New Methods in Quantum Theory

NATO ASI Series

Advanced Science Institutes Series

A Series presenting the results of activities sponsored by the NATO Science Committee, which aims at the dissemination of advanced scientific and technological knowledge, with a view to strengthening links between scientific communities.

The Series is published by an international board of publishers in conjunction with the NATO Scientific Affairs Division

A	**Life Sciences**	Plenum Publishing Corporation
B	**Physics**	London and New York
C	**Mathematical and Physical Sciences**	Kluwer Academic Publishers
D	**Behavioural and Social Sciences**	Dordrecht, Boston and London
E	**Applied Sciences**	
F	**Computer and Systems Sciences**	Springer-Verlag
G	**Ecological Sciences**	Berlin, Heidelberg, New York, London,
H	**Cell Biology**	Paris and Tokyo
I	**Global Environmental Change**	

PARTNERSHIP SUB-SERIES

1.	**Disarmament Technologies**	Kluwer Academic Publishers
2.	**Environment**	Springer-Verlag / Kluwer Academic Publishers
3.	**High Technology**	Kluwer Academic Publishers
4.	**Science and Technology Policy**	Kluwer Academic Publishers
5.	**Computer Networking**	Kluwer Academic Publishers

The Partnership Sub-Series incorporates activities undertaken in collaboration with NATO's Cooperation Partners, the countries of the CIS and Central and Eastern Europe, in Priority Areas of concern to those countries.

NATO-PCO-DATA BASE

The electronic index to the NATO ASI Series provides full bibliographical references (with keywords and/or abstracts) to more than 50000 contributions from international scientists published in all sections of the NATO ASI Series.
Access to the NATO-PCO-DATA BASE is possible in two ways:

– via online FILE 128 (NATO-PCO-DATA BASE) hosted by ESRIN, Via Galileo Galilei, I-00044 Frascati, Italy.

– via CD-ROM "NATO-PCO-DATA BASE" with user-friendly retrieval software in English, French and German (© WTV GmbH and DATAWARE Technologies Inc. 1989).

The CD-ROM can be ordered through any member of the Board of Publishers or through NATO-PCO, Overijse, Belgium.

Series 3: High Technology – Vol. 8

New Methods in Quantum Theory

edited by

Constantinos A. Tsipis

Laboratory of Applied Quantum Chemistry,
Faculty of Chemistry,
Aristotle University of Thessaloniki,
Thessaloniki, Greece

Vladimir S. Popov

Institute of Theoretical and Experimental Physics,
Moscow, Russia

Dudley R. Herschbach

Department of Chemistry,
Harvard University,
Cambridge, MA, U.S.A.

and

John S. Avery

Department of Chemistry,
H.C. Oersted Institute,
University of Copenhagen,
Copenhagen, Denmark

Kluwer Academic Publishers

Dordrecht / Boston / London

Published in cooperation with NATO Scientific Affairs Division

Proceedings of the NATO Advanced Research Workshop on
New Methods in Quantum Theory
Halkidiki, Greece
May 14–19, 1995

A C.I.P. Catalogue record for this book is available from the Library of Congress

ISBN-13: 978-94-010-6585-6 e-ISBN-13: 978-94-009-0227-5
DOI: 10.1007/978-94-009-0227-5

Published by Kluwer Academic Publishers,
P.O. Box 17, 3300 AA Dordrecht, The Netherlands.

Kluwer Academic Publishers incorporates the publishing programmes of
D. Reidel, Martinus Nijhoff, Dr W. Junk and MTP Press.

Sold and distributed in the U.S.A. and Canada
by Kluwer Academic Publishers,
101 Philip Drive, Norwell, MA 02061, U.S.A.

In all other countries, sold and distributed
by Kluwer Academic Publishers Group,
P.O. Box 322, 3300 AH Dordrecht, The Netherlands.

Printed on acid-free paper

CONTENTS

PREFACE

This volume contains all of the invited lectures presented at the NATO ARW on *"New Methods in Quantum Theory"* held in Halkidiki, Greece from May 14th to May 19th, 1995. This survey of new perspectives, techniques and results in quantum theory contains 26 chapters by leading quantum chemists and physicists from 14 countries. The book covers a wide range of topics, though the emphasis throughout is on new approaches and their interrelationships. Topics covered include dimensional scaling, the hyperspherical method applied both to reactive scattering theory and to bound state problems, chaotic behaviour, large-order perturbation theory, complex eigenvalues and quasistationary states, semiclassical methods, cusps in hyperaccurate wave functions, density functional theory, relativistic quantum theory, and quantum Monte Carlo methods. I hope very much that the various lectures presented at this Advanced Research Workshop will be of as great interest to the reader as they were to the participants.

The organization of the ARW would have been impossible without the generous funding by NATO, which is gratefully acknowledged. I sincerely thank Drs. J.-M. Cadiou, Assistant Secretary General for Scientific and Environmental Affairs, and J. A. Rausell-Colom, Programme Director for Priority Area on High Technology, North Atlantic Treaty Organization, who helped me throughout the various stages of the ARW.

It is my pleasure to acknowledge the essential financial support and technical assistance provided by the Aristotle University of Thessaloniki (Research Committee), the Ministry of Macedonia-Thrace, the Greek Secretariat of Research and Technology, the Greek National Tourist Organisation, the ALICO (American Life Insurance Company), the APEX A.B.E.E.-High Technology Systems, the Athenian Brewery S.A., the DOW Hellas A.B.E.E., the EXIN A.B.E.E.-High Technology Systems, the Hellenic Sugar Industry S.A., the Macedonia-Thrace Bank, the MEVGAL S.A.-Dairy Product Industry, the I. Boutaris and Son Ltd. (Quality Wines since 1879) and the Public Petrol Corporation.

Moreover, I would like to express my sincere gratitude to my esteemed colleagues on the organizing committee, Profs. Vladimir S. Popov, Dudley R. Herschbach and John S. Avery, for their help and advice to make this ARW to be the success it turned out to be.

Many thanks go also to the members of the local committee, Profs. C.L. Tsiamis, G.A. Katsoulos, M.P. Sigalas and E.G. Bakalbassis, for their valuable advice on various matters relating to the ARW.

I am very grateful to Dr. Maria-Aglaia D. Stiakaki for the invaluable help and assistance she rendered as the ARW Secretary.

Finally, I am particularly thankful to the participants, the truly essential actors, who made this meeting a success through their cooperation, enthusiasm and, above all, their piercing critique and critical inquiries.

Constantinos A. Tsipis

INTRODUCTORY REMARKS

C. A. TSIPIS
Laboratory of Applied Quantum Chemistry,
Faculty of Chemistry,
Aristotle University of Thessaloniki,
540 06 Thessaloniki,
Greece

Ladies and Gentlemen:

It is my pleasure to welcome you to this NATO Advanced Research Workshop on "New Methods in Quantum Theory", particularly hoping that this week-long meeting will create lively interactions and vigorous correlations among this elite gathering of dynamic chemists and physicists. In all, there are seventy one participants and speakers from 16 countries. I am happy to welcome also participants coming from Latvia, Lithuania, Russia, Ukraine, Slovakia and Poland, exhibiting a significant impact in the development of Quantum Theory.

The broad spectrum of the work to be presented at the workshop is the reflection of a time of intensive search for new ways of solving some of the most fundamental problems in quantum theory.

I will try to avoid restricting my introductory remarks to detailed facts; instead, I will try to paint a more general picture basing my reflections on the scientific results.

The development of new materials for high-technology applications (e.g. high density computer memories, molecular switches, solar cells, etc.) requires the ability to predict, through the techniques of quantum theory, the electronic structure, electromagnetic properties, and chemical reactivity of materials even before they are synthesized.

The subject of electronic structure calculations is a very timely one. We are often reminded how the power of computers is growing exponentially. *Is our subject of developing new methods and techniques in quantum theory also growing exponentially in its contribution to the advancement of electronic structure calculations ? What is its role ?*

Much of the following week will be devoted to the technical intricacies of making such calculations, and before we get wholly immersed in those I want to set the work in the wider contex; *what do we hope to achieve ?* In all of science, achieving a real advance in *understanding* often depends on selecting just the right experiment or calculation to shed light on a phenomenon and not merely in heaping up more facts. Clearly such understanding is not going to come without intelligent computation, which establishes an essential role for our subject at the core of modern facets of quantum theory.

An accurate and consistent treatment of electron correlation is one of the great challenges currently confronting electronic structure calculations in theoretical chemistry, atomic and molecular physics, and condensed matter theory. Such a treatment is critical for many aspects of the ab initio determination of atomic and molecular structure.

In the last few years, there has been an upsurge in the evolution of novel many-body correlation methods with hitherto unsurpassed degrees of sophistication. A number of post-Hartree-Fock methods, either configurational or perturbational, have been developed adding many-body correlation at various levels of approximation, but all these conventional correlation methods scaling in the N^6 regime become arduous or intractable even with the mighty computing power now available or in prospect. Other important formalisms that treat exactly the correlation energy are density functional theory, electron propagator and quantum Monte Carlo methods. The former approach is gaining momentum and is useful in a variety of applications.

Besides, there has been a fresh approach of solving directly the many-particle Schrödinger equation in a hyperspace, without making use of the usual approximations such as the Born-Oppenheimer or the Hartree-Fock approximations. In the emerged new technique, known as dimensional scaling, the dimensionality of space, D, is treated as a variable parameter, and perturbation expansions in $1/D$ are employed to calculate the wavefunctions and energetics of a system at the physical 3-dimensional world. The new procedure provides a convenient and intuitively meaningful way of treating electron correlation.

A related method of solving directly the many-particle Schrödinger equation, also involving hyperspaces, makes use of hyperspherical coordinates and hyperspherical harmonics. The hyperspherical method yields very accurate solutions both to bound-state problems in many-body quantum theory and to problems involving reactive scattering.

Last, but not least, breakthrous have been made in the development of ab initio relativistic electronic structure theory. The Dirac's relativistic theory provides a very natural and convenient way of treating magnetic effects in quantum chemistry and solid state theory. Since both the relativistic and electron correlation effects are expected to be very pronounced, it is mandatory to treat both these effects *accurately* for systems of heavy elements. Therefore, there is an urgent need for a concerted effort by leading scientists working in both areas, to delineate and difuse the state-of-the-art theoretical and computational strategies so as to enable scientists to mount a joint attack on these gargantuan problems. This ARW was, therefore, arranged to allow scientists from the interdisciplinary areas of non-relativistic and relativistic quantum chemistry, atomic, molecular and solid-state physics, etc., to assemble and discuss the various aspects of the effects of relativity and electron correlation on the electronic structure, bonding, and dynamics of molecules and solids.

I believe that this very diversity of topics represents one of the most attractive features of the workshop, which I hope will provide a good orientation in the abundant flow of ideas in modern quantum theory. In addition, apart from the usual exchange of results and views that takes place in an international gathering of learned scholars like this, I particularly hope that all participants will experience its generous benefits and taste the sweet nectar of the scientific progress.

A title such as "New Methods in Quantum Theory" may remind of discussions on "Foundations of Quantum Mechanics" which often indulge to futile speculations, influenced by philosophical prejudices which are irrelevant for the understanding of a scientific theory. As a result, the controversial statements which are inevitably to be found in any meeting on "foundations" or on "fundamental aspects" (of anything) form the seasoning and not the main course of the present workshop. Remember that we are in the place, where the great philosopher Aristotle was born in 384 BC and laid down his views on natural philosophy and science.

Let me extract some of his terse formulations related to our modern philosophical thoughts about nature " the whole of the heavens is harmony and number"; that is, the manifest order evinced by the motions of the heavenly bodies, which in itself already constituted an element of beauty appealing to the Greek aesthetic sense, can be defined by numerical proportions, and on this account is included with acoustic euphony in the concept of harmony; the harmony of the spheres need not therefore be audible: the mind's eye here takes the place held by the ear in music.

The very discovery of the Kepler's laws represented an outstanding act of intuitive "harmonization". Analyzing all mystical numerical ratios established in the planetary system, documented at least since Pythagoras, in terms of many-body systems coupled by feedback, harmonies are established all by themselves in this world when, under certain feedback conditions, chaos is allowed to "play itself out".

Of great interest is also Aristotle's view on the concept of infinity which strongly resembles the ideas that Gauss had later on the same subject. Aristotle rejected the idea of an actual infinity and replaced it by something that can be made as large as we wish. In his opinion the infinity of a straight line does not mean that it actually becomes infinite but that it can always be prolonged, no matter how far out we are. Similarly, a sum which would grow beyond any conceivable quantity would be mathematically meaningless. We encounter here the first demands towards convergence- a concept which became of such leading importance in the investigations of Gauss, Abel and Cauchy around the beginning of the nineteenth century.

I am sure that if Aristotle and his teacher Plato were around today would enjoy all lectures on hyperspaces, as the space was the cornerstone in their explanation of the whole cosmos. Entering the philosopher's chora, our goal is not to prove, but to improve, to become better able to think and to see the world.

Our world is an island, an island of order, an island of physical lows, an island of ideas, an island of trust. We live on our island. Since the world is harmonic, ideas and matter exist in a creative relationship with each other, the world is finite, love is a strong, integrating power in it, we live in a beautiful world.

It was my pleasure to act as the Director of the ARW and I am most grateful to my esteemed collegues on the organising committee and the invited lecturers for their enthusiastic support and valuable advice on various matters relating to the ARW. My special heartfelt thanks go to the participants without whose interest and participation the ARW would not have been possible.

Finally, to all participants I hope that, amid their deliberations on the latest results and theoretical thinking, you may find time to see something of the beauty of Halkidiki and Macedonia and the city of Thessaloniki which, I may say not without some pride experience the pulsation of a dynamic growing city. Halkidiki, the birthplace of Aristotle, offers a pleasant environment and inspires people to explore their own capabilities. I hope that new collaborations will emerge and new links will be forged among the participants. The thirst for new methods in Quantum Theory would be enhanced by the food for thought that the lecturers of the ARW will provide in abundance in this meeting.

Let me say welcome again and wish you all the best in your endeavours.

DIMENSIONAL SCALING

J. G. LOESER
Department of Chemistry
Oregon State University
Corvallis, OR 97331-4003
United States

AND

D. R. HERSCHBACH
Department of Chemistry
Harvard University
Cambridge, MA 02138
United States

Abstract. This chapter reviews three kinds of dimensional scaling techniques, for each one commenting on motivation, strengths and weaknesses, as well as current and prospective applications. (1) Perturbation *expansions* in powers of $1/D$ exploit the simplicity of the $D \to \infty$ limit, which is generally characterized by some kind of localization; these expansions can be developed systematically, but diverge strongly, so resummation techniques are usually required in order to extract highly accurate results. (2) *Dimensional interpolation* takes advantage of the fact that many problems simplify not only for $D \to \infty$, but also at low D, and vary smoothly between these limits; interpolation is simple and often surprisingly accurate, but less systematic and more limited in applicability than $1/D$ expansions. (3) *Dimensional bypass* methods call on complementary physical correlates which often enable dimensional scaling to be used to correct the errors of other approximate treatments; the bypass approach is quite effective, but inherently nonuniform. The expansion, interpolation, and bypass techniques are illustrated with calculations for three model problems: the asphericity of a random walk, the third virial coefficient of a hard sphere fluid, and the ground-state energy of a helium atom. We give also an annotated bibliography for D-scaling and briefly survey conjugal approximations for many-electron correlation energies.

C. A. Tsipis et al. (eds.), New Methods in Quantum Theory, 1–32.

1. Introduction

Dimensional scaling as applied to chemical physics embraces a number of distinct techniques[1]. What these techniques have in common is the use of dimensionally generalized versions of real-world problems, wherein D, the number of translational degrees of freedom per particle, is taken as a variable parameter. The goal is to exploit simple $D \neq 3$ solutions to construct or improve approximate $D=3$ results. Most current D-scaling techniques may be assigned to three categories: $1/D$ *expansions*, in which solutions are developed as perturbation expansions about the high-D limit; *dimensional interpolation*, in which solutions are approximated using results obtained at both high- and low-D; and various *dimensional bypass* methods, in which a dimensional scaling approximation is combined with some other approximation having complementary strengths.

In this chapter, we review basic features of the D-scaling techniques and briefly discuss some recent and prospective applications. For reference, we also provide an annotated bibliography[2–45]. This extends a previous listing[1] to mid-'95 and includes a few earlier papers that had been overlooked. In order to compare and contrast the three basic techniques, we will apply each of them to three prototypical problems. These are the asphericity of a random walk, the third virial coefficient of a hard sphere fluid, and the energy of a helium atom. These problems are simple enough to be treated in detail, yet complex enough to bring out characteristic strengths and limitations of the methods.

First we define the quantities to be computed in the three model problems and note the accurate reference values available for $D=3$. The asphericity of a random walk [46, 47], usually denoted $\langle A_D \rangle$, is a number between 0 and 1 that characterizes the average disparity of the moments of inertia for an ensemble of random walks. It is defined by

$$\langle A_D \rangle = \frac{1}{D-1} \left\langle \sum_{\mu=1}^{D} \sum_{\substack{\nu=1 \\ \mu<\nu}}^{D} \left(R_\mu{}^2 - R_\nu{}^2 \right)^2 \Big/ \left(\sum_{\mu=1}^{D} R_\mu{}^2 \right)^2 \right\rangle , \tag{1}$$

where the $R_\mu{}^2$ are eigenvalues of the radius of gyration tensor (also referred to as the shape tensor or moment of inertia tensor), whose components are $Q_{\mu\nu} \equiv \overline{(r_\mu - \overline{r}_\mu)(r_\nu - \overline{r}_\nu)}$. Here bars denote averaging over the segments of a single random walk, and angle brackets denote averaging over an ensemble of walks. We consider only very long walks ($N \to \infty$, where N is the number of steps), which renders the asphericity for any D a universal constant, independent of the lattice type. (For finite N the value does show rather strong dependence on the kind of lattice[48].) The $D=3$ value is known to be[49]

$$\langle A_3 \rangle = 0.39427\ldots \qquad \boxed{\text{reference value}} \,. \tag{2}$$

For a hard sphere fluid, the third virial coefficient, denoted here by C, appears in the quadratic term of the low-density expansion for the compressibility factor,

$$z = \frac{p}{\rho kT} = 1 + B\rho + C\rho^2 + \dots \,. \tag{3}$$

It is convenient to work with unitless coefficients (often, but imprecisely, called dimensionless!), obtained by dividing the higher virial coefficients by the appropriate powers of B, the second coefficient. Therefore we will actually compute

$$\tilde{C}_D = C_D/{B_D}^2 = \frac{1}{2}\left(\frac{\partial^2 z}{\partial \rho^2}\right) \Big/ \left(\frac{\partial z}{\partial \rho}\right)^2 \Bigg|_{\rho=0} . \tag{4}$$

For hard spheres, the nth virial coefficient can be expressed in terms of integrals describing simultaneous overlaps among n spheres. In particular, $\tilde{C}_D$ is given by

$$\tilde{C}_D = \frac{4}{3} \iint_{\triangle} d^D\mathbf{r}_1\, d^D\mathbf{r}_2 \Big/ \iint_{\wedge} d^D\mathbf{r}_1\, d^D\mathbf{r}_2 \,. \tag{5}$$

Here $\triangle$ denotes the domain where spheres 1 and 2 overlap one another and each overlaps a third sphere located at the origin (that is, $|\mathbf{r}_1 - \mathbf{r}_2|$, $|\mathbf{r}_1|$, and $|\mathbf{r}_2|$ are each less than one sphere diameter), while $\wedge$ denotes the domain where one of these three constraints is removed. The $D=3$ value is known to be[50]

$$\tilde{C}_3 = \tfrac{5}{8} \qquad \boxed{\text{reference value}} \,. \tag{6}$$

For the helium atom, we consider just the ground state energy. All small corrections are ignored, so this is the lowest eigenvalue of the Hamiltonian (in atomic units)

$$\mathcal{H}_D = -\tfrac{1}{2}{\nabla_1}^2 - \tfrac{1}{2}{\nabla_2}^2 - Z/r_1 - Z/r_2 + 1/r_{12} \,, \tag{7}$$

where $Z=2$, and the Laplacians and distances are all D-dimensional,

$$\nabla^2 = \sum_{\mu=1}^{D} \frac{\partial^2}{\partial {x_\mu}^2} \,, \qquad r = \left(\sum_{\mu=1}^{D} {x_\mu}^2\right)^{1/2} . \tag{8}$$

We use a system of dimensionally generalized atomic units, chosen so as to eliminate the dominant dimension dependence from the solutions[51]. In conventional (hartree/bohr) atomic units, the hydrogen atom ground state solution yields $E=-2/(D-1)^2$ and $\langle r\rangle=D(D-1)/4$, so reduced quantities are defined by[19, 52]

$$\widetilde{E}_D = \frac{(D-1)^2}{4} E_D\,, \qquad \tilde{r}_D = \frac{6}{D(D-1)} r_D\,. \tag{9}$$

(The numerical factors 4 and 6 are chosen so that the scalings have no effect at $D=3$.) A frequently used alternative scaling for the distance is $\hat{r} = 4(D-1)^{-2} r$. We have chosen the scaling in Eq. 9 because it leads to finite expectation values for distances over the full range of D that we will be considering, from $D=1$ to $D \to \infty$. For the helium atom, the $D=3$ energy is known to be[53]

$$\widetilde{E}_3 = -2.90372\ldots \qquad \boxed{\text{reference value}}\,. \tag{10}$$

For comparison with numbers cited below, we also make note of the expectation values for the electron-nucleus and electron-electron distances[53], which are $\langle\tilde{r}_1\rangle = \langle\tilde{r}_2\rangle = 0.92947$ and $\langle\tilde{r}_{12}\rangle = 1.42207$.

2. Dimensional perturbation expansions

One might expect that as D is increased, it would become ever more difficult to characterize the likely configurations of a system. In fact the opposite is true. Quite generally, the internal structure of a many-body system will become increasingly well-defined as D is increased[54, 55]. The large-D limit is in fact a singular limit characterized by complete localization, in the sense that for any pair of particles the relative uncertainty in their separation vanishes; that is, $\Delta r_{ij}/\langle r_{ij}\rangle \to 0$ as $D\to\infty$.

Localization arises from the fact that different regions of configuration space are weighted ever more disparately as the dimensionality is increased. The simplest example of this is found in a two-body system, where the pertinent weighting is just the radial volume element factor, r^{D-1}. Localization results when such factors compete against restraining influences imposed by the physics of the problem, such as hard walls or exponential tails.

Localization allows the $D \to \infty$ limit solution to be evaluated easily for a wide range of problems[56, 57, 58]. Often this can be done analytically, even for many-body problems[52, 59, 60]. With a bit more work, one can utilize the limiting solutions as a platform for perturbation expansions[36, 61, 62] in powers of $1/D$. (Note that these have also been called $1/d$, $1/N$, and $1/n$ expansions, and that the denominator parameter is often defined as a combination of D with quantum numbers[63].) High-order $1/D$ expansions have now been derived for a wide range of problems, including the anharmonic oscillator[64], the Yukawa potential[65] and other central potentials[61, 66, 67], the hydrogen atom in an electric field[68] or a magnetic field[35, 37, 69], the hydrogen molecule ion[70, 71], two-electron atoms in both ground[12, 72] and excited[24] states, and the Lennard-Jones potential[26].

Because of the singular character of the $D \to \infty$ limit, $1/D$ expansions are usually divergent, and it requires some effort to extract accurate results from them. The summation problem can be attacked at many levels. The crudest is simple asymptotic truncation, wherein the series is terminated before the term smallest in magnitude. Better, the series can be recast in continued fraction form[73],

$$a_0 + \frac{a_1}{D} + \frac{a_2}{D^2} + \frac{a_3}{D^3} + \ldots = \cfrac{c_0}{1 + \cfrac{c_1/D}{1 + \cfrac{c_2/D}{1 + \cfrac{c_3/D}{1 + \ldots}}}} = \cfrac{c_0}{1 + \cfrac{c_1}{D + \cfrac{c_2}{1 + \cfrac{c_3}{D + \ldots}}}}, \quad (11)$$

and then this series can be truncated at successive levels. This procedure is equivalent to constructing the alternating sequence of $[N, N]$ and $[N, N{+}1]$ Padé approximants. Unlike the a_n coefficients, which typically grow as $n!$ and thus become huge, the c_n coefficients remain within modest bounds even at high order; that is a marked advantage for tabulation.

It is also useful to invoke the Borel representation of the series[73],

$$a_0 + \frac{a_1}{D} + \frac{a_2}{D^2} + \frac{a_3}{D^3} + \ldots = \int_0^\infty \left(a_0 + \frac{t\, a_1}{1!\, D} + \frac{t^2\, a_2}{2!\, D^2} + \frac{t^3\, a_3}{3!\, D^3} + \ldots \right) e^{-t}\, dt\,. \quad (12)$$

The Borel sum is obtained by first treating the series in parentheses (which is much nicer than the original series), for example by means of continued fractions, and then integrating. The combined use of continued fractions and the Borel representation is a special case of what is commonly known as Padé-Borel summation.

Consideration of the explicit singularity structure of a problem leads to additional summation tools. Among these are removal of known poles[10, 74], quadratic Padé (or Hermite-Padé) summation[75], Darboux-Borel summation[24, 76], and Levin-Borel summation[77]. Still better means for taming the divergences may emerge from recent discoveries concerning the analytic origins of the asymptotic behavior[22].

2.1. RANDOM WALK

The origin and consequences of the localization which occurs at large D can be seen most simply in the random walk problem. Since the asphericity is independent of lattice type for very long walks, we consider a hypercubic lattice, conceptually the easiest. For any dimensionality D and path length N, there will be a large number of shapes for the walk. As $D \to \infty$, however, walks in which every new step is perpendicular to every previous step become overwhelmingly most probable. This is simply because at each step, there are only finitely many choices for the next step in directions already used in the walk, and infinitely many choices that are perpendicular to all of those directions[47].

One way to see graphically the extent of localization with increasing D is to consider the probability distributions[78, 79] for end-to-end distances in long random walks. For finite D there will be a range of values, but for $D \to \infty$ the end-to-end distance becomes fixed at $\sqrt{N}$; this is just the body diagonal of the hypercube defined by the N-step walk. The probability distributions for end-to-end distances at several values of D are plotted in Figure 1a.

The $D \to \infty$ limit for the asphericity is readily evaluated. There is only one walk to consider, and its simple all-perpendicular structure permits the radius of gyration tensor to be diagonalized exactly. In fact, the eigenvalues turn out to have a "hydrogenic" spectrum, with relative values $1 : \frac{1}{4} : \frac{1}{9} : \ldots$; from these one obtains the simple result[47, 80]

$$\langle A_\infty \rangle = \tfrac{2}{5} \; . \tag{13}$$

The $D \to \infty$ limit solution may be considered to be the zeroth-order term in a $1/D$ expansion. To evaluate further terms in this expansion, one needs to consider what happens when D is large, but not infinite. In this case a small fraction of the walks will *not* be of the all-perpendicular form. For example, the leading correction, of order D^{-1}, arises from walks in which just two steps are along the same axis. The leading terms of the $1/D$ series for the asphericity are[49, 81]

$$\langle A_D \rangle \sim \tfrac{2}{5} - \tfrac{12}{175} D^{-1} + \tfrac{48}{175} D^{-2} - \ldots \; . \tag{14}$$

The divergent character of the series at $D=3$ is already apparent, since the third term is larger in magnitude than the second. This suggests that the optimal truncation, which for an asymptotic series is made just before the smallest term, consists of only the zeroth-order term. (Actually there are terms of smaller magnitude in the series[81], but only due to the existence of an oscillatory "envelope", so this is the correct truncation.) Therefore the $1/D$ series approximation to the asphericity at $D=3$ is just

$$\langle A_3 \rangle = \tfrac{2}{5} \qquad \boxed{1/D \text{ series: } +1.45\% \text{ error}} \; . \tag{15}$$

Much more accurate values can be obtained by employing resummation procedures[81]. We consider just two of these here. Figure 2 shows the

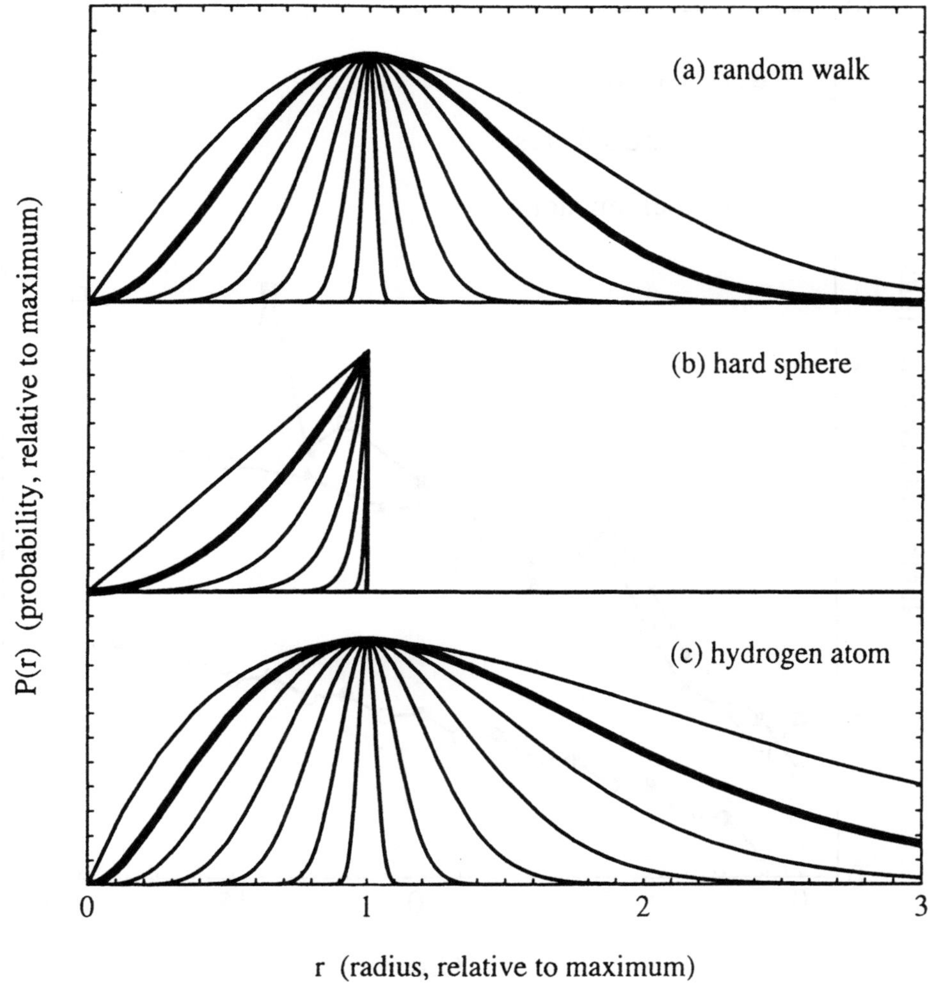

Figure 1. Illustration of localization with increasing D for the three dimensionally generalized problems. Plotted are probability distributions for (a) end-to-end distances in long random walks, (b) radii of random points in a sphere, and (c) electronic radii in a hydrogen atom. In each case curves plotted are for $D=2$ (most diffuse), $D=3$ (bold line), $D=5$, $D=10$, $D=25$, $D=100$, and $D=1000$ (most localized). All distributions are normalized by the position and height of the maximum. Functions plotted are all of the form $\{r \exp[-(r^n-1)/n]\}^{D-1}$, but with $n=2$ for the random walk, $n\to\infty$ for the hard sphere, and $n=1$ for the hydrogen atom.

accuracy of values for $\langle A_3\rangle$ obtained by converting the $1/D$ series to continued fraction form and truncating at successive orders. This yields 7-digit accuracy by 20th order. Figure 3 shows the corresponding values obtained by taking the Borel transform of the original $1/D$ series, summing this by means of the continued fraction representation, and then inverting the

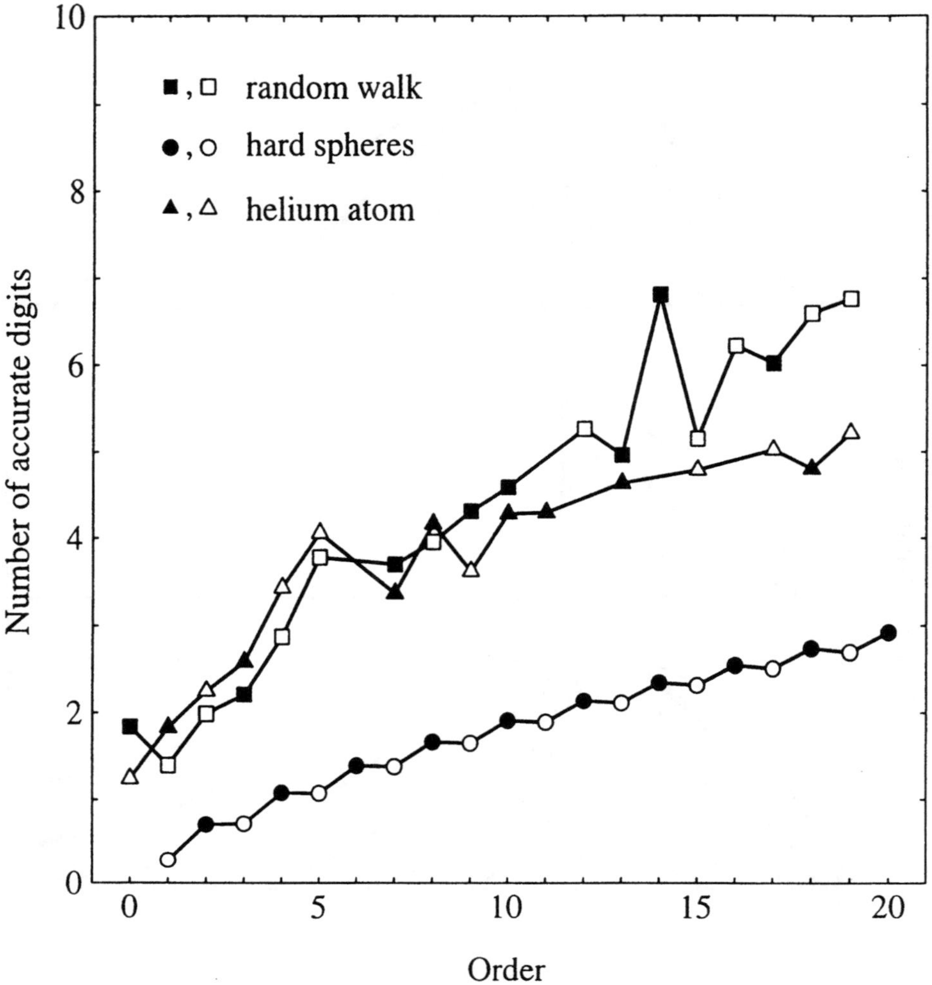

Figure 2. Accuracy of continued fraction partial sums for the three model problems. The number of accurate digits is defined as $-\log_{10}|\text{error/value}|$; positive and negative errors are indicated by filled and open symbols, respectively. Values were obtained by truncating the continued fraction representation of the $1/D$ series at successive orders, omitting those which generated poles whose distance from $\frac{1}{3}$ in the complex $1/D$ plane was less than $\frac{1}{3}$.

transform by numerical integration. This yields roughly 9-digit accuracy by 20th order.

2.2. HARD SPHERES

For the hard sphere problem the phenomenon of localization is really just a consequence of the fact that, due to the r^{D-1} factor in the hyperspherical

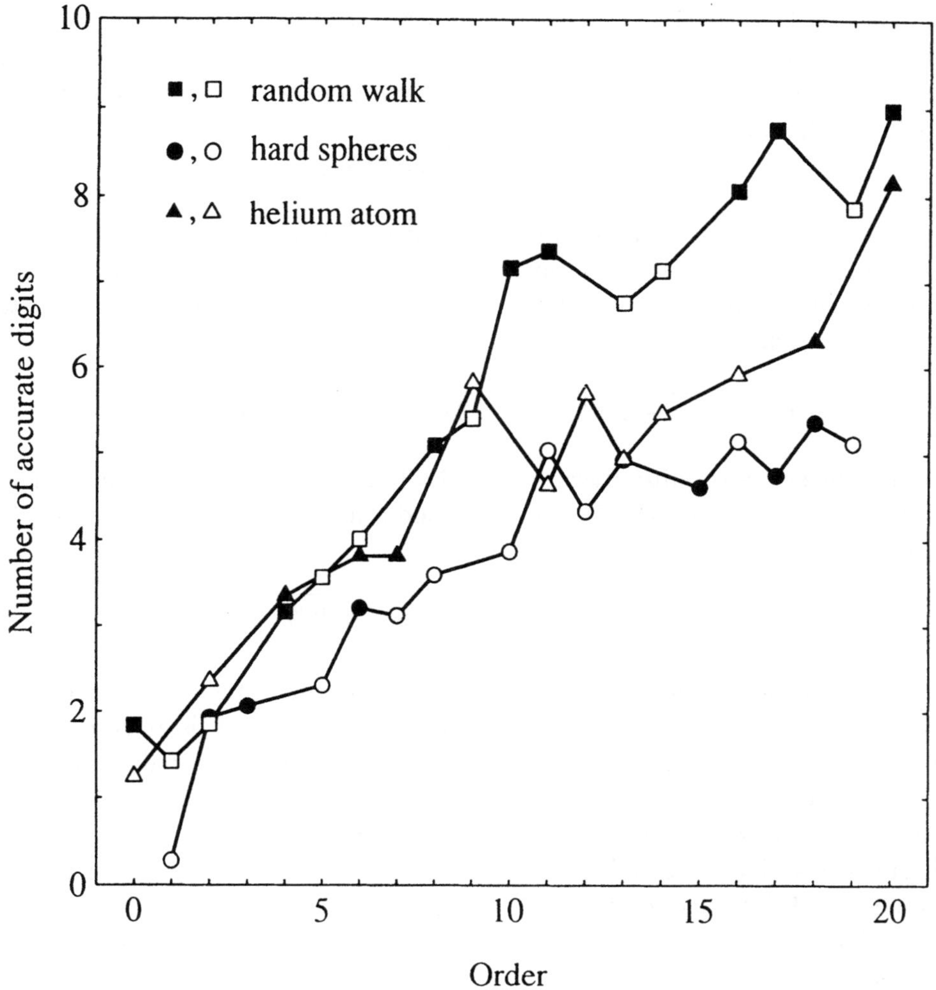

Figure 3. Accuracy of Padé-Borel summation for the three model problems. Notation as in previous figure. Values were obtained by numerically integrating truncated continued fraction representations of the $1/D$ Borel series at successive orders, omitting those which generated poles along the positive real axis in the $1/D$ plane.

volume element, the entire volume of a high-dimensional sphere is concentrated at its surface. The localization is shown graphically in Figure 1b, which plots probability distributions for radii of random points within hyperspheres of various dimensionalities.

A consequence of this localization is that when D is large, integrals describing the simultaneous overlaps of a number of spheres are determined entirely by configurations where spheres just touch each other[60, 82]. For example, the $D \to \infty$ limit behavior of the $\triangle$ integral in Eq. 5 is governed by the equilateral triangular geometry of three spheres in mutual contact. The fact that twice the dimensionless area of this triangle is $\sqrt{3/4}$ is all that one needs to characterize the limiting behavior of the third virial coefficient at large D, which is

$$\lim_{D\to\infty}(\widetilde{C}_D)^{1/D} = \sqrt{3/4}\,. \tag{16}$$

We can proceed to higher order by considering what happens when D is large but not infinite, so that some of the sphere volume "leaks" inward from its surface. For purposes of writing down the series it is most convenient to use the logarithm. For the third virial coefficient one obtains[77]

$$\ln\widetilde{C}_D \sim D\ln\sqrt{3/4} - \tfrac{1}{2}\ln D + \ln\sqrt{24/\pi} - \tfrac{15}{4}D^{-1} + 32D^{-2} - \ldots\,. \tag{17}$$

Again the series is quite divergent at $D=3$, and simple asymptotic truncation, before the term smallest in magnitude, means that we are limited to the first term. Thus we obtain

$$\widetilde{C}_3 = 0.64952 \qquad \boxed{1/D \text{ series: } +3.92\% \text{ error}}\,. \tag{18}$$

Using resummation techniques, one can extract much more accurate values from this and other virial coefficient series. For this purpose, it is preferable to treat the first three terms of the series (the only irrational ones) separately, and resum to the remaining terms with the help of continued fractions, the Borel transform, etc. Results for $\widetilde{C}_3$ are plotted in Figures 2 and 3. It can be seen that by 20th order, continued fractions alone give 3-digit accuracy , while using the Borel transform improves this to 5-digit accuracy. Even better values can be obtained by using Levin-Borel summation[77].

2.3. HELIUM ATOM

Finally we turn to electronic structure. Here it is the competition between volume element factors (which vanish for small interparticle separations) and the exponential tails of the wavefunction (which vanish for large separations) that results in localization[83]. The localization process is illustrated in Figure 1c, which shows hydrogen atom radial probability distributions for several values of D. For $D \to \infty$ the scaled electron-nucleus distance becomes well-defined.

More generally, any electronic structure problem in the $D \to \infty$ limit yields a classical "electronic structure", with interparticle distances fixed relative to one another[51, 55]. These structures are somewhat similar to those of the old quantum theory, though they have a quite different status and geometry[33, 84]. The structures and their energies are obtained by considering the singular limit of the Hamiltonian for the probability amplitude within the space of internal coordinates. Quite generally, this has the form Ω^{-1}(second derivatives) + (non-differential terms), where Ω is of order D^2. Therefore the $D\to\infty$ limit solution can be obtained as the minimum of the non-differential part of the Hamiltonian. For the helium atom this is just[52]

$$\tilde{\mathcal{H}}_\infty = \frac{9(\tilde{r}_1^2+\tilde{r}_2^2)}{4(\tilde{r}_1^2\tilde{r}_2^2+\tilde{r}_2^2\tilde{r}_{12}^2+\tilde{r}_{12}^2\tilde{r}_1^2)-2(\tilde{r}_1^4+\tilde{r}_2^4+\tilde{r}_{12}^4)} - \frac{3Z}{2\tilde{r}_1} - \frac{3Z}{2\tilde{r}_2} + \frac{3}{2\tilde{r}_{12}} \ . \quad (19)$$

Minimizing Eq. 19 gives $\tilde{r}_1 = \tilde{r}_2 = 0.91045$ and $\tilde{r}_{12} = 1.34572$ (which are a few percent smaller than the corresponding $D=3$ expectation values), and

$$\tilde{E}_\infty = -2.73777 \ . \quad (20)$$

As in the other problems, one can also utilize the $D\to\infty$ solutions to develop a systematic perturbation expansion in powers of $1/D$. For helium, the expansion for the energy begins[12]

$$\tilde{E}_D \sim -2.73777 - 0.58205D^{-1} + 0.51523D^{-2} - 2.23386D^{-3} + \dots \ . \quad (21)$$

In this case it is the third term which is smallest in magnitude when $D=3$, so the optimal asymptotic truncation is after two terms. Therefore

$$\tilde{E}_3 = -2.93179 \qquad \boxed{1/D \text{ series: } +0.97\% \text{ error}} \ . \quad (22)$$

Again, much better energies can be obtained by using resummation methods on higher-order series. From Figures 2 and 3 one sees that simple continued fraction truncation yields 5-digit accuracy for $\tilde{E}_3$ by 20th order, while use of the Borel transform yields 7 digits. Removal of poles yields about two more digits[12].

3. Dimensional interpolation

One of the most important and useful characteristics of dimensionality as a parameter is that for many problems, one finds dramatic simplification at *both* high D and low D, with smooth variation of the physics in between. This invites a dimensional interpolation strategy for obtaining approximate $D=3$ solutions from the dimensional limit solutions.

Note that there are relatively few parameters for which the physical value is bracketed like this. In those cases where one can find two simplifying limits, there is often discontinuous behavior between them. Examples are the nuclear charge in an atom, where the atom falls apart as one heads for the low-charge limit, and the density in the hard sphere problem, where a phase transition separates the high-density and low-density limits. An example of a parameter which *does* show smooth behavior between simplifying limits is the bond length of a diatomic molecule, which yields simplification in both the united atom limit and the separated atom limit. Likewise, not all physical quantities vary smoothly with D, so dimensional interpolation may sometimes be inappropriate. An example of such an uncooperative quantity will be pointed out below.

Dimensional interpolation is less straightforward than a $1/D$ expansion. In particular, the choice of the scale for interpolation usually requires care. Suppose, for example, that one can solve a problem for both $D \to \infty$ and $D=1$. Then one might be able to use $1/D$ (the perturbation parameter) as the interpolation scale. For many electronic structure problems, this is in fact an appropriate and effective choice. However, if one seeks to utilize a solution obtained at $D=0$ (a frequently useful value), then clearly another parameter is needed. For example, one might employ a shifted parameter $(D+a)^{-1}$, where a is some constant[85]. This is a somewhat arbitrary choice, but because of its simplicity and generality, it is the method that we will use here. However, we note that there are more effective strategies available, just as there are better approaches to $1/D$ series than simple asymptotic truncation. In general these involve analysis of the singularities of the problem[10].

Input for an interpolation function consists of values and possibly derivatives at both high and low D. Because the interpolant is constrained to have the correct behavior at one or more low values of D, it is often possible to employ more terms from the $1/D$ expansion than were used for the optimal asymptotic truncation of the series. (Recall that we were able to use only one term for random walks, one term for hard spheres, and two terms for the helium atom.) However, the asymptotic character of the $1/D$ expansion still limits the number of terms that can be used for the interpolation.

3.1. RANDOM WALK

Again we start with the random walk. One can consider various low-D versions of this problem, but for purposes of evaluating the asphericity it is easiest to consider $D = 0$. Clearly, as in the large-D limit, a walk in zero-dimensions can no longer be random! Here one finds that

$$\langle A_0 \rangle = \tfrac{1}{2} \ . \tag{23}$$

Since the $1/D$ series in Eq. 14 diverges at $D=0$, we shift the expansion parameter to $(D+a)^{-1}$. If we use all three terms shown in the series, the interpolation function becomes

$$\langle A_D \rangle \approx \tfrac{2}{5} - \tfrac{12}{175}(D+a)^{-1} + \left(\tfrac{48}{175} - \tfrac{12}{175}a\right)(D+a)^{-2} \ . \tag{24}$$

The coefficient for the third term insures that the $1/D$ expansion agrees with Eq. 14 to order D^{-2}. The parameter a can now be fixed by using the $D = 0$ value for $\langle A_D \rangle$; this gives $a = 1.10679$. Finally, evaluating the expression at $D = 3$ yields

$$\langle A_3 \rangle = 0.39507 \qquad \boxed{\text{interpolation: } +0.20\% \text{ error}} \; . \tag{25}$$

This is significantly better than the truncated series result.

We note that dimensional interpolation would probably be much less appropriate for modeling the asphericities of *self-avoiding* (or excluded volume) random walks[86]. This is an example of a problem for which the physics does not vary as smoothly with D. Although still continuous, $\langle A_D \rangle$ shows more complex behavior[87] below $D = 4$. Dimensionality-based techniques of a different kind (expansions about $D = 4$, known as $4-\epsilon$ expansions) are useful for treating this problem[87, 88], but simple dimensional interpolation probably is not.

3.2. HARD SPHERES

The hard sphere problem becomes[60, 89] the hard disk problem for $D = 2$, the hard rod problem for $D = 1$, and the hard point problem for $D = 0$. Once again the $D = 0$ version is especially easy to solve. The equation of state[90] is $z_0 = -\ln(1-\rho)/\rho$, which yields

$$\tilde{C}_0 = \tfrac{1}{3}/(\tfrac{1}{2})^2 = \tfrac{4}{3} \; . \tag{26}$$

$D = 1$ is also very simple. The equation of state[91] is $z_1 = 1/(1-\sigma\rho)$, where σ is the sphere diameter, which yields

$$\tilde{C}_1 = \sigma^2/(\sigma)^2 = 1 \; . \tag{27}$$

(Note that the $D = 0$ and $D = 1$ equations of state also yield all higher virial coefficients.)

In order to interpolate, we again take several terms from the large-D series, but shift the parameter wherever necessary in order to prevent divergence at $D = 0$. Since we have two independent constraints at low D, we take four terms from the series, and introduce independent parameter shifts:

$$\ln \tilde{C}_D \approx D \ln \sqrt{3/4} - \tfrac{1}{2} \ln(D+a) + \ln \sqrt{24/\pi} - \tfrac{15}{4}(D+b)^{-1} \; . \tag{28}$$

Using the $D = 0$ and $D = 1$ values to fix the two constants, one obtains $a = 2.70491$ and $b = 16.20249$. The resulting $D = 3$ value is then

$$\tilde{C}_3 = 0.61828 \qquad \boxed{\text{interpolation: } -1.07\% \text{ error}} \; . \tag{29}$$

This is again better than the value obtained by truncating the $1/D$ series.

We note that dimensional interpolation has also been applied to the computation of the next two virial coefficients[60, 82]. The procedure used for these coefficients was somewhat different, in that only the first terms from the large-D expansions were used, but the interpolation scale was based on the known behavior of $\widetilde{C}_D$ as a function of D.

3.3. HELIUM ATOM

Electronic structure problems are generally characterized by singular behavior in the limit $D \rightarrow 1$. As in the case of the $D \rightarrow \infty$ limit, however, this singular behavior can be tamed and turned into a source of simplicity. For the helium atom, the $D \rightarrow 1$ limit of the scaled dimensionally generalized Hamiltonian is just the one-dimensional delta-function model[19, 92],

$$\widetilde{\mathcal{H}}_1 = -\tfrac{9}{2}(\partial^2/\partial\tilde{r}_1^2) - \tfrac{9}{2}(\partial^2/\partial\tilde{r}_2^2) - 3Z\delta(\tilde{r}_1) - 3Z\delta(\tilde{r}_2) + 3\delta(\tilde{r}_1 - \tilde{r}_2)\ . \quad (30)$$

The unique bound state eigenvalue is readily evaluated (though not in closed form), and found to be[93, 94]

$$\widetilde{E}_1 = -3.15539\ . \quad (31)$$

We construct an interpolating function by taking a couple of terms from the $1/D$ expansion, shifting the argument,

$$\widetilde{E}_D \approx -2.73777 - 0.58205(D+a)^{-1}\ , \quad (32)$$

and then fixing the parameter a by using of the $D \rightarrow 1$ energy; this gives $a = 0.39373$. (The reason for truncating the series after the first-order term is that the attempt to use more terms leads to complex-valued solutions for a.) The resulting $D=3$ energy is

$$\widetilde{E}_3 = -2.90928 \qquad \boxed{\text{interpolation: } +0.19\% \text{ error}}\ . \quad (33)$$

Once again the result is better than that obtained by truncating the $1/D$ series.

Dimensional interpolation has been applied to a number of simple electronic structure problems, including confined and shielded hydrogen atoms, the hydrogen molecule and molecule ion, and the helium atom in electric and magnetic fields[19]. The method generally provides at least rough estimates for $D=3$ energies and properties, assuming that there is smooth variation of the physics with respect to D. But, relative to $1/D$ expansions, interpolation is less systematic and more difficult to generalize to many-electron problems.

4. Dimensional bypass methods

Dimensional scaling is usually not the only avenue available for constructing approximate solutions to a problem. Moreover, when another approximation method can be found, it will frequently be much better than a simple dimensional expansion or interpolation. If one can make an approximation by some means not involving dimensionality, however, then one can usually construct an overall approximation which is better than either dimensional scaling or the alternative method by itself. The scheme, which we will call a dimensional "bypass"[95], can be diagramed[96] as shown in Figure 4.

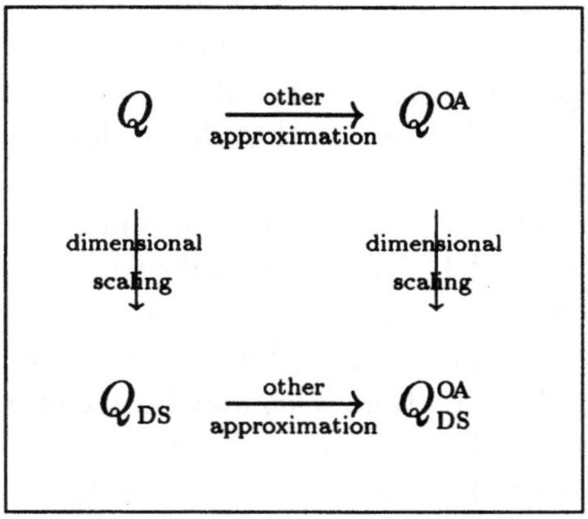

Figure 4. Scheme for the bypass approximation. See text for explanation.

The goal is to model Q, and we assume that we have available some other approximation to it, denoted Q^{OA}. This might be a mean-field approximation, for example. We also have available the possibility of a dimensional scaling approximation (either an expansion or an interpolation), denoted Q_{DS}. Finally, if we can compute both Q^{OA} and Q_{DS}, then we will almost certainly be able to compute the double approximation $Q^{\mathrm{OA}}_{\mathrm{DS}}$ as well.

The three approximate values just considered can be used to construct a better one, namely

$$Q \approx Q^{\mathrm{OA}} - Q^{\mathrm{OA}}_{\mathrm{DS}} + Q_{\mathrm{DS}} \,. \tag{34}$$

This is the dimensional bypass approximation. Geometrically, this equation amounts to saying that the four values in Figure 4 define an approximately flat surface (within the two-dimensional space of approximations), so that any three corner values can be used to model the fourth. Of course we use the three easy ones to model the hard one.

One reason why the bypass approach may be expected to be better than either underlying approximation by itself is that it amounts to using one approximation to model not the whole quantity Q, but rather only that part of Q which is not modeled by the second approximation. To see this, rewrite the equation for the bypass approximation as

$$Q - Q^{\text{OA}} \approx Q_{\text{DS}} - Q_{\text{DS}}^{\text{OA}} \,. \tag{35}$$

Thus dimensional scaling is being used to approximate only the error in the other approximation, $\Delta = Q - Q^{\text{OA}}$, which is presumably a much smaller quantity than Q itself. Alternatively, one can rewrite the bypass equation as

$$Q - Q_{\text{DS}} \approx Q^{\text{OA}} - Q_{\text{DS}}^{\text{OA}} \,, \tag{36}$$

and see that the other approximation is being used to model only the error $\delta = Q - Q_{\text{DS}}$ in the dimensional scaling approximation.

The two rearrangements of Eq. 34 make it clear that the dimensional bypass will be effective to the extent that dimensional scaling doesn't "interact" with the other approximation. That is, the other approximation needs to be in some sense orthogonal to the dimensional scaling approximation. (Thus the label "OA" might be taken to be an abbreviation for "orthogonal approximation".) It is certainly not assured that this will be the case. However, it is often true that dimensional scaling methods possess this property of limited interaction. This may be attributed to the quite different physical content of dimensional variation, compared to that of many common approximations. It is for this reason that, even though a bypass can be constructed out of any pair of approximations, it is particularly appropriate when one of them is based upon dimensional variation.

4.1. RANDOM WALK

An alternative approximation to the asphericity of a random walk is the quantity[47]

$$\overline{A}_D = \frac{1}{D-1} \left\langle \sum_{\substack{\mu=1 \\ \mu<\nu}}^{D} \sum_{\nu=1}^{D} \left({R_\mu}^2 - {R_\nu}^2 \right)^2 \right\rangle \Big/ \left\langle \left(\sum_{\mu=1}^{D} {R_\mu}^2 \right)^2 \right\rangle \,. \tag{37}$$

Comparing this with Eq. 1, one sees that $\overline{A}_D$ is the ratio of two averages, while $\langle A_D \rangle$ was the average of the ratio. The alternative approximation is readily evaluated[47] for any D, giving

$$\overline{A}_D = \frac{2D+4}{5D+4} \,. \tag{38}$$

Having an explicit expression for $\overline{A}_D$, we now seek to model the difference $\Delta_D = \langle A_D \rangle - \overline{A}_D$ by means of dimensional scaling. Since we know the behavior of this quantity for both $D \to \infty$ and $D=0$ (from the limiting behavior of $\langle A_D \rangle$ and the exact expression for $\overline{A}_D$), we will approximate it by dimensional interpolation. Here we take only the leading-order large-D behavior, but shift the parameter in order to keep the expression finite at $D=0$:

$$\Delta_D \approx -\tfrac{96}{175}(D+a)^{-1} . \tag{39}$$

Evaluating a using the $D=0$ result gives $a = \frac{192}{175}$. Then the bypass approximation $\langle A_D \rangle \approx \overline{A}_D + \Delta_D$ at $D=3$ is

$$\langle A_3 \rangle = 0.39242 \qquad \boxed{\text{bypass: } -0.47\% \text{ error}} . \tag{40}$$

This value is better than the series result, but not as good as the interpolation result.

The reason for not including one more term in Eq. 39, as we did for the simple interpolation of Eq. 24, is that doing so leads to a complex value for a. This difficulty, already mentioned in connection with Eq. 32, appears often when one attempts to tie down a truncated and shifted large-D expansion using low-D results. In this case, however, the values obtained when an extra term is included are still fairly meaningful. In fact, the two-term version of Eq. 39 yields complex conjugate pairs of values for $\langle A_D \rangle$ for any D (stemming from a complex conjugate pair of a values), but the imaginary components are quite small. At $D=3$ the values are $\langle A_3 \rangle = 0.39484 \pm 0.00040\,i$, whose error of $(0.14 \pm 0.10\,i)\%$ might be considered smaller than that for the dimensionally interpolated result.

4.2. HARD SPHERES

An alternative approach for the hard sphere fluid is the Percus-Yevick approximation. Virial coefficients are readily extracted from this approximation[97]. The values obtained actually depend on the route taken to the equation of state; here we consider the coefficients obtained by using the compressibility equation. For $D=3$ the equation of state can be evaluated in closed form,

$$z_3 = \frac{1+\eta+\eta^2}{(1-\eta)^3} , \tag{41}$$

where $\eta = \frac{1}{6}\pi\sigma^3\rho$ (with σ again denoting the sphere diameter). The third virial coefficient obtained from this equation is

$$\widetilde{C}_3^{\rm PY} = \tfrac{5}{18}\pi^2\sigma^6 \Big/ \left(\tfrac{2}{3}\pi\sigma^3\right)^2 = \tfrac{5}{8} . \tag{42}$$

Comparing this with Eq. 6, one sees that the Percus-Yevick approximation is exact for the third virial coefficient at $D=3$. In fact, this can be shown to be true for any D. Thus there isn't much content to the bypass procedure here. Formally, we can evaluate the difference $\Delta_D = \widetilde{C}_D - \widetilde{C}_D^{\mathrm{PY}}$ in the $D \to \infty$ limit and at $D=0$ and $D=1$, in all cases finding that it vanishes. Then the only reasonable interpolating function is

$$\Delta_D = 0 \ , \tag{43}$$

and the dimensional bypass approximation is identical to the Percus-Yevick approximation. Thus

$$\widetilde{C}_3 = \tfrac{5}{8} \qquad \boxed{\text{bypass: exact}} \ . \tag{44}$$

Of course, since there was no dimensional scaling contribution to the answer, this doesn't say anything about the bypass approach, except perhaps that it seems not to destroy a good approximation.

The bypass approximation becomes much more interesting for higher virial coefficients, since here the Percus-Yevick approximation is not exact. Using dimensional interpolation with a scale based on $\widetilde{C}_D$ to model the errors in the Percus-Yevick approximation, one obtains the following $D=3$ results[82] for the fourth through eighth coefficients:

	$\widetilde{D}_3$	$\widetilde{E}_3$	$\widetilde{F}_3$	$\widetilde{G}_3$	$\widetilde{H}_3$	
Percus-Yevick[97]	0.29688	0.12109	0.04492	0.01562	0.00519	
D-interpolation	0.27680	0.10219	0.03348	0.00993	0.00268	(45)
PY / DI bypass	0.28338	0.10797	0.03691	0.01169	0.00350	
Accurate values[98]	0.28695	0.11025	0.03881	0.01307	0.00432	

This demonstrates how the bypass approximation tends to be superior to either of the two underlying approximations. These computations can easily be taken further, but beyond the eighth coefficient there are no literature values for comparison.

4.3. HELIUM ATOM

For electronic structure problems, an obvious alternative approach is the Hartree-Fock approximation. With this choice, $\Delta_D = \widetilde{E}_D - \widetilde{E}_D^{\mathrm{HF}}$ is just the correlation energy. Thus a simple dimensional bypass approximation starts with the Hartree-Fock approximation, which for a $D=3$ helium atom is[99]

$$\widetilde{E}_3^{\mathrm{HF}} = -2.68168 \ , \tag{46}$$

and combines this with a dimensional scaling approximation to the correlation energy.

Again we will use dimensional interpolation to model the Δ_D. For this purpose we need the $D\to\infty$ and $D\to 1$ limiting behaviors for both the exact energy $\tilde{E}_D$ and Hartree-Fock energy $\tilde{E}_D^{\mathrm{HF}}$. For the exact energy these were given by Eq. 20 and Eq. 31. The corresponding results for the Hartree-Fock energy are are[100, 101, 102]

$$\tilde{E}_D^{\mathrm{HF}} = -(Z-c)^2 + 2\Big[(Z-c)^{3/2}(Z-\tfrac{5}{2}c)^{1/2} - (Z-c)(Z-\tfrac{11}{8}c)\Big]D^{-1} + \ldots$$
$$\tilde{E}_1^{\mathrm{HF}} = -Z^2 + \tfrac{1}{2}Z - \tfrac{1}{12}\,, \tag{47}$$

where $c=2^{-3/2}$, and for helium $Z=2$. Taking just the first two terms from the resulting $1/D$ series for Δ_D and shifting the parameter, we get

$$\Delta_D \approx -0.02698 - 0.06088(D+a)^{-1}\,. \tag{48}$$

Now using the $D\to 1$ results to evaluate a, one finds $a=0.35060$. Finally, evaluating the bypass approximation $\tilde{E}_D \approx \tilde{E}_D^{\mathrm{HF}} + \Delta_D$ at $D=3$ yields

$$\tilde{E}_3 = -2.90683 \qquad \boxed{\text{bypass: } +0.11\% \text{ error}}\,. \tag{49}$$

This is much better than the Hartree-Fock approximation itself, which is in error by -1.45% for helium.

Once again we have limited the interpolation function to two terms because the attempt to use more yields complex values for a. In this case it is even better to cut down further on the input to the calculation, using *only* the $D\to\infty$ and $D\to 1$ limiting results. (That is, it is better not to use the D^{-1} terms from the $1/D$ series.) When the limiting values are combined using a simple linear interpolation in $1/D$, one obtains quite accurate approximations for the correlation energy Δ_D, and consequently for the total energy[103]. For helium one obtains three-digit accuracy for the correlation energy, and five-digit accuracy for the total energy.

The bypass approximation has also been applied to the computation of correlation energies of larger atoms and ions[33]. For these calculations only the $D\to\infty$ solutions were used, but in conjunction with a generalized (multiple) scaling. Specifically, the total energy was partitioned into shell energies (ionization energies), and that for the nth shell was scaled by $(D+2n-3)^2/4$; these are just the factors which remove all dimension-dependence for a one-electron atom. This procedure yields the following energies for the noble gases:

	He	Ne	Ar	Kr
Hartree-Fock[99]	−2.86168	−128.547	−526.818	−2752.05
D-scaling	−2.73777	−128.205	−530.405	−2806.82
HF / DS bypass	−2.88866	−128.855	−527.443	−2753.74
Accurate values [99, 104, 105]	−2.90372	−128.938	−527.540	−2753.95

(50)

(For reference, we note that the dimensional scaling approximation used here is equivalent to the correlated "post-scaling" approximation of Ref. [33], and also to the "1-term approximation" of Ref. [59], except that the configurations used here are the experimental ones, rather than those obtained from the $D\to\infty$ limit.) Again, one can easily go further with the HF/DS bypass, but there is a lack of adequate literature values to compare to beyond those in the table.

5. Frontiers and prospects

In the foregoing, our main aim has been to sketch the chief D-scaling techniques and to show how they work in some very simple contexts. We turn now to the research frontiers. Table 1, patterned after its counterpart in Ref. [1], provides a concise guide to the research reported in the papers of our annotated bibliography[2–45] and in subsequent chapters of this book. We will focus here on efforts to develop practical ways to apply D-scaling to the daunting problem of many-body correlation effects in electronic structure. Although some of these approaches do not fit neatly into the three categories we've thus far considered, all seek to exploit the ability of the large-D limit to treat nonseparable many-body interactions.

Subhamiltonian techniques[82, 33] offer such an approach. Subhamiltonians are essentially $D\to\infty$ limit Hamiltonians which have been internally modified to simulate dominant finite-D effects. The simplest way to do this is to "pre-scale" the kinetic (centrifugal) terms of the limiting Hamiltonian using the same hydrogenic factors $(D+2n-3)^2/4$ mentioned above in connection with the "post-scaling" approach. That is, the subhamiltonian approach differs only in that the scalings are performed *before* the hamiltonian is minimized. Although the energies given by the two procedures are of comparable accuracy, the structures obtained from subhamiltonians constitute significantly more realistic representations of the true solutions at $D=3$. One advantage of this is that it is much easier to apply the model to the calculation of quantities other than energies. Another advantage is that the closer connection with real electronic structure opens a path to more

TABLE 1. Inventory of D-scaling studies of two-electron atoms (TEA), many-electron atoms (MEA), few-electron molecules (FEM), many-electron molecules (MEM), and other systems; citations to annotated bibliography and to other chapters in this book (Ch. ...).

Large-D limit: MEA[33, Ch. 2]; FEM[38]; MEM[20]; [13]
Low-D limit: [11, 19]
Dimensional interpolation: [19]
Low-order $1/D$ expansions: [2, 3, 4, 5, 6, 7, 14, 45]
Large-order $1/D$ expansions: TEA[12, 24, 40, Ch. 5]; FEM[22, 28]; [26, 36, 37, Ch. 4]
Summation methods: [3, 10, 12, 22, 24, 28, 36]
Subhamiltonians: MEA[33, Ch. 2]
Renormalization: TEA[23, 41]; MEA[25, 31, 32, 41]; MEM[41, 42, 44, Ch. 3]
Variant D-scaling schemes: TEA[40]; FEM[45]; [18, 19]
Complex D-scaling: [17, 21, 26]
Quasistationary states: [21, 26]
Properties other than energy: [19, 36, Ch. 2]
Electric or magnetic fields: [28, 35]
Inter-D degeneracies: TEA[9, 16]
Density functionals: [8, 23, 30]
Large N and Z limit: MEA[31, Ch. 3]
Other topics: [15, 18, 23, 29, 34, 45]
Reviews: [1, 7, 27, 39, 43]

drastic modifications of $D \to \infty$ hamiltonians. These "pre-structuring" approaches, which lead to representations wherein each lobe of an orbital is represented by a fractional point charge, are described in Chapter 2 of this book.

5.1. RENORMALIZATION

Another versatile approach employs dimensional *renormalization*. This is akin to the bypass strategy, in that it combines D-scaling with another approximation having complementary strengths. However, instead of employing a differencing scheme, as in Eq. 34, renormalization "injects" one approximation into the other via a suitably chosen parameter. The key idea is to renormalize the parameter with the aid of one approximation, then use the adjusted value in the other approximation to enhance its accuracy. The method can be implemented in either of two ways. Figure 5 shows these in their simplest or zeroth-order forms. The two versions may be called dimensional renormalization (DR) and orthogonal renormalization (OR), depending on whether the adjusted parameter is actually used in the dimensionality-based approximation or in the other one, the "orthogonal" approximation. (We use that awkward name only for consistency with the notation of Figure 4.)

DR-0 approximation	OR-0 approximation
$$\begin{array}{ccc} Q(Z) & \rightarrow & Q^{OA}(Z) \\ \wr & & \parallel \\ Q_{DS}(Z^{OA}) & \rightarrow & Q^{OA}_{DS}(Z^{OA}) \end{array}$$	$$\begin{array}{ccc} Q(Z) & \approx & Q^{OA}(Z_{DS}) \\ \downarrow & & \downarrow \\ Q_{DS}(Z) & = & Q^{OA}_{DS}(Z_{DS}) \end{array}$$

Figure 5. Schemes for renormalization approximations. See text for explanation.

Here Z is the parameter, and Z_{DS} and Z^{OA} are the renormalized values for the two schemes. As indicated in Figure 5, the modified values are *defined* by the respective equalities (right column for DR-0, bottom row for OR-0),

$$Q^{OA}(Z) = Q^{OA}_{DS}(Z^{OA}) \qquad\qquad Q_{DS}(Z) = Q^{OA}_{DS}(Z_{DS}) \, . \tag{51}$$

The renormalized parameters are then *used* in the two approximate equalities (left column for DR-0, top row for OR-0), which constitute the actual approximations:

$$Q(Z) \approx Q_{DS}(Z^{OA}) \qquad\qquad Q(Z) \approx Q^{OA}(Z_{DS}) \, . \tag{52}$$

Note that if one formally inverts Eqs. 51, then the two approximations may be summarized as

$$Q(Z) \approx Q_{DS}(1/Q^{OA}_{DS})Q^{OA}(Z) \qquad\qquad Q(Z) \approx Q^{OA}(1/Q^{OA}_{DS})Q_{DS}(Z) \, . \tag{53}$$

These schematic expressions make it clear that renormalization is utilizing one approximation to undo the errors of the other. Thus, one can expect that dimensional renormalization, like the dimensional bypass approximation, will be effective only if there is minimal interaction between the two underlying approximations.

As an example of the renormalization approach, we again consider many-electron atoms, and again utilize Hartree-Fock as the other approximation. Results obtained with the two renormalization schemes[25, 41, 106] are compared with those obtained with the bypass approach in the first three lines of the following table:

	He	Ne	Ar	
HF / DS bypass	−2.88866	−128.85457	−527.44307	
DR-0 approximation[25]	−2.88876	−128.85474	−527.44275	
OR-0 approximation[41, 106]	−2.88928	−128.85045	−527.43109	(54)
DR-1,2 approximation[32]	−2.90418	−128.90896	−527.56731	
Accurate values[99, 104, 105]	−2.90372	−128.93757	−527.53967	

The three basic ways of combining approximations (bypass, DR-0, and OR-0) give similar results. In each case, injecting the large-D limit provides much of the correlation energy (about 65%, 80%, 85%, respectively, for He, Ne, Ar) that is lacking in the Hartree-Fock approximation.

Figure 6 gives a geometrical construction that shows why, despite substantial differences in formulation, the HF/DS bypass, DR-0, and OR-0 approximations are virtually equivalent (as seen in the numerical results above). The renormalized charges, $Z+\delta Z$, differ appreciably for the DR and OR procedures, but for both the displacement from the actual charge is quite small. The correlation energy, Δ_D, thus is nearly proportional to the increment in the renormalized charge,

$$\Delta_D \equiv E_D - E_D^{\mathrm{HF}} = S_D^{\mathrm{HF}}\, \delta Z_D \ . \tag{55}$$

The slope $S_D^{\mathrm{HF}} \equiv \partial E_D^{\mathrm{HF}}/\partial Z$, and here we denote by δZ_D the increment (indicated in Figure 6 by solid points) that yields an exact value for Δ_D. For either procedure, the correlation energy at $D=3$ then is related to that at the large-D limit by

$$\Delta_D = (S_3^{\mathrm{HF}}/S_\infty^{\mathrm{HF}})\,(\delta Z_3/\delta Z_\infty)\,\Delta_\infty \ . \tag{56}$$

In both zeroth-order approximations, the unknown increment δZ_3 is merely replaced by δZ_∞, and since the slope does not change much with D, both procedures give $\Delta_3 \approx \Delta_\infty$ in accord with the HF/DS bypass.

The invocation of an auxiliary parameter as the means for combining the two underlying approximations may seem at first like a complication, compared to the bypass approach. In fact, however, the parametrized approach has the advantage of providing a natural route to improvements. Since the renormalization procedure is limited by the interaction of its component approximations, it is clear that improvements can be made only if this interaction is addressed. In general one can expect this to be a difficult task.

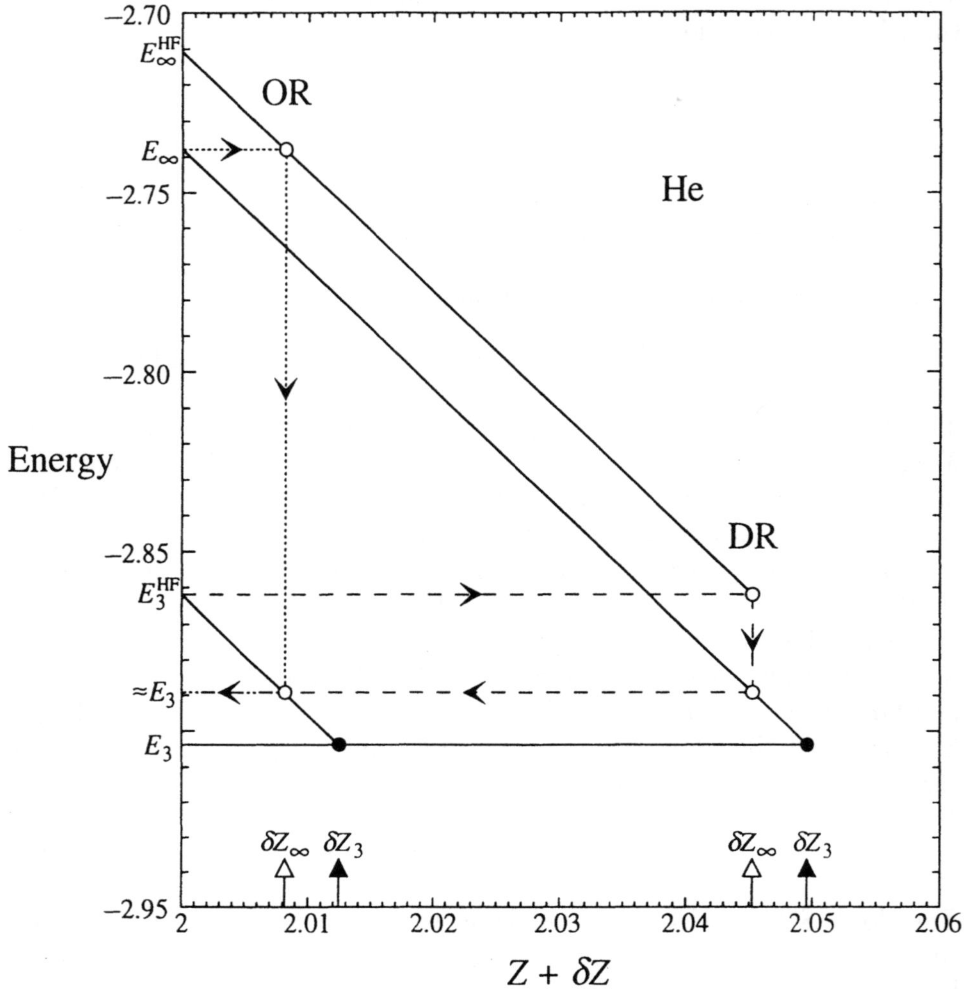

Figure 6. Charge renormalization procedures utilizing as input Hartree-Fock energies at $D = 3$ and $D \to \infty$ and exact energy at $D \to \infty$ limit to construct approximations to the $D = 3$ energy for helium (in hartree units). *Cf.* Eq. 53 and Figure 5. For the OR procedure: $E_3(Z) \approx E_3^{\text{HF}}(Z + \delta Z)$; for the DR procedure: $E_3(Z) \approx E_\infty(Z + \delta Z)$. Criteria for determining δZ are, for the OR procedure: $E_\infty^{\text{HF}}(Z{+}\delta Z) = E_\infty(Z)$; for the DR procedure: $E_\infty^{\text{HF}}(Z{+}\delta Z) = E_3^{\text{HF}}(Z)$. Open circles denote values of charge increments (δZ_∞) determined from these criteria, full circles values (δZ_3) that render the approximations exact.

However, it may yield to attack by perturbation theory in the renormalization parameter. For many-electron atoms, at least, this approach has proven to be very effective[32]. This is demonstrated by the energies labeled DR-1,2 in the table, which were obtained by augmenting the zeroth-order renormalization scheme with the leading order correction in $1/Z$. In particular, when electronic configurations of the same symmetry become degenerate for $Z\to\infty$, as often occurs, it becomes important to include the zeroth-order mixing[107]. That correction (included in DR-1 and OR-1) only requires diagonalizing a small matrix in a hydrogenic basis. Again, the accuracy obtained is much better than that of the input approximation; incorporating the leading $1/Z$ term in the renormalization was found[32, 41] to reduce the mean error in the correlation energy to only about 5%, for atoms and cations with N = 2–18 electrons and Z = 2–28.

Although the HF/DS bypass and DR procedures have not been tractable for many-electron molecules, happily the OR approximation has now provided a promising approach[42, 44]. In essence, this combines renormalized charge increments δZ_α determined for each constituent atom α with slopes, $\sigma_\alpha \equiv \partial E^{\rm HF}/\partial Z_\alpha$, computed from the *molecular* Hartree-Fock energy. The charge increments are related to the *atomic* correlation energies and slopes by

$$\Delta_\alpha = S_\alpha \, \delta Z_\alpha \; . \tag{57}$$

The corresponding approximation for the correlation energy of a heteronuclear diatomic molecule is

$$\Delta_{\alpha\beta} \approx \left(\sigma_\alpha + \frac{Z_\beta}{R}\right) \delta Z_\alpha + \left(\sigma_\beta + \frac{Z_\alpha}{R}\right) \delta Z_\beta + \frac{\partial E}{\partial R}\, \delta R \; . \tag{58}$$

The Z/R terms arise from the nuclear-nuclear repulsion. The third term represents a contribution from renormalizing R, the internuclear distance. Its contribution is estimated[44] to be an order of magnitude smaller than those from renormalizing the nuclear charges; also, at the equilibrium bond distance and for large R the third term vanishes. For a polyatomic molecule, the correlation energy can be obtained in the same way, just by adding analogous terms to Eq. 58. Kais and coworkers[42, 44] have tested this approach by determining Δ at the equilibrium R for about 20 molecules, including Be_2 ... F_2; HF, LiF, BeO; H_2O, CH_2, and CH_4. Comparison with high level correlated *ab initio* calculations, MP4 and CISD(T), indicates that their renormalization method typically gives 70% or more of the correlation energy. This is remarkable, since beyond large-D results for the constituent atoms, the method requires only slight modifications of the standard Hartree-Fock algorithm. As expected, better results are obtained with input approximations which model nondynamical electron correlation, such as CAS (complete active space multiconfigurational HF) and unrestricted HF. Further aspects are discussed by Kais and Bleil in Chapter 3 of this book, including prospects for renormalizing expectation values and wavefunctions.

5.2. CONTINUATION AND CONJUGACY

The key question in any D-scaling treatment is how to connect the dimensional limit to $D = 3$. This requires some means, explicit or implicit, for dimensional continuation. There are many options, since the only rigorous constraint on the scaling scheme is that it reduce to an identity at $D=3$. All D-scaling methods benefit from the fact that many-body effects, when expressed in units calibrated by prototype one- or few-body problems, tend to have only mild dependence on dimensionality. The subhamiltonian and renormalization procedures indeed rely entirely on such scaling, without attempting dimensional continuation for the actual many-body system. However, $1/D$ perturbation expansions and summation techniques enable explicit evaluation of the dimension dependence, and the recent work has substantially extended the scope of this forthright method. Here we comment on this, again emphasizing prospects for conjugal links[95] with other D-scaling or conventional methods.

The state of the art has been markedly advanced by a linear algebraic method for accurate computation of the $1/D$ perturbation coefficients[36]. The algorithm is formulated completely in terms of tensor arithmetic. That makes it easier to treat many-body systems and excited states, as well as simplifying the computer codes and memory management. Furthermore, it is particularly well suited for implementation on parallel computer architectures. A general parallel code is now available for calculation of energy levels (either real or complex), wavefunctions, and expectation values for an arbitrary many-body system[37].

The scope of rigorous, unadulterated $1/D$ calculations is of course limited, even with today's supercomputers. An analysis of the processing and memory requirements[36] indicates that large-order (20th or higher) expansions will be feasible for no more than 6 nonseparable degrees of freedom, lower order (say, 10th) for about 10 degrees of freedom. Practical applications to larger systems thus will require use of separability approximations. This compares favorably with conventional methods, however, which employ the Born-Oppenheimer and self-consistent field approximations to reduce computations to differential equations with a single degree of freedom. In analogous fashion, a systematic procedure might be formulated to approximately separate a large system into subsets of strongly coupled degrees of freedom. The ability of the $1/D$ method to handle such subsets might then be exploited, for instance, to incorporate correlations between electron pairs or triples in a generalized Hartree-Fock method. Goodson is developing that approach, as described in Chapter 4 of this book.

Although the perturbation expansions are usually asymptotic, accurate exponentially convergent results can be obtained (as suggested by Figures

2 and 3) by means of summation approximants designed to model singularity structure in the complex $1/D$ plane. Such results have been obtained for a variety of electronic structures, as noted above and in Table 1. Of particular interest are results for large $|m|$ Rydberg states of the H atom in a magnetic field[35], a notorious challenge for conventional methods. In contrast to other approaches, the $1/D$ expansion is applicable to the entire range of field strengths and copes well with myriad avoided crossings. Also gratifying is the applicability of $1/D$ expansions to complex energy eigenvalues, corresponding to quasibound or resonant states. The extension to quasistationary states, first made by Popov and coworkers[68], merely involves allowing the radial coordinate for the large-D limit to be complex. For several model systems, even a first or second-order $1/D$ expansion was found to give good results for both the location and width of the resonant states[21]. The calculations are much simpler than conventional methods. High-order $1/D$ expansions (~20–30) markedly improve the accuracy of the resonance positions and lifetimes[26]. Among the quasibound states treated are very broad rotational resonances of H_2 which lie far above the centrifugal barrier and have not yet been evaluated by other methods.

Another recent development has neatly solved a basic problem that had often handicapped $1/D$ expansions. Since these are Taylor expansions about the $D \to \infty$ limit, the zeroth-order term is evaluated at the minimum of an effective potential surface for that limit. However, for systems that are weakly bound at $D=3$, such as the hydride ion, the effective potential surface for large D often exhibits a saddle point rather than a minimum. One way to deal with this is to use complex D-scaling; it enabled Rost to develop an analytic first-order $1/D$ expansion for feebly bound three-particle Coulomb systems which gave good estimates for the ground-state binding energies[17]. A more convenient method has been employed by Watson and Goodson[40]. They simply replace Z in the potential by a nominal D-dependent charge, $\zeta(D)$. This is chosen to be large enough at $D \to \infty$ to endow the potential with a decent minimum there, whereas at $D = 3$ the charge assumes the actual physical value, $\zeta(D) = Z$. At intermediate points, $1/\zeta$ was defined to be linear in $1/D$. Since the modified potential has a stable large-D limit, the $1/D$ expansion can be carried out and summed in the usual way. For the hydride ion, a difficult case for conventional methods, this gave excellent accuracy[40]. The $1/D$ method and results are reviewed by Dunn, *et al.* in Chapter 5 of this book.

Beyond its computational utility, the $\zeta(D)$ device exemplifies an important aspect of D-scaling[18, 19], also seen in the subhamiltonian and renormalization methods. Only the Hamiltonian at $D=3$ is sacred; at any other dimension, we are free to modify either or both the kinetic and potential energy terms. Thereby we can retain the simplicity offered at special D

values such as the large-D and low-D limit, while adjusting these in ways that better model the $D=3$ solution and facilitate dimensional continuation. In analogy to the situation familiar in electrodynamics, we refer to this flexibility as the *gauge property.*

This property has a key role in D-scaling treatments of molecules. The scaling usually used for atoms (the alternative noted under Eq. 9) is not adequate for molecules. With that scaling, the potential curves at $D\rightarrow\infty$ are purely repulsive because the sum of the scaled centrifugal term and the nuclear repulsion outweighs chemical bonding[38]. In contrast, the scaling chosen in Eq. 9 is akin to using the $\zeta(D)$ gauge[18]. The scaled potential curves at both $D\rightarrow\infty$ and $D\rightarrow 1$ then have stable minima qualitatively like $D=3$. This scaling hence fulfills the basic aim of removing the major, generic D-dependence and is suitable both for interpolation[19] and for carrying out the $1/D$ expansion. A careful study of further aspects, with detailed application to H_2^+, is given by Cohen and Goodson[45].

The gauge property implies that for some ideal choice of scaling, the energy at $D\rightarrow\infty$ would become identical to that at $D=3$. Thereby the $1/D$ expansion would shrink to just the zeroth-order term. Of course, we have no way to find that ideal scaling without determining the exact D-dependence for the problem, something harder to do than solving it for $D=3$ alone! The same situation pertains for the subhamiltonian and renormalization methods. Yet, to the extent that an approximation to the ideal scaling can be found with reasonable effort, use of that approximation to define the scaling will remove much of the D-dependence. The rest can then be more easily determined from a $1/D$ expansion, which will become shorter as the starting approximation is improved.

This situation suggests melding the $1/D$ expansion with the subhamiltonian or renormalization approaches. Several ways to do this seem promising; here we note just one as an example. For a many-electron atom or molecule the exact D-dependence of Hartree-Fock results can readily be determined. It requires only a straightforward modification of standard $D=3$ computer codes, to include a D-dependent centrifugal potential in the Hamiltonian; this has already been done for the two-electron atom[102]. The Hartree-Fock results can be used to define the scaling, essentially just by replacing the H atom results used in Eq. 9 (or those for higher orbitals in many-electron systems[33, 59]). Then the $1/D$ expansion would need only account for the D-dependence of the correlation energy. From the extant renormalization and HF/DS bypass studies, we already know that the zeroth-order term would include most of the correlation energy. The same strategy could be systematically pursued using *ab initio* input or perhaps density functional theory to refine the D-dependence for the scaling and the $1/D$ expansion to obtain improved approximations to the correlation energy.

Many such liaisons of D-scaling with other methods are now in prospect; we expect further conjugal unions will emerge and be fruitful.

Acknowledgments

This paper represents the work of many coworkers, now not at all localized. We are especially grateful to Gerardo Beltrame for his derivations of $1/D$ expansions for random walks and hard spheres; to Martin Dunn, Tim Germann, David Goodson, John Morgan III, and Deborah Watson for their extensive development of $1/D$ expansions for electronic structure; and to Sabre Kais for his work on renormalization.

Annotated bibliography

The annotated references [2–45] are intended to extend the list given in [1], to include all papers (up to mid-1995) treating applications of dimensional scaling or equivalent methods to chemical physics. Table 1 and the text provide a guide to the major topics and also supply some cross references not made apparent by the brief annotations.

References

1. D. R. Herschbach, J. Avery, and O. Goscinski, eds., *Dimensional Scaling in Chemical Physics* (Kluwer Academic, Dordrecht, 1993).
2. A. A. Belov and Yu. E. Lozovik, *Sov. Phys. JETP* **67**, 2413 (1988). *First order $1/D$, bound states for clusters of N identical particles.*
3. A. V. Sergeev, *Sov. J. Nucl. Phys.* **50**, 589 (1989). *$1/D$ expansions for analytic three-body problems.*
4. A. A. Belov, Yu. E. Lozovik, and A. Gonzalez, *Phys. Lett. A* **142**, 389 (1989). *First-order $1/D$ for electrons in a harmonic oscillator potential.*
5. A. A. Belov and Yu. E. Lozovik, *Theor. Math. Phys.* **81**, 1294 (1989). *First order $1/D$ for 3 and 4 particle clusters, dynamical symmetry.*
6. A. Gonzalez, *Few Body Syst.* **8**, 73 (1990). *First-order $1/D$ for N bosons, central power-law potential.*
7. A. Gonzalez, *Few Body Syst.* **10**, 43 (1991). *Quantum problems and $1/D$ expansions*
8. A. Holas, P. M. Kozlowski, and N. H. March, *J. Phys. A* **24**, 4249 (1991). *Gradient expansion for kinetic energy density in D-dimensions.*
9. J. M. Rost, S. M. Sung, D. R. Herschbach, and J. S. Briggs, *Phys. Rev. A* **46**, 2410 (1992). *Inter-D degeneracies, two-electron doubly excited states.*
10. D. Z. Goodson and D. R. Herschbach, *Phys. Rev. A* **46**, 5428 (1992). *Approximate $1/D$ summation methods.*
11. C. M. Bender, S. Boettcher, and L. Lipatov, *Phys. Rev. Lett.* **68**, 3674 (1992). *Zero-D limit, expansions in powers of D.*
12. D. Z. Goodson, M. López-Cabrera, D. R. Herschbach, and J. D. Morgan III, *J. Chem. Phys.* **97**, 8481 (1992). *High-order $1/D$ perturbation expansions for two-electron atoms.*
13. Y. Y. Goldschmidt, *Nucl. Phys. B* **393**, 507 (1993). *Path integral form for large-D limit, N-body problem.*

14. A. Gonzalez and D. Leal, *J. Phys. B* **26**, 1253 (1993). *First order 1/D for binding energies of rare gas clusters.*
15. S. Kais and G. Beltrame, *J. Phys. Chem.* **97**, 2453 (1993). *D-scaling for Regge trajectories.*
16. M. Dunn and D. K. Watson, *J. Phys. Chem.* **97**, 2457 (1993). *Wrong parity states of doubly excited two-electron atoms.*
17. J. M. Rost, *J. Phys. Chem.* **97**, 2461 (1993). *Complex D-scaling.*
18. C. A. Traynor and D. Z. Goodson, *J. Phys. Chem.* **97**, 2464 (1993). *D-scaling without the Born-Oppenheimer approximation.*
19. M. López-Cabrera, A. L. Tan, and J. G. Loeser, *J. Phys. Chem.* **97**, 2467 (1993). *D-scaling and interpolation.*
20. S. M. Sung and J. M. Rost, *J. Phys. Chem.* **97**, 2479 (1993). *Chemical binding in the large-D limit.*
21. S. Kais and D. R. Herschbach, *J. Chem. Phys.* **98**, 3990 (1993). *D-scaling for quasistationary states.*
22. V. S. Popov and A. V. Sergeev, *Phys. Lett. A* **172**, 193 (1993). *Analytic results for 1/D expansion for* H_2^+.
23. S. Kais, D. R. Herschbach, N. C. Handy, C. W. Murray, and G. J. Laming, *J. Chem. Phys.* **99**, 417 (1993). *Density functionals and D-renormalization for exactly solvable Hooke's law atom.*
24. D. Z. Goodson and D. K. Watson, *Phys. Rev. A* **48**, 2668 (1993). *High-order 1/D expansions for excited two-electron atoms.*
25. S. Kais, S. M. Sung, and D. R. Herschbach, *J. Chem. Phys.* **99**, 5184 (1993). *D-renormalization for many-electron atoms.*
26. T. C. Germann and S. Kais, *J. Chem. Phys.* **99**, 7739 (1993). *High-order 1/D expansions for complex eigenvalues.*
27. D. R. Herschbach, *Proc. Am. Phil. Soc.* **137**, 532 (1993). *Review including D-interpolation of virial coefficients.*
28. V. S. Popov and A. V. Sergeev, *Phys. Lett. A* **193**, 165 (1994). *Large order 1/n expansion for many degrees of freedom.*
29. C. M. Bender, S. Boettcher, and L. Mead, *J. Math. Phys.* **35**, 368 (1994). *Random walk on a D-dimensional spherical lattice.*
30. S. M. Valone, *Int. J. Quantum Chem.* **49**, 591 (1994). *D-scaling for constrained search energy density functionals.*
31. S. Kais, S. M. Sung, and D. R. Herschbach, *Int. J. Quantum Chem.* **49**, 657 (1994). *D-renormalization for large-Z and large-N atoms.*
32. S. Kais and D. R. Herschbach, *J. Chem. Phys.* **100**, 4367 (1994). *D-renormalization and 1/Z expansion for many-electron atoms.*
33. J. G. Loeser, J. H. Summerfield, A. L. Tan, and Z. Zheng, *J. Chem. Phys.* **100**, 5036 (1994). *Correlated models suggested by large-D limit.*
34. C. M. Bender, S. Boettcher, and M. Moshe, *J. Math. Phys.* **35**, 4941 (1994). *Spherically symmetric random walks in noninteger D.*
35. T. C. Germann, D. R. Herschbach, M. Dunn, and D. K. Watson, *Phys. Rev. Lett.* **74**, 658 (1995). *D-scaling for* H *atom in magnetic field.*
36. M. Dunn, T. C. Germann, D. Z. Goodson, C. A. Traynor, J. D. Morgan III, D. K. Watson, and D. R. Herschbach, *J. Chem. Phys.* **101**, 5987 (1994). *Matrix method for 1/D perturbation expansions.*
37. T. C. Germann, D. R. Herschbach, and B. M. Boghosian, *Comput. Phys.* **8**, 712 (1994). *Parallel computation for 1/D expansions.*
38. S. Kais, T. C. Germann, and D. R. Herschbach, *J. Phys. Chem.* **98**, 11015 (1994). *Corresponding states for D=3 and large-D potential curves of two-electron molecules.*
39. J. Avery, in *Structure and Dynamics of Atoms and Molecules: Conceptual Trends*, J.-L. Calais and E. S. Kryachco, eds. (Kluwer Academic, Dordrecht, 1995), p. 135. *Review, large-D limit for 3-particle systems.*

40. D. K. Watson and D. Z. Goodson, *Phys. Rev. A* **51**, R5 (1995). *1/D expansion for hydride ion improved by Z-variation.*
41. S. Kais and R. Bleil, *J. Chem. Phys.* **102**, 7472 (1995). *D-renormalization of Hartree-Fock energies for atoms.*
42. R. Bleil, A. Faliks, M. Miletic, and S. Kais, *J. Chem. Phys.* (in press, 1995). *D-renormalization for homonuclear diatomic molecules.*
43. D. R. Herschbach, *Int. J. Quantum Chem.* (in press, 1995). *Review, D-scaling and renormalization.*
44. R. Bleil and S. Kais, *Int. J. Quantum Chem.* (in press, 1995). *D-renormalization for heteronuclear diatomic and polyatomic molecules.*
45. J. M. Cohen and D. Z. Goodson, *J. Chem. Phys.* (in press, 1995). *D-scaling for* H_2^+ *without Born-Oppenheimer approximation.*
46. D. N. Theodorou and U. W. Suter, *Macromolecules* **18**, 1206 (1985).
47. J. Rudnick and G. Gaspari, *J. Phys. A* **19**, L191 (1986); *Science* **237**, 384 (1987).
48. G. Zifferer and O. F. Olaj, *J. Chem. Phys.* **100**, 636 (1994).
49. H. W. Diehl and E. Eisenriegler, *J. Phys. A* **22**, L87 (1989).
50. J. E. Kilpatrick, *Adv. Chem. Phys.* **20**, 39 (1971).
51. D. R. Herschbach, *J. Chem. Phys.* **84**, 838 (1986).
52. A. L. Tan and J. G. Loeser, Ref. [1], p. 230.
53. C. L. Pekeris, *Phys. Rev.* **115**, 1216 (1959).
54. D. R. Herschbach, Ref. [1], p. 7.
55. E. Witten, *Phys. Today* **33** (no. 7), 38 (1980).
56. J. Avery, D. Z. Goodson, and D. R. Herschbach, *Theor. Chim. Acta* **81**, 1 (1991).
57. J. G. Loeser, Ref. [1], p. 389.
58. D. Vollhardt, in *Correlated Electron Systems*, V. J. Emery, ed. (World Scientific, Singapore, 1993), p. 57.
59. J. G. Loeser, *J. Chem. Phys.* **86**, 5635 (1987).
60. J. G. Loeser, Z. Zhen, S. Kais, and D. R. Herschbach, *J. Chem. Phys.* **95**, 4525 (1991).
61. A. Chatterjee, *Phys. Rep.* **186**, 249 (1990).
62. S. Coleman, *Aspects of Symmetry* (Cambridge, Cambridge, 1985), p. 351.
63. V. S. Popov, Ref. [1], p. 217.
64. A. D. Dolgov and V. S. Popov, *Phys. Lett. B* **185**, 556 (1979).
65. G. Moreno and A. Zepeda, *J. Phys. B* **17**, 21 (1984).
66. L. D. Mlodinow and M. P. Shatz, *J. Math. Phys.* **25**, 943 (1984).
67. C. H. Lai, *J. Math. Phys.* **28**, 1801 (1987).
68. V. M. Vaĭnberg, V. D. Mur, V. S. Popov, and A. V. Sergeev, *Pis'ma Zh. Eksp. Teor. Fiz.* **44**, 9 (1986) [*JETP Lett.* **44**, 9 (1986)].
69. C. M. Bender, L. D. Mlodinow, and N. Papanicolaou, *Phys. Rev. A* **25**, 1305 (1982).
70. V. D. Mur, V. S. Popov, and A. V. Sergeev, *Zh. Eksp. Teor. Fiz.* **97**, 32 (1990) [*Sov. Phys. JETP* **70**, 16 (1990)].
71. M. López-Cabrera, D. Z. Goodson, D. R. Herschbach, and J. D. Morgan III, *Phys. Rev. Lett.* **68**, 1992 (1992).
72. D. Z. Goodson and D. R. Herschbach, *Phys. Rev. Lett.* **58**, 1628 (1987).
73. C. M. Bender and S. A. Orszag, *Advanced Mathematical Methods for Scientists and Engineers* (McGraw-Hill, New York, 1978), pp. 381 and 395.
74. D. J. Doren and D. R. Herschbach, *Chem. Phys. Lett.* **118**, 115 (1985).
75. V. S. Popov, V. D. Mur, A. V. Sergeev, and V. M. Weinberg, *Phys. Lett. A* **149**, 418 (1990).
76. D. Z. Goodson and M. López-Cabrera, Ref. [1], p. 275.
77. G. Beltrame, Ph.D. dissertation (Harvard University, 1995).
78. S. Chandrasekhar, *Rev. Mod. Phys.* **15**, 1 (1943).
79. G. F. Lawler, *Intersections of Random Walks* (Birkhäuser, Boston, 1991), p. 12.
80. W. C. Forsman and R. E. Hughes, *J. Chem. Phys.* **38**, 2118 (1963).

81. G. Beltrame, unpublished work.
82. Z. Zhen and J. G. Loeser, Ref. [1], p. 429.
83. D. R. Herschbach, Ref. [1], p. 61.
84. D. R. Herschbach, J. G. Loeser, and D. K. Watson, *Z. Phys. D* **10**, 195 (1988).
85. U. Sukhatme and T. Imbo, *Phys. Rev. D* **28**, 418 (1983).
86. G. Slade, *Math. Intelligencer* **16**, 29 (1994).
87. J. A. Aronovitz and D. R. Nelson, *J. Phys. (France)* **47**, 1445 (1986).
88. O. Jagodzinski, E. Eisenriegler, and K. Kremer, *J. Phys. I (France)* **2**, 2243 (1992).
89. M. Baus and J. L. Colot, *Phys. Rev. A* **36**, 3912 (1987).
90. A. Baram and M. Luban, *J. Phys. A* **19**, L585 (1986).
91. L. Tonks, *Phys. Rev.* **50**, 955 (1936).
92. D. R. Herrick and F. H. Stillinger, *Phys. Rev. A* **11**, 42 (1975).
93. C. M. Rosenthal, *J. Chem. Phys.* **55**, 2474 (1971).
94. D. J. Doren and D. R. Herschbach, *J. Chem. Phys.* **87**, 433 (1987).
95. Z. A. Melzak, *Bypasses* (Wiley, New York, 1983).
96. Z. Zhen and J. G. Loeser, Ref. [1], p. 83.
97. D. A. McQuarrie, *Statistical Mechanics* (Harper & Row, New York, 1976), p. 279.
98. E. J. Janse van Rensburg, *J. Phys. A* **26**, 4805 (1993).
99. C. F. Bunge, J. A. Barrientos, A. V. Bunge, and J. A. Cogordan, *Phys. Rev. A* **46**, 3691 (1992); C. F. Bunge, J. A. Barrientos, and A. V. Bunge, *At. Data Nucl. Data Tables* **53**, 113 (1993).
100. D. Z. Goodson and D. R. Herschbach, *J. Chem. Phys.* **86**, 4997 (1986).
101. Y. Nogami, M. Vallières, and W. van Dijk, *Am. J. Phys.* **44**, 886 (1976); L. L. Foldy, *Am. J. Phys.* **44**, 1192 (1976).
102. J. G. Loeser and D. R. Herschbach, *J. Chem. Phys.* **84**, 3893 (1986).
103. J. G. Loeser and D. R. Herschbach, *J. Phys. Chem.* **89**, 3444 (1985); *J. Chem. Phys.* **86**, 3512 (1987).
104. S. J. Chakravorty, S. R. Gwaltney, E. R. Davidson, F. A. Parpia, and C. Froese Fischer, *Phys. Rev. A* **47**, 3649 (1993).
105. V. Termath, W. Klopper, and W. Kutzelnigg, *J. Chem. Phys.* **94**, 2002 (1991).
106. S. Kais, unpublished work.
107. E. R. Davidson, S. A. Hagstrom, and S. J. Chakravorty, *Phys. Rev. A* **44**, 7071 (1991).

SUBHAMILTONIAN METHODS

J. G. LOESER
Department of Chemistry
Oregon State University
Corvallis, OR 97331-4003
United States

Abstract. We describe a formalism for generating high-dimensional point-charge representations of electronic structure. The pseudoclassical structures arise as the minima of subhamiltonians, or large-D limit hamiltonians internally modified (without sacrificing their derivative-free form) to reflect certain finite-D effects. Simple subhamiltonians are obtained by scaling the kinetic terms (generalized centrifugal potentials) of a $D \to \infty$ limit hamiltonian, while more realistic ones utilize additional potentials which reflect explicit nodal structure. For molecules, subhamiltonians typically yield distinct structures for alternative assignments of charges to the atoms. A simple approximation for the interaction of these structures, as well as pathways to better approximations, are described. Subhamiltonians are very rough models, but their simplicity and the dramatically different physical content of the underlying approximations, compared to that of orbital-based treatments, render them potentially useful for modeling correlation effects in large systems.

1. One atom

For an electronic structure problem treated in the Born-Oppenheimer approximation, the large-dimension limit — that is, the hypothetical problem in which each electron is assigned unlimited Cartesian degrees of freedom — is a singular limit of classical character. The limiting solution describes a geometry of point charges with fixed interparticle distances. These "electronic structures" bear a superficial resemblance to pre-quantal models, except for the fact that they are inherently high-dimensional. That is, the set

C. A. Tsipis et al. (eds.), New Methods in Quantum Theory, 33–54.

of interparticle distances typically cannot be realized in a three-dimensional figure.

The classical structure may be obtained by taking the $D \to \infty$ limit of the D-dependent probability distribution in the space of internal coordinates, $|\Phi(\{r_{ij}\})|^2$, which collapses to the singular limiting structures (with widths proportional to $D^{-1/2}$). But it is much simpler to work directly with the $D \to \infty$ limit of the hamiltonian for the probability amplitude, $\mathcal{H}_\Phi(\{r_{ij}\})$. The singular limit is characterized by the disappearance of derivative terms (which have coefficients proportional to D^{-2}), so that the solution is obtained simply by minimization of the derivative-free portion of the hamiltonian.[1, 2, 3, 4, 5]

Although the $D \to \infty$ limit solutions are very easy to obtain, they are usually not quantitatively useful as models for $D=3$ electronic structure. At least three different routes have been taken, either alone or in combination, in addressing this problem. First is the method of dimensional interpolation, in which the $D \to \infty$ limit solutions are combined with solutions obtained at other non-physical but simplifying values of D.[6, 7, 8] Second is the method of $1/D$ expansions, or dimensional perturbation theory, in which the missing derivatives are reintroduced within a perturbation framework.[9, 10, 11, 12, 13, 14] And third is the method of subhamiltonians, in which the effects of the derivatives are simulated in a rough way by careful internal modification of the $D \to \infty$ limit hamiltonian (but without reintroducing derivatives).[15, 16] We will be concerned here only with the third approach.

1.1. SIMPLEST SUBHAMILTONIANS

Consider an N-electron atom or ion of nuclear charge Z. Probably the simplest subhamiltonian is (in atomic units, which are used throughout this presentation)[15, 16]

$$\mathcal{H} = \sum_{i=1}^{N} \frac{{n_i}^2}{2{r_{n_i}}^2} - \sum_{i=1}^{N} \frac{Z}{r_{n_i}} + \sum_{\substack{i=1 \\ i<j}}^{N} \sum_{j=1}^{N} \frac{1}{\sqrt{{r_{n_i}}^2 + {r_{n_j}}^2}} . \tag{1}$$

This is just the $D \to \infty$ limit of either the Hartree or Hartree-Fock hamiltonian, but with the centrifugal terms scaled on the basis of the atomic principal quantum numbers n_i. It is also implicitly assumed that solutions will have maximal symmetry. For neon, for example, Eq. 1 becomes

$$\mathcal{H} = \frac{2(1)^2}{2{r_1}^2} + \frac{8(2)^2}{2{r_2}^2} - \frac{2Z}{r_1} - \frac{8Z}{r_2} + \frac{1}{\sqrt{2}\,r_1} + \frac{16}{\sqrt{{r_1}^2 + {r_2}^2}} + \frac{28}{\sqrt{2}\,r_2} , \tag{2}$$

where $Z=10$ and r_1 and r_2 are radii associated with the two shells.

Minimization of Eq. 1 yields a high-dimensional point-charge structure. Each electron sits at a fixed distance from the nucleus, and at 90° from every other electron. (This mutual orthogonality is tantamount to the orbital approximation; all angles would be larger than 90°, though not by much, without this approximation.[15]) All electrons in a given shell sit at the same distance from the nucleus, so that the geometry is completely characterized by a set of shell radii. Comparing these radii with values obtained from accurate calculations (such as maxima of radial probability distributions for the SCF atomic orbitals), one finds that the values obtained from Eq. 1 are realistic for inner shells, but too large for outer shells. Total energies are too low by 1–6%.

Like the better subhamiltonians described below, that in Eq. 1 can be thought of as an approximation to a certain conventional variational calculation. In this case, the conventional calculation is itself highly approximate, namely a single-ζ Hartree treatment of an atom or ion with hypothetical configuration $1s^2\,2p^8\,3d^{18}\ldots$ (that is, an approximation obtained by combining hydrogenic pieces, by neglecting both exchange or correlation, and by forgetting about inner lobes). In fact the structures and energies obtained from Eq. 1 are in reasonable agreement with those given by a conventional calculation, as can be seen in the first two lines of Table 1 (shell radii for argon) and Table 2 (total energies for all noble gases).

TABLE 1. Comparison of shell or subshell radii for argon obtained from subhamiltonian (S) and conventional Hartree (H) or Hartree-Fock (HF) calculations. Values are in atomic units (bohrs).

Kind	Source	r_1 or r_{1s}	r_2 or r_{2s}, r_{2p}	r_3 or r_{3s}, r_{3p}
S	Eq. 1	0.0568	0.2947	1.5085
H	Eq. 3	0.0567	0.2951	1.6814
S	Eq. 12	0.0570	0.2945	1.1366
H	Ref. [17], $r_n \equiv n/\zeta_n$	0.0569	0.293	1.19
S	Eq. 6	0.0570	0.2928, 0.2948	1.0280, 1.1841
HF	Ref. [18], $r_{n\ell} \equiv \langle r^{-1}\rangle_{n\ell}^{-1}$	0.0570	0.2813, 0.2899	1.0395, 1.2284

Explicit comparison with the conventional calculation just mentioned provides some insight into the workings of Eq. 1. For a normalized radial Slater orbital $\chi = Nr^{n-1}e^{-\zeta_n r}$ with $N^2 = (2\zeta_n)^{2n+1}/(2n)!$, one finds

TABLE 2. Comparison of total energies for noble gases obtained from subhamiltonian (S) and conventional Hartree (H) and Hartree-Fock (HF) calculations. Values are in atomic units (hartrees).

Kind	Source	He	Ne	Ar	Kr	Xe	Rn
S	Eq. 1	−2.71079	−123.808	−509.579	−2688.44	−7064.05	−21476.7
H	Eq. 3	−2.84766	−123.760	−507.204	−2652.43		
S	Eq. 12	−2.71079	−125.666	−520.535	−2751.52	−7263.54	−22101.9
H	Ref. [17]	−2.84766	−125.226	−521.558			
S	Eq. 6	−2.71079	−125.668	−520.587	−2752.03	−7264.51	−22105.9
HF	Ref. [18]	−2.86168	−128.547	−526.818	−2752.05	−7232.14	−21866.8

$\langle\chi| - \frac{1}{2}\nabla^2|\chi\rangle = \frac{1}{2}\zeta_n^2$ and $\langle\chi|r^{-1}|\chi\rangle = \zeta_n/n$. Replacing each ζ_n by the corresponding characteristic radius $r_n = n/\zeta_n$, one therefore obtains

$$\mathcal{H} = \sum_{i=1}^{N} \frac{{n_i}^2}{2{r_{n_i}}^2} - \sum_{i=1}^{N} \frac{Z}{r_{n_i}} + \sum_{\substack{i=1 \\ i<j}}^{N} \sum_{j=1}^{N} J_{n_i n_j} \,, \tag{3}$$

where

$$\begin{aligned} J_{11} &= 5/(8r_1) \\ J_{12} &= (32r_1^4 + 80r_1^3 r_2 + 40r_1^2 r_2^2 + 10r_1 r_2^3 + r_2^4)/(2r_1 + r_2)^5 \\ J_{22} &= 93/(512r_2) \\ J_{13} &= (2187r_1^6{+}5103r_1^5 r_3{+}2835r_1^4 r_3^2{+}945r_1^3 r_3^3{+}189r_1^2 r_3^4{+}21r_1 r_3^5{+}r_3^6)/ \\ &\quad (3r_1{+}r_3)^7 \\ &\vdots \end{aligned} \tag{4}$$

(These are just the spherically symmetric, or Slater-Condon F^0, contributions to the two-electron integrals. Only these were used for the variational calculations in the second line of Table 1, since $D=3$ angular momentum algebra cannot be applied to the hypothetical configurations $1s^2\,2p^8\,3d^{18}\ldots$.) The reason that Eq. 1 works as well as it does is that the simple approximation implicit within it, namely

$$J_{n_i n_j} \approx \frac{1}{\sqrt{{r_{n_i}}^2 + {r_{n_j}}^2}} = \frac{1}{\sqrt{(n_i/\zeta_{n_i})^2 + (n_j/\zeta_{n_j})^2}} \,, \tag{5}$$

is fairly effective. The approximation in Eq. 5 constitutes an overestimate for inner shells (the worst error being for the $1s$, where $\frac{5}{8}$ is approximated as $2^{-1/2}$) and an underestimate for outer shells, particularly for orbitals with radial nodes. However, the approximation is usually accurate to within 10%, and becomes exact when $r_{n_i} \ll r_{n_j}$ or $r_{n_i} \gg r_{n_j}$.

A slightly different perspective on the radii r_n is obtained by considering what happens as one transforms the $D=3$ hamiltonian smoothly into Eq. 1 (by gradually assigning each electron additional degrees of freedom, while simultaneously scaling the problem to preserve basic hydrogenic expectation values). As the problem is transformed, each electron makes ever less use of the first three coordinates, and ever more use of the auxiliary coordinates (which is really just a matter of statistics, since the axes are treated democratically), until finally each electron sits at the origin in the physical subspace, but overall at a finite distance from the nucleus. Furthermore, the dispersions in both the electron-nucleus distances and the electron-electron angles diminish smoothly to zero as $D \to \infty$ (again just a matter of statistics). Thus, one can think of the process as being one in which each electron collapses to a single most-representative point (the origin), while simultaneously assuming a finite distance along an auxiliary axis, with the distance along this axis encoding the dispersion which was lost during the collapse. This "encoding" might be considered as somewhat analogous to the way in which the imaginary component of an energy codes for its Lorentzian line width, although in the subhamiltonian treatment everything remains real-valued.

1.2. REPRESENTING RADIAL STRUCTURE

From an energetic standpoint, the most severe limitation of Eq. 1 is the lack of representation of inner lobes of orbitals, and therefore of penetration effects. Perhaps the simplest way to represent these effects is by means of a subhamiltonian of the form

$$\mathcal{H} = \sum_{i=1}^{N} \frac{{n_i}^2}{2r_{n_i\ell_i}{}^2} - \sum_{i=1}^{N} \frac{Z}{r_{n_i\ell_i}} + \sum_{\substack{i=1\\i<j}}^{N} \sum_{j=1}^{N} \sum_{\sigma=1}^{n_i-\ell_i} \sum_{\tau=1}^{n_j-\ell_j} \frac{\phi_{n_i\ell_i\sigma}\,\phi_{n_j\ell_j\tau}}{\sqrt{(\xi_{n_i\ell_i\sigma}\, r_{n_i\ell_i})^2 + (\xi_{n_j\ell_j\tau}\, r_{n_j\ell_j})^2}} \,. \tag{6}$$

Here each electron has been partitioned into pieces, one for each radial lobe. Thus $1s$ and $2p$ electrons remain "intact", each $2s$ electrons is represented by two pieces, and so forth. The positions of the pieces are given by $\xi_{n\ell\sigma}\, r_{n\ell}$, and the weights (or fractions of an electron) by $\phi_{n\ell\sigma}$, where $r_{n\ell}$ is a single overall radial position and the $\xi_{n\ell\sigma}$ and $\phi_{n\ell\sigma}$ are the zeros of the polynomials given in Table 3.

TABLE 3. Hydrogenic radial positions and weights. Given are the polynomials whose zeros are the radial multipliers $\xi \equiv \xi_{n\ell\sigma}$ and weights $\phi \equiv \phi_{n\ell\sigma}$.

$n\ell$	$\xi_{n\ell\sigma}$ are zeros of ...	$\phi_{n\ell\sigma}$ are zeros of ...
1s	$\xi - 1$	$\phi - 1$
2s	$4\xi^2 - 6\xi + 1$	$20\phi^2 - 20\phi + 1$
2p	$\xi - 1$	$\phi - 1$
3s	$18\xi^3 - 36\xi^2 + 15\xi - 1$	$783\phi^3 - 783\phi^2 + 81\phi - 1$
3p	$9\xi^2 - 15\xi + 4$	$81\phi^2 - 81\phi + 8$
3d	$\xi - 1$	$\phi - 1$
4s	$256\xi^4 - 640\xi^3 + 432\xi^2 - 84\xi + 3$	$137984\phi^4 - 137984\phi^3 + 19872\phi^2 - 624\phi + 3$
4p	$32\xi^3 - 72\xi^2 + 40\xi - 5$	$6016\phi^3 - 6016\phi^2 + 860\phi - 25$
4d	$16\xi^2 - 28\xi + 9$	$208\phi^2 - 208\phi + 27$
4f	$\xi - 1$	$\phi - 1$
5s	$1250\xi^5 - 3750\xi^4 + 3500\xi^3 - 1200\xi^2 + 135\xi - 3$	$10834375\phi^5 - 10834375\phi^4 + 1898750\phi^3 - 93750\phi^2 + 1275\phi - 3$
5p	$250\xi^4 - 700\xi^3 + 585\xi^2 - 165\xi + 12$	$683725\phi^4 - 683725\phi^3 + 119583\phi^2 - 5823\phi + 72$
5d	$250\xi^3 - 600\xi^2 + 385\xi - 63$	$186625\phi^3 - 186625\phi^2 + 31955\phi - 1323$
5f	$25\xi^2 - 45\xi + 16$	$425\phi^2 - 425\phi + 64$
6s	$10368\xi^6 - 36288\xi^5 + 43200\xi^4 - 21600\xi^3 + 4500\xi^2 - 330\xi + 5$	$1776707136\phi^6 - 1776707136\phi^5 + 355460400\phi^4 - 23176800\phi^3 + 531900\phi^2 - 3780\phi + 5$
6p	$2592\xi^5 - 8640\xi^4 + 9576\xi^3 - 4284\xi^2 + 735\xi - 35$	$655852464\phi^5 - 655852464\phi^4 + 131118792\phi^3 - 8515584\phi^2 + 191835\phi - 1225$
6d	$72\xi^4 - 216\xi^3 + 204\xi^2 - 70\xi + 7$	$214704\phi^4 - 214704\phi^3 + 42624\phi^2 - 2660\phi + 49$
7s	$470596\xi^7 - 1882384\xi^6 + 2722734\xi^5 - 1800750\xi^4 + 565950\xi^3 - 79380\xi^2 + 4095\xi - 45$	$1892028276775\phi^7 - 1892028276775\phi^6 + 416877348951\phi^5 - 32928694575\phi^4 + 1050017325\phi^3 - 13196925\phi^2 + 55125\phi - 45$

Alternatively, and more generally, the $\xi_{n\ell\sigma}$ are the roots of the equation

$$n(1-\xi)L_{n+\ell}^{2\ell+1}(2n\xi) - (n+\ell)L_{n+\ell-1}^{2\ell+1}(2n\xi) = 0\ , \tag{7}$$

and each $\phi_{n\ell\sigma}$ is given in terms of the corresponding $\xi_{n\ell\sigma}$ by

$$\phi = (\xi^2/n)\Big/\Big[-\xi^2 + 2\xi - \ell(\ell+1)/n^2\Big]. \tag{8}$$

By way of example, the neon atom is described by the function

$$\begin{aligned} \mathcal{H} \;=\; & \frac{2(1)^2}{2{r_{1s}}^2} + \frac{2(2)^2}{2{r_{2s}}^2} + \frac{6(2)^2}{2{r_{2p}}^2} - \frac{2Z}{r_{1s}} - \frac{2Z}{r_{2s}} - \frac{6Z}{r_{2p}} \\ & + \frac{1}{\sqrt{2}\,r_{1s}} + \frac{\frac{7}{20}\sqrt{2}+\frac{1}{35}\sqrt{7}}{r_{2s}} + \frac{15}{\sqrt{2}\,r_{2p}} + \frac{12}{\sqrt{{r_{1s}}^2 + {r_{2p}}^2}} \\ & + \sum_{\pm} \frac{4(\frac{1}{2}\pm\frac{1}{5}\sqrt{5})}{\sqrt{{r_{1s}}^2 + (\frac{3}{4}\pm\frac{1}{4}\sqrt{5})^2 {r_{2s}}^2}} + \sum_{\pm} \frac{12(\frac{1}{2}\pm\frac{1}{5}\sqrt{5})}{\sqrt{(\frac{3}{4}\pm\frac{1}{4}\sqrt{5})^2 {r_{2s}}^2 + {r_{2p}}^2}} \;. \end{aligned} \tag{9}$$

Eq. 6 corresponds to assuming that each electron behaves hydrogenically, so that both the relative locations and the integrated probabilities for the various radial lobes are fixed. The values for these locations and lobes (the zeros of the polynomials in Table 3) are obtained from a hydrogenic subhamiltonian which accounts for radial structure, namely[16]

$$h_{n\ell} = \frac{(\ell+1)^2}{2r^2} + \sum_{\kappa=1}^{n-\ell-1} \frac{1}{2(r-c_\kappa)^2} - \frac{Z}{r}\,, \tag{10}$$

where the c_κ are the radial nodal positions. This function has $n-\ell$ minima of equal energy, namely $E = -\frac{1}{2}Z^2/n^2$. The details of this derivation are described elsewhere.[16] Here we note only the following points: First, the relative positions $\xi_{n\ell\sigma}$ agree exactly with the maxima of the true hydrogenic radial probability distribution, measured relative to the characteristic radius n^2/Z. Second, the $\phi_{n\ell\sigma}$, which are obtained by weighting each position in inverse proportion to the radial kinetic energy at the minimum, agree very closely (within $\frac{1}{10}\%$) with true integrated probabilities for the different lobes. And third, the reason that the kinetic terms in Eq. 6 do not involve the $\xi_{n\ell\sigma}$ and $\phi_{n\ell\sigma}$ is that the hydrogenic values permit collapse to the simpler forms shown. That is,

$$\begin{aligned} \frac{n^2}{2r^2} \;&=\; \sum_{\sigma} \phi_{n\ell\sigma} \left[\frac{(\ell+1)^2}{2(\xi_{n\ell\sigma} r)^2} + \sum_{\kappa=1}^{n-\ell-1} \frac{1}{2(\xi_{n\ell\sigma} r - c_\kappa)^2} \right], \\ -\frac{Z}{r} \;&=\; \sum_{\sigma} \phi_{n\ell\sigma} \left[-\frac{Z}{\xi_{n\ell\sigma} r} \right]. \end{aligned} \tag{11}$$

A slightly simplified version of the subhamiltonian in Eq. 6 is obtained by replacing $r_{n_i\ell_i}$ with r_{n_i},

$$\mathcal{H}=\sum_{i=1}^{N}\frac{{n_i}^2}{2{r_{n_i}}^2}-\sum_{i=1}^{N}\frac{Z}{r_{n_i}}+\sum_{\substack{i=1\\i<j}}^{N}\sum_{j=1}^{N}\sum_{\sigma=1}^{n_i-\ell_i}\sum_{\tau=1}^{n_j-\ell_j}\frac{\phi_{n_i\ell_i\sigma}\,\phi_{n_j\ell_j\tau}}{\sqrt{(\xi_{n_i\ell_i\sigma}\,r_{n_i})^2+(\xi_{n_j\ell_j\tau}\,r_{n_j})^2}}\,. \tag{12}$$

That is, all subshells of a given shell are assigned the same radius. This is the equivalent, within the framework of subhamiltonians, to equalizing the effective nuclear charges for all electrons in a shell. Taking neon as the example again,

$$\mathcal{H}=\frac{2(1)^2}{2{r_1}^2}+\frac{8(2)^2}{2{r_2}^2}-\frac{2Z}{r_1}-\frac{8Z}{r_2}+\frac{1}{\sqrt{2}\,r_1}+\frac{12}{\sqrt{{r_1}^2+{r_2}^2}} \tag{13}$$
$$+\frac{\frac{7}{20}\sqrt{2}+\frac{1}{35}\sqrt{7}+\frac{15}{2}\sqrt{2}+\frac{4}{5}\sqrt{75+6\sqrt{5}}}{r_2}+\sum_{\pm}\frac{4(\frac{1}{2}\pm\frac{1}{5}\sqrt{5})}{\sqrt{{r_1}^2+(\frac{3}{4}\pm\frac{1}{4}\sqrt{5})^2{r_2}^2}}\,.$$

Structural and energetic results for the minima of Eqs. 6 and 12 are given in Tables 1 and 2, along with comparison values. The subhamiltonians correspond most closely to a Hartree calculation employing pure hydrogenic orbitals. Because of the recognized limitations of such a treatment (including lack of exchange and nonorthogonality of orbitals), calculations of this kind are not readily available. Those of Plato,[17] which utilized one variational parameter per shell and therefore correspond to Eq. 12, are listed in the tables. Also listed for reference are accurate Hartree-Fock calculations.[18]

1.3. ANGULAR STRUCTURE AND OTHER ENHANCEMENTS

There are several avenues by which the subhamiltonian representations of isolated atoms can be improved further. First, one can in principle represent effects of dynamical electron correlation.[15, 16, 19] In the $D\rightarrow\infty$ limit this is accomplished by keeping the interelectron angles or distances as dynamical variables (rather than pre-orthogonalizing the electrons, as has been done in the models described above). The angles are then determined by a balance between kinetic effects (which are orthogonalizing) and Coulombic effects (which are obtusifying). In the $D\rightarrow\infty$ limit the orthogonalizing potential is expressed in terms of Gramian determinants involving the interelectron angles, but the optimal means for modifying the many-body centrifugal potentials for use at finite D is still an open question.

Second, one can avoid the assumption of purely hydrogenic form for the radial structuring. This is desirable because the hydrogenic form overrepresents the inner lobes, and therefore overestimates penetrations effects, in many-electron atoms. Ideally this problem should be addressed by means of a subhamiltonian transcription of the requirement of orbital orthogonality. Although the attempt to do this quickly destroys the simplicity of the approach, one can implement approximate versions without much complication. For example, a very simple model is suggested by the observation that the nodal positions for different orbitals in an atom are nearly coincident.[20] Thus, one can to a first approximation set radii representing inner lobes equal to other basic radii of the problem. (For example, the inner lobe positions for all s orbitals would be set equal to the $1s$ radius.) Note that changes in the relative positions of the lobes also require changes in the relative weights.[16]

Finally, one can introduce angular structure into the problem.[16] This is essentially a matter of continuing the process of breaking up the overall centrifugal potential into terms representing individual nodal effects. For the one-electron atom, the final result is a subhamiltonian in which each kinetic term represents the effect of a single node, and whose minima correspond to the individual lobes of the wave function. The process may be represented as a stepwise breakup:

$$\frac{n^2}{2r^2}\left\{\begin{array}{l} \dfrac{(\ell+1)^2}{2r^2}\left\{\begin{array}{l} \dfrac{1}{2{r_\perp}^2} \\[2ex] \dfrac{\ell^2}{2{r_\parallel}^2}\left\{\begin{array}{l} \displaystyle\sum_\lambda \frac{1}{2{r_\parallel}^2(f_\lambda\cos\theta - g_\lambda)^2} \\[2ex] \dfrac{m^2}{2{r_\parallel}^2\sin^2\theta}\left\{\displaystyle\sum_\mu \frac{1}{2{r_\parallel}^2\sin^2\theta(a_\mu\cos\phi + b_\mu\sin\phi)^2}\right. \end{array}\right. \end{array}\right. \\[2ex] \displaystyle\sum_\kappa \frac{1}{2(r-c_\kappa)^2} \end{array}\right. \tag{14}$$

Here $r_\parallel = |\mathbf{r}_\parallel|$ and $r_\perp = |\mathbf{r}_\perp|$ refer to the original and auxiliary components of the electronic position,

$$\mathbf{r} = (\underbrace{\ldots\, u,\, v,\, w,}_{\mathbf{r}_\perp}\, \underbrace{x,\, y,\, z}_{\mathbf{r}_\parallel}) \,. \tag{15}$$

Also, the angular coefficients are normalized by ${f_\lambda}^2 - {g_\lambda}^2 = 1$ and ${a_\mu}^2 + {b_\mu}^2 = 1$. The nodal positions ($c_\kappa$, g_λ/f_λ, and b_μ/a_μ) are those which equalize the energies of the several minima, or equivalently minimize the weighted average of the energies of the minima. (Alternatively, the positions can be read off from true hydrogenic wave functions.)

By way of example, for a $5f_{x(5z^2-r^2)}$ (or $5f_{\pi c}$[21]) electron, the fully structured hydrogenic subhamiltonian (the final stage of the breakup) is

$$\frac{1}{2{r_\perp}^2} + \frac{1}{2(r-20/Z)^2} + \sum_\pm \frac{1}{2{r_\parallel}^2(\frac{1}{2}\sqrt{5}\cos\theta \pm \frac{1}{2})^2} + \frac{1}{2{r_\parallel}^2\sin^2\theta\cos^2\phi} - \frac{Z}{r} \,. \tag{16}$$

This subhamiltonian has twelve minima, at each of which the value is $-\frac{1}{50}$. In general, the minima of a fully structured hydrogenic subhamiltonian correspond one-to-one with the various lobes of a real-valued hydrogenic orbital. For each minimum the physical components of the position ($r_\parallel, \theta, \phi$ or x, y, z) describe the location of the lobe, while the radius within the space of all other components ($r_\perp$) corresponds roughly to the size of the lobe. These interpretations of the meaning of subhamiltonian minima remain valid for the less structured forms, except that the positions now correspond to averages of one or more lobes.

Within the hydrogenic approximation, there is a simple transcription between solutions at various levels of structuring. Thus, one finds that an overall radius r decomposes into

$$r_\perp = \sqrt{\tfrac{1}{\ell+1}}\, r\,, \qquad\qquad r_\parallel = \sqrt{\tfrac{\ell}{\ell+1}}\, r\,, \tag{17}$$

while the angular minima are just the positions of the extrema of real-valued linear combinations spherical harmonics. Note that the natural combinations may depend on the symmetry of the problem. (For example, the f orbitals appropriate to linear and cubic symmetry differ.[21]) Here we utilize the simplest real-valued forms, which are just the m-diagaonal combinations $P_\ell^{|m|}(\cos\theta)\,\{\cos m\phi, \sin m\phi\}$. The resulting angular positions and weights are given in Table 4. The minima with respect to the azimuthal angle are arranged symmetrically and weighted equally.

The weights with respect to the polar angle were fixed by demanding that the lower-order multipoles of the "wave function" vanish. Specifically, if the angular positions are denoted θ_σ, then the corrsponding weights ϕ_σ were determined by symmetry and by the requirement that the multipoles along the quantization axis, namely $\sum_\sigma \alpha_\sigma P_k(\cos\theta_\sigma)$ where $\alpha_\sigma = \pm\sqrt{\phi_\sigma}$ (with signs chosen in alternating fashion), vanish for $k < \ell$. For $m = 0$, where the subhamiltonian does not vary with the azimuthal angle, a simple point-charge representation can be achieved by choosing two azimuthal positions 180° apart for each polar angle other than 0° or 180° (so that the electron will be represented by 2ℓ angular positions overall). There is some arbitrariness in this choice. However, it is a natural and simple representation, and is the one we have used. (We note that alternative representations for $m = 0$ electrons seem to lead to very similar results in derived quantities.) With this choice, all angular states considered except f_0 and $f_{\pm 1}$ are represented by point-charge distributions in which each angular position is assigned an equal weight. For f_0 the positions have relative weights of 4:5:5:5:5:4 (the smaller values being along the z axis), while for $f_{\pm 1}$ the positions have relative weights 16:15:15:15:15:16 (the larger values being in the xy plane).

TABLE 4. Hydrogenic angular positions and weights. Given are the cosines of the polar angle and the corresponding weights ϕ. (The corresponding values for the azimulthal angle are determined by symmetry.)

	p electrons		d electrons		f electrons	
	$\cos\theta$	ϕ	$\cos\theta$	ϕ	$\cos\theta$	ϕ
$m=0$	± 1	$\frac{1}{2}, \frac{1}{2}$	± 1	$\frac{1}{4}, \frac{1}{4}$	± 1	$\frac{2}{14}, \frac{2}{14}$
			0	$\frac{1}{2}$	$\pm\sqrt{\frac{1}{5}}$	$\frac{5}{14}, \frac{5}{14}$
$m=\pm 1$	0	1	$\pm\sqrt{\frac{1}{2}}$	$\frac{1}{2}, \frac{1}{2}$	$\pm\sqrt{\frac{11}{15}}$	$\frac{15}{46}, \frac{15}{46}$
					0	$\frac{16}{46}$
$m=\pm 2$			0	1	$\pm\sqrt{\frac{1}{3}}$	$\frac{1}{2}, \frac{1}{2}$
$m=\pm 3$					0	1

By way of example, for the $5f_{x(5z^2-r^2)}$ example, one can read off from Tables 3 and 4, and the fact that the simplest subhamiltonian for this electron ($n^2/2r^2 - Z/r$ where $n=5$) yields $r=25/Z$, that

$$
\begin{array}{ccl|ll}
 & & & \underline{\text{position}} & \underline{\text{weight}} \\
r & = & \left\{ \begin{array}{l} \\ \\ \end{array} \right. & \begin{array}{l} 12.19/Z \\ 32.81/Z \end{array} & \begin{array}{l} 0.1847 \\ 0.8152 \end{array} \\
\theta & = & \left\{ \begin{array}{l} \\ \\ \\ \end{array} \right. & \begin{array}{l} 31.1^\circ \\ 90.0^\circ \\ 148.9^\circ \end{array} & \begin{array}{l} 0.3261 \\ 0.3478 \\ 0.3261 \end{array} \\
\phi & = & \left\{ \begin{array}{l} \\ \\ \end{array} \right. & \begin{array}{l} 0^\circ \\ 180^\circ \end{array} & \begin{array}{l} 0.5000 \\ 0.5000 \end{array}
\end{array}
\qquad (18)
$$

The weights for the twelve individual lobes are the products of the r, θ, and ϕ weights.

Thus, for many purposes one can work with subhamiltonians which are radially structured (for example, Eq. 6 or 12), and then transcribe to fully structured solutions. It should be noted, however, that some aspects of the physics only appear when angular structure is introduced. For a given configuration of a many-electron atom or ion, for example, each term corresponds to a different weighting of relative positions for the electrons in the unfilled shells.[16]

There are still other "correct" forms for hydrogenic subhamiltonians — that is, forms which yield meaningful structures and correct energies. In particular, in the first couple of stages of the sequence of structuring described above, one can replace all occurrences of r in the kinetic terms by $r_\perp$ (obtaining for example $n^2/2{r_\perp}^2 - Z/r$ in place of $n^2/2r^2 - Z/r$). The forms with $r_\perp$ in the centrifugal terms force $x{=}y{=}z{=}0$, or $r = r_\perp$, at each minimum. The forms with r in the centrifugal terms, on the other hand, are less specific, in that they provide r values without in any way restricting the breakdown among $r_\parallel$ and $r_\perp$. It is appropriate to utilize subhamiltonians with perpendicular distances in the centrifugal terms whenever the physics of the problem selects out some axis or set of axes.[4, 8, 22] For example, when an atom is placed in an electric field, the natural subhamiltonians utilize perpendicular distances from the axis defining the field direction. And for a three-dimensional molecule, the natural subhamiltonians utilize perpendicular distances from the subspace defined by the nuclei, namely (x, y, z)-space.

2. More than one atom

We now turn to electronic structure in a collection of atoms. This might be an isolated molecule, or a small piece of a solid with or without periodic boundary condiditions. We work within the Born-Oppenheimer approximation, so the nuclear framework is considered to be fixed and strictly three-dimensional.[22]

2.1. SIMPLEST SUBHAMILTONIANS

Subhamiltonians for molecules and solids are constructed in essentially the same way as for isolated atoms. The simplest subhamiltonians are therefore those of the form

$$\mathcal{H} = \sum_i \frac{{n_i}^2}{2{r_{i\perp}}^2} + \sum_{I<J}\sum \frac{Z_I Z_J}{r_{IJ}} - \sum_I \sum_j \frac{Z_I}{r_{Ij}} + \sum_{i<j}\sum \frac{1}{r_{ij}} \,. \tag{19}$$

It is only the electronic structure which is dimensionally continued, so only the electronic positions have auxiliary components. Thus in the preceding equation

$$\begin{aligned} r_{IJ} &= \left[(x_I - x_J)^2 + (y_I - y_J)^2 + (z_I - z_J)^2\right]^{1/2} \\ r_{iJ} &= \left[(x_i - x_J)^2 + (y_i - y_J)^2 + (z_i - z_J)^2 + {r_{i\perp}}^2\right]^{1/2} \\ r_{ij} &= \left[(x_i - x_j)^2 + (y_i - y_j)^2 + (z_i - z_j)^2 + {r_{i\perp}}^2 + {r_{j\perp}}^2\right]^{1/2} . \end{aligned} \tag{20}$$

Minimization of Eq. 19 will in general yield multiple minima. At normal nuclear geometries, these will correspond to different assignments of electrons to the atoms. (One might also call these configuration state functions or valence bond structures.) Thus, for a diatomic molecule AB one will in general obtain minima corresponding to A^0B^0, A^+B^-, A^-B^+, etc. The principle quantum numbers utilized are those appropriate to the associations of electrons with individual atoms. Note that in some cases these quantum numbers will change with the charge state. In Li^+H^-, for example, all electrons would be assigned $n = 1$, while in Li^0H^0 one electron would have to be assigned $n = 2$.

Calculations can be carried out at various levels of approximation. For instance, one can eliminate degrees of freedom for the electrons within (x, y, z)-space by fixing the x, y, z coordinates of each electron i to be those of some nucleus I, so that its projection into physical space coincides with that nucleus; this corresponds to eliminating polarization effects. One can also consider less approximate representations that that in Eq. 19. Thus, one can introduce radial structure into the problem in precisely the same manner as in atoms, or one can introduce short-range dynamical correlation effects by means of the interelectron angles (which would render the $r_\perp$ components no longer perpendicular to each other).

As an example, consider F_2. For simplicity we equivalence electronic and nuclear x, y, z coordinates, as described in the last paragraph, obtaining

$$\begin{aligned}\mathcal{H} = {} & \frac{2(1)^2}{2{r_1}^2} + \frac{m(2)^2}{2{r_2}^2} - \frac{2Z}{r_1} - \frac{mZ}{r_2} - \frac{2Z}{\sqrt{{r_1}^2 + R^2}} - \frac{mZ}{\sqrt{{r_2}^2 + R^2}} \\ & + \frac{2(1)^2}{2{s_1}^2} + \frac{n(2)^2}{2{s_2}^2} - \frac{2Z}{s_1} - \frac{nZ}{s_2} - \frac{2Z}{\sqrt{{s_1}^2 + R^2}} - \frac{nZ}{\sqrt{{s_2}^2 + R^2}} + \frac{Z^2}{R} \\ & + \frac{1}{\sqrt{2}\,r_1} + \frac{2m}{\sqrt{{r_1}^2+{r_2}^2}} + \frac{\frac{1}{2}m(m-1)}{\sqrt{2}\,r_2} + \frac{1}{\sqrt{2}\,s_1} + \frac{2n}{\sqrt{{s_1}^2+{s_2}^2}} + \frac{\frac{1}{2}n(n-1)}{\sqrt{2}\,s_2} \\ & + \frac{4}{\sqrt{{r_1}^2+{s_1}^2+R^2}} + \frac{2m}{\sqrt{{r_2}^2+{s_1}^2+R^2}} + \frac{2n}{\sqrt{{r_1}^2+{s_2}^2+R^2}} + \frac{mn}{\sqrt{{r_2}^2+{s_2}^2+R^2}} \,.\end{aligned} \tag{21}$$

Here $Z=9$, R is the internuclear separation, and m and n are valence shell electron counts. (In particular, $m=n=7$ for a covalent assignment, while $m=6$ and $n=8$ or vice versa for an ionic assignment.) The shell radii used as variables are really the components in the auxiliary space ($r_\perp$ values), but for simplicity have been abbreviated as r_1, r_2 for the left atom and s_1, s_2 for the right atom.

A special case of Eq. 21 is that obtained by setting $r_1=s_1$ and $r_2=s_2$, which corresponds to equivalencing the atomic orbitals on the two atoms. Denoting this by a bar, the SCF hamiltonian is then

$$\mathcal{H}^{\mathrm{SCF}} = \tfrac{1}{2}\left[\overline{\mathcal{H}}_{m=n} + \overline{\mathcal{H}}_{m=n\pm2}\right] , \tag{22}$$

TABLE 5. Energies for F_2 at various levels of treatment.

Level of treatment	Covalent minimum	Ionic minimum		Eigen-value	Differ-ence	Refer-ence[23]
SCF	−191.422	−190.893	HF	−191.464		
					−0.099	−0.073
			CASSCF	−191.562		
					−0.042	−0.032
+ breathing	−191.423	−191.087		−191.604		
					−0.007	−0.006
+ distortion	−191.428	−191.101		−191.611		

corresponding to equal weighting of covalent and ionic contributions. On the other hand, an extension of Eq. 21 is that obtained by accounting for polarization effects, or displacements of the electrons along the internuclear axis. This would be modeled by rewriting the subhamiltonian with electronic positions not tied to nuclear coordinates (so that, for example, the first Coulomb term would be replaced by $-2Z/\sqrt{r_1{}^2 + z_1{}^2}$).

The minima of Eq. 21 and its variants are given in the first two number columns of Table 5. The SCF values are those obtained by evaluating the covalent and ionic subhamiltonians at the minimum of the SCF hamiltonian (which at the experimental internuclear separation of $R = 2.68$ gives $r_1 = s_1 = 0.116$ and $r_2 = s_2 = 0.811$). The remaining energies were obtained by independently minimizing the relevant covalent and ionic forms of the subhamiltonian. The geometries of the minima are in accord with physical expectations. In particular, one finds that for the ionic charge assignments the electrons on the negatively charged atom move radially outward and toward the opposite nucleus, while on the positive atom they move inward and away. Note that in a conventional treatment these movements would be called dynamical relaxation or polarization effects, and would be modeled by single excitations out of the valence shell. Specifically, the two kinds of relaxation, which have been called breathing and distortion, are those represented through ℓ-conserving and ℓ-nonconserving excitations.[23]

2.2. INTERACTION OF SUBHAMILTONIAN MINIMA

We now turn to the problem of treating those aspects of the physics which have no representation within the framework of subhamiltonian minimizations. Roughly speaking, these are the less "classical" aspects of the physics.

Formally, the best way to address these features is probably that of a perturbation expansion, with the subhamiltonian as the initital approximation. Such an expansion is completely analogous to a conventional $1/D$ expansion. Recent developments in the algorithms for deriving and summing $1/D$ series for larger systems suggest that it may be possible to carry out this program.[10, 12, 24, 25] However, this endeavor is bound to be far more complicated than the procedures we have been considering so far.

Another avenue, significantly less complex but still challenging, involves the transcription of subhamiltonian minima into conventional atomic orbitals,[16] followed by explicit computation of matrix elements. For hydrogenic orbitals, all of the required one-electron integrals can be expressed[26] in terms of tabulated[27] overlap integrals.

We consider here a third route to the approximation of at least some of the integrals that are needed. This is a bold and rough approximation, stimulated by the desire to complete the picture of the physics without having to worry about the complexities of the better procedures. To motivate it, consider what happens when two orbitals are brought together. If the orbitals (or pieces thereof) are out of phase, then there is a node whose effects need to be modeled. The subhamiltonians for atoms suggest that this should be accomplished by means of a simple repulsive potential of the form $\frac{1}{2}(\text{distance})^{-2}$. On the other hand, if the orbitals (or pieces thereof) are out of phase, then there is an antinode whose effects must be modeled. There is nothing in the atomic subhamiltonians corresponding to the influence of an antinode, but since the long-range energetic effect is just the opposite of a node, one can at least consider the use of a simple attractive potential of the form $-\frac{1}{2}(\text{distance})^{-2}$.

In general, we will represent an interaction via a sum of such nodal and antinodal potentials. The contributions need to be weighted, however. By analogy with a conventional treatment, the natural way to do this is with the probability amplitudes $\alpha = \pm\sqrt{\phi}$ mentioned above. Thus, the approximation to the interaction is taken to be

$$t = \tfrac{1}{2} \sum_{\mu} \sum_{\nu} \frac{\alpha_\mu \alpha'_\nu}{(x_\mu - x'_\nu)^2 + (y_\mu - y'_\nu)^2 + (z_\mu - z'_\nu)^2}\,, \tag{23}$$

where the sum is over all pairs of lobes, one on each atom. (For example, a $2p$-$2p$ interaction would have four terms in the sum, two positive and two negative.) Note that there are no $r_\perp$ components present in this expression because it pertains to a single electron whose auxiliary coordinates on the two atoms are assumed to be identical. Eq. 23 is analogous to that for the Coulomb interaction of point-charge multipoles, except for the fact that the terms being summed are of kinetic (inverse distance squared) form.

Following this analogy, one can assume that $r_{n\ell} \ll R$ and $r_{n'\ell'} \ll R$, or roughly speaking that the overlap is small, expand Eq. 23 in inverse power of R, and keep only the leading-order term. The values obtained in this way have the general form

$$t_{n\ell\, n'\ell'\, \lambda} = a_{n\ell\, n'\ell'\, \lambda} \frac{(r_{n\ell})^{\ell}\,(r_{n'\ell'})^{\ell'}}{R^{\ell+\ell'+2}} \ . \tag{24}$$

Explicit expressions for interactions of the radially nodeless orbitals $1s$, $2p$, $3d$, and $4f$ are given in Table 6.

TABLE 6. Approximations to interaction matrix elements (hopping integrals) for radially nodeless orbitals.

$n\ell\, n'\ell'\, \lambda$	$t_{n\ell\, n'\ell'\, \lambda}$	$n\ell\, n'\ell'\, \lambda$	$t_{n\ell\, n'\ell'\, \lambda}$
$1s\, 1s'\, \sigma$	$2\, R^{-2}$	$3d\, 3d'\, \sigma$	$448/9\, r_{3d}{}^2\, r_{3d'}{}^2\, R^{-6}$
$1s\, 2p\, \sigma$	$4\, r_{2p}\, R^{-3}$	$3d\, 3d'\, \pi$	$320/9\, r_{3d}{}^2\, r_{3d'}{}^2\, R^{-6}$
$1s\, 3d\, \sigma$	$16/3\, r_{3d}{}^2\, R^{-4}$	$3d\, 3d'\, \delta$	$64/9\, r_{3d}{}^2\, r_{3d'}{}^2\, R^{-6}$
$1s\, 4f\, \sigma$	$16\sqrt{189}/35\, r_{4f}{}^3\, R^{-5}$	$3d\, 4f\, \sigma$	$704\sqrt{21}/35\, r_{3d}{}^2\, r_{4f}{}^3\, R^{-7}$
$2p\, 2p'\, \sigma$	$12\, r_{2p}\, r_{2p'}\, R^{-4}$	$3d\, 4f\, \pi$	$352\sqrt{621}/115\, r_{3d}{}^2\, r_{4f}{}^3\, R^{-7}$
$2p\, 2p'\, \pi$	$4\, r_{2p}\, r_{2p'}\, R^{-4}$	$3d\, 4f\, \delta$	$16\sqrt{2}\, r_{3d}{}^2\, r_{4f}{}^3\, R^{-7}$
$2p\, 3d\, \sigma$	$64/3\, r_{2p}\, r_{3d}{}^2\, R^{-5}$	$4f\, 4f'\, \sigma$	$5184/25\, r_{4f}{}^3\, r_{4f'}{}^3\, R^{-8}$
$2p\, 3d\, \pi$	$32/3\, r_{2p}\, r_{3d}{}^2\, R^{-5}$	$4f\, 4f'\, \pi$	$110352/575\, r_{4f}{}^3\, r_{4f'}{}^3\, R^{-8}$
$2p\, 4f\, \sigma$	$16\sqrt{189}/7\, r_{2p}\, r_{4f}{}^3\, R^{-6}$	$4f\, 4f'\, \delta$	$84\, r_{4f}{}^3\, r_{4f'}{}^3\, R^{-8}$
$2p\, 4f\, \pi$	$88\sqrt{621}/115\, r_{2p}\, r_{4f}{}^3\, R^{-6}$	$4f\, 4f'\, \phi$	$81/8\, r_{4f}{}^3\, r_{4f'}{}^3\, R^{-8}$

The crude (one might even say *ad hoc*) nature of the argument given for the representation of an orbital interaction renders the very meaning of the result open to question. However, we will take the values in Eq. 23 and Table 6 to approximate (up to a sign) the one-electron matrix elements or hopping integrals between the two orbitals. (We designate the values by the usual symbol for a hopping integral in solid state applications, t.)

Although the derivation is questionable, it turns out that the general form for the matrix elements, as given by Eq. 24, is quite similar to that provided by empirical fitting[28] and by other (also quite rough) theoretical arguments,[29] most of which are of the form $(r_{n\ell})^{\ell-\frac{1}{2}}\,(r_{n'\ell'})^{\ell'-\frac{1}{2}}/R^{\ell+\ell'+1}$. In addition, relative values for different λ are in good agreement with those obtained by other means[31] (for example, 7 : 5 : 1 compared to 6 : 4 : 1 for the $\sigma : \pi : \delta$ ratio among $3d\,3d$ integrals), and the actual numerical values obtained are at least of the right order of magnitude. (It should be noted, however, that both the present approximation and the similar ones in the literature have at most a quite limited range of applicability. In particular, they are clearly incorrect in both limiting regimes: for small R the actual values remain finite, and for large R the actual behavior is exponential.)

Now consider two subhamiltonian minima which are connected by a one-electron transfer, such as one covalent and one ionic minimum in F_2. Continuing in the spirit of bold simplification, we will ignore all overlap effects for now. Then the overall matrix element between the states corresponding to the two minima can be approximated by the one-electron matrix elements.

Returning to F_2, and restricting attention to the four states connected by transfers of the two bonding electrons, we obtain

$$\mathbf{H} = \begin{pmatrix} E_{\text{cov}} & 0 & -t & -t \\ 0 & E_{\text{cov}} & -t & -t \\ -t & -t & E_{\text{ion}} & 0 \\ -t & -t & 0 & E_{\text{ion}} \end{pmatrix} . \tag{25}$$

Here the four states are those which in a conventional treatment might be labeled $A\overline{B}$, $B\overline{A}$, $A\overline{A}$, and $B\overline{B}$. (Of course this can be block-diagonalized by using $A\overline{B} \pm B\overline{A}$ and $A\overline{A} \pm B\overline{B}$ as a basis, but we prefer to keep the treatment as close as possible to the classical picture.) The diagonal elements, which depend on the level of subhamiltonian representation, are the individual covalent and ionic energies discussed above, while the single off-diagonal value is approximated with the $2p\,2p\,\sigma$ expression from Table 6. Using $R = 2.68$ (the equilibrium bond length) and $r_{2p} = 0.811$ (from the SCF minimum), one finds $t = 0.153$.

Finally, the total ground state energy can be computed in terms of the three independent parameters of the matrix, or equivalently in terms of $\overline{E} = \frac{1}{2}(E_{\text{cov}} + E_{\text{ion}})$, $U = E_{\text{ion}} - E_{\text{cov}}$, and t. In particular, the Hartree-Fock energy is given by $\overline{E} - 2t$ (which corresponds to equal weighting of covalent and ionic contributors), while the energy at any higher level of approximation is given by $\overline{E} - 2\sqrt{t^2 + U^2/16}$ (where the parameters depend on the level). The resulting total energies are given in the third numerical column of Table 5. Note that for simplicity we have utilized a single value for t in all calculations, since we do not yet know how characterize the influence of dynamical relaxation effects on the hopping integral itself.

Because the subhamiltonian in Eq. 21 is of the simplest possible kind, with no explicit radial or angular structure, the total energies are fairly distant from the true total energy (-191.6 compared to -198.9). However, energy differences may be significant. In particular, we consider the energy differences between the successive levels of approximation in Table 5. These correspond to correlation effects of three different kinds, namely nondynamical correlation and dynamical relaxation effects associated with breathing and distortion of the individual structures.[23, 32] The values obtained are compared with literature values in the last two columns. One can see that the subhamiltonian treatment is at least roughly simulating the various correlation effects. Of course, the several approximations embedded in this treatment (in particular the crudeness of the representation of the interactions, and the lack of representation of overlap) mean that it is probably too early to try to interpret the directions and sizes of the discrepancies. But the partial agreement is encouraging, and suggests that this approach, when carried out using better subhamiltonians and hopefully an improved description of interactions, will provide a useful means for studying both nondynamical and dynamical correlation effects.

It is possible to carry the correlation treatment one step further by considering the contribution of short-range dynamical correlation. This is the part associated with cusp behavior, represented in a conventional treatment by double excitations out of the valence shell,[23] and thought to be the part modeled by a local density approximation.[33, 34] In a subhamiltonian treatment this component of the correlation energy is represented by angle openings between the electrons.[15, 16, 22] These effects are largely atomic in character, and so may be approximated by considering the individual atoms or ions. (For reference, the dynamical correlation energies for F^+, F, and F^- at this level of calculation are respectively -0.242, -0.308, and -0.379.) This contribution to the correlation energy has been considered in detail elsewhere both for isolated atoms and for a model solid (metallic hydrogen), and will not be considered further here.

It is also possible to look at other energy differences. For example, one can in principle compute bond strengths by comparing the total energies in Table 5 to that of two fluorine atoms at the same level of treatment (Eq. 1), namely -191.516. The values obtained are of the right order of magnitude,

	Subhamiltonian	Reference [23]	
HF	−0.053	−0.057	
CASSCF	0.046	0.018	(26)
+ breathing	0.088	0.050	
+ distortion	0.095	0.056	

However, compared to the correlation energies, the dissociation energies are much more sensitive to the approximation for the hopping integral, and may also be expected to be more sensitive to facets of the physics which have not even been considered here (like overlap) Therefore it is probably unwise to attach much significance to dissociation energies computed at this level of treatment.

2.3. DIRECTIONS FOR IMPROVEMENT

Although there are many deficiencies in the procedures described, the most serious one at this point is the treatment of the matrix elements connecting alternative charge states. Several avenues are available for addressing this shortcoming. First is the possibility of transcribing atomic subhamiltonian minima into conventional orbitals,[16] followed by utilization of tabulated matrix elements.[27] Note that if these two steps are combined, one will obtain a representation of the matrix elements directly in terms of the geometric parameters of the subhamiltonians (as in Eq. 4, except that now the quantities of primary interest are two-center one-electon integrals rather than one-center two-electron integrals).

It may also be possible to develop less arbitrary representations of nodal and antinodal influences. In the case of H_2^+, for example, one can utilize an LCAO approximation to the wavefunction, analytic motivation for the forms of subhamiltonians,[16] and some simple geometric reasoning to arrive at the forms

$$h_{\rm g} = \frac{1}{2r_\perp{}^2} - \frac{z^2}{r_\perp{}^2+z^2}\sinh^2(z/r_\perp) - \sum_\pm \frac{1}{\sqrt{r_\perp{}^2+(z\pm\frac{1}{2}R)^2}} + \frac{1}{R}\,, \quad (27)$$

$$h_{\rm u} = \frac{1}{2r_\perp{}^2} + \cosh^2(z/r_\perp) - \sum_\pm \frac{1}{\sqrt{r_\perp{}^2+(z\pm\frac{1}{2}R)^2}} + \frac{1}{R}\,. \quad (28)$$

Here the hyperbolic functions arise as the second derivatives of the log of the wavefunction, and the additional ratio in $h_{\rm g}$ is a projection factor. The minima of these subhamiltonians reproduce the potential energy curves for both states very closely over the entire range of R. Unfortunately, this approach quickly becomes more complex as one proceeds to other interactions.

Finally, it may be possible to improve the crude model for the interaction of subhamiltonian minima. As already mentioned, the form given in Eq. 24 is manifestly incorrect at both large and small R, yet it is very close to phenomenological forms and to those obtained by alternative theoretical arguments. Comparing values, one finds that the simple model generally gives smaller values than the other treatments, and that this may be attributed to smaller values of the characteristic radii $r_{n\ell}$. This suggests the possibility of scaling the $r_{n\ell}$ values. This might be justified if it could be argued that the appropriate quantities were some kind of expectation value, rather than probability maxima. (This is in fact how they are interpreted in the other treatments.[29]) One finds, for example, that augmentation of each $r_{n\ell}$ by a factor of $\sqrt{2}$ brings the matrix elements into much better agreement with accepted values for hopping integrals involving d orbitals.[35] It is curious that for orbitals with equal weighting of different lobes (which includes everything except the $4f_0$ and $4f_{\pm1}$ among those considered above), utilization of this scaling is completely equivalent to ignoring the amplitudes α_μ in Eq. 23 (but still keeping the alternation of signs).

3. Perspective and some future directions

With the help of physical reasoning and some analytic results for the dimensionally generalized one-electron atom, one can motivate parameter-free internal modifications of $D \to \infty$ limit hamiltonians for many-electron problems which render them much more realistic as models for $D=3$ electronic structure. The resulting "subhamiltonians" often yield results which are more accurate than might be anticipated, especially when carried to the level of radial or angular structuring. It appears that the basic ingredient of the formalism, which is the representation of each electron by a point or set of points in a space of inflated dimensionality, preserves much of the energetic balance of the actual three-dimensional physics. Of course there are inherent limitations in any collapsed representation, but it is perhaps too soon to draw conclusions regarding the ultimate quantitative potential of the approach.

We believe that one area where subhamiltonian representations may find practical use is in the treatment of electronic structure in highly correlated solids. The primary appeal of the method is its ability to model correlation effects, especially dynamical relaxation or polarization effects associated with charge carrier motions and interactions,[36, 37] within a scheme which makes no use of basis sets. Thus, breathing effects are model without utilization of diffuse functions, distortion effects without polarization functions, and local correlation effects without configuration interaction. These correlation effects all tend to stabilize preferentially the ionic contributor states, so the subhamiltonian approach does not discriminate against such states the way orbital-based treatments tend to. Of course, another advantage of subhamiltonian methods in any solid-state application would be its computational simplicity.

Another area for future investigation is that of perturbation expansions based on subhamiltonians. Such expansions would be completely analogous to standard $1/D$ expansions, though the perturbation is somewhat more complex.[15] We have not pursued this avenue because of the rather large step involved in going beyond simple subhamiltonian minimizations. However, recent advances in techniques for both evaluating and utilizing $1/D$

expansions[3, 12, 38] suggest that this may be an area worth exploring. At a minimum, it is bound to provide better understanding of the relationship between high-dimensional point-charge representations and real-world electronic structure.

Acknowledgments. I thank Agnes Tan, John Summerfield, and Zhen Zheng for their valuable contributions to the research reported here. Funding from the National Science Foundation for the earlier stages of this research is also gratefully acknowledged.

References

1. E. Witten, *Phys. Today* **33**(7), 38 (1980).
2. J. Avery, D. Z. Goodson, and D. R. Herschbach, *Theor. Chim. Acta* **81**, 1 (1991).
3. D. R. Herschbach, J. Avery, and O. Goscinski (eds.), *Dimensional Scaling in Chemical Physics* (Kluwer, Dordrecht, 1993).
4. A. L. Tan and J. G. Loeser, Ref. [3], p. 230.
5. A. A. Belov, Yu. E. Lozovik, and A. González, *Phys. Lett. A* **142**, 389 (1989).
6. D. R. Herschbach, *J. Chem. Phys.* **84**, 838 (1986).
7. J. G. Loeser, Z. Zhen, S. Kais, and D. R. Herschbach, *J. Chem. Phys.* **95**, 4525 (1991).
8. M. López-Cabrera, A. L. Tan, and J. G. Loeser, *J. Phys. Chem.* **97**, 2467 (1993).
9. S. S. Stepanov and R. S. Tutik, *Sov. Phys. JETP* **73**, 227 (1991).
10. M. López-Cabrera, D. Z. Goodson, D. R. Herschbach, and J. D. Morgan III, *Phys. Rev. Lett.* **68**, 1992 (1992).
11. T. C. Germann and S. Kais, *J. Chem. Phys.* **99**, 7739 (1993).
12. M. Dunn, T. C. Germann, D. Z. Goodson, C. A. Traynor, J. D. Morgan III, D. K. Watson, and D. R. Herschbach, *J. Chem. Phys.* **101**, 5987 (1994).
13. V. S. Popov and A. V. Sergeev, *Sov. Phys. JETP* **78**, 303 (1994).
14. D. K. Watson and D. Z. Goodson, *Phys. Rev. A* **51**, R5 (1995).
15. Z. Zheng and J. G. Loeser, Ref. [3], p. 83.
16. J. G. Loeser, J. H. Summerfield, A. L. Tan, and Z. Zheng, *J. Chem. Phys.* **100**, 5036 (1994).
17. G. Plato, *Ann. d. Phys.* [5] **21**, 745 (1935).
18. C. F. Bunge, J. A. Barrientos, and A. V. Bunge, *At. Data Nucl. Data Tables* **53**, 113 (1993).
19. J. G. Loeser, *J. Chem. Phys.* **86**, 5635 (1987).
20. C. Froese Fischer, *The Hartree-Fock Method for Atoms* (Wiley-Interscience, New York, 1977), p. 90.
21. C. K. Jørgensen, *Oxidation Numbers and Oxidation States* (Springer, New York, 1969), p. 15.
22. J. G. Loeser, Ref. [3], p. 389.
23. R. Ghailane, M.-B. Lepetit, and J.-P. Malrieu, *J. Phys. Chem.* **97**, 94 (1993).
24. D. Z. Goodson and D. R. Herschbach, *Phys. Rev. A* **46**, 5428 (1992).
25. T. Germann, D. R. Herschbach, and B. M. Boghosian, *Comput. Phys.* **8**, 712 (1994).
26. T. Koga and T. Matsuhashi, *J. Chem. Phys.* **88**, 1110 (1988).
27. R. S. Mulliken, C. A. Rieke, D. Orloff, and H. Orloff, *J. Chem. Phys.* **17**, 1248 (1949); W. Urland, *J. Chem. Phys.* **70**, 5335 (1979).
28. S. Maekawa, *Comments Cond. Mat. Phys.* **16**, 79 (1992).
29. S. Froyen and W. A. Harrison, *Phys. Rev. B* **20**, 2420 (1979); W. A. Harrison and S. Froyen, *Phys. Rev. B* **21**, 3214 (1980); W. A. Harrison, *Phys. Rev. B* **218**, 550 (1983).
30. O. K. Andersen, *Solid State Comm.* **13**, 133 (1973).
31. A. P. Sutton, *Electronic Structure of Materials* (Oxford, Oxford, 1993), p. 183.

32. P. C. Hiberty, J. P. Flament, and E. Noizet, *Chem. Phys. Lett.* **189**, 259 (1992).
33. T. Ziegler, *Chem. Rev.* **91**, 651 (1991).
34. A. D. Becke, *J. Chem. Phys.* **98**, 1372 (1993).
35. W. A. Harrison, *Electronic Structure and the Properties of Solids* (Dover, New Yorkm, 1989), p. 552.
36. G. J. M. Janssen and W. C. Nieuwpoort, *Phys. Rev. B* **38**, 3449 (1988); *Int. J. Quantum Chem. Symp.* **22**, 679 (1988).
37. G. A. Sawatzky, in *Narrow-Band Phenomena — Influence of Electrons with Both Band and Localized Character*, ed. by J. C. Fuggle, G. A. Sawatzky, and J. W. Allen (Plenum, New York, 1988), p. 117.
38. T. C. Germann, doctoral dissertation, Harvard University, 1995.

DIMENSIONAL RENORMALIZATION FOR ELECTRONIC STRUCTURE CALCULATIONS

S. KAIS AND R. BLEIL
Department of Chemistry
Purdue University
West Lafayette, IN 47907 USA

Abstract. By utilizing the analytically known solutions for the Hartree-Fock and exact energies of N-electron atoms in the large-D limit, we have developed a systematic way to determine an effective nuclear charge $Z_D^{\mathrm{R}} = Z + \delta Z$ such that the Hartree-Fock energies will be significantly closer to the exact energies. This expansion is systematic and self-consistent, and when utilized in standard Hartree-Fock calculations significantly reduces the HF calculation error compared with exact results for atoms, weakly bound systems, and diatomic and polyatomic molecules at their equilibrium geometry. In this approach, we recover more than 2/3 of the correlation energy with no additional computational cost over standard HF calculations since the leading term for the correlation energy, $\frac{\partial E^{\mathrm{HF}}}{\partial Z}\,\delta Z$, is readily available using the Hellmann-Feynman theorem.

1. Introduction

One of the main goals of modern quantum chemistry is to perform highly accurate calculations at the lowest computational cost. The Hartree-Fock approximation[1], which is based on the idea that we can approximately describe an interacting fermion system in terms of an effective single-particle model, remains the major approach for quantitative calculations. This self-consistent field approach usually yields good zeroth order approximation results, and in the asymptotic limit of large molecules scales approximately as $N_{\mathrm{B}}^{2.2-2.3}$, where N_{B} is the number of basis functions involved in the calculation[2]. Recovering the remaining error in the total energy, which is the correlation energy, is the main driving force for introducing new

C. A. Tsipis et al. (eds.), New Methods in Quantum Theory, 55–70.

methods for calculating electronic structure. A wide variety of techniques are currently available for predicting accurate correlation energies, including configuration interaction (CI), many-body perturbation theory (MP2, MP3, MP4, etc.), multi-configuration Hartree-Fock (MCSCF), and coupled cluster methods (CC). Unfortunately, most of these methods are quite computationally expensive relative to Hartree-Fock calculations, and scale at least as N_{B}^5[3].

Dimensional scaling offers a new approach to treat the many-body correlation problem. This involves generalizing the problem to D-dimensional space and treating D as a free parameter (for reviews see Ref. [4]). Typically, the limit $D \to \infty$ yields dramatic simplification in the analysis of a wide class of problems, and often an analytic solution can be obtained in that limit. Finite-D corrections can then be taken into account by introducing a systematic perturbation expansion in $1/D$. At the end of the calculation, one sets $D = 3$ to obtain the "real-world" solution. Among recent applications of dimensional scaling are large order dimensional perturbation expansions for few-body systems[5], including complex dimensional scaling for resonances and unstable states[6], correlated electronic structure models for atoms and solids based on the subhamiltonian approximation[7], and dimensional renormalization for atoms[8] and molecules[9].

Herschbach[4] suggested the use of charge renormalization as another means to exploit the simplicity of the $D \to \infty$ limit. In this approach, a renormalized nuclear charge is found which renders the dimensionally scaled energy at $D \to \infty$ a good approximation to that for $D = 3$ with the actual Z. The renormalized charge is readily evaluated by comparing the HF energy with its $D \to \infty$ limit. With only HF input in addition to the $D \to \infty$ limit, the renormalization procedure yields about 2/3 or more of the correlation energy for all neutral atoms[8].

In the present approach, our goal is to find a systematic way to determine an effective nuclear charge which will have the effect of reducing the Hartree-Fock calculation error to give exact results at $D = 3$. A means to find this effective charge has been developed utilizing the large-D limit where both the exact and Hartree-Fock energies are known analytically. Using the renormalized charge at $D \to \infty$ as a zeroth order approximation for the renormalized charge in three dimensional Hartree-Fock calculations recovers well over 80% of the correlation energy of atoms, ions and molecules at equilibrium.

Although the use of this renormalized nuclear charge is shown to correct for the correlation energy at equilibrium for molecules, our approach still gives the wrong dissociation limit whenever the HF energy dissociates to the wrong limit. In order to correct the energy at the dissociation limit, it becomes necessary to solve for δZ as a function of internuclear distance

R, $\delta Z(R)$, or renormalize simultaneously both the nuclear charge and the internuclear distance.

In this chapter, we provide the details for the determination of δZ, as well as for the use of the renormalized charge in the calculation of the correlation energy. We provide several examples where this method predicts the correlation energy for atoms, ions, weakly bound ionic systems, as well as homonuclear and heteronuclear diatomic and polyatomic molecules with a degree of accuracy camparable to high level theories but using only a standard Hartree-Fock calculation.

2. Dimensional Renormalization

In the Hartree-Fock theory one approximates the wave function, Ψ, of the nonseparable N-electron problem as a single Slater determinant, Ψ^{SD}. The Slater determinant formed will consist of a set of spin orbitals such that this determinant is the best possible approximation to the ground state of the N-electron system described by an exact electronic Hamiltonian H. In dimensional renormalization, we would like to take this approximation a step further by generalizing it to D-dimensional space and allowing the nuclear charge Z to be a free parameter which will vary with the dimensionality of the system. The effective nuclear charge, or the renormalized nuclear charge Z_D^{R}, will be fixed by requiring that

$$E_D^{\mathrm{HF}}(Z_D^{\mathrm{R}}, N) = E_D^{\mathrm{exact}}(Z, N) \; . \tag{1}$$

The goal is to find a nuclear charge renormalization procedure in which the Hartree-Fock eigenvalues will yield the exact energy which is systematic and self-consistent.

In order to accomplish this goal, we take advantage of the fact that both the Hartree-Fock and the exact energy of N-electron atoms have known analytical expressions in the large-D limit. By assuming that the effective charge is a function of the dimensionality D, expanding both sides of Eq. (1) in powers of $\frac{1}{D}$ allows us to obtain the renormalized charge Z_D^{R} in a systematic way. We begin by first expanding the renormalized charge, Z_D^{R}, which gives

$$Z_D^{\mathrm{R}} = Z_\infty + \frac{1}{\kappa} Z_1 + \frac{1}{\kappa^2} Z_2 + \ldots ; \qquad \kappa = \frac{D-1}{2} \; . \tag{2}$$

Substitution of this expansion for Z_D^{R} in Eq. (1) and expanding both sides in powers of $\frac{1}{D}$, leads to a set of equations for the expansion coefficients of the renormalized charge Z_D^{R}. At the large-D limit, the zeroth order approximation, Z_∞, is readily found from the condition

$$E_\infty(Z, N) - E_\infty^{\mathrm{HF}}(Z_\infty, N) = 0 \; . \tag{3}$$

For two electron atoms the solution has been shown to give the analytical expression[10]

$$Z_\infty = \frac{1}{2^{\frac{3}{2}}} + Z\left[1 + \frac{5}{32Z^2} - \frac{1}{2048}\left(\frac{1}{Z^4} + \frac{1}{Z}(128 + \frac{1}{Z^2})^{\frac{3}{2}}\right)\right]^{\frac{1}{2}} \tag{4}$$

for Z_∞, which can be expanded in powers of $\frac{1}{Z}$ to give

$$\delta Z_\infty = Z_\infty - Z = \frac{1}{64}\frac{1}{Z} + \frac{1}{512\sqrt{2}}\frac{1}{Z^2} + \frac{1}{8192}\frac{1}{Z^3} + O(\frac{1}{Z^4}) . \tag{5}$$

This result for δZ_∞ can be generalized to N-electron atoms by solving

$$E_\infty^{\text{exact}}(Z, N) = E_\infty^{\text{HF}}(Z+\delta Z_\infty, N) , \tag{6}$$

where both the exact and HF energies are known analytically at the $D \to \infty$ limit[11]. By limiting consideration to closed-shell atoms[12], $E_\infty^{\text{exact}}(Z, N)$ and $E_\infty^{\text{HF}}(Z+\delta Z, N)$ can be expanded in powers of $Z^{-1/3}$, where the expansion coefficients are given in Ref. [12]. After a little algebra δZ_∞ can be obtained asymptotically for the large-atom limit, this gives

$$\delta Z_\infty = \frac{1}{96}\left(\frac{1}{2}\nu^3 + \nu^2 + \nu - 1\right)\frac{1}{Z} + O(\frac{1}{Z^2}) , \tag{7}$$

where ν is given by

$$\nu = \mu\left[1 - \frac{1}{2\mu} + \frac{1}{12\mu^2} - \frac{1}{5184\mu^6} + O(\frac{1}{\mu^7})\right] \tag{8}$$

and $\mu = (\frac{3N}{2})^{\frac{1}{3}}$. Figure 1 shows the exact δZ_∞ from solving Eq. (6) along with the asymptotic formula for neutral atoms with $N = Z = 2$–48.

3. N-electron Atoms and Weakly Bound Systems

For two electron atoms the renormalized nuclear charge, Z_D^{R}, which makes the Hartree-Fock solution the same as the exact solution in all dimensions, can be obtained analytically to first order in powers of $\frac{1}{D}$. This first order renormalized charge, $Z_\infty + Z_1$ is within only 0.001% error compared with the exact renormalized charge which makes the HF solution equal the exact at $D = 3$[10]. The percent error shows that the Hartree-Fock energies utilizing the renormalized charges differ from the exact energies by less than 0.2% for small Z, corresponding to recovery of at least 85% of the correlation energy[10].

For N-electron atoms we find that renormalization of the nuclear effective charge by means of Z_∞ yields much improved Hartree-Fock energies for

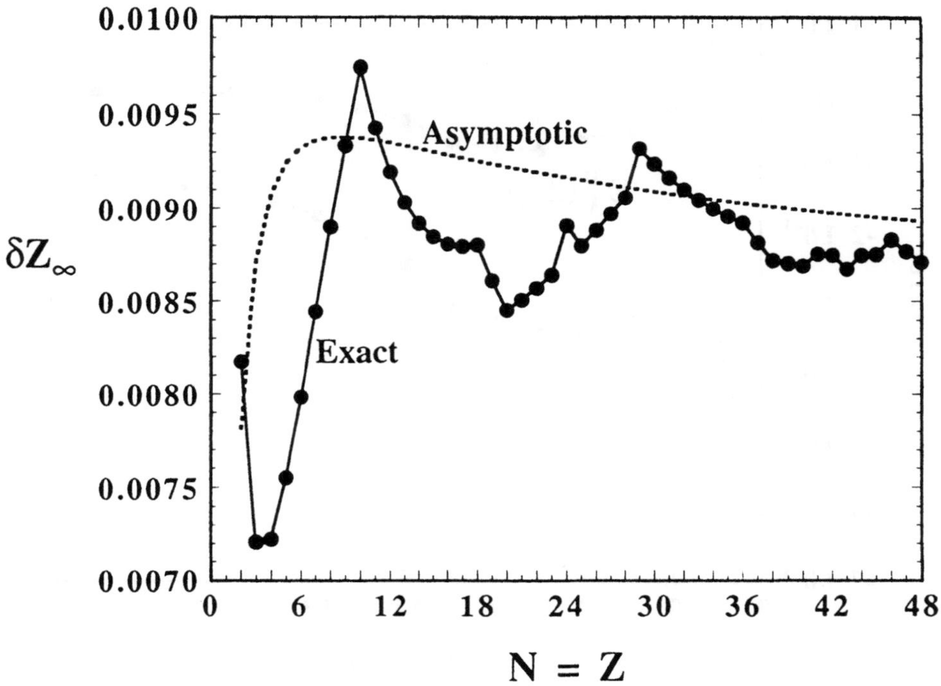

Figure 1. Charge renormalization parameter δZ_∞ (solid line) as a function of the number of electrons N. The asymptotic limit (dashed line) was calculated via Eq. (7).

N-electron atoms in $D = 3$. The renormalized charges used in the Hartree-Fock calculations have been obtained through the use of Eq. (6) as the condition to fix Z_∞. Comparing Z_∞ with $Z_3^{\rm R}$, the effective nuclear charges which force the Hartree-Fock energy to be equal to the exact energies at $D = 3$, typically yields percent error of less than $\sim 0.03\%$ for small atoms, and $\sim 0.002\%$ for large N.

Evaluating the Hartree-Fock energies with the effective charge, Z_∞ at $D = 3$, we obtain a typical accuracy of $\sim 0.003\%$ for large atoms[10]. These results are shown graphically in Figure 2. This figure also shows the percentage error of a typical Hartree-Fock energy calculation using the actual nuclear charge.

Ultimately, we would like to be able to correct the Hartree-Fock approximation, which means that we would like to be able to calculate the correlation energy. Inclusion of some small perturbation resulting in the renormalized charge for any given atom $Z_D^{\rm R} = Z + \delta Z$ allows us to estimate the leading term of the correlation energy by expansion of the renormalized

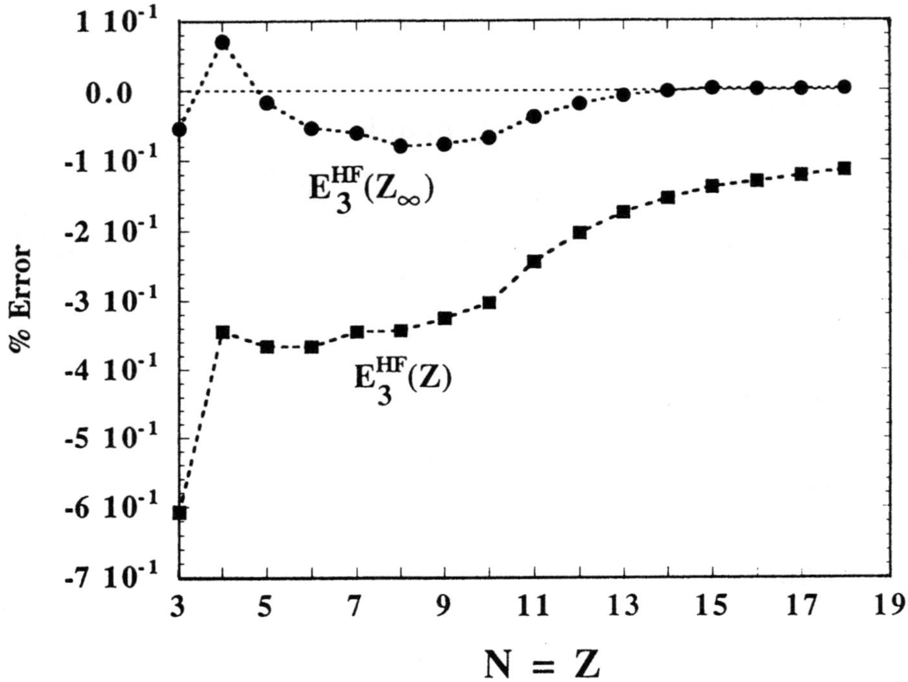

Figure 2. Percent error of the total Hartree-Fock (CAS) energies compared with the exact known energies both with (dotted line) and without (dashed line) renormalized nuclear charges as a function of the number of electrons N.

Hartree-Fock energy to the first order in δZ, giving

$$E^{\text{corr}}(Z) \approx \frac{\partial E^{\text{HF}}}{\partial Z}\,\delta Z \tag{9}$$

as our approximate expression for the correlation energy of an atom. Using the Hellmann-Feynmann theorem, one can obtain the partial derivative of the Hartree-Fock energy with respect to the nuclear charge Z, which is simply the average electron-nuclear potential energy.

For atoms with known near degeneracy mixing (such as beryllium, boron and carbon), the Hartree-Fock calculations were modified to approximate the complete active valence space multiconfigurational Hartree-Fock energy, $E^{\text{CAS}}(Z)$. Following Davidson *et al.*[13], this energy can be written to a good approximation as

$$E^{\text{CAS}}(Z) \approx E^{\text{HF}}(Z) + \Delta B_1(N)\,Z + \Delta B(N)\ , \tag{10}$$

where $\Delta B_1(N)$ has been tabulated in Ref. [13] and $\Delta B(N)$ was fixed to give the exact CAS energies. Substituting $E^{\rm CAS}(Z)$ for $E^{\rm HF}(Z)$ in Eq. (9), we arrive at the corrected correlation energy for this special case as

$$E^{\rm corr}(Z) \approx \left(\frac{\partial E^{\rm HF}}{\partial Z} + \Delta B_1(N)\right) \delta Z \, . \tag{11}$$

In order to verify that Eq. (9) does indeed give the leading term for the correlation energy, we estimated this term at the asymptotic limit, $N \to \infty$ for fixed $\frac{N}{Z}$, where the Thomas-Fermi model gives exact results ($E^{\rm TF} = -0.768745 Z^{7/3}$)[14]. The next two leading corrections, Z^2 and $Z^{5/3}$, were estimated as the contribution of the inner electrons and electron exchange, which produces a remarkably simple formula for the electronic binding energy of a neutral atom with Z electrons[15] as

$$E = -0.768745\, Z^{7/3} + 0.5\, Z^{6/3} - 0.2699\, Z^{5/3} \, . \tag{12}$$

At the asymptotic limit, Eq. (5) gives $\delta Z_\infty \sim 2^{-7} N\, Z^{-1}$ and $\partial E^{\rm HF}_\infty / \partial Z \sim -2(\frac{3}{2})^{1/3} N^{1/3} Z$, which gives the leading correction to the correlation energy

$$E^{\rm corr} \sim -2^{-6} (\tfrac{3}{2})^{1/3}\, Z^{4/3} = -0.017886\, Z^{4/3} \, . \tag{13}$$

This result is in complete agreement with our previous finding, where we have shown that the electron correlation contribution to the total energy is proportional to $Z^{4/3}$[12].

For highly correlated systems such as open-shell systems, multiply excited states, or anions, the Hartree-Fock theory fails as a zeroth-order model. The ground state energy of the hydride ion, H^-, is a prototype example of this inadequacy. Our results show that by using charge renormalization, one can drastically improve the Hartree-Fock energy for these systems. For H^-, Eq.(3) gives $Z_\infty = 1.01714$. Using this effective charge instead of the actual nuclear charge in the Hartree-Fock equations reduces the error from 7.6% to 3%, and gives $E^{\rm HF}(Z_\infty) = -0.5115601$. To carry out the first order correction to the renormalized charge, Z_1, special treatment is needed. Although the HF large-D limit solution remains stable for $Z = 1$ ($Z_{\rm c}^{\rm HF} = 0.8839$), it becomes unstable for the exact large-D solution. Since $Z = 1$ is below the critical charge Z_c, the antisymmetric stretching vibrational mode becomes negative and the first order correction to the energy $E_1(Z)$ becomes complex. Thus, the symmetric configuration corresponds to a saddle point. Use of the complex dimensional scaling procedure[16] allows us to obtain the first order correction to the renormalized charge, Z_1, which gives a stable corrected Hartree-Fock energy of $E_3^{\rm HF}(Z_\infty + Z_1) = -0.5179018$, reducing the percentage error for H^- to just 1.9% [10].

4. Diatomic Molecules

For molecular systems, we can vary the set of the nuclear charges Z_i on each atom i and the internuclear distances R_{ij} between each pair of atoms i and j. As before, we still vary the nuclear charges by the addition of some small correction term resulting in renormalized nuclear charges $Z_D^{\mathrm{R}} = Z+\delta Z$, but now we also vary the internuclear distances $R_{ij}^{\mathrm{R}} = R_{ij}+\delta R_{ij}$. Expanding the renormalized Hartree-Fock energy to first order, this time in both δZ and δR, gives

$$E^{\mathrm{corr}}((R_{ij}),(Z_i)) = \sum_i \frac{\partial E^{\mathrm{HF}}}{\partial Z_i}\delta Z_i + \sum_{i<j} \frac{\partial E^{\mathrm{HF}}}{\partial R_{ij}}\delta R_{ij} \tag{14}$$

as our approximate expression for the correlation energy.

At the asymptotic limit as well as at equilibrium distances, $\frac{\partial E^{\mathrm{HF}}}{\partial R} = 0$. We therefore expect that the leading contribution to the correlation energy for diatomic molecules will be[17],

$$E^{\mathrm{corr}}(Z_a, Z_b, R) \approx \frac{\partial E^{\mathrm{HF}}}{\partial Z_a}\delta Z_a + \frac{\partial E^{\mathrm{HF}}}{\partial Z_b}\delta Z_b\ . \tag{15}$$

The increment in the effective nuclear charge, δZ, is taken to be the large-D limit, δZ_∞, for each individual atom in the molecule, which has been determined in the previous section. To calculate the term $\frac{\partial E^{\mathrm{HF}}}{\partial Z}$, necessary for the calculation of the correlation energy, we use the Hellmann-Feynman theorem, which yields[17]

$$E^{\mathrm{corr}}(R_{ab}, Z_i) \approx -\sum_i \delta Z_i \mathbf{U}^{\mathrm{t}}\mathbf{D}_{\mathrm{T}} + \frac{Z_a\delta Z_b + Z_b\delta Z_a}{R_{ab}}\ , \tag{16}$$

where $\mathbf{U}^t$ is the Hermitian matrix of the electron-nucleus potential energy term in the Hamiltonian, and $\mathbf{D}_{\mathrm{T}}$ is the total electron density matrix[18].

We have calculated the correlation energy from Eq. (15) using $\delta Z \sim \delta Z_\infty$ at the large-D limit as calculated in the previous section for several diatomic molecules in their ground states. These results are shown graphically in Figure 3 as a plot of the correlation energy at the equilibrium distance for diatomic molecules as a function of the number of electrons ranging from $N = 4$ to 18. We have included for comparison purposes higher level theories (such as MP2, MP4, and QCISD(T)). We have recovered about 70% or better of the correlation energy for all molecules examined. This is an encouraging result, since this has been done without modifications to δZ as a result of a molecular rather than atomic environment.

As an example, we present the energy curves for the FH($X^1\Sigma^+$) in Figure 4. As can be seen in this figure, our correlation energy significantly

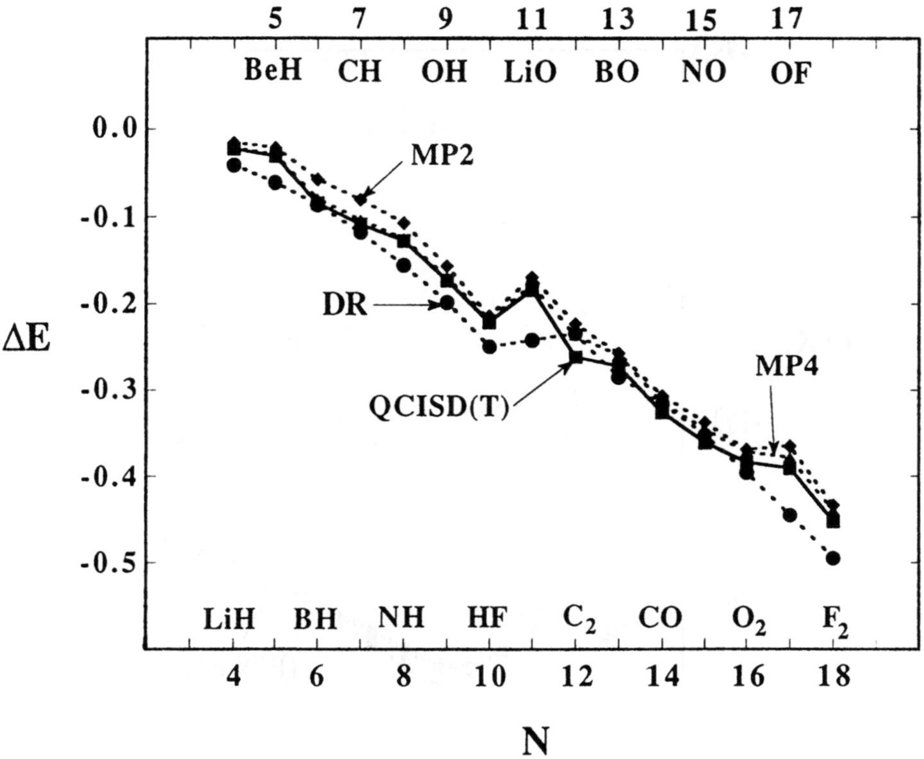

Figure 3. Correlation energy calculated by dimensional renormalization (DR) and higher level theories (MP2, MP4 and QCISD(T)) as a function of the number of electrons in the molecule. The molecules used in the making of this figure are listed along the axes.

reduces the Hartree-Fock error near equilibrium, making the Hartree-Fock results comparable with the best high level theory calculations such as QCISD(T) and CCSD(T). The only shortcoming of this approach is that our energy values tend towards the Hartree-Fock limit. However, in the dissociation limit, many of the higher level theories (such as MP4) diverge. In this graph, we present the Hartree-Fock energy curve, MP4, CCSD(T) and the QCISD(T) curve as the exact curve, as well as the dimensional renormalization curve (DR) calculated in the present work, which is the Hartree-Fock energy as a function of internuclear distance R plus our correlation energy.

The unrestricted Hartree-Fock calculation has been shown to go to the correct dissociation limit for many cases where the restricted Hartree-Fock calculation does not. To verify that the unrestricted Hartree-Fock calculation does give the correct dissociation limit behavior with the dimensional renormalization procedure, we have executed both restricted (RHF) and unrestricted (UHF) Hartree-Fock calculations for $N_2(^1\Sigma_g^+)$ and compared

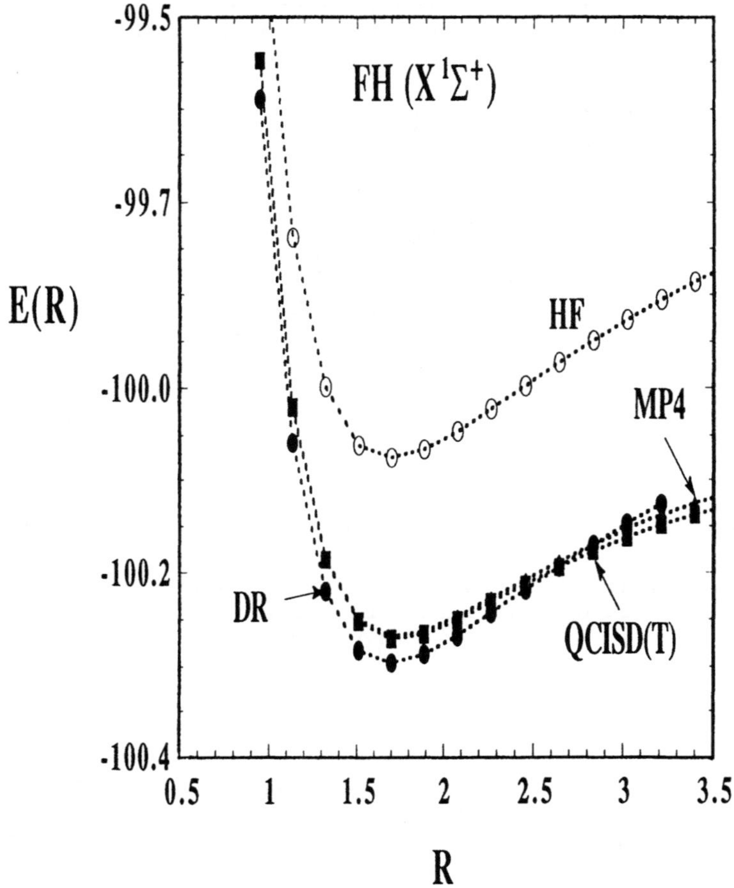

Figure 4. Dimensional renormalization of hydrogen flouride as compared with restricted Hartree-Fock, MP4 and QCISD(T) methods versus internuclear separation distances.

the results with MP4, CCSD(T) and QCISD(T) results, as shown in Figure 5. In this figure, it is seen that the restricted Hartree-Fock method diverges from the unrestricted method around 2.5 bohr, where the UHF method continues to a more realistic dissociation limit than the RHF method. The current method, labeled as RDR and UDR for restricted dimensional renormalization and unrestricted dimensional renormalization, shows a similar behavior, where the UDR method shows the correct dissociation limit behavior but the RDR method does not. It should be kept in mind that this less expensive method gives better results in this dissociation limit than the more expensive higher level theories (such as MP4), which tend to diverge. Although the use of unrestricted Hartree-Fock calculations does improve

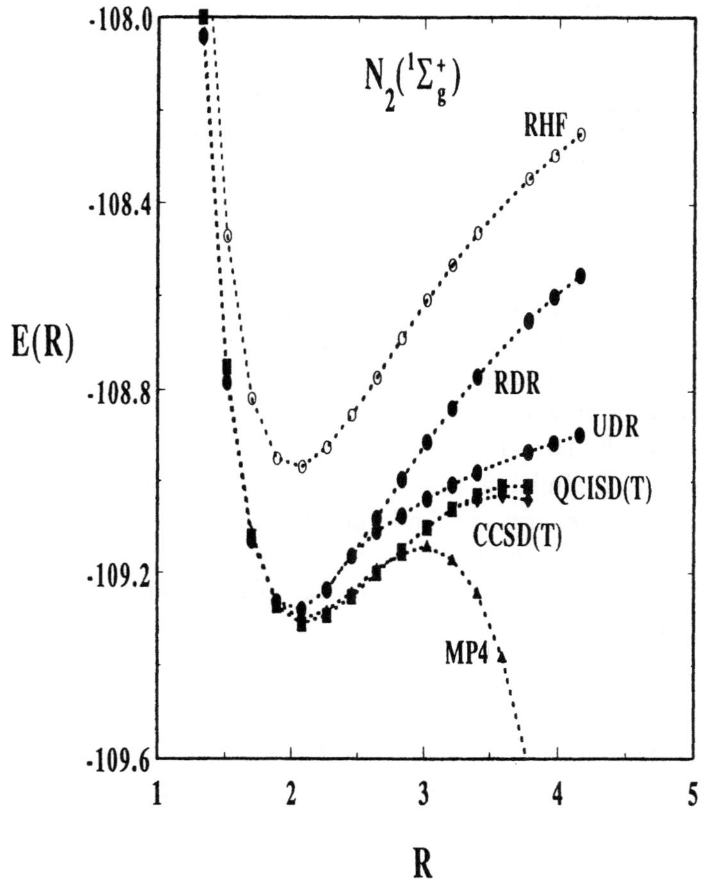

Figure 5. Restricted and Unrestricted dimensional renormalization of diatomic nitrogen as compared with restricted and unrestricted Hartree-Fock, MP4 and QCISD(T) methods versus internuclear separation distances.

the correlation energy, it would be best if we could find a more rigorously correct method for calculating the correlation energy for all internuclear distances, preferrably along the same lines as we have calculated it at equilibrium.

An alternative approach would be to calculate the charge renormalization parameter, δZ, as a function of internuclear distance R, $\delta Z(R)$, from the large-D limit of molecules. For example, let us consider the four electron diatomic molecule, where the effective potential in the large-D limit using D-dimensional cylindrical coordinates has the form [19]

$$V_{\text{eff}} = \sum_{i=1}^{N} \left(\frac{\Gamma^i}{2\rho_i \Gamma} - \frac{Z}{\sqrt{\rho_i^2 + (z_i + \frac{1}{2}R)^2}} - \frac{Z}{\sqrt{\rho_i^2 + (z_i - \frac{1}{2}R)^2}} + \sum_{j=i+1}^{N} \frac{1}{\sqrt{\rho_i^2 + \rho_j^2 - 2\rho_i\rho_j\gamma_{ij} + (z_i - z_j)^2}} \right) + \frac{Z^2}{R}, \tag{17}$$

where the two nuclei are located on the z-axis at $-\frac{R}{2}$ and $+\frac{R}{2}$, with equal charges, Z, and the Γ is the Gramian determinant $| \hat{\mathbf{r}}_i \cdot \hat{\mathbf{r}}_j |=| \gamma_{ij} |$ for all N electrons, and Γ_i is the Gramian determinant for all but the ith electron. The ith electron is located at (ρ_i, z_i), and γ_{ij} is the cosine of the dihedral angle defined by the two electrons and the z-axis.

We begin by minimizing this potential with respect to $(\rho_i, z_i, \gamma_{ij})$ to determine the exact energy at the large-D limit. Then, by setting $\gamma_{ij} = 0$, we solve for (ρ_i, z_{ij}) to find the Hartree-Fock energy. As we have done previously, we then vary δZ for a given value of R until the large-D Hartree-Fock energy is equal to the exact energy,

$$E_{\infty}^{\text{exact}}(R, Z) = E_{\infty}^{\text{HF}}(R, Z+\delta Z_{\infty}) . \tag{18}$$

Solving Eq. (18) for all R, we find $\delta Z_{\infty}(R)$.

We have carried out this procedure for He_2 as an example. Although Hatree-Fock calculations do not give a minimum in the potential energy curve for the helium dimer, using $\delta Z_{\infty}(R)$ in a standard HF calculation results in an energy curve with a clear minimum at about 6.0 a.u. and well depth of -7.88 cm^{-1} compared with the exact of a minimum at 5.73 a.u. and well depth of -7.428 cm^{-1}[20].

5. Polyatomic Molecules

For polyatomic molecules, the generalization of Eq. (16) is straightforward and gives the leading contribution to the correlation energy to be[21]

$$E^{\text{corr}}(R_{ab}, Z_i) \approx -\sum_a \delta Z_a \mathbf{U}^t \mathbf{D}_{\text{T}} + \sum_{a,b} \frac{Z_a \delta Z_b + Z_b \delta Z_a + \delta Z_a \delta Z_b}{R_{ab}} . \tag{19}$$

Using this formula we have examined a variety of polyatomic molecules, including H_2O, H_3O^+, CH_2, CH_4, BeH_2, BH_2 and others[21]. These results are presented in Figure 6, which shows that our model predicts the correlation energy at equilibrium with high accuracy, typically within 20% error, compared with higher level theories such as MP4 and QCISD(T). Special complications arise for polyatomic molecules that are not at equilibrium.

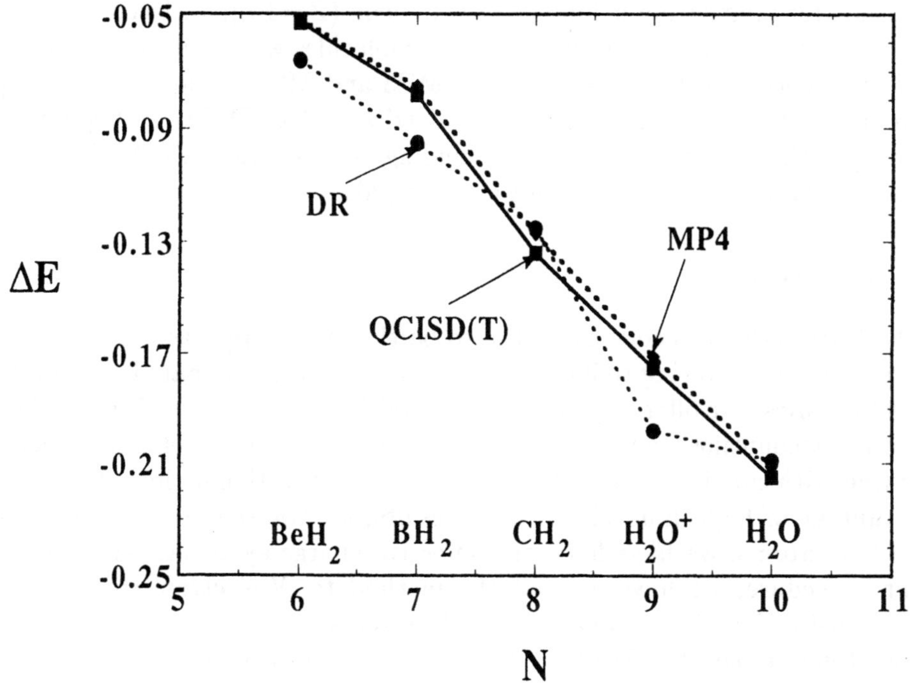

Figure 6. Correlation energy calculated by dimensional renormalization (DR) and higher level theories (MP4 and QCISD(T)) as a function of the number of electrons in the molecule. The molecules used in the making of this figure are listed along the axes.

Therefore, these results all represent the molecular energies only in their equilibrium geometries.

6. Details of the Hartree-Fock Calculation

For atoms, the correlation energies were calculated on a modified algorithm originally designed by Roothaan and Bagus[22]. The modifications allowed us to vary the nuclear charge as described, and also modified the basis set to account for the new nuclear charges. For all molecules, the correlation energies were calculated using a modified program originally designed by Gomperts and Clementi[23] in order to find correlation energy independently from the Hartree-Fock energy in accordance with Eq. (14). All higher level theory calculations and geometry minimizations were performed using Gaussian92[24].

For all calculations, we used the well-tempered Gaussian type functions as our basis set based upon the Roothaan-Hartree-Fock atomic orbital expansion for atoms from helium ($Z=2$) to neon ($Z=10$)[25]. To compare our correlation energy with exact results, for molecules we calculated all Møller-Plessett perturbation theories (MP2, MP3 and MP4), coupled cluster approximation with singles, doubles and triples (CCSD(T)) and quadratic configuration interaction theories including triples terms (QCISD(T)) with the standard 6-311G* triple zeta basis set[26].

7. Discussion

The Hartree-Fock theory continues to be the major approach for quantitative calculations owing to its relative inexpense when compared with higher level theories. We offer a systematic method for correcting the Hartree-Fock energy, accounting for the correlation energy by taking an effective nuclear charge with no significant addition to the computational time. By taking advantage of high dimensional limit results, known analytically for all N-electron atoms, we have fixed this effective charge systematically. To test our procedure, we have applied the method to N-electron atoms (N = 2–18, with Z = 2–28), weakly bound systems such as H^-, diatomic and polyatomic molecules. We have found that this procedure significantly reduces the error of the Hartree-Fock calculations and recovers more than 2/3 of the correlation energy without additional computational cost compared with standard Hartree-Fock calculations.

Previous methods have been proposed to include correlation energy into Hartree-Fock calculations. Examples include the g-Hartree approach which has been used to correct for the correlation energy of atoms and excited states of $H_2(^1\Sigma_g^+)$[27]. In this approach, a free parameter, g, was introduced into the Fock operator and fixed by requiring that the second order correction, MP2, be zero.

A more refined approach makes use of a soft-Coulomb hole effective interaction of the form $e^{-\omega_{ij}r_{ij}^2}/r_{ij}$ whose magnitude was calibrated semi-empirically to account for the correlation energies of atoms and molecules. Although these are two very different approaches, each of these methods resulted in the introduction of arbitrarily adjustable parameters which required either experimental or higher-level theory results. The model we propose in this chapter is for correcting the Hartree-Fock calculations to include correlation energy without increasing the cost of the computation by including higher level theories or introducing arbitrarily adjustable parameters. All of these approaches, however, fall under the same generalization of finding an effective Hamiltonian such that the eigenvalues of the effective Hamiltonian approaches the exact Hamiltonian eigenvalues[29].

Currently, exact solutions in the large-D limit exist only for N-electron atoms and a few small molecular systems. However, in principle, the solution of molecules at the large-D is similar to that of N-electron atoms, which involves finding the global minimum of a molecular effective potential. The problems to overcome in addition to finding a global minimum of a multivariable function is that the large-D limit does not exhibit shell-structure, and typically, the effective potential acquires multiple minima when nuclear charges or geometrical parameters are varied, some means to incorporate tunneling among such minima is also required[30]. Currently, work is in progress to develop methods to overcome these difficulties and obtain the large-D limit solution for molecules.

Acknowledgment

We are very grateful to Dudley Herschbach for initiating the idea of dimensional renormalization and for many constructive and fruitful discussions. We would like to thank Aaron Stanton, Marina Miletic and Avial Faliks for their contributions to this project. This work has been supported by the Chemistry Department of Purdue University and the Pittsburgh Supercomputing Center for their help and for making their computing facilities available to us, grant DMR900015P.

References

1. A. Szabo, and N.S. Ostlund, Modern Quantum Chemistry, (MacMillan Publishing Co., NY 1982), Chap. 3 and references therein.
2. D.L. Strout and G.E. Scuseria, J. Chem. Phys. 102, 8448 (1995).
3. I. Shavitt, "The Method of Configuration Interaction", in The Electronic Structure of Atoms and Molecules, Ed. H.F. Schaefer III (Plenum, NY 1977); R.J. Bartlett, Ann. Rev. Phys. Chem. 32, 359 (1981).
4. D.R. Herschbach, J. Avery and O. Goscinski, Dimensional Scaling in Chemical Physics (Kluwer Academic: Dordrecht, 1993).
5. M. Dunn, T.C. Germann, D.Z. Goodson, C.A. Traynor, J.D. Morgan III., D.K. Watson and D.R. Herschbach, J. Chem. Phys. 101, 5987 (1994).
6. S. Kais, and D.R. Herschbach, J. Chem. Phys. 3990, 98 (1993); T.C. Germann and S. Kais, J. Chem. Phys. 99, 7739 (1993).
7. J.G. Loeser, J.H. Summerfield, and A.L. Tan, and Z. Zheng, J. Chem. Phys. 5036, 100 (1994); J.G. Loeser, Chapter 9 of Ref. [4]
8. S. Kais, S.M. Sung, and D.R. Herschbach, J. Chem. Phys. 99, 5184 (1993); S. Kais, and D.R. Herschbach, J. Chem. Phys. 100, 4367 (1994)
9. S. Kais, T.C. Germann and D.R. Herschbach, J. Phys. Chem. 98, 11015 (1994).
10. S. Kais and R. Bleil, J. Chem. Phys. 102, 7472 (1995).
11. J.G. Loeser, J. Chem. Phys. 86, 5635 (1987).
12. S. Kais, S.M. Sung, and D.R. Herschbach, Int. J. Quantum Chem. 49, 657 (1994).
13. E.R. Davidson, S.A. Hagstrom and S.J. Chakravorty, Phys. Rev. A 44, 7071 (1991); S. J. Chakravorty, S.R. Gwaltney and E.R. Davidson, Phys. Rev. A 47, 3649 (1993).
14. E.H. Lieb and B. Simon, Adv. Math. 23, 22 (1977).
15. N.H. March and R.J. White, J. Phys. B 5, 466 (1972); R. Shakeshaft, L. Spruch and J.B. Mann, J. Phys. B14, L121 (1981); B.G. Englert and J. Schwinger, Phys. Rev. A32, 47 (1985).
16. J.M. Rost, J. Phys. Chem. 97, 2461 (1993).

17. R. Bleil, A. Faliks, M. Miletic and S. Kais, J. Chem. Phys. (submitted).
18. P.O. Löwdin, J. Mol. Spectroscopy, 3, 46 (1959).
19. A. L. Tan and J.G. Loeser, Chap. 6 of Ref[4].
20. C.A. Parish and C.E. Dykstra, J. Chem. Phys. 101, 7618 (1994); J. B. Anderson, C.A. Traynor, J. Chem. Phys. 99, 345 (1993).
21. R. Bleil and S. Kais, Int. J. Quantum Chem. (in press).
22. C.C.J. Roothaan and P.S. Bagus, Methods in Computational Physics, Ed. B. Alder (Academic Press, 1963).
23. E. Clementi and D.R. Davis, J. Comp. Phys. 2, 223 (1967); E. Ortoleva, G. Castiglione and E. Clementi, Comp. Phys. Comm. 19, 337 (1980).
24. M.J. Frisch, G.W. Trucks, M. Head-Gordon, P.M. W. Gill, M.W. Wong, J.B. Foresman, B.G. Johnson, H.B. Schlegel, M.A. Robb, E.S. Replogle, R. Gomperts, J.L. Andres, K. Raghavachari, J.S. Binkley , C. Gonzalez, R.L. Martin, D.J. Fox, D.J. DeFress, J. Baker, J.J.P. Stewart and J.A. Pople, GAUSSIAN 92 (Gaussian, INC., Pittsburgh, PA 1992).
25. S. Huzinaga and M. Klobukowski, J. Mol. Structure, 167, 1 (1988).
26. R. Krishnan, J.S. Binkley, R. Seeger and J.A. Pople, J. Chem. Phys. 72, 650 (1980).
27. B.A. Hess and M. Ohno, Phys. Rev. A39, 5637 (1989); R. Sawatzki and L.S. Cederbaum, Phys. Rev. A38, 6059 (1988); R. Sawatzki and L.S. Cederbaum, Chem. Phys. Lett. 126, 430 (1986).
28. E. Clementi and D.W.M. Hofmann, Int. J. Quantum Chem. 52, 849 (1994); S.J. Chakravorty and E. Clementi, Phys. Rev. A 39, 2290 (1989).
29. V. Hurtubis and K.F. Freed, Adv. Chem. Phys. 80, 465 (1993).
30. S. Kais, J.D. Morgan III and D.R. Herschbach, J. Chem. Phys. 95, 9028 (1991).

GENERALIZED SELF-CONSISTENT-FIELD APPROXIMATIONS

DAVID Z. GOODSON
Department of Chemistry
Southern Methodist University
Dallas, Texas 75275
USA

Abstract

A major motivation for using dimensional perturbation theory is to avoid the use of the standard independent-particle separability assumptions. Large-order perturbation theory in $1/D$ seems capable of providing highly accurate solutions of the exact Schrödinger equation for systems with as many as 10 coupled internal degrees of freedom. For larger systems it will be necessary to introduce separability approximations that uncouple at least some of the degrees of freedom. I will formulate the $1/D$ expansion within a self-consistent-field theory. Separability can be invoked in terms of the large-D normal coordinates (each of which describe correlated motions of many particles) or in terms of clusters of up to 10 coupled normal coordinates. This approach removes the size constraint that has until now limited applications of the large-order expansions to sytems with only a few particles.

Dimensional continuation has yielded intriguing qualitative descriptions of the dynamics of electrons in atoms and molecules [1]. The first-order perturbation theory in $1/D$ is mathematically very simple yet it gives a surprisingly good description of very complicated correlation and tunneling effects. However, in order to obtain highly accurate results for energies of Coulombic systems from dimensional perturbation theory (DPT) it is apparently necessary to carry out the perturbation expansion to rather large order [2-7]. Computations of expansion coefficients to sufficiently high order, *with no approximations* made in the Schrödinger equation, will probably only be possible for systems with a relatively small number of coupled internal degrees of freedom. To treat larger systems it will be necessary to introduce approximations. If the errors caused by the approximations are less severe than those from the standard quantum chemistry approximations then the method will still be viable.

Before considering approximate treatments, let us estimate the size constraints on the exact theory. Suppose we assume that the accuracy at a given

C. A. Tsipis et al. (eds.), New Methods in Quantum Theory, 71–82.

order in the expansion for the ground state of an arbitrary atom or molecule will be comparable to that for two-electron systems [6,8],

$$|E_{DPT} - E_{exact}| \approx 10^{-(0.5k+0.2)}, \tag{1}$$

where k is the order in the perturbation theory and the energies are in atomic units. Eq. (1) implies that 12th-order perturbation theory will be needed in order to obtain an accuracy of 10^{-6} a.u. ("spectroscopic" accuracy). Figure 1 shows a rough estimate of the accuracy that will be obtainable from large-order DPT as a function of the number of coupled degrees of freedom. The Figure is based on Eq. (1) and the analysis of computational costs and roundoff error given in Ref. [10].

For a D-dimensional system of N particles the number of internal degrees of freedom is

$$\tau = N(N-1)/2, \tag{2}$$

after separating out D translational coordinates and $(N-1)D-\tau$ rotations. Therefore, highly accurate calculations of this type using the exact Schrödinger equation are probably not be feasible for systems with more than 5 particles using the computer workstations that are currently available.

I will consider here approximations introduced through the use of separability assumptions in the wavefunction. For example, one could assume complete separability in terms of the large-D normal coordinates q_1, q_2, ...,

$$\Psi(q_1, q_2, q_3, \ldots) = \psi^{(1)}(q_1)\psi^{(2)}(q_2)\psi^{(3)}(q_3)\cdots. \tag{3}$$

This would reduce the problem to a system of equations each of which involves only one degree of freedom and thereby remove completely the DPT size constraint. Since the q_i each represent coupled motions of the particles, this separability assumption can be expected to account for correlation effects that are left out by the Hartree-Fock approximation. Less drastic separability assumptions could also be used, for example,

$$\Psi = \psi^{(a)}(q_1, q_2)\psi^{(b)}(q_3, q_4)\psi^{(c)}(q_5, q_6)\cdots. \tag{4}$$

Groupings of up to 10 normal modes will in principle be feasible with current computational resources. The best solutions for the ψ_i can be determined from the variational principle for the energy [11]. This approach leads to a *generalized self-consistent-field* (GSCF) approximation [12] that yields a set of coupled differential equations, each of which has fewer degrees of freedom than the original Schrödinger equation.

The differential equations can be uncoupled and solved recursively order by order in $D^{-1/2}$. To demonstrate this, I will consider an arbitrary system with two degrees of freedom and zero total angular momentum. After appropriate dimensional scalings [6,9,10,13-16], the Hamiltonian takes the form

$$\mathcal{H} = T + U + V. \tag{5}$$

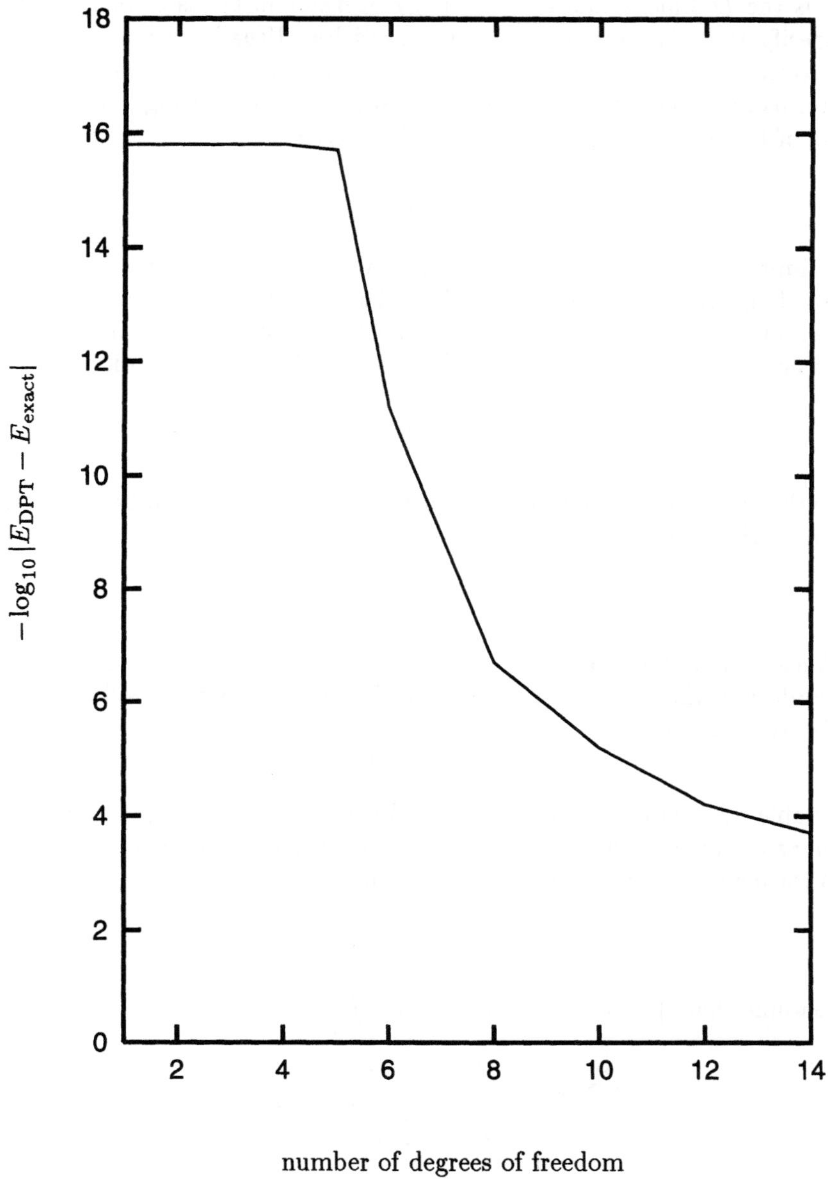

Figure 1. Estimate of accuracy obtainable from DPT, as a function of the number of coupled internal degrees of freedom. The energies are in atomic units.

$T+U$ is the D-dimensional kinetic energy operator in terms of the coordinates that specify the relative positions of the particles. Usually, the wavefunction is scaled so that this operator does not contain first derivatives. Then T consists of those terms that contain second derivatives while U is a centrifugal potential that consists of the terms that do not contain derivatives. Let

$$\delta = 1/D. \tag{6}$$

In the limit $\delta \to 0$ the operator T goes to zero while U and V remain finite and nonzero. U generally has a quadratic dependence on δ. It is advantageous to also include a quadratic δ dependence in V, subject to the constraint that V reduce to the physical potential at $\delta = 1/3$ [7,16]. The Schrödinger equation is

$$(\mathcal{H} - \widetilde{E})\Psi = 0. \tag{7}$$

$\widetilde{E}(\delta)$ is the energy in dimension-scaled distance units [15]. It will be calculated as an asymptotic expansion,

$$\widetilde{E} = \sum_{k=0}^{\infty} \widetilde{E}_k \delta^k. \tag{8}$$

Consider a system with only two degrees of freedom. Let r_1 and r_2 be the two coordinates that describe the configuration of the system. I will assume that the effective potential,

$$W = U + V, \tag{9}$$

has a stable minimum at $r_1 = \bar{r}_1$, $r_2 = \bar{r}_2$. In the limit $\delta \to 0$ the system becomes stationary at this configuration. To take into account displacements, at nonzero δ, it is convenient to transform to the coordinates x_i,

$$r_i = \bar{r}_i + \delta^{1/2} x_i. \tag{10}$$

I will assume that W can be expanded in the form

$$W = \sum_{k=0}^{\infty} \delta^{k/2} \sum_{m=0}^{\min(2,k/2)} \sum_{j=0}^{k-2m} w_{k,m,j} x_1^{k-2m-j} x_2^j \tag{11a}$$

and that T can be expanded in the form

$$T = -\frac{1}{2}\delta \left(t_{2,0} \frac{\partial^2}{\partial x_1^2} + t_{2,1} \frac{\partial^2}{\partial x_1 \partial x_2} + t_{2,2} \frac{\partial^2}{\partial x_2^2} \right) + \sum_{k=3}^{\infty} \delta^{k/2} \sum_{j=0}^{k-2} t_{k,j} x_1^{k-j-2} x_2^j \frac{\partial^2}{\partial x_1 \partial x_2}. \tag{11b}$$

Eqs. (11) are sufficiently general for most problems of interest.

Substituting Eqs. (8) and (11) into Eq. (7) and collecting terms of order 0 in $\delta^{1/2}$ gives

$$\widetilde{E}_0 = w_{0,0,0}. \tag{12}$$

Terms of order $(\delta^{1/2})^2$ give a harmonic oscillator Schrödinger equation with eigenvalues

$$\widetilde{E}_1 = w_{2,1,0} + (n_1 + \tfrac{1}{2})\omega_1 + (n_2 + \tfrac{1}{2})\omega_2 \tag{13}$$

and eigenvectors [10]

$$\Psi_0 = \psi^{(1)}_{n_1,0}(q_1)\psi^{(2)}_{n_2,0}(q_2), \tag{14}$$

$$\psi^{(i)}_{n,0}(q) = \omega_i^{1/4} h_n(\omega_i^{1/2} q), \tag{15}$$

$$h_n(q) = \left(\frac{1}{2^n n! \pi^{1/2}}\right)^{1/2} H_n(q)\, e^{-q^2/2}, \tag{16}$$

where H_n is the Hermite polynomial of degree n and the q_i are normal coordinates,

$$q_i = b_{i,1}x_1 + b_{i,2}x_2. \tag{17}$$

The ω_i and the $b_{i,j}$ can be derived using the standard techniques of molecular vibration analysis [17]. The operators W and T in normal coordinates have the forms

$$W = w_{0,0,0} + \delta\left(w_{2,1,0} + \frac{1}{2}\omega_1^2 q_1^2 + \frac{1}{2}\omega_2^2 q_2^2\right) + \sum_{k=3}^{\infty} \delta^{k/2} \sum_{m=0}^{\min(2,k/2)} \sum_{j=0}^{k-2m} W_{k,m,j}\, q_1^{k-2m-j} q_2^j, \tag{18a}$$

$$T = -\frac{1}{2}\delta\left(\frac{\partial^2}{\partial q_1^2} + \frac{\partial^2}{\partial q_2^2}\right) + \sum_{k=3}^{\infty} \delta^{k/2} \sum_{j=0}^{k-2} T_{k,j}\, q_1^{k-j-2} q_2^j \frac{\partial^2}{\partial q_1 \partial q_2}. \tag{18b}$$

Up to this point in the analysis no approximations have been introduced. To order $(\delta^{1/2})^2$ (which is first order in the energy expansion), Ψ is exactly separable in terms of the normal coordinates. Let us now assume that this separability will hold at all orders of $\delta^{1/2}$, that is,

$$\Psi(q_1, q_2) = \psi^{(1)}_{n_1}(q_1)\psi^{(2)}_{n_2}(q_2). \tag{19}$$

The subscripts n_i in Eq. (19) are the large-D harmonic quantum numbers. Each eigenstate of the system will correspond to a different choice of (n_1, n_2) [18]. Eq. (19) introduces an approximation into the analysis, since higher-order terms in the expansion of $\mathcal{H}$ couple q_1 and q_2.

The goal now is to determine the best wavefunction ("best" in the sense of the variational principle for the energy) that has the form of Eq. (19). This is the self-consistent-field (SCF) solution [11]. With conventional SCF methods this solution is determined iteratively starting with a guess. DPT yields the asymptotic expansion of the SCF solution directly, order by order, without using an iterative procedure.

The SCF equations are

$$\left\langle \psi_{n_2}^{(2)}(q_2) \middle| \left(\mathcal{H} - \widetilde{E}^{(1)}\right) \middle| \psi_{n_2}^{(2)}(q_2) \right\rangle_{q_2} \psi_{n_1}^{(1)}(q_1) = 0, \tag{20a}$$

$$\left\langle \psi_{n_1}^{(1)}(q_1) \middle| \left(\mathcal{H} - \widetilde{E}^{(2)}\right) \middle| \psi_{n_1}^{(1)}(q_1) \right\rangle_{q_1} \psi_{n_2}^{(2)}(q_2) = 0. \tag{20b}$$

Note that Slater determinants [11] are not used in Eqs. (20), because the separable wavefunction, Eq. (19), already has the correct permutation symmetry. The normal-mode energies $\widetilde{E}^{(1)}$ and $\widetilde{E}^{(2)}$ are related to $\widetilde{E}$, the SCF approximation for the total energy, by

$$\begin{aligned} \widetilde{E} &= \langle \psi_{n_2}^{(2)} \psi_{n_1}^{(1)} | \mathcal{H} | \psi_{n_1}^{(1)} \psi_{n_2}^{(2)} \rangle \\ &= \frac{1}{2}\left(\widetilde{E}^{(1)} + \widetilde{E}^{(2)}\right). \end{aligned} \tag{21}$$

Let

$$\widetilde{E}^{(i)} = \widetilde{E}_0 + \delta \widetilde{E}_1 + \delta \sum_{k=1}^{\infty} \delta^{k/2} \epsilon_k^{(i)}, \tag{22}$$

where $\epsilon_k^{(i)} = 0$ for odd k, and let

$$\psi_n^{(i)}(q) \propto \sum_{k=0}^{\infty} \delta^{k/2} \psi_{n,k}^{(i)}(q), \tag{23}$$

with

$$\left\langle \psi_{n,0}^{(i)} | \psi_{n,k}^{(i)} \right\rangle = \delta_{0,k}. \tag{24}$$

Each of the $\psi_{n,k}^{(i)}$ can be expressed as *finite* linear combinations of the harmonic eigenfunctions [10],

$$\psi_{n,k}^{(i)}(q) = \sum_{j=0}^{3k+n} {}^{j}a_{n,k}^{(i)} \psi_{j,0}^{(i)}(q). \tag{25}$$

Refs. [10] and [19] have demonstrated the utility of representing the wavefunction in terms of vectors in the Hilbert space spanned by the harmonic eigenfunctions. Thus, $\psi_{n,k}^{(i)}$ for $k > 0$ will be represented in the Hilbert space of the $\psi_{n,0}^{(i)}$ by the column vector $\mathbf{a}_{n,k}^{(i)}$, which has 0 in row $n+1$ (on account of Eq. (24)) and the number ${}^{j-1}a_{n,k}^{(i)}$ in the jth row for $j \neq n+1$, while $\psi_{n,0}^{(i)}$ will be represented by the column vector with 1 in row $n+1$ and zeros in all the other rows. From here on, for the sake of notational simplicity, I will omit the subscripts n that label the eigenstate under consideration.

We will need to calculate normalized moment integrals of the form

$$c_m^{(i)} = \left\langle \psi^{(i)}(q) | q^m | \psi^{(i)}(q) \right\rangle, \tag{26a}$$

$$d_m^{(i)} = \left\langle \psi^{(i)}(q) \left| q^{m-1} \frac{d}{dq} \right| \psi^{(i)}(q) \right\rangle. \tag{26b}$$

Let $\hat{c}_m^{(i)}$ be the unnormalized moment

$$\hat{c}_m^{(i)} = \sum_{k=0}^{\infty} \delta^{k/2} \sum_{j=0}^{k} \left(\mathbf{a}_{k-j}^{(i)} \right)^T \mathbf{q}_i^m \mathbf{a}_j^{(i)}, \tag{27}$$

where $\mathbf{q}_1$ and $\mathbf{q}_2$ are the matrix representations of the coordinate operators q_1 and q_2 in their respective Hilbert spaces. These matrices were derived in Ref. [10]. They are

$$\mathbf{q}_i = (2\omega_i)^{-1/2} \mathbf{q}, \tag{28}$$

where

$$\mathbf{q} = \begin{pmatrix} 0 & \sqrt{1} & 0 & 0 & & \\ \sqrt{1} & 0 & \sqrt{2} & 0 & & \\ 0 & \sqrt{2} & 0 & \sqrt{3} & \cdots & \\ 0 & 0 & \sqrt{3} & 0 & & \\ & & \vdots & & \ddots & \end{pmatrix}. \tag{29}$$

Formally, the normalized moments are

$$c_m^{(i)} = \hat{c}_m^{(i)} / \hat{c}_0^{(i)}. \tag{30}$$

It follows that $c_0^{(i)} = 1$. In general, however, Eq. (30) is a ratio of power series, which must be expanded into a single power series. Let

$$\frac{1}{\hat{c}_0^{(i)}} = \sum_{p=0}^{\infty} \delta^{p/2} \mathcal{N}_p^{(i)}. \tag{31}$$

Then

$$\begin{aligned} 1 &= \hat{c}_0^{(i)} \sum_{p=0}^{\infty} \delta^{p/2} \mathcal{N}_p^{(i)} \\ &= \sum_{k=0}^{\infty} \delta^{k/2} \sum_{p=0}^{k} \mathcal{N}_p^{(i)} \sum_{j=0}^{k-p} \left(\mathbf{a}_{k-p-j}^{(i)} \right)^T \cdot \mathbf{a}_j^{(i)}. \end{aligned} \tag{32}$$

Collecting terms according to powers of $\delta^{k/2}$ and then solving for $\mathcal{N}_k^{(i)}$ gives

$$\mathcal{N}_0^{(i)} = \left[\left(\mathbf{a}_0^{(i)}\right)^T \cdot \mathbf{a}_0^{(i)}\right]^{-1} = 1, \tag{33a}$$

$$\mathcal{N}_1^{(i)} = -2\left(\mathbf{a}_1^{(i)}\right)^T \cdot \mathbf{a}_0^{(i)} = 0, \tag{33b}$$

$$\mathcal{N}_{k>1}^{(i)} = -\sum_{p=0}^{k-2} \mathcal{N}_p^{(i)} \sum_{j=1}^{k-p-1} \left(\mathbf{a}_{k-p-j}^{(i)}\right)^T \cdot \mathbf{a}_j^{(i)}. \tag{33c}$$

The normalized moments are

$$c_m^{(i)} = \sum_{k=0}^{\infty} c_{k,m}^{(i)} \delta^{k/2}, \tag{34}$$

$$c_{k,m}^{(i)} = (2\omega_i)^{-m/2} \sum_{p=0}^{k} \mathcal{N}_p^{(i)} \sum_{j=0}^{k-p} \left(\mathbf{a}_{k-p-j}^{(i)}\right)^T \mathbf{q}^m \mathbf{a}_j^{(i)}. \tag{35}$$

Similarly,

$$d_m^{(i)} = \sum_{k=0}^{\infty} d_{k,m}^{(i)} \delta^{k/2}, \tag{36}$$

$$d_{k,m}^{(i)} = \omega_i (2\omega_i)^{-m/2} \sum_{p=0}^{k} \mathcal{N}_p^{(i)} \sum_{j=0}^{k-p} \left(\mathbf{a}_{k-p-j}^{(i)}\right)^T \mathbf{q}^{m-1} \frac{\mathbf{d}}{\mathbf{dq}} \mathbf{a}_j^{(i)}, \tag{37}$$

and $\frac{\mathbf{d}}{\mathbf{dq}}$ is the matrix [10]

$$\frac{\mathbf{d}}{\mathbf{dq}} = \begin{pmatrix} 0 & \sqrt{1} & 0 & 0 & \\ -\sqrt{1} & 0 & \sqrt{2} & 0 & \\ 0 & -\sqrt{2} & 0 & \sqrt{3} & \cdots \\ 0 & 0 & -\sqrt{3} & 0 & \\ & & \vdots & & \ddots \end{pmatrix}. \tag{38}$$

In addition, we will need expectation values of the harmonic component of $\mathcal{H}$. Note that

$$\left[-\frac{1}{2}\frac{d^2}{dq^2} + \frac{1}{2}\omega^2 q^2 - \left(n + \frac{1}{2}\right)\omega\right] h_j(\omega^{1/2} q) = (j - n)\omega h_j(\omega^{1/2} q). \tag{39}$$

Therefore,

$$\left\langle \psi^{(i)} \middle| \left[-\frac{1}{2}\frac{d^2}{dq^2} + \frac{1}{2}\omega_i^2 q^2 - \left(n_i + \frac{1}{2}\right)\omega_i\right] \middle| \psi^{(i)} \right\rangle = \mathcal{K}^{(i)}, \tag{40}$$

where n_i is the harmonic quantum number corresponding to $\psi^{(i)}$ in Eq. (19) and

$$\mathcal{K}^{(i)} = \sum_{k=1}^{\infty} \delta^{k/2} \mathcal{K}_k^{(i)}, \tag{41}$$

$$\mathcal{K}_k^{(i)} = \omega_i \sum_{p=0}^{k-2} \mathcal{N}_p^{(i)} \sum_{j=1}^{k-p-1} \left(\mathbf{a}_{k-p-j}^{(i)}\right)^T \mathbf{K}^{(i)} \mathbf{a}_j^{(i)}, \tag{42}$$

with the diagonal matrix $\mathbf{K}^{(i)}$ defined by

$$\left(\mathbf{K}^{(i)}\right)_{j,k} = (j - n_i)\delta_{j,k}. \tag{43}$$

It is now straightforward to derive the expectation values in Eqs. (20). Collecting terms according to powers of $\delta^{1/2}$ leads to the set of equations

$$\sum_{k=0}^{p} \left(\mathbf{H}_k^{(i)} - \epsilon_k^{(i)} \mathbf{I}\right) \mathbf{a}_{p-k}^{(i)} = \mathbf{0}, \quad p = 1, 2, 3, \ldots, \tag{44}$$

where $\mathbf{I}$ is the identity matrix, $\mathbf{0}$ is the vector with all elements equal to zero, and the $\mathbf{H}_k^{(i)}$ are the matrix representations of the $\mathcal{H}_k^{(i)}$,

$$\begin{aligned}\mathcal{H}_k^{(1)}(q_1) = \mathcal{K}_k^{(2)} &+ \sum_{p=0}^{k-1} \sum_{j=0}^{k-p} T_{k-p+2,k-p-j} d_{p,k-p-j}^{(2)} q_1^j \frac{d}{dq_1} \\ &+ \sum_{p=0}^{k-1} \sum_{m=0}^{\min(2,(k-p+2)/2)} \sum_{j=0}^{k-p-2m+2} W_{k-p+2,m,k-p-2m-j+2} c_{p,k-p-2m-j+2}^{(2)} q_1^j,\end{aligned} \tag{45a}$$

$$\begin{aligned}\mathcal{H}_k^{(2)}(q_2) = \mathcal{K}_k^{(1)} &+ \sum_{p=0}^{k-1} \sum_{j=0}^{k-p} T_{k-p+2,j} d_{p,k-p-j}^{(1)} q_2^j \frac{d}{dq_2} \\ &+ \sum_{p=0}^{k-1} \sum_{m=0}^{\min(2,(k-p+2)/2)} \sum_{j=0}^{k-p-2m+2} W_{k-p+2,m,j}\, c_{p,k-p-2m-j+2}^{(1)} q_2^j,\end{aligned} \tag{45b}$$

It follows that [10]

$$\mathbf{a}_p^{(i)} = \omega_i^{-1} \widetilde{\mathbf{K}}^{(i)} \sum_{k=1}^{p} \left(\epsilon_k^{(i)} \mathbf{I} - \mathbf{H}_k^{(i)}\right) \mathbf{a}_{p-k}^{(i)}, \quad p = 1, 2, 3, \ldots, \tag{46}$$

where the elements of $\widetilde{\mathbf{K}}^{(i)}$ are

$$\left(\widetilde{\mathbf{K}}^{(i)}\right)_{j,k} = \begin{cases} (j-n_i)^{-1}\delta_{j,k}, & j \neq n_i, \\ 0, & j = n_i, \end{cases} \tag{47}$$

and

$$\epsilon_p^{(i)} = \left(\mathbf{a}_0^{(i)}\right)^T \sum_{k=1}^{p} \mathbf{H}_k^{(i)} \mathbf{a}_{p-k}^{(i)}. \tag{48}$$

The generalization to more than two degrees of freedom, either with complete separability as in Eq. (3) or with groupings of coupled normal coordinates, is straightforward. With groupings of coordinates the $\mathbf{a}_k^{(i)}$ will be tensors of order τ_i, where τ_i is the number of coordinates in the ith group, rather than column vectors as in the present case [10].

Eqs. (46) and (48) have the same form as the matrix equations for a one-degree-of-freedom problem [10,19]. However, there is a significant difference: $\mathbf{H}_j^{(i)}$ depends on the $\mathbf{a}_p^{(i')}$ via the constants $c_{k,m}^{(i')}$ and $d_{k,m}^{(i')}$, where $i = 1$, $i' = 2$ or vice versa. $\mathbf{H}_j^{(i)}$ depends only on the $\mathbf{a}_p^{(i')}$ with $p < j$. Therefore, Eqs. (46) and (48) are solved in the following order:

$$\begin{array}{cccccc} & & \mathbf{a}_1^{(1)}, & \mathbf{a}_1^{(2)}, & \epsilon_2^{(1)}, & \epsilon_2^{(2)}, \\ \mathbf{a}_2^{(1)}, & \mathbf{a}_2^{(2)} & \mathbf{a}_3^{(1)}, & \mathbf{a}_3^{(2)}, & \epsilon_4^{(1)}, & \epsilon_4^{(2)}, \\ & & & \vdots & & \\ \mathbf{a}_{p_{max}-2}^{(1)}, & \mathbf{a}_{p_{max}-2}^{(2)}, & \mathbf{a}_{p_{max}-1}^{(1)}, & \mathbf{a}_{p_{max}-1}^{(2)}, & \epsilon_{p_{max}}^{(1)}, & \epsilon_{p_{max}}^{(2)}. \end{array}$$

If the expansion of $\widetilde{E}$, Eq. (8), is to be calculated through order k_{max}, then

$$p_{max} = 2k_{max} - 2. \tag{49}$$

This approach will be viable even when τ, the total number of internal degrees of freedom, is rather large. The size limitation in the exact case [10] was due to the size of the $\mathbf{a}_k$, which was a tensor of order τ. The number of tensor elements therefore increased exponentially with τ. With separability approximations, the total number of elements in the $a_k^{(i)}$ for all i increases only linearly with τ, which implies an N^2 increase, where N is the number of particles. The only quantities that increase at a faster than linear rate are the expansion coefficients $t_{k,j_1,\ldots,j_{\tau-1}}$ and $w_{k,m,j_1,\ldots,j_{\tau-1}}$, but these are easy to calculate and will easily be accommodated by the disk storage of current desktop computer workstations. In fact, the number of these expansion coefficients that will need to be stored is much smaller than one might expect, on account of the high symmetry of the large-D configuration. For example, in the case of Loeser's Hamiltonian for the N-electron atom [9], with S_N symmetry, the storage requirement for 12th-order perturbation theory with quadruple-precision (128-bit) arithmetic will be less than 100 MB. The time needed to evaluate the $c_{k,m}^{(i)}$ and $d_{k,m}^{(i)}$ will impose a constraint, but this will certainly be much less severe than the time constraint without separability approximations.

Although the primary motivation for GSCF approximations is to allow calculations for larger systems, there other potential uses for the technique. For example, one could use the separability assumptions as a probe of particle correlations. The degree of coupling between normal modes could provide a valuable physical picture of the correlated dynamics. Another possible use would be to improve the convergence of the large-order expansions for highly excited states. Converged δ expansions for the energies of doubly excited states of the He atom, for example, are quite difficult to obtain on account of singularities at complex values of δ very close to $\delta = 0$ [20]. These singularities are caused by avoided crossings between the doubly excited state and an infinite number of singly excited states [21]. All of the singly excited states lie below the doubly excited states at $\delta = 1/3$, but most lie above them at $\delta = 0$. GSCF approximations would eliminate this class of singularities by removing any couplings between the normal modes. This would presumably improve the convergence of the δ expansions.

GSCF calculations could also be used to analyze molecular vibrations with empirical Born-Oppenheimer nuclear potentials. These calculations would bear a strong resemblance to the vibrational SCF approach of Bowman and coworkers [22]. However, the δ expansion yields a much more direct solution of the SCF equations, requiring fewer simplifying approximations and avoiding the need for iterative procedures, and allows for the selective coupling of groups of normal modes. The δ expansion could also be used to solve the molecular Schrödinger equation without the Born-Oppenheimer approximation [16]. Those normal coordinates that reduce to purely electronic coordinates in the limit of infinite nuclear masses could be separated from the remaining coordinates with a GSCF approximation. This would retain the partitioning of the problem into separate "vibrational" and "electronic" analyses but would include non-Born-Oppenheimer effects, since for finite nuclear masses many of the "electronic" normal coordinates would represent coupled electronic and nuclear motions.

Acknowledgments

I am grateful to Dr. Melchior Elout for helpful discussions and for assistance in estimating the computational requirements of the GSCF calculations. This work was supported by grants from the Robert A. Welch Foundation and from the U.S. National Science Foundation.

References

1. For an excellent review see D. R. Herschbach (1993), in D. R. Herschbach, J. Avery, and O. Goscinski (eds.), *Dimensional Scaling in Chemical Physics*, Kluwer Academic Publishers, Dordrecht, pp. 7-80.

2. D. Z. Goodson and D. R. Herschbach (1987), *Phys. Rev. Lett.* **58**, 1628; D. Z. Goodson, M. López-Cabrera, D. R. Herschbach, and J. D. Morgan III (1992), *J. Chem. Phys.* **97**, 8481; M. López-Cabrera, D. Z. Goodson, D. R. Herschbach, and J. D. Morgan III (1992), *Phys. Rev. Lett.* **68**, 1992.
3. V. S. Popov, V. D. Mur, A. V. Scheblykin, and V. M. Weinberg (1987), *Phys. Lett.* **124 A**, 77; V. D. Mur, V. S. Popov, and A. V. Sergeev (1990), *Zh. Eksp. Teor. Fiz.* **97**, 32 [*Sov. Phys. JETP* **70**, 16].
4. T. C. Germann and S. Kais (1993), *J. Chem. Phys.* **99**, 7739.
5. D. Z. Goodson and D. K. Watson (1993), *Phys. Rev. A* **48**, 2668.
6. D. K. Watson and D. Z. Goodson (1995), *Phys. Rev. A* **51**, R5.
7. D. Z. Goodson, C. D. Elliston, and D. K. Watson (1995), unpublished.
8. I am assuming that the accuracy at a given order in the perturbation expansion will be no worse than that found for the two-electron atom ground state isoelectronic sequence. The rate of convergence of summation approximants for those systems is exponential. The convergence improves with increasing nuclear charge, which implies that the rate of convergence for H^- is a lower bound. Eq. (1) is the error in the ground state energy for H^- from Ref. [6]. Refinements in dimensional continuation techniques [7] will probably improve on this somewhat. It is not certain that this estimate will apply to larger systems. However, the accuracy of Loeser's perturbation theory for the N-electron atom [9] at zeroth order is essentially constant over the range of atomic number for which his symmetric large-D structure is valid (H through Si), which is consistent with my assumption of constant accuracy at given order.
9. J. G. Loeser (1987), *J. Chem. Phys.* **86**, 5635.
10. M. Dunn, T. C. Germann, D. Z. Goodson, C. A. Traynor, J. D. Morgan III, D. K. Watson, and D. R. Herschbach (1994), *J. Chem. Phys.* **101**, 5987.
11. J. C. Slater (1963) *Quantum Theory of Molecules and Solids*, vol. 1, McGraw-Hill, New York, pp. 41-45; and refs. therein.
12. D. Z. Goodson and D. R. Herschbach (1987), *J. Chem. Phys.* **86**, 4997.
13. J. Avery, D. Z. Goodson, and D. R. Herschbach (1991), *Theor. Chim. Acta* **81**, 1.
14. A. L. Tan and J. G. Loeser (1993), in Ref. [1], pp. 230-255.
15. M. López-Cabrera, A. L. Tan, and J. G. Loeser (1993), *J. Phys. Chem.* **97**, 2467.
16. J. M. Cohen and D. Z. Goodson (1995) A Unified Approach to Molecular Structure and Molecular Vibrations, submitted to *J. Chem. Phys.*
17. L. A. Woodward (1972) *Introduction to the Theory of Molecular Vibrations and Vibrational Spectroscopy*, Oxford University Press, London, pp. 168-182, 233-238.
18. For many-electron systems the Pauli principle will restrict the choices of the harmonic quantum numbers. See, for example, Ref. [9].
19. V. M. Vainberg, V. D. Mur, V. S. Popov, A. V. Sergeev, and A. V. Shcheblykin (1988), *Teor. Mat. Fiz.* **74**, 399 [*Theo. Math. Phys.* **74**, 269].
20. D. Z. Goodson and D. K. Watson (1993), in Ref. [1], pp. 359-374.
21. A. V. Sergeev (1994), unpublished; T. C. Germann and M. Dunn (1994), unpublished.
22. J. M. Bowman (1986), *Acc. Chem. Res.* **19**, 202; and refs. therein.

DIMENSIONAL EXPANSIONS FOR ATOMIC SYSTEMS

D. K. WATSON, M. DUNN, T. C. GERMANN*, D. R. HERSCHBACH*,
D. Z. GOODSON† AND J. R. WALKUP

Department of Physics and Astronomy
University of Oklahoma
Norman, OK 73019
United States

Abstract.

Dimensional perturbation theory seeks physical insight into three dimensional systems by studying such systems from the perpective of higher dimensions. Early results have yielded a wealth of insight into quite diverse phenomena, including geometrical insight into electron correlation, an explanation of patterns in atomic spectra and similarities in hyperspherical potential curves and approximate quantum numbers for highly-correlated atomic states.

In this work, we use dimensional perturbation theory to determine very accurate properties for two atomic systems. In particular we have studied some circular Rydberg and other large $|m|$ states of diamagnetic H. Taking advantage of an exact interdimensional degeneracy between states with high $|m|$ values at $D = 3$ and states with high D values for $|m| = 0$, we find that dimensional perturbation theory yields accurate energies and expectation values for the entire range of magnetic field strengths. In addition, the large dimension quantum numbers provide physically meaningful labels for these states at all magnetic field strengths.

We have also investigated an extension of dimensional perturbation theory to H^-, a weakly bound system, which is unstable in the limit $D \to \infty$ under the conventional dimensional continuation of the Schrödinger equation. By attenuating the repulsive part of the potential with a D-dependent factor, we are able to obtain accurate energies from large-order perturbation theory. Our result for the ground state of H^- is accurate to 8 digits, slightly better than that obtained for the ground state of He using conventional dimensional continuation. We expect that using dimension-dependent po-

C. A. Tsipis et al. (eds.), New Methods in Quantum Theory, 83–97.

tentials will also improve the convergence even for systems that are not weakly bound.

1. Introduction

During the last ten years dimensional perturbation theory has emerged as a viable new approach for studying atomic and molecular systems [1, 2, 3]. In this approach the Cartesian dimensionality of space, D, is used as the perturbation parameter. The solution of a dimensionally-continued Schrödinger equation as $D \to \infty$ is obtained exactly, and used as a starting point to develop an expansion for physical observables as a function of $1/D$. Although this method can be viewed as strictly a mathematical technique to obtain a perturbation series in a particular parameter, the advantages of the method are rooted in both the qualitative physical pictures and the quantitative results that emerge from the low perturbation orders which include a significant amount of correlation. The $D \to \infty$ results bear a close physical similarity to $D = 3$ results as verified by the comparison of expectation values at these two values of D [4]. This has prompted the formulation of several alternatives to dimensional perturbation theory that take advantage of the information and insight from the large D limit. These include *dimensional interpolation* between the $D = 1$ and $D \to \infty$ limits [5], *dimensional renormalization*[6, 7] of large D parameters and *subhamiltonian formulations*[8] which use the nondifferential form of the $D \to \infty$ hamiltonian at $D = 3$.

The present chapter will focus on the use of the more traditional dimensional perturbation theory to study two systems - diamagnetic hydrogen and the ground state of the hydride ion. Diamagnetic hydrogen is an example of a simple system with two nonseparable degrees of freedom, while the ground state of the hydride ion is an example of a weakly bound system for which very accurate results can be obtained by stabilizing the large D limit.

These two systems serve to demonstrate the well-known advantages of dimensional perturbation theory in addition to revealing some new types of insight from the large dimension limit. In particular the Fermi resonances at harmonic order help to illuminate the analysis of the fully converged states for diamagnetic hydrogen.

2. Diamagnetic Hydrogen

Understanding the hydrogen atom in a magnetic field remains a major unsolved problem of atomic physics. The diamagnetic term in the potential results in a nonseparable Hamiltonian in two degrees of freedom for which no general solution has been obtained. Even at low fields where the Coulomb potential dominates, the diamagnetic interaction has a profound effect since it breaks the symmetry of the Coulomb field resulting in l-mixing of states in a given n manifold. The evolution of states from pure hydrogenic at $B = 0$ to the harmonic Landau states as $B \to \infty$ is a nontrivial problem as the two potentials compete at intermediate fields to govern the shape of wavefunctions and the structure of spectra.

Many experimental and theoretical investigations, including semiclassical methods, have been carried out leading to numerous insights (for reviews see References [9, 10, 11]). Various methods have been tried using basis sets appropriate to either the low or high field limits. In contrast dimensional perturbation theory is applicable to the entire range of magnetic field strengths. The zeroth-order starting point and the basis set of harmonic functions, which are obtained from the large D limit, not only automatically adapt as the magnetic field strength changes, but they adapt in a way that is sensitive to the interplay between the Coulomb and diamagnetic potentials.

The Schrödinger equation for a hydrogen atom in a constant magnetic field, B, (units of ${m_e}^2 e^3 c/\hbar^3 = 2.35 \times 10^5\, T$), aligned in the z direction is given in atomic units by :

$$\left\{ -\frac{1}{2}\left[\frac{1}{\rho}\frac{\partial}{\partial\rho}\left(\rho\frac{\partial}{\partial\rho}\right) + \frac{\partial^2}{\partial z^2} - \frac{m^2}{\rho^2}\right] + \frac{mB}{2} + \frac{B^2\rho^2}{8} - \frac{Z}{\sqrt{\rho^2+z^2}} \right\} \Psi(\rho,z) = E\,\Psi(\rho,z)\,. \qquad (1)$$

The trivial paramagnetic Zeeman term, $mB/2$, can be ignored since it does not affect the dynamics. This equation has two good quantum numbers, the azimuthal quantum number, m, and the z parity, π_z. The dimensional continuation of this equation may be derived using well known techniques[1, 12], where ρ becomes the radius of a hypersphere in a $D-1$ dimensional Euclidean space. By introducing the substitution

$$\Phi(\rho,z) = \rho^{(D-2)/2}\Psi(\rho,z), \qquad (2)$$

we obtain an equation which has a simplified derivative term plus new centrifugal terms which keep the system bound as $D \to \infty$. After defining a new expansion parameter, κ,

$$\kappa = D + 2|m| - 1 \qquad (3)$$

and introducing dimensionally scaled variables

$$\rho = \kappa^2 \tilde{\rho}, \qquad z = \kappa^2 \tilde{z}, \qquad \tilde{B} = \kappa^3 B, \qquad \epsilon = \kappa^2 E, \tag{4}$$

we obtain the dimension-dependent scaled Schrödinger equation which now has a finite energy ϵ as $\kappa \to \infty$[12, 13]:

$$\left\{ -\frac{1}{2}\delta^2 \left(\frac{\partial^2}{\partial \tilde{\rho}^2} + \frac{\partial^2}{\partial \tilde{z}^2} \right) + \frac{1 - 4\delta + 3\delta^2}{8\tilde{\rho}^2} \right.$$
$$\left. + \frac{\tilde{B}^2 \tilde{\rho}^2}{8} - \frac{Z}{\sqrt{\tilde{\rho}^2 + \tilde{z}^2}} \right\} \Phi(\tilde{\rho}, \tilde{z}) = \epsilon\, \Phi(\tilde{\rho}, \tilde{z}) . \tag{5}$$

In this equation $\delta = 1/\kappa$ is treated as a continuous perturbation parameter. The same form was obtained by Bender and coworkers [14] by replacing $|m|$ in Eq. (1) by $(\kappa - 2)/2$.

To obtain a zeroth-order starting point we let $\kappa \to \infty$. In this limit, the problem reduces to finding the minimum of the effective potential, V_{eff}, for $\tilde{\rho}_m$ and $\tilde{z}_m$

$$V_{\text{eff}}(\tilde{\rho}, \tilde{z}) = \frac{1}{8\tilde{\rho}^2} + \frac{\tilde{B}^2 \tilde{\rho}^2}{8} - \frac{Z}{\sqrt{\tilde{\rho}^2 + \tilde{z}^2}}. \tag{6}$$

This potential is clearly minimized for $\tilde{z}_m = 0$. The minimum value for $\tilde{\rho}_m$ depends on the value of the magnetic field. The unscaled values vary from $\rho_m = (1 + |m|)^2$ at zero field to $2^{1/2}(1 + |m|)^{1/2} B^{-1/2}$ at large fields. Note that the zero field value is equal to the average bohr radius, n^2, for circular Rydberg states ($|m| = n - 1$), i.e. for the lowest state in an m manifold, while for high fields ρ_m decreases as the magnetic field increases. Thus, these positions, $\tilde{\rho}_m$ and $\tilde{z}_m$, which, for a given value of B, define frozen positions about which the basis functions will be expanded, adapt smoothly from low field to high field values. The resulting zeroth-order energy varies from $E_0 = -Z^2/(2(|m| + 1)^2)$, the correct ground state energy in each m manifold, at zero field to $E_0 = \frac{1}{2}B(|m| + 1)$, which is simply the continuum threshold in each m^{π_z} subspace at high field. We expand the solution about $\tilde{\rho}_m$ and $\tilde{z}_m$ by defining dimension-scaled displacement coordinates x_1 and x_2

$$\tilde{\rho} = \tilde{\rho}_m + \delta^{1/2} x_1, \qquad \tilde{z} = \delta^{1/2} x_2. \tag{7}$$

and expanding the wavefunction and energy as

$$\Phi(x_1, x_2) \;=\; \sum_{j=0}^{\infty} \Phi_j(x_1, x_2)\, \delta^{j/2}, \tag{8}$$

$$\epsilon = V_{\text{eff}}(\tilde{\rho}_m, \tilde{z}_m) + \delta \sum_{j=0}^{\infty} \epsilon_{2j} \delta^j. \tag{9}$$

The $\Phi_j(x_1, x_2)$ are expanded in terms of the large dimension harmonic oscillator eigenfunctions, h_ν

$$\Phi_j(x_1, x_2) = \sum_{\nu_1} \sum_{\nu_2} {}^{\nu_1,\nu_2}a_j h_{\nu_1}(\omega_1^{1/2} x_1) h_{\nu_2}(\omega_2^{1/2} x_2), \tag{10}$$

which are solutions to the first-order perturbation equation:

$$\left\{ -\frac{1}{2} \left(\frac{\partial^2}{\partial x_1^2} + \frac{\partial^2}{\partial x_2^2} \right) + \left(\frac{1}{2}\omega_1^2 x_1^2 + \frac{1}{2}\omega_2^2 x_2^2 + v_{0,0} \right) \right\} \Phi_0(x_1, x_2) = \epsilon_0 \ \Phi_0(x_1, x_2), \tag{11}$$

$$\epsilon_0 = \left(\nu_1 + \tfrac{1}{2}\right) \omega_1 + \left(\nu_2 + \tfrac{1}{2}\right) \omega_2 + v_{0,0}\,, \tag{12}$$

$$\Phi_0(x_1, x_2) = h_{\nu_1}(\omega_1^{1/2} x_1) h_{\nu_2}(\omega_2^{1/2} x_2)\,, \tag{13}$$

$$v_{0,0} = -\frac{1}{2\tilde{\rho}_m^2}.$$

The harmonic frequencies, ω_1 and ω_2 automatically adapt as the magnetic field changes:

$$\omega_1^2 = \frac{3}{4\tilde{\rho}_m^4} - \frac{2Z}{\tilde{\rho}_m^3} + \frac{\tilde{B}^2}{4}, \tag{14}$$

$$\omega_2^2 = \frac{Z}{\tilde{\rho}_m^3}. \tag{15}$$

For $B = 0$, $\omega_1 = \omega_2$. As B increases, $\omega_1 > \omega_2$ and both ω's increase resulting in basis functions that are squeezed inward responding appropriately to the effect of larger magnetic fields which squeeze the states of the H atom toward the origin. Note also that the basis functions are sensitive to the interplay between the Coulomb and diamagnetic potentials through the quantity $\tilde{\rho}_m$, obtained from the solution of Eq. (6), which determines both ω_1 and ω_2 and the expansion center for the basis functions.

2.1. CIRCULAR RYDBERG STATES

We chose to investigate circular ($|m| = n - 1 >> 1$) or near circular Rydberg states[15, 16]. There are several reasons why dimensional perturbation theory should yield excellent results for these states.

TABLE 1. Partial sums for the $m = -24$ circular state energy and expectation values $\langle z^2 \rangle$ and $\langle \rho^2 \rangle$ at a field strength of $B = 3 \times 10^{-4} \simeq 70.5$ T ($\tilde{B} = 37.5$). The series for $\langle z^2 \rangle$ begins with the first order term, and all series are unchanged (to the number of digits shown) beyond 12th order. Variational result (var) is taken from [17].

order	E (10^{-3} Rydbergs)	$\langle z^2 \rangle$	$\langle \rho^2 \rangle$
0	-5.076 181 876 237 13		113 234.573 995 659 7
1	-5.014 796 998 599 69	3086.414 795 836 26	114 603.520 931 015 4
2	-5.015 918 889 892 29	3207.951 676 065 63	114 556.338 735 429 2
3	-5.015 899 510 826 58	3209.095 452 458 65	114 556.253 103 047 4
4	-5.015 899 705 161 41	3209.077 579 145 55	114 556.258 353 394 3
5	-5.015 899 713 911 55	3209.078 138 849 00	114 556.258 311 853 3
6	-5.015 899 712 980 01	3209.078 120 371 57	114 556.258 311 121 1
7	-5.015 899 713 037 12	3209.078 120 433 78	114 556.258 311 315 1
8	-5.015 899 713 034 87	3209.078 120 527 51	114 556.258 311 342 3
9	-5.015 899 713 034 82	3209.078 120 512 36	114 556.258 311 341 8
10	-5.015 899 713 034 84	3209.078 120 514 11	114 556.258 311 341 8
11	-5.015 899 713 034 84	3209.078 120 513 95	114 556.258 311 341 8
12+	-5.015 899 713 034 84	3209.078 120 513 96	114 556.258 311 341 8
var	-5.015 899 7		

The dimensionally continued Schrödinger equation, Eq. (5), is invariant under changes $|m|, D \leftrightarrow |m|-1, D+2$ due to the fact that κ doesn't change under this transformation. This results in a complete set of interdimensional degeneracies. Thus large $|m|$ states at $D = 3$ are degenerate with large D states with $|m| = 0$ which we expect to converge rapidly since our expansion parameter $\delta = 1/\kappa$ is small. For example in Table 1 we give results for the $|m| = 24, D = 3$ state at a field strength of $B = 70.5T$ which we obtained by calculating the $|m| = 0$, $D = 51$ state. The expansion parameter, $\delta = 1/\kappa = 1/50 = .02$ results in a highly convergent expansion, which agrees with the variational results of Wunner, Ruder and Kost[17] by fourth order and has reached machine accuracy by tenth order. The expansion is so strongly convergent that we were able to use simple straightforward summation to obtain these results rather than Padé summation. A second reason for the rapid convergence is that as $|m|$ increases the wavefunction becomes more and more circular and thus closer and closer to the zeroth-order dimensional perturbation wavefunction which defines a circle in the $z = 0$ plane (as

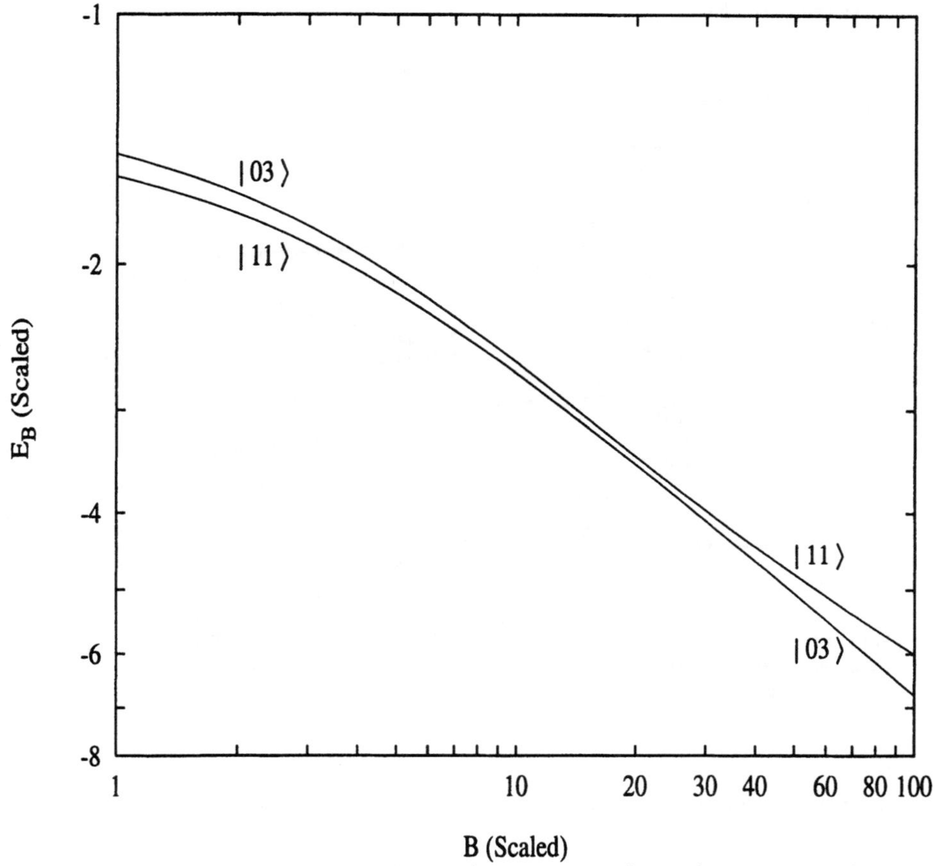

Figure 1. Converged energy curves for the $|\nu_1 = 1, \nu_2 = 1\rangle$ and $|\nu_1 = 0, \nu_2 = 3\rangle$ states showing an avoided crossing with the quantum numbers crossing diabatically

$\delta \to 0$, $\rho \to \tilde{\rho}_m$, $z \to 0$).

We studied the change in the energy as a function of B for several states in both even and odd $z-$parity manifolds. From an analysis of the expectation values of $\langle z^2 \rangle$ and $\langle \rho^2 \rangle$[15] we found that we could label these states with the harmonic quantum numbers, ν_1 and ν_2, from the solution of the first-order perturbation equation (Eq. (11)). As seen in Figure 1 which shows an avoided crossing for the $|\nu_1 = 1, \nu_2 = 1\rangle$ and $|\nu_1 = 0, \nu_2 = 3\rangle$ states, the harmonic quantum numbers pass through the avoided crossings diabatically, reflecting the exchange of character that occurs at an avoided crossing between states of the same symmetry.

These quantum numbers provide physically meaningful labels at all magnetic field strengths and correlate with the low B and high B quantum numbers. The ν_1 quantum number, which measures the excitation in the ρ direction, correlates with N, the Landau quantum number at high field since both quantum numbers count the number of nodes in the ρ direction.

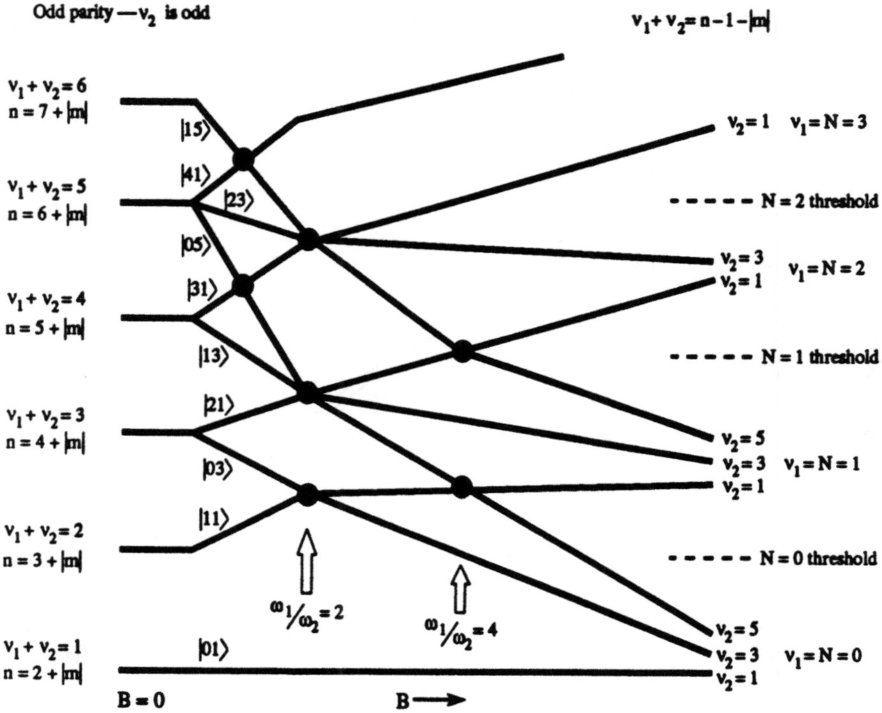

Figure 2. Correlation diagram for odd parity states showing Fermi resonances

At low field ν_1 correlates with $n-k-|m|-1$, where the quantum number, k, counts the states in an n manifold that are l-mixed due to the diamagnetic term. The highest state is labelled with $k=0$ down to $k=n-|m|-1$ for the lowest state. The ν_2 quantum number, which measures excitation in the z direction, correlates with 2ν for even parity states ($\pi_z = 1$) and with $2\nu-1$ for odd parity states ($\pi_z=-1$) where ν is the one-dimensional Coulomb quantum number from the solution of the Landau problem as $B \to \infty$. (The factor of 2 comes from the degeneracy of even and odd parity states as $B \to \infty$.) At low fields $\nu_2 \to k$.

The avoided crossings originate from the changing order of the states as B increases from low field to high field. At the low field limit, the ordering is governed by the Coulomb potential while in the high field limit the ordering of states is governed by the diamagnetic potential. Figure 2 shows a correlation diagram between low field and high field. The crossings have been placed at the magnetic field strengths where the ratio of ω_1/ω_2 becomes a rational number. At these positions the harmonic energies, i.e. first-order energies, become degenerate for certain pairs of states. These

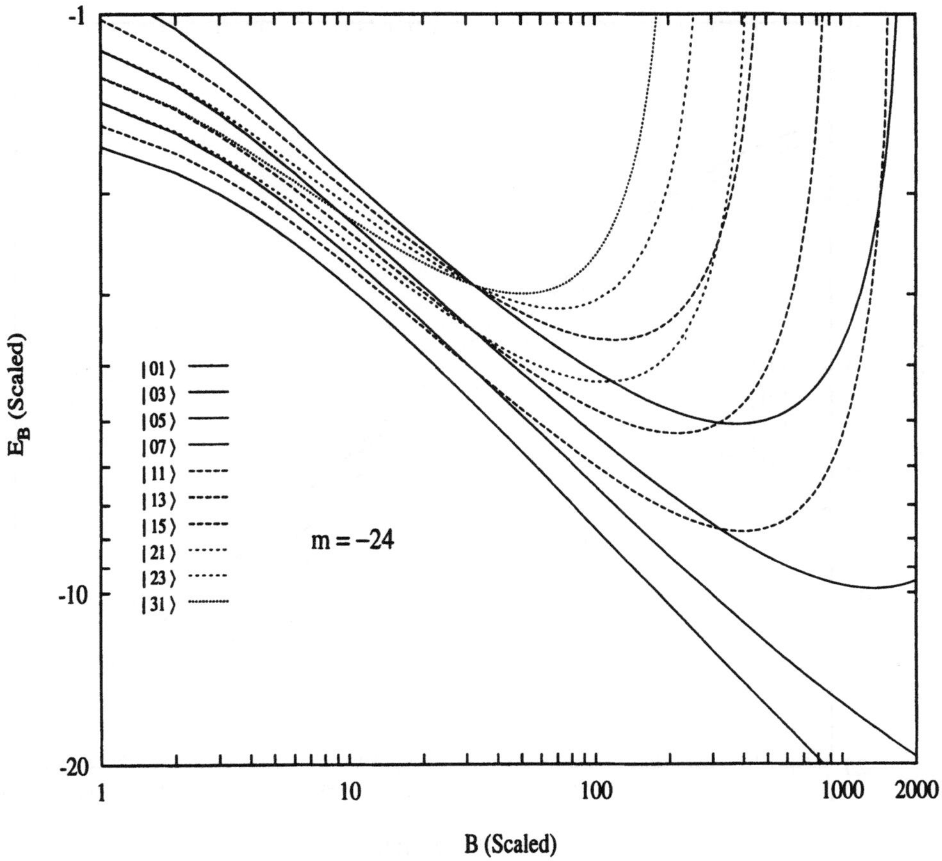

Figure 3. **Harmonic (first-order) energy curves for some odd parity states**

degenerate energies are called Fermi resonances. In Figure 3 we have plotted the harmonic energies for some odd parity states which display several Fermi resonances where these curves cross. The fully converged curves (Figure 4) show **avoided** crossings at B fields near these harmonic crossings due to the Wigner-von Neumann no-crossing rule for states of the same symmetry. By first graphing the harmonic energy levels which have simple crossings as a function of B, one can discern the overriding pattern for the converged energy levels which present a very complicated picture since they undergo many avoided crossings from low B to high B. The ordering of states at zero field and high field plus the position of the Fermi resonances essentially pins down the basic structure of the extremely complicated intermediate region.

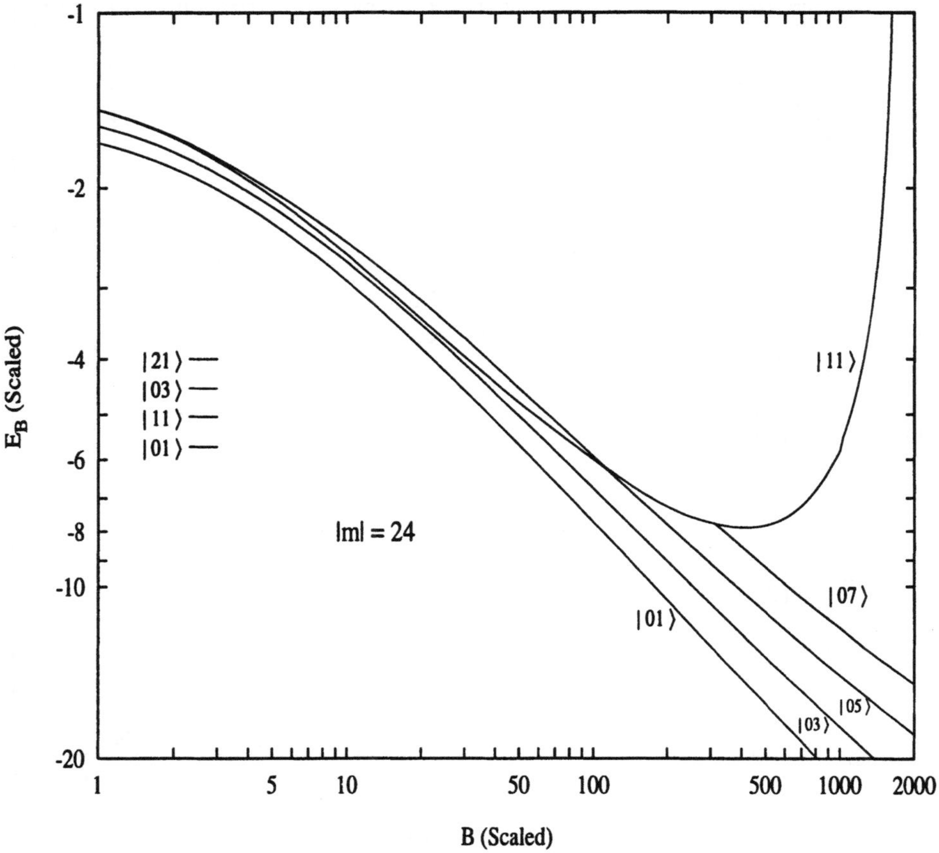

Figure 4. **Converged energy curves for some odd parity states. The ordering of four of the states at low *B* field is listed on the left.**

2.2. SINGULARITY ANALYSIS OF THE ENERGY SERIES

Analysis of the poles of the energy series using simple Padé analysis reveals that certain pairs of states have identical square-root branch points. For example the states $|\nu_1 = 1, \nu_2 = 1\rangle$ and $|\nu_1 = 0, \nu_2 = 3\rangle$ share a pair of branch points whose position varies as a function of B, while the states $|\nu_1 = 1, \nu_2 = 1\rangle$ and $|\nu_1 = 0, \nu_2 = 5\rangle$ share a different pair of square-root branch points whose behavior as a function of B is distinctly different from the first pair. This distinctive behavior of the branch points as a function of B is due to the type of interaction that couples the two states and will be explored in another paper. Both of these behaviors suggest that each of the states in a pair represents a different branch of the same multi-sheeted energy function, E(δ), which, for a given value of B, can be represented by the form [13]

TABLE 2. Energy expansion coefficients of the $|11\rangle$ and $|03\rangle$ states for scaled magnetic field $\tilde{B} = 20$. The numbers in parenthesis are the powers of ten by which each entry should be multiplied.

order j	ϵ_{2j}, $\lvert 11\rangle$		ϵ_{2j}, $\lvert 03\rangle$	
V_{eff}	−0.1125239354663	(00)	−0.1125239354663	(00)
0	+0.4106435280968	(02)	+0.4524609416892	(02)
1	−0.1555378224115	(03)	−0.1781736368247	(03)
2	− 0.9443167285383	(03)	+ 2.697780701731	(03)
3	+ 0.9102214876078	(04)	− 2.095374272998	(04)
4	+0. 6564553608858	(06)	−0. 5716205621973	(06)
5	−0. 7406130298139	(07)	+0. 6716141872573	(07)
6	−0.54 24078205954	(09)	+0.54 70817407763	(09)
7	+0.78 42529494748	(10)	−0.78 82728712806	(10)
8	+0.554 3801764684	(12)	−0.554 1619245327	(12)
9	−0.981 8192837103	(13)	+0.981 5949345757	(13)
10	−0.6275 772390791	(15)	+0.6275 825404210	(15)
11	+0.1328 691513258	(17)	−0.1328 704028611	(17)
12	+0.755131 3414370	(18)	−0.755131 8140234	(18)
13	−0.187987 6371776	(20)	+0.187987 5485558	(20)
14	−0.944717 6901749	(21)	+0.944717 5924958	(21)
15	+0.2734999 427303	(23)	−0.2734999 521390	(23)
16	+0.12126831 01541	(25)	−0.12126831 13109	(25)
17	−0.40541402 99715	(26)	+0.40541402 88027	(26)
18	−0.15831112 70058	(28)	+0.15831112 68897	(28)
19	+0.608850104 6078	(29)	−0.608850104 7401	(29)
20	+0.2088186863 258	(31)	−0.2088186863 363	(31)

$$E^{\pm}(\delta) = E^{0}(\delta) \pm E'(\delta)\sqrt{(1-\delta/\delta_{s1})(1-\delta/\delta_{s2})}. \tag{16}$$

Because these branch points, δ_{s1} and δ_{s2}, are very close to the origin, i.e. $\delta_{s1}, \delta_{s2} << 1$, the expansion of the square root in Eq. (16) in powers of δ should determine the large-order coefficients for both series. This is verified in Table 2 in which the large-order coefficients for the $|\nu_1 = 1, \nu_2 = 1\rangle$ and $|\nu_1 = 0, \nu_2 = 3\rangle$ states are seen to be identical in size but opposite in magnitude. This square-root branch point behavior is a direct consequence of the avoided crossings that exist for the fully converged states. The Fermi resonances at the harmonic level where the two states are exactly degenerate become avoided crossings for the fully converged states which cannot cross. However the states can and do become degenerate for complex values of δ where the two Riemann sheets of the multi-valued energy function are connected. These complex branch points are the solutions to the 2×2 secular equation that couples these two almost degenerate states.

3. The Ground State of the Hydride Ion

The hydride ion is a two-electron system which is weakly bound supporting only two bound states, the ground $1s^2\ {}^1S$ state and the $2p^2\ {}^3P^e$ state which is embedded in the continuum. The failure of restricted Hartree-Fock theory to predict these two bound states points to the importance of correlation effects which have proved difficult to describe theoretically. For high accuracy, wavefunctions with both radial and angular correlation are required. This type of correlation is included in the large D limit of dimensional perturbation theory since r_1, r_2 and θ_{12} are simultaneously determined from the minimum of the $D \to \infty$ effective potential. However, the conventional dimensional continuation of the Schrödinger equation, in which the kinetic energy is generalized to arbitrary D while the potential term remains unchanged, results in an unstable effective potential for H^- which allows the outer electron to escape. Thus conventional dimensional perturbation theory is not a possibility for the hydride ion. Several methods have attempted to use other types of dimensional analysis to study H^-. Doren and Herschbach devised a dimensional interpolation procedure[5] which uses information from both the $D = 1$ and $D \to \infty$ limits. Their results were within a few tenths of a percent for both states or about 2 digits of accuracy. In another approach to the hydride ion and other weakly bound three-body systems[18], Rost has expanded about a saddle point of the effective potential V_{eff} in the large dimension limit.

Here we report on a new method of using dimensional analysis to study weakly bound systems[19]. This method appears to be quite flexible, easily extendible to a variety of systems and amenable to systematic improvements. We prescribe a new method for the dimensional continuation of the Schrödinger equation which results in a stabilized large D limit. This is accomplished by carefully adding D dependence to the potential so that the interelectronic potential, $1/r_{12}$, is attenuated as D increases. The resulting effective potential is required to have a stable minimum about which an expansion can be performed and the correct Schrödinger equation must result at $D = 3$. Within these restrictions the dimensional continuation is arbitrary.

The potential energy term in the D-scaled Schrödinger equation for a two-electron atom in units of Z^2 atomic units is

$$V(r_1, r_2, r_{12}) = -\frac{1}{r_1} - \frac{1}{r_2} + \frac{\lambda}{r_{12}}. \tag{17}$$

where $\lambda = 1/Z$. For this initial study we chose to replace the λ/r_{12} term by

$$\lambda/r_{12} \rightarrow \lambda(\delta)/r_{12} \tag{18}$$

where $\lambda(\delta) = \lambda_0(1+\beta\delta)$. As $D \rightarrow \infty$ (i.e. $\delta \rightarrow 0$) , $\lambda \rightarrow \lambda_0$ which must be chosen to give a stable minimum. Doren and Herschbach found that if $\lambda_0 >$ .8144 the effective potential has a saddle point rather than a symmetric minimum. After λ_0 is chosen, β must be chosen so $\lambda(\delta = 1/3) = 1/Z = 1$. Thus we chose:

$$\beta = 3({\lambda_0}^{-1} - 1). \tag{19}$$

We calculated the coefficients, E_k, in the energy expansion

$$E = \delta^2 \sum_{k=0}^{\infty} \delta^k E_k \tag{20}$$

through fourteenth order using the moment method[20] with double-precision arithmetic. Several different values of λ_0 in the range $.45 < \lambda_0 < .8$ were tried. To sum the series we first factored out the first and second order poles at $D = 1$ determining the residue, a_{-2}, from Rosenthal's series for the eigenvalue solution of the one-dimensional Schrödinger equation [21] and the residue, a_{-1}, from Padé analysis. The resulting series takes the form

$$E = \delta^2 \left[\frac{a_{-2}}{(1-\delta)^2} + \frac{a_{-1}}{1-\delta} + \sum_{k=0}^{\infty} \delta^k {E_k}' \right] \tag{21}$$

where ${E_k}' = E_k - (k+1)a_{-2} - a_{-1}$. Our result shown in Table 3 for the ground state of H^- is accurate to 8 digits. We found that a λ_0 close to .5, i.e. the helium value, gave the best results. Smaller values of λ_0 resulted in poorer values for a_{-2} from Rosenthal's series due to the value of λ at $D = 1$ becoming too large for good convergence.

4. Conclusions

Dimensional perturbation theory has the advantage of dual strengths. First it offers physical insight into properties of atomic systems from the low orders of perturbation theory while offering very accurate results from high order calculations. The low-order terms include significant correlation and are still simple enough (in fact harmonic) to draw pictures and obtain approximate quantum numbers. The pictures of the harmonic energy levels as a function of magnetic field which show crossings at the Fermi resonances

TABLE 3. Comparison of ground state energies of H^-

Energy (a.u.)	Method	No. of basis functions	No. of perturbation terms
-0.527751014	Dimensional perturbation[19]	–	14
-0.527751017	Variational[22]	616	–
-0.527751017	$1/Z$ Variational perturbation[23]	476	401

aid tremendously in the analysis of the complicated picture presented by the converged energy levels at intermediate fields. The zeroth-order correlation for diamagnetic hydrogen is sensitive to the interplay between the Coulomb potential and the diamagnetic potential at any magnetic field strength, while the correlation included at zeroth-order for the hydride ion includes both radial and angular types.

Secondly, dimensional perturbation theory offers an extremely flexible formalism since the continuation of the Schrödinger equation to $D \to \infty$ is subject only to 2 constraints: the $D \to \infty$ limit is required to be stable and the correct Schrödinger equation must be recovered at $D = 3$. This was taken advantage of in the hydride study in which we stabilized the large D limit by adding D dependence to the $1/r_{12}$ term so that it was attenuated as $D \to \infty$. The accuracy (8 digits) of our result for the ground state of the hydride ion is impressive. This is due in part to the zeroth-order starting point which includes both radial and angular correlation but also to the extraction of the first and second order poles at $D = 1$ which results in a more highly convergent energy expansion.

In the diamagnetic hydrogen problem one is able to take advantage of a complete set of interdimensional degeneracies which enabled us to calculate very high $|m|$ states at $D = 3$ by calculating $|m| = 0$ states at large values of D resulting in a rapidly convergent and very accurate expansion.

We plan to continue our investigation of both these systems. For H^- we are extending our study to quadruple precision and plan to study whether stabilizing the $D \to \infty$ limit for systems such as He which are not weakly bound will improve the convergence. For diamagnetic hydrogen we plan a more thorough study of the circular Rydberg states as well as states with smaller values of $|m|$. We are hopeful that the combination of insight from harmonic order and high accuracy from large-order perturbation expansions will combine to illuminate this long-studied fundamental problem of atomic physics.

Acknowledgements

This work was supported by grants from the National Science Foundation (PHY-9123199) and the Office of Naval Research (NOOO14-94-1-0998. T.C.G. acknowledges a DOE Computational Science Graduate Fellowship and D.Z.G. acknowledges support from the R.A. Welch Foundation.

*Permanent address: Department of Chemistry, Harvard University, Cambridge, MA, USA 02138.

†Permanent address: Department of Chemistry, Southern Methodist University, Dallas, TX USA 75275.

References

1. *Dimensional Scaling in Chemical Physics*, edited by D.R. Herschbach, J. Avery, and O. Goscinski (Kluwer, Dordrecht, 1992).
2. M. Dunn, T.C. Germann, D.Z. Goodson, C.A. Traynor, J.D. Morgan III, D.K. Watson, and D.R. Herschbach, *J. Chem. Phys.* **101**, 5987 (1994).
3. V.S. Popov, V.D. Mur, A.V. Shcheblykin, and V.M. Weinberg, *Phys. Lett.* **124A**, 77 (1987); V.S. Popov, V.D. Mur, A.V. Sergeev, and V.M. Weinberg, *ibid.* **149A**, 418 (1990); V.S. Popov, V.D. Mur, and A.V. Sergeev, *ibid.* **149A**, 425 (1990); V.M. Vaĭnberg, V.S. Popov, and A.V. Sergeev, *Sov. Phys. JETP* **71**, 470 (1990). See also V.S. Popov in *Dimensional Scaling in Chemical Physics*, edited by D.R. Herschbach, J. Avery, and O. Goscinski (Kluwer, Dordrecht, 1992), Sections 5.3 and 6.2.
4. D. R. Herschbach, J.G. Loeser and D.K. Watson, *Z. Phys. D.* **10**, 195 (1988).
5. D.J. Doren and D.R. Herschbach, *J. Am. Chem. Soc.* **92**, 1816 (1988).
6. S. Kais, S.M. Sung and D.R. Herschbach, *J. Chem. Phys.* **99**, 5184 (1993); *Int. J. Quantum Chem.* **49**, 657 (1994).
7. S. Kais and D.R. Herschbach, *J. Chem. Phys.* **100**, 4367 (1994).
8. J.G. Loeser, J.H. Summerfield, A.L. Tan, and Z. Zheng, *J. Chem. Phys.* **100**, 5036 (1994).
9. H. Friedrich and D. Wintgen, *Phys. Rep.* **2**, 37 (1989).
10. D. Delande, F. Biraben and J.C. Gay, *New Trends in Atomic Physics*, Les Houches Session XXXVIII(Elsevier, Amsterdam, p.351 (1984).
11. C.W. Clark, K.T. Lu and A.F. Starace, *Progress in Atomic Spectroscopy* Part C, eds H.J. Beyer and H. Kleinpoppen(Plenum, New York, 1984) p. 247.
12. T.C. Germann, D.R. Herschbach, and B.M. Boghosian, Comput. Phys. **8**, 712 (1994).
13. T.C. Germann, *Dimensional Perturbation Theory for Atomic Systems*, Ph.D. thesis, Harvard University, 92 (1995).
14. C.M. Bender, L.D. Mlodinow, and N. Papanicolaou, *Phys. Rev. A* **25**, 1305 (1982).
15. T.C. Germann, D.R. Herschbach, M. Dunn, and D.K. Watson, *Phys. Rev. Lett.* **74**, 658 (1995).
16. See also A.V. Sergeev, to be published.
17. G. Wunner, M. Kost, and H. Ruder, *Phys. Rev. A* **33**, 1444 (1986).
18. J.M. Rost, *J. Phys. Chem.* **97**, 2461 (1993).
19. D.K. Watson and D.Z. Goodson, *Phys. Rev. A* **51**, R5 (1995).
20. D.Z. Goodson and D.R. Herschbach, *Phys. Rev. Lett.* **58**, 1628 (1987).
21. C.M. Rosenthal, *J. Chem. Phys.* **55**, 2474 (1971).
22. G.W.F. Drake, *Nucl. Inst. Meth. Phys. Res.* **B31**, 7 (1988).
23. J.D. Baker, D.E. Freund, R.N. Hill and J.D. Morgan III, *Phys. Rev. A* **41**, 1247 (1990).

THE MANIFESTATION OF CHAOTIC BEHAVIOUR OF THE HYDROGEN ATOM IN THE PRESENCE OF UNIFORM ELECTRIC FIELD

R. DAMBURG AND B. BARANOVSKY

Institute of Atomic Physics and Spectroscopy, Latvian University, Raina bulv., Riga LV-1586, Latvia

Abstract

We analyze the problem of a hydrogen atom in the presence of a uniform electric field and show that it provides one of the simplest examples of chaotic situation in atomic physics.The explanation of our understanding of a "chaos" is provided. We show that in the presence of a uniform electric field the unique for $F = 0$ ionization threshold at $E = 0$ splits in two kind of subthresholds depending on quantum numbers n and m. The energies of the subthresholds can be calculated exactly for any value of F. The obtained theoretical predictions are compared with the experimental data so far available.

1. Introduction

With the development of laser technique the detailed study of highly excited Rydberg states of atoms in the presence of the electric field began. In the paper of Kleppner *et.al.* [1] the behaviour of Stark manifolds of Li and Na for total quantum numbers n from 12 up to 20 and quantum number $m = 0, 1, 2$ was presented and analysed. The energies of all states of Stark manifolds for $m = 1$ and $m = 2$, except regions of very small F, can be quantitatively described by the Schrödinger perturbation expansion for hydrogen. The terms $\sim F^3$ are too small to be verified for values of F studied in [1]. The authors [1] claim that they found some essential deviations from hydrogenic case.

"In spite of their similarities, however the lithium and hydrogenic structures fundamentally differ from each other. None of the lithium's levels cross" [1].

C. A. Tsipis et al. (eds.), New Methods in Quantum Theory, 99–134.

This experimental observation was explained in the following way. "The degenerate 'sea' of levels has no effect on the dynamics of hydrogen because the Stark states do not mix as a result of the dynamic symmetry...The symmetry is only exact, however, for a pure Coulomb potential. Any perturbation to the potential, for instance, the core interaction in an alkali-metal atom, upsets the symmetry and causes the hydrogenic Stark levels to mix. If a stable state mixes with a decaying state, it also decays" [1]. The experimental studies of alkali-metal spectra started in [1] were extended later on to much larger values of F. Mc Nicholl *et.al.* [2] observed for three states of Na that at certain values of F and E (see review [3] where earlier papers are mentioned) the signal which gives the evidence of the presence of the Stark level is sharply decreasing and than at slightly larger value of F regenerating again. The phenomenon got a name "stabilization of Stark state at the saddle point". It was also explained by the influence of atomic core [2]. By saying that the results for the Stark states of alkali-metals are different from hydrogen ones the authors [1] do not even mention that there are no up to now similar experimental studies for the hydrogen. They relied purely on the theory of the Stark effect for the hydrogen atom in the presence of the uniform electric field. In no case we are going to deny this theory. But we shall show that it should be fundamentally supplemented. Such supplemented theory for the hydrogen atom explains also the experimental results obtained for Stark states of alkali-metals.

But the most dramatic example which immediately shows that the existing theory for the Stark effect is incomplete can be traced in the experimental data on the microwave ionization of highly excited states of the hydrogen atom [4]. Such atoms are effectively ionized at so small values of maxima of oscillating field and so small frequencies that the decay of atoms cannot be explained by quantum tunneling or by the absorption of reasonably small number of photons which can transfer the energy of excited atoms above the ionization threshold in the absence of field. In explaining the phenomenon theoreticians in any case did not assume that microwave ionization can be understood solely by examining the behaviour of an atom in a constant field. Hence model time-dependent (classical and quantum) equations were solved and attempts to find a correlation between the obtained numerical and analytical results and features of microwave ionization were made. References to the theoretical papers can be found in [5]. We can note only that no simple and physically straightforward mechanism of the phenomena occurred in the microwave ionization of atoms was suggested. We consider also the photoionization of the hydrogen atom in the presence of uniform electric field and show that here the experimental data contain also some unexpected features [6,7]. The experimental data obtained in different studies of highly excited states clearly show that there is some feature in the theory of the nonrelativistic hydrogen atom in the presence of the uniform electric field which had not been taken into account.

In interpreting the experimental data both experimenters and theoreticians assumed, without even discussing the problem, that all behaviour of atoms in a field can be understood by considering only Stark levels. Very general considerations show that this is not the case. The results of the action of photons on an atom at $F = 0$ cannot be understood only by considering discrete states, since if the photon energy exceeds the ionization energy, the atom becomes ionized. Stark states, however, are simply a natural continuation of the discrete levels of an atom at $F = 0$ into the region with $F \neq 0$. The presence of a field lifts the degeneracy of the states in the parabolic quantum numbers n_1, n_2 and m. A method for such a continuation was suggested by Schrödinger, who used a perturbation theory expansion in the small parameter F [8,9]. This continuation is asymptotically exact as $F \to 0$. The question of what happens with the ionization limit of an atom when the field F is switched on, which at $F = 0$ has $E = 0$ and is infinitely degenerate in n_1, n_2, and $|m|$, has not even been fully posed. The Stark states alone provide a fairly full picture of the behaviour of an atom in a field only up to energies at which direct ionization of the atom becomes possible. But for $F \neq 0$ we cannot even introduce the notion of an ionization limit determined by a natural value of energy E. Only one mechanism of direct ionization of an atom in a field has so far been suggested in Sec. 54 of Bethe and Salpeter's monograph [9]. These authors did not consider it in detail but did show that it was impossible to identify it using the experimental data available at the time. The possibility emerged only when it became possible to gather data on the behaviour of highly excited Rydberg states in Li and Na atoms at $m = 1, 2$ in the presence of a constant electric field [1,10].

Bethe and Salpeter [9] defined the threshold for disappearance of line, i.e. starting point for subthreshold for the ionization, in terms of classical mechanics. Certain hindrances to the understanding of this circumstance could have been caused by the fact that Bethe and Salpeter referred to the different phenomenon, i.e. to the quantum tunneling, as ionization by wave mechanics. Evidently, for such reason, the Sec. 54 of their book was not properly understood for many years, since one might think that the ionization by wave mechanics is just more rigorous approach than the ionization by classical mechanics. But actually, Bethe and Salpeter were considering two different phenomena. We shall consider below four different mechanisms for "direct ionization" in the presence of a constant electric field. The energies of "subthresholds" for ionization at $F \neq 0$ can be calculated exactly in two cases. The theoretical consideration in all cases will be supplemented by comparisons with the experimental data.

2. Ionization by Electric Field. Disappearance of Lines in Stark Effect

The title of this Chapter coincides with that of Sec.54 of the book by Bethe and Salpeter [9], and it should be therefore clear that we consider here the effect which was predicted by them.

Consider the Schrödinger equation for the Stark effect for the hydrogen atom in the uniform electric field

$$\left(\Delta + \frac{2}{r} - 2Fz + 2E\right)\Psi = 0. \tag{1}$$

Equation (1) is separable in coordinates

$$x = (\xi\eta)^{1/2}\cos\varphi, y = (\xi\eta)^{1/2}\sin\varphi, z = \frac{1}{2}(\xi - \eta). \tag{2}$$

Defining the wave function Ψ as the product

$$\Psi = (\xi\eta)^{-1/2}V(\xi)U(\eta)e^{\pm m\varphi} \tag{3}$$

and substituting (3) into (1) we obtain

$$\frac{d^2V}{d\xi^2} + \left(\frac{E}{2} + \frac{Z_1}{\xi} + \frac{1-m^2}{4\xi^2} - \frac{F}{4}\xi\right)V = 0, \tag{4}$$

$$\frac{d^2U}{d\eta^2} + \left(\frac{E}{2} + \frac{Z_2}{\eta} + \frac{1-m^2}{4\eta^2} + \frac{F}{4}\eta\right)U = 0, \tag{5}$$

$$Z_1 + Z_2 = 1. \tag{6}$$

The behaviour of $V(\xi)$ and $U(\eta)$ at the origin and at infinity is determined by the formulae

$$V(\xi) \simeq_{\xi\to 0} \xi^{(m+1)/2}, \tag{7}$$

$$V(\xi) \simeq_{\xi\to\infty} \frac{A}{\xi^{1/4}}\exp\left(-\frac{F^{1/2}}{3}\xi^{3/2} + \frac{E}{F^{1/2}}\xi^{1/2}\right), \tag{8}$$

$$U(\eta) \simeq_{\eta\to 0} \eta^{(m+1)/2}, \tag{9}$$

$$U(\eta) \simeq_{\eta\to\infty} \frac{B}{\eta^{1/4}}\sin\left(\frac{F^{1/2}}{3}\eta^{3/2} + \frac{E}{F^{1/2}}\eta^{1/2} + \Phi\right), \tag{10}$$

where asymptotic amplitudes A and $B(Z_2)$ and phase shift $\Phi(Z_2)$ are constants for fixed F and E.

Since we are able to separate variables in (1) there is no problem to obtain the exact solution of it. But often this statement is not understood properly, i.e. by assuming that it is possible for such reason to get exact and unique values of Stark energy E_0 and level width Γ for any state and for any value of F. It goes without saying that there is no need to perform calculations for very large F only for the reason that it is impossible to study experimentally very diffused lines. But it is clear from Sec. 54 [9] that Bethe and Salpeter had a quite different opinion on this subject. They determine the threshold for disappearance of line by using the notion of classical mechanics. According to classical mechanics a level exists and is stable as long as it remains under the barrier. Above the barrier it disappears immediately.

The condition of disappearance of level in equation (5) is given in [9]

$$4Z_2 = \frac{E_t^2}{F_t}. \tag{11}$$

Let us turn now to a discussion of what can be obtained as a result of numerical solution of equations (4) and (5). For the fixed F and E we can in principle obtain the exact eigenvalues Z_1 for a given quantum number n_1. In the limit $F \to 0$ we go over to the field free case. By integrating equation (5) for fixed F and E we can calculate exactly the amplitude $B_E(Z_2)$. Consequently, we can determine exactly the values of Z_2 where $B_E(Z_2)$ reach minima (and maxima). As $F \to 0$ the minima gradually go to unperturbed eigenvalues of equation (5) at $F = 0$. We can parameterize the amplitude $B_E(Z_2)$ in the vicinity of the minima by using the Breit-Wigner formula

$$B(Z) = B\left[(Z - Z_0)^2 + \frac{\gamma^2}{4}\right], \tag{12}$$

and determine values Z_0 and γ. Similarly, solving the system of equations (4)–(6) simultaneously we can determine Stark energy E_0 and level width Γ. Such a procedure is completely justified for small values of F where Γ is small and E_0 can differ from minimum of amplitude by an amount $\sim \Gamma^2$. But when F is increasing the situation is gradually changing. The magnitude of Γ (and also E_0) starts to depend significantly on the points of amplitude which were chosen for the parameterization. Is there some reasonable procedure which allows to choose E_0 and Γ in a unique way?

A procedure has been proposed of summing the perturbation theory (PT) series for the energy in Borel sense [11] which converges for all values of F. It allows one, in principle, to determine E_0 and Γ, as accurately as one may like, for any values of quantum numbers n_1, n_2, m and any value of F. But in no case can it be said that this approach is uniquely rigorous from mathematical point of view. The PT series diverges asymptotically and terms $\sim \Gamma^2$ which are essentially smaller than the irreducible error of the main series can be sensibly added to it. For large F, and this is the case that interests us, we get a different answer.

It would seem that for large F it should be possible to implement a procedure which makes use of the absence of a diverging wave in the solution. In this case we seek a pole in the complex E-plane, whose real part corresponds to the Stark energy, and whose imaginary part corresponds to Γ. As $F \to 0$, the pole converges to the unperturbed solution of the problem, i.e., the approach is well founded. We can require that the absolute value of the coefficient of the diverging wave approach zero, and seek the values of E_0 and Γ in this way. But this approach, although it would seem to be the unique approach, is in no way

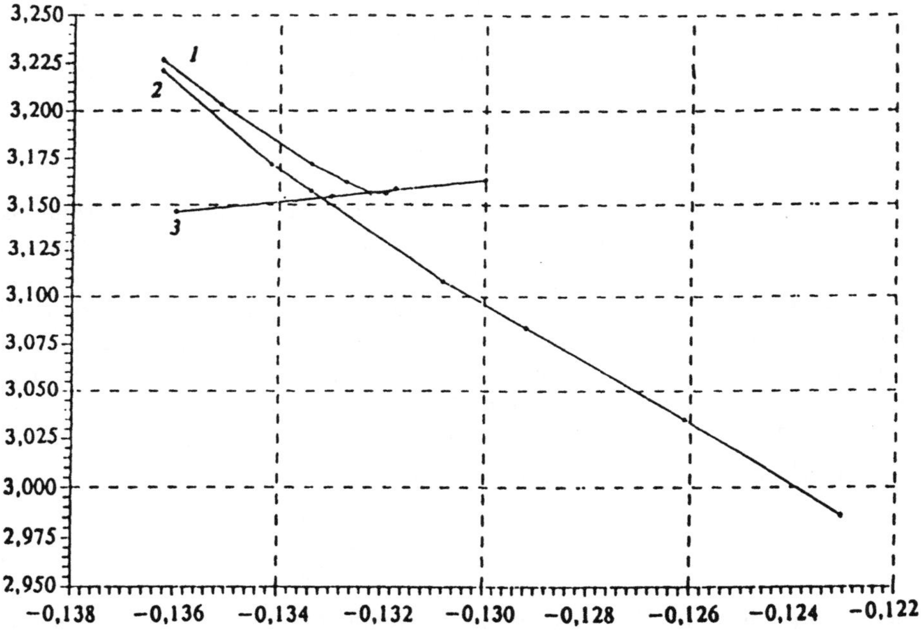

Figure 1. Dependence of values $4-4Z_1$ and $4Z_2$ of energy E: 1- values $4Z_2$ corresponding to the minimum $B_E(Z_2)$, 2- values $4Z_2$ corresponding to the minimum $B_{Z_2}(E)$, 3- $4-4Z_1$, where Z_1 corresponds to the eigenvalue of equation (4) at $E = const$. The crossing point of the curves 3 and 1 is corresponding to the "stabilization" of the state $(n_1, n_2, m) = (3, 18, 1)$. $F = 0.599 \times 10^{-6}$a.u.

unique. We can require that the convergence to zero of the real and imaginary parts of the coefficient in question converge to zero in a different way. In such a case we would obtain different values of E_0 and Γ.

Naturally, significant differences between the results of the approaches arise only for large values of F. The fact that the pole in the complex plane cannot be used in a "rigorous" determination of the quasistationary state (QS) for any F was implicitly noted in the first edition of the monograph by Landau and Lifshitz [8]. The procedure of introducing a complex energy to describe it has been called the "peculiar" method. Its usefulness is obvious since by this approach the connection between linewidth and the lifetime of the level is revealed at once. But in the later edition of that book, prepared for publication at the time when Landau stopped his activities as the theoretician because of illness, this method came to be called the "formal" method. Considerable "improvements" were also introduced into chapters of the book devoted to the definition of QS.

It can be verified by careful reading of the first and the next editions of the book. This served as a reason to assume that the method of "complex poles" was rigorously founded and that one can use it to find a unique solution, which one cannot because the problem does not have a unique solution.

Therefore, at large F we cannot define value of Γ both experimentally and theoretically. If it were not so the theory would be inadequate. This important feature of QS was discarded in attempts to define them in a unique way for large F.

The minima of $B_E(Z_2)$ – if they even exist, which, as we will show, is not always the case – can be found exactly. By solving Eqs. (4) and (5) simultaneously, i.e., satisfying condition (6), where Z_1 corresponds to the exact eigenvalue of Eq. (4), and Z_2 corresponds to the minimum of $B_E(Z_2)$, we can determine the Stark energy E_0. We consider in detail the case $m = 1$, since there is the possibility to compare results of the calculation with the experimental data on Na [2]. We performed the calculations of $B_E(Z_2)$ for one value of F and different values of E. The minimum of amplitude denoted as Z_2 originated from the unperturbed eigenvalues of equation (5) at $F = 0$ (or $F = const$, $E \to -\infty$). There is no need to perform the calculations for different F. We can obtain new curves $Z_2(E)$ by obvious scaling procedure $F \to A^3F$, $E \to A^2E$, $Z_2 \to AZ_2$, where A – is an arbitrary positive parameter.

One of the results of calculations of $Z_2(E)$ is shown at Fig. 1. For large negative E the minimum $B_E(Z_2)$ is very sharp, but when E is increasing (decreasing in magnitude) it becomes more and more diffused. Consequently, the Breit-Wigner parameterization becomes less and less reliable. But the positions of minima and corresponding maxima can be obtained (in principle) exactly. When E is increasing the value of Z_2 is decreasing as it can be expected from perturbation expansion for Z_2. But at some value of $E = E_t$ the exact value of Z_2 starts increasing and then corresponding minimum of $B_E(Z_2)$ rapidly disappears. It is clearly seen from the results of calculation that at some E minimum sticks together with the upper maximum $B_E(Z_2)$. For larger values of E we are able to find only bending points of curve $B_E(Z_2)$. Such disappearances of minima $B_E(Z_2)$ are sharp for all n_2. For $E > E_t$ we cannot more talk about the level (for given n_2), it disappeared.

In Tab. 1 we present the results obtained numerically for different n_2. We supply it also by the comparison with the formula (11).

It is clear that the formula (11) is asymptotically correct at $n_2 \to \infty$ as it can be expected. By solving the Schrödinger equation we have obtained the results which correspond to classical mechanics. In our approach we calculated the value Z_2 which is strictly defined and repudiated the notion of level width. Such an approach is natural, if only for the reason that in classical mechanics there is no concept of width, i.e. state exists and is stable or there is no state. Thus, the transition from quantum mechanics to classical mechanics does not have to take the form of simple washing out of QS. Actually, such idea is hidden in Sec. 54 [9].

TABLE 1. $m = 1$, $F = 0.638544\times10^{-6}$, E_t – value of E where Z_2 starts increasing.

n_2	$4Z_2$	$\times10^{-4}E_t$	$\frac{E^2}{F} = 4Z_2'$	$\frac{Z_2}{Z_2'}$	$\times10^{-4}R$	$6f$	$\frac{Z_2(E_t,n_2)}{\sqrt{-2E_t}(n_2+1)}$
15	2.5461	-12.17	2.318	1.099	13.302	0.1277	0.8065
16	2.7672	-12.71	2.523	1.094	12.456	0.1271	0.8071
17	2.9930	-13.25	2.748	1.089	11.707	0.1264	0.8076
18	3.2234	-13.77	2.970	1.086	11.045	0.1259	0.8082
19	3.4582	-14.29	3.197	1.082	10.450	0.1254	0.8086
21	3.9404	-15.30	3.664	1.075	9.435	0.1246	0.8096
23	4.4386	-16.28	4.148	1.070	8.596	0.1238	0.8104
29	6.0206	-19.07	5.694	1.057	6.778	0.1220	0.8131

$\frac{E^2}{F} = 4Z_2'$ – Bethe-Salpeter asymptotic formula (11) [9].

On the basis of the results presented in Tab. 1, we can introduce a quantum-mechanical correction in formula (11) which follows from classical mechanics:

$$4Z_2 = \frac{E^2}{F}\left(1 + \frac{A}{n_2}\right), \tag{11'}$$

where $A \approx 1.6\ldots$

The disappearance of the minima of $B_E(Z_2)$ appear paradoxical if we proceed in an unrigorous way from perturbation theory. It is just in this way that physicists usually proceed, and in the majority of cases such an approach is justified. Indeed, we can find Z_2 from the formula

$$Z_2 = \sqrt{-2E}\left[n_2 + \frac{m+1}{2} - (6n_2^2 + 6n_2 m + 6n_2 + m^2 + 3m + 2)R - \cdots\right], \tag{13}$$

where

$$R = \frac{F}{4}(-2E)^{-3/2}, \tag{14}$$

including in it a large enough number of decreasing (in magnitude) terms and discarding the growing ones. For the case shown in Fig. 1, for all E, including $E = E_t$, around ten terms in formula (13) decrease. It would seem that the series (13) is entirely suitable for the calculation of Z_2 since the problem in question does not generally require great accuracy. But in this case, how are we to understand the disappearance of the level, taking into account that all the terms in Eq. (13) except the first are negative? The answer to this question is contained in the rigorous formalism of Large Order PT, which has been developed only in recent years [12]. We will not delve deeply here in the far-from-trivial mathematical subtleties of this question. We will only point out that the series (13) is sign-constant and asymptotically divergent, i.e., for any arbitrarily small R its terms, starting at some smallest term, grow. Therefore, the complete expansion of Z_2, analogous to the expansion of the centre of a doublet for symmetric two-well problems [13], together with the power-law terms in R, contains some exponentially small terms (different from the imaginary terms which characterize the width of the level). These terms do not manifest themselves at small values of the field, especially for small n_2. For this reason their influence cannot be noticed if we

compare the results of very accurate numerical calculations of Z_2 with the results obtained using formula (13) for small F. However, starting at some value of F, these terms begin to grow very rapidly, causing the disappearance of minima. The situation with sign-changing series, for example, the expansion of Z_1, is different.

To calculate the Stark energies E_0, we must solve simultaneously equations (4)-(6). We will not describe this simple procedure. One example of the numerical determination of the critical values of F_d and E_d at which stabilization of the level $n_1 = 3$, $n_2 = 18$, $m = 1$ occurs is presented on Fig. 1. For $F < 0.599 \times 10^{-6}$a.u. this state exists, while for larger F the condition $Z_1 + Z_2 = 1$, where Z_2 corresponds to the minimum of $B_E(Z_2)$ at $n_2 = 18$, cannot be satisfied and consequently the level disappears.

We present below the simple semiempirical formulae that allow to obtain values F_d and E_d which correspond to the "stabilization" points. These formulae are based on the results presented in Tab. 1. We see that for all n_2:

$$\frac{Z_2(E_t, n_2)}{(n_2+1)\sqrt{-2E_t}} \approx 0.8086\cdots. \tag{15}$$

This result can be understood on the basis of formula (13) which for $m = 1$ can be represented in the form:

$$\frac{Z_2(E_t, n_2)}{(n_2+1)\sqrt{-2E_t}} = 1 - 6(n_2+1)R - [68(n_2+1)^2 + 10]R^2 - \cdots. \tag{16}$$

Of course, we will not use formula (16) to calculate Z_2 but only to clarify at which R the level disappears for large n_2. It is clear that the left side of expression (16) depends significantly only on the quantity

$$f = (n_2+1)R. \tag{17}$$

We see that at the turning points E_t the value of f is approximately constant and equal to:

$$f_t = 0.021\cdots. \tag{18}$$

Basing ourselves on these observations and using the PT formula for $Z_1(E)$, which is unquestionably valid for the fields under the consideration and $n_1 \ll n_2$, and also condition (6), we obtain

$$E_d = -\frac{1}{2(Z_1' + Z_2')^2}, \tag{19}$$

TABLE 2. Comparison of experimental and theoretical values F_d and E_d for "stabilization" points.

		E_d	F_d	n_1	n_2	n
A	*a*	-12.79×10^{-4}	0.6457×10^{-6}			
	b	-12.51×10^{-4}	0.6185×10^{-6}	16	5	23
B	*a*	-12.77×10^{-4}	0.6378×10^{-6}			
	b	-12.86×10^{-4}	0.6087×10^{-6}	17	4	23
C	*a*	-13.00×10^{-4}	0.5873×10^{-6}			
	b	-13.20×10^{-4}	0.5995×10^{-6}	18	3	23

a- experimental data [2] converted to a.u.
b- results obtained by formulae (19)–(22)

$$F_d = \frac{0.084}{(n_2+1)}(-2E_d)^{3/2}, \tag{20}$$

where

$$Z_2' = 0.8086(n_2+1), \tag{21}$$

$$Z_1' = (n_1+1)\left[1+\left(\frac{n_1+1}{n_2+1}\right)0.126 - \frac{17}{9}\left(\frac{n_1+1}{n_2+1}\right)^2 0.126^2 + \cdots\right]. \tag{22}$$

Formulae (19)–(22) not only produce quite reliable results for F_d and E_d ($n_1 \sim 0-6$, $n_2 \sim 12-25$) which was checked by the comparison with exact numerical calculations for several cases but also provide possibility to "decipher" the results of [2], where the "stabilization" points were obtained experimentally for three lines of Na. It is evident that quantum numbers n_1, n_2 cannot be deduced from the experimental data. The quantum number $m = 1$, on the contrary, is determined by the experimental procedure.

The experimental technique of Mc Nicholl *et.al.* [2], in principle developed already in earlier investigations of the dependence of the behaviour of the Stark levels of alkali atoms on the magnitude of the field F, is described in [1]. The atomic beam technique was used. The Rydberg states of the Na atom were populated by three-step photoexcitation with the help of dye lasers. The Na^+ ions that form by the decay of Na atoms give evidence of the presence of the Stark level. If the field F is varied slowly by scanning the laser frequency, i.e., varying the energy E, the evolution of the Stark line can be traced out. In their studies of the Stark levels, Mc Nicholl *et.al.* [2], were able to achieve significantly larger values of F than those achieved in previous studies [1,10]. They observed that at certain values of F_d and E_d characteristic of the three levels the Na^+ signal abruptly decreased.

In Tab. 2 we present the comparison of the experimental and theoretical values of E_d and F_d. Formulas (19)-(22) show at once that there are no other quantum numbers n_1, n_2 that can produce some other values F_d and E_d which are close enough to the experimental in "reasonably" close intervals in E and F simultaneously. By discussing results, first of all, we should mention that our calculation were performed for hydrogen but experimental data obtained for Na. Even for such reason it is difficult to expect the better agreement between calculations and experimental data. We should mention here one even more important fact. The experimental accuracy of measurements of F is not higher than 2–3%, which is typical for all of experiments of such kind. It was clearly noted in [2]. The "exact" experimental values of F_d presented in Tab. 2 are in fact not such (as it was clearly noted by the authors of [2]), but were "refined" by comparison with the theoretical data of a model which takes the atomic core into account [1,2]. According to this theory and in contradiction with Sec. 54 [9] the "stabilization" for Stark level in hydrogen is impossible.

Bethe and Salpeter [9] made an attempt to explain the disappearance of spectral Stark lines of the Balmer series in H with the increase of F studied by Traubenberg by using Formula (11) but failed. They presented the explanation: the ionization of levels was not directly studied in this experiment, but spectral lines were disappearing at much lower fields, because of quantum tunneling. (They were using for tunneling the terminology: ionization by the wave mechanics, which might cause confusion. Someone might think that classical ionization is only an approximate description of the phenomenon. But the authors [9] meant different, i.e. the states can decay in the field by tunneling but at some field F_d they simply disappear.)

In the experiment [2] the ionization was studied directly. Notwithstanding that experimentally was studied the sodium the results of hydrogenic calculations are in a quite convincing agreement with the experimental data.

We obtained E_d and F_d by exact solution of Schrödinger equation. But we can determine them approximately much easier. For this we should satisfy simultaneously slightly improved formula of Bethe and Salpeter, i.e. the (11′) and formula for PE for the energy by accounting in it 4-5 terms. Such calculations produce quite reliable results for large n_2 ($n_2 > 12$) and much smaller n_1. Instead of direct calculation of Z_2 which enters in formula (11′) by using PT it is better to calculate Z_1 for which PE is converging rapidly. Then we can obtain Z_2 by using (6). The values E_d and F_d obtained by such procedure differ from exact ones not more than by 2-3% for all the considered by us cases. It is evident that for the practical use such accuracy is satisfactory. Thus we arrived to a quite paradoxical conclusion: the PE is practically valid up to the point where the Stark level disappears.

We considered in detail the case $m = 1$. Exactly this case was discussed in [9]. Similar results can be obtained for $m \geq 2$. The case $m = 0$ is special. The minimum $B_E(Z_2)$ does not disappear when we are passing the top of the potential barrier. It just becomes more and more diffused when we increase F. Mathematically this case differs from the others by the fact, that "centrifugal" potential $(1 - m^2)/(4\eta^2)$ in equation (5) has the different sign. The particle is falling on the center in the classical mechanics for such potential.

Though the agreement between experimental and theoretical values F_d and E_d is quite convincing, in our opinion, still we have not explained all the details of the experimental data. Indeed, the experimental studies [2] clearly show that at further increase of F, starting from $F = F_d$ the signals from the "stabilized" levels appear again, though as weaker ones.

Thus, the arguments of Sec. 54 [9], though they are undoubtedly correct, do not give the complete picture of the behaviour of Stark levels in strong electric field. It seems that partially for such reason the experimenters were not able to connect the observations with the predictions of Sec. 54 [9].

The situation can be elucidated again by the exact solution of equations (4)–(6). What did we miss in foregoing consideration?

The traditional way to solve these equations started from the PT first applied by Schrödinger. Here we fix the value of E and solve equations (4) and (5) independently. Values Z_1 and Z_2 are the eigenvalues in such approach, since we consider that wavefunctions are decreasing at infinity exponentially and the decay of the atom is impossible. After Z_1 and Z_2 are determined for arbitrary E, its true value is found from condition (6).

But there is another possibility to solve the same problem. We can fix first the values Z_1 and Z_2 and after the eigenvalue E is found to satisfy the condition (6). It is evident that when our calculations are based on PT, as in the case $F = 0$, we can not find anything new by using such approach.

But, solving out the problem exactly according to traditional scheme, we departed from the framework of PT. And only such procedure allowed us to find the "stabilization" points.

Quite similarly by considering the second way of solution of equations (4)-(6), we should go beyond the boundaries of PE. The equation (4) at $F \neq 0$ describe the bound problem and therefore the second approach would not give us anything new. The situation with the equation (5) is different. Now by solving it we are fixing the value Z_2 and looking for the minima $B_{Z_2}(E)$ for chosen value of n_2. Again we are able to find them exactly. For small F we obtain by both procedures almost coinciding values Z_2 and E. But only at $F = 0$ they merge together in one single eigenvalue. For large F positions of minima corresponding to the same n_2 are diverging noticeably. The example of calculations is shown on Fig. 1. The small scale of the draft should be taken into account. From practical point of view the difference in the positions of minima is important only in a very close vicinity of the disappearance of minimum $B_E(Z_2)$. The calculations show that the minima $B_{Z_2}(E)$ do not disappear with the increase of E, they only become more and more diffused. The presence of a series of minima in $B_{Z_2}(E)$ reflects the other face of quantum mechanics, namely its wave character. Using the value of Z_2 which corresponds to the minimum $B_{Z_2}(E)$ we can determine what can be, by stipulation, defined as a "second" component of a Stark line. We remind here some basic facts of Copenhagen interpretation of quantum mechanics. "It is in principle impossible, however, to formulate the basic concepts of quantum mechanics without using classical mechanics" [8]. Since there is no notion of QS in classical mechanics, in quantum mechanics it can not be rigorous and valid for all F. "For such systems with a small disintegration probability, we can introduce the concept of QS, in which the particles move 'inside the system' for a considerable period of time,leaving it only when a fairly long time interval τ has elapsed" [8]. Therefore it is clear that the extrapolation of the notion of QS by Borel summation or some other procedure to arbitrary large values of F violates the basic principles of quantum mechanics.But, two possibilities for the extension of the energy of the bound state at $F = 0$ into region $F \neq 0$ by calculating positions of minima of amplitude $B_{Z_2}(E)$ and $B_E(Z_2)$ are not only legitimate, but also necessary. By introducing the "rigorous" notion of complex poles for the energy valid for all F one eliminates not only the link between quantum and classical mechanics, but also the wave character of quantum particle.

After such deviation we continue to consider our problem. We omit the Stark energies for any concrete levels as functions of the magnitude of the field F, calculated using our technique. The quantitative difference between them and the results which can be obtained using the standard PT for the energy, taken out to 4–5 terms, proves to be insubstantial even for the region where "stabilization" is observed. The two components of the Stark line, from the practical point of view, coincide.

The presence of the "second" component manifests itself in the experiment as a "revival" of Stark line after its "disappearance" in the stabilization point.

The two solutions of equation (1) are exact and explicitly manifest themselves in experiment [2]. We emphasize here that we are talking about not just qualitative, but also about quantitative agreement between calculations and experimental data. By quantitative here we mean that an accuracy of 2–3% in the determination of F_d and E_d is sufficient for error-free identification of the "stabilization" points.

In no case we claim that the solutions we have found are the unique solutions of equation (1). It is possible with some unnecessary technical efforts to obtain the whole series of quite "reasonable" solutions which would also satisfy (1), at least for the reason that the spectrum of the problem is continuous and by defining the notion "exact solution" we rely only on the "common sense". But in the case of small F, where the "usual" notion of QS is valid, it is also introduced into physics basing on "common sense" and in no case it represents some additional postulate of quantum mechanics.

At first sight, we contradict here to the commonly accepted view on the solution of the problems in physics, where we have the Schrödinger equation with the necessary boundary conditions and a unique solution. Even when we have no possibility to solve it exactly, we are sure that such solution exists.

When we consider QS and not only in a Stark effect we have a different situation. The energy E_0 and level width Γ are not the values which can be determined, that is to say, calculated with arbitrary accuracy [8]. For better understanding of the Stark problem we can compare it with the problem of auto-ionization states of He or H^-. Further, we can consider the process of the elastic scattering of electron by He^+ or H. By solving this problem, which can be rigorously formulated in terms of quantum mechanics, we are able, at least in principle, to calculate exactly the phaseshift for the elastic scattering and also the partial scattering cross-section. But, if the phaseshift is rapidly increasing by π in small interval of energy we can introduce the notion of auto-ionization states of He or H^-. Usefulness of such concept is evident and not only for the reason that it is much easier to perform the calculations by the solution of the approximate Schrödinger equation than to calculate the phaseshift. But in no way does it follow that one should ascribe the concept of an exact value to an approximate value in attempting to determine the energy and level width with the excessive accuracy, as is sometimes done. In those cases in which Γ turn out to be too large, the concept of auto-ionization states simply loses meaning, but the scattering problem from which it originated continues to be rigorous.

When we are considering the Stark problem, by repudiating the notion Γ, we can advance slightly more further, i.e. we can find the place where level disappears completely. For autoionization states similar consideration would present just formal interest, and we would not able to connect it with some experimental verification.

3. Another Manifestation of the Mechanism of Direct Ionization Suggested by Bethe and Salpeter [9]

In this section we will consider the experimental data on the behaviour of the Stark states of Li for $m = 1$ and $n = 18-20$ [1]. First, we describe schematically the experimental procedure. The atoms in the uniform electric field F were excited by tunable laser. At small F the excited atoms were practically stable, since lifetimes for the quantum tunneling for them were too large. After some delay, during which the excited atoms were not essentially decaying radiatively, they were ionized by high-voltage pulse. The "stable" Rydberg atoms produced the signals, i.e. the ions Li^+. Such technique allowed to study the Stark energies for small F. The obtained energies for Li appeared to be coinciding practically with the hydrogenic ones for $m = 1$.

However, at some critical, and still very small F, the signals which were manifesting that the states exist suddenly disappear. The authors [1] made a conclusion that at such fields the corresponding states were suddenly decaying. (It is clear, that there is a possibility of another conclusion. At such fields the amount of atoms which are excited into mentioned states are sharply decreasing and signals disappear for such reason.)

According to the hydrogenic theory the states, which were studied in [1] have enormously large lifetimes for the quantum tunneling. Besides, according to hydrogenic theory the the shortest-lived states for a given total quantum number n are those deepest in energy. The experiment for Li and, at first sight, has shown a quite different behaviour of the states.

In the paper [10] the authors discussed also one important observation, which was later on omitted in review paper [1], i.e. nonmonotonic behaviour of the lifetime for some states with increase of F.

It is necessary to note an important feature in the experimental observations. The mentioned above states do not disappear at the critical field completely. It was verified by using different approach to the measurements, when the high-voltage pulse was not more applied, but signal arise only from quantum tunneling in the static fields. This method has shown that Stark levels continue to exist for F larger than critical value and behave also like hydrogenic. But in fields which are somewhat below critical, by such approach it is impossible to register directly the signals from the Stark states.

The authors [1] presented "almost" correct explanation for the obtained experimental results, basing on the simple "saddle-point" model. According to this model, atom can ionize when its energy reaches the critical value

$$E = -2F^{1/2}. \tag{23}$$

The authors [1] presented also another correct interpretation of the obtained data, which do not contradict to the mentioned model. When critical values of F are reached, then practically stable Stark levels with large n_1 and small n_2 start crossing extremely broad levels from higher manifolds (with larger total quantum number n) with large n_2. It causes in an observation the "disappearances" of the states which are in fact stable. These "disappearances" are quite different from those which were predicted in Sec. 54 [9], though both phenomena, as it will be shown below, are connected.

Basing on absolutely correct from point of view of physics arguments, the authors [1] made a strange conclusion that such behaviour of Stark levels is typical only for complex atom with the core. In their opinion in the case of hydrogen levels would not mix, since the potential is pure Coulomb, and therefore the mechanism of decay is different.

Up to now the experimental studies for H similar to [1] were not performed. It is possible that the authors [1] came to the mentioned above conclusion because they were influenced by the experimental data of Traubenberg [9], who studied the disappearance of the spectral Balmer lines of H with the increase of F. At first sight, the results of Traubenberg are in a drastic disagreement with the data obtained in [1]. We note at once that such disagreement is caused first of all by the fact that states which were studied in two cases had different total quantum numbers n.

Let us then refine the conclusions reached in [1]. According to Sec. 54 [9] and our treatment

$$E = -2(Z_2F)^{1/2}, \tag{24}$$

where $Z_2 = 1 - Z_1$.

It is useful to give a physical interpretation of formula (24) which is asymptotically exact at $F \to 0$. At $F = 0$ the ionization spectrum starts at $E = 0$. At $F \neq 0$ the threshold for ionization is decreasing and described by formula (24). In the presence of the electric field the ionization threshold splits as it does the energy level of a bound atomic state, characterized at $F = 0$ only by total quantum number $n = n_1 + n_2 + m + 1$. The ionization "subthresholds" are characterized by quantum number n_1, since it exists at $F \neq 0$ (m is fixed, quantum number n_2 does not exist). The disappearances of signals from practically stable levels with the increase of F is caused by the fact that the energies have reached the first subthresholds for the ionization: formula (24) at $n_1 = 0$. At such fields the excitation of atomic states in the experiment [1] is essentially suppressed by the possibility of direct ionization of an atom. But unlike the case $F = 0$ some amount of the atoms are nevertheless excited, which was verified by the experiment done in the second manner [1]. With the further increase of F the next subthresholds at $n_1 = 1, 2, \cdots$. are reached and situation is repeated. It explains seemingly nonmonotonic character of the ionization rates for Stark sublevels. We do not present the numerical data which can be obtained by (24), since there are no direct measurements for H. But they practically coincide with the data which were obtained by (23) in [1] (with some additional small semiempirical corrections) and compared there with the experimental data for Li at $m = 1$. Now we explain the reason for the disagreement between the experimental data of Traubenberg [9] for hydrogen and the experimental data for Li [1]. For this we should compare the hydrogenic formula for Γ [14]

$$\Gamma = \left(\frac{4}{Fn^3}\right)^{2n_2+m+1} \exp\left[3(n_1 - n_2) - \frac{2}{3Fn^3}\right] \frac{1}{n^3 n_2!(n_2+m)!}[1 - O(F)]. \tag{25}$$

with the formula (24). It can be easily verified that with the increase of total quantum number n the Stark levels are reaching the "region of ionization" more and more stable in the respect of the tunneling in the field. Thus, in the limiting case $n_1 \to \infty$ for small n_2 in the intersection point $F \sim \frac{1}{16n^4}$ and consequently the linewidth behaves on the "ionization curve" like

$$\Gamma \sim \exp\left(-\frac{23}{3}n_1\right). \tag{26}$$

At finite values of n_1 and n_2 the calculation of Γ at intersection points can be performed accurately by using formula (73) from [14]. Still it is clear from (26) that at small values of n Stark levels decayed essentially in the process of tunneling at fields which are less than critical, i.e. before they reach the region of direct ionization. At large n the situation is different. The experiments of Kleppner *et.al.* [1] were done for $n = 18 - 20$ but optical measurements for H [9] for $n = 4 - 7$. It explains so drastic but only deceptive disagreement in the results of two experiments. Though the quantum tunneling still operates in the same way, i.e., for the same quantum numbers n it is greater for deeper states, it cannot reveal itself in the experiments [1].

We see now, why Bethe and Salpeter [9] by introducing the notion of the ionization of the atomic states in the presence of the electric field had no in mind the process of quantum tunneling. Some misunderstanding of this fact might be caused by the terminology. The process of the tunneling was named by Bethe and Salpeter [9] as an ionization by wave mechanics.

We see that side by side with the quantum tunneling, QS at some values F and E simply do not exist, similarly to the case of $F = 0$ where bound states do not exist at $E > 0$. QS spectrum at $F \neq 0$, similarly to bound states spectrum at $F = 0$, can be described by quantum numbers n_1, n_2, m. But in contrast to bound states spectrum, which does not penetrate into region where the ionization is possible, QS spectrum can penetrate partially into this region. On the other side, at any value of F at some negative values of E ($m \geq 1$) some QS with finite values of n_2 simply do not exist, which is impossible for the bound states at $F = 0$.

Thus, the "continuous" and "discrete" spectra of a hydrogen atom at $F \neq 0$ are mutually penetrating.

4. Splitting of the Threshold for the Ionization which is not Connected Directly with Classical Mechanics

4.1. GENERAL CONSIDERATION

The Schrödinger equation for the Stark effect for the hydrogen separates and for such reason there is no difficulty to obtain for it the exact solution. But this equation by itself does not describe any physical process. And therefore the problem how to connect the solution of such equation with the real physical process arises. At small F everything, at first sight, seems quite trivial. In such case we excite the atoms with slightly changed energies, which can be calculated by using PT. When we increase the value of F, the energies of Stark states are changing, besides the states get the possibility to decay by quantum tunneling. At further increase of F the states under the consideration are gradually practically disappear. Is it all? In such consideration we assume tacitly that the direct ionization threshold is irrelevant to the problem. Only one indication that it cannot be always so has been made by Bethe and Salpeter [9] and we have considered this case in detail above.

The other mechanism for "direct" ionization of the hydrogen atom in the presence of the uniform electric field are "technically" much simpler than the considered above.

We start from perturbation theory and limit our consideration by just two terms for the constants of separation Z_1 and Z_2, which allows us to obtain the energy E

$$Z_1 = \sqrt{-2E}\left[n_1 + \tfrac{m+1}{2} + \left(6n_1^2 + 6n_1 m + 6n_1 + m^2 + 3m + 2\right)\right.$$
$$\left.\times \tfrac{F(-2E)^{-3/2}}{4} - \cdots\right], \tag{27}$$

$$Z_2 = \sqrt{-2E}\left[n_2 + \tfrac{m+1}{2} - \left(6n_2^2 + 6n_2 m + 6n_2 + m^2 + 3m + 2\right)\right.$$
$$\left.\times \tfrac{F(-2E)^{-3/2}}{4} - \cdots\right], \tag{28}$$

$$Z_1 + Z_2 = 1. \tag{29}$$

We can find the energy of quasi-stationary states $E_{n_1 n_2 m}(F)$ at small F by satisfying (27)–(29) simultaneously

$$E_{n_1 n_2 m}(F) = -\frac{1}{2n^2} + \frac{3}{2}n(n_1 - n_2)F - \cdots \tag{30}$$

where $n = n_1 + n_2 + m + 1$ is the principal quantum number.

Consider now the continuous spectrum. It is evident that at $F = 0$ we can approach the discrete spectrum,starting from continuum, i.e. $Z_1 > 1$, $Z_2 < 0$ or $Z_2 > 1$, $Z_1 < 0$ by satisfying (27)–(29) simultaneously only in the limit $Z_1 = 1$, i.e. $E = 0$, $n_1 \to \infty$ (n_2, m are fixed) or $Z_2 = 1$, i.e. $E = 0$, $n_2 \to \infty$ (n_1, m are fixed). But at $F \neq 0$ the situation immediately becomes different. We can satisfy $Z_1 = 1$ by choosing the energy

$$E^+_{n_1 m}(F) = -\frac{1}{2\left(n_1 + \frac{m+1}{2}\right)^2} + \frac{6n_1^2 + 6n_1 m + 6n_1 + m^2 + 3m + 2}{4}F - \cdots, \tag{31}$$

or $Z_2 = 1$ by

$$E^-_{n_2 m}(F) = -\frac{1}{2\left(n_2 + \frac{m+1}{2}\right)^2} - \frac{6n_2^2 + 6n_2 m + 6n_2 + m^2 + 3m + 2}{4}F - \cdots. \tag{32}$$

Since the whole spectrum of the problem at $F \neq 0$ is, strictly speaking, continuous we can satisfy (29) at all energies. But since the condition $Z_1 = 1$ or $Z_2 = 1$ were the starting points for ionization at $F = 0$, their natural continuations into region $F \neq 0$ by formulae (31) and (32) provide us with the starting points for the ionization at $F \neq 0$. We see that in the presence of the electric field the ionization boundary splits, similarly to the splitting of the energy of the bound states. The situation with the ionization curves (31) and (32) is quite similar to the one for the Stark sublevels (30). Excitation of "discrete" and "continuous" states of the spectrum at $F \neq 0$ is most effective for the values of E obtained by PT, but experimentally they manifest themselves differently. At energies E satisfying condition (30) the atom is the most stable with the respect to decay in field. The states corresponding to such energies became known as quasistationary. To characterize such states the concept of lifetime in relation to decay in fields was introduced in addition to the concept of energy. The quantum number n_2 does not exist for the "ionization" states (31) and n_1 for the states (32). For such reason we may speak of ionization states. At such energies one quantum number disappears, hence ionization states decay instantly. The "ionization states" in no way are connected with the quantum tunneling. Their presence is the consequence of the two-dimensionality of the problem.

We note that curves (31) and (32) are similar only for small F. The symmetry between n_1 and n_2 suggested, at first sight, by formulae (31) and (32) is only approximate, as it should be at $F \neq 0$. Similarly to the quasi-stationary states, curve (32) is gradually loosing its clear physical meaning with growing F. Curve (31) continues to be physically meaningful at all F. But perturbation expansion (31), where much more terms can be included, becomes inconvenient at large F. The values $E^+_{n_1 m}(F)$ can be calculated exactly at all F for different n_1 and m by numerical solution of the corresponding separated Schrödinger equation (4) by putting in it $Z_1 = 1$.

When $E^+_{n_1 m}(F)$ are close to $E = 0$ then for large n_1 they can be also calculated accurately by using semiclassical formula at $Z_1 = 1$ [15].

As the field F becomes stronger, the Stark lines $E_{n_1 n_2 m}(F)$ may reach the ionization curves $E^+_{n_1 m}(F)$ or $E^-_{n_2 m}(F)$. The process of direct atomic ionization occurs at the points where these curves intersect in addition to excitation of Stark states. Hence the strength of the signal indicating the presence of Stark states drops sharply, so that the states seems to simply disappear. For small principle quantum numbers the Stark states reach the intersection points with the ionization curves at values of F for which they already practically decayed due to the quantum tunneling. It can be seen from (30)-(32) that at large $n_1 \sim n$, or $n_2 \sim n$ that at intersection points $F \sim 1/(3n^5)$. Looking at (26) we can conclude that in such case the Stark state reaches the intersection point being practically stable. Hence clear experimental observation of direct ionization of atoms in the presence of electric field is possible for high Rydberg states. For low excited states it can be completely masked by the tunneling effect.

4.2. COMPARISON WITH EXPERIMENTAL DATA ON STARK STATES

There is still no experimental data on the behaviour of highly excited Stark levels of the hydrogen atom caused by variations in F. We therefore examine the results obtained for Li and Na [1,10], which for $m \geq 1$ are close to those for hydrogen.

Comparing (31) and (32) with (30), we see that the ionization curves behave differently for odd and even values of m. For instance, for m odd the ionization curves $E^+_{n'_1 m}(F)$ and $E^-_{n'_2 m}(F)$ encompass the set of Stark sublevels $E_{n_1 n_2 m}(F)$

belonging to the same principal quantum number $n = n_1 + n_2 + m + 1$. Hence, in the intersections of Stark levels whose principal quantum numbers differ by unity, the strength of the signal indicating the presence of a level drops sharply. This subtle effect was clearly identified in the experiment reported in [1] for Stark lines of Li at $m = 1$. It was interpreted, however, as a rule according to which the Stark levels of all atoms except hydrogen do not cross. Hydrogen states, however, must behave in exactly the same manner. For m even and a field increasing in strength, the ionization curves begin to intersect the Stark states at a certain field strength, and at this field strength the signal from the Stark state drops off sharply and in some cases can practically disappear.

Kleppner and his collaborators [10] began their studies of the behaviour of Stark levels caused by varying F for Na at $m = 2$, since this case resembles most closely the situation with hydrogen. They discovered that the level with quantum numbers $n_1 = 6$, $n_2 = 3$, and $m = 2$, which is almost completely stable with respect to decay in the field, suddenly disappears at an energy $E \sim 730\text{cm}^{-1}$ and a field strength $F \sim 15.2\text{kVcm}^{-1}$ and then, as the field grows in strength, reappears. This phenomenon was discussed in [10], where the researchers explained the observed effect by the intersection of the $E_{6,3,2}(F)$ state with the $E_{0,11,2}(F)$ state, which, however, did not manifest itself directly in the experiment owing to its extremely short lifetime with respect to decay in the field. Actually, the temporary disappearance of the $E_{6,3,2}(F)$ state is caused by the intersection of this state with the ionization curve $E^{-}_{13,2}(F)$.

When the authors of [1,10] moved from Na atomic states with $m = 2$ to Li atomic states with $m = 1$, they found no similarity in the behaviour of the states. Not expecting the states with $m = 1$ and $m = 2$ to behave differently, Littman, Zimmerman, and Kleppner apparently arrived at the conclusion that there was an error of some kind in the measurements for Na, with the result that in a detailed review (see [1]) there is no reference to the case with $m = 2$. The most complete experimental information about ionization states can be found by examining the Stark curves of the Li atom for $m = 0$ and the principal quantum number n equal to 15. We cannot a priori expect very good agreement between the theoretical formulas obtained for hydrogen and the experimental data, since the case with $m = 0$ for Li is decidedly not hydrogenlike. Nevertheless, here too it is possible to easily verify that the ionization curves $E^{-}_{15,0}(F)$ and $E^{+}_{14,0}(F)$ qualitatively describe the most important feature in the behaviour of the set of Stark sublevels $E_{n_1 n_2 0}(F)$ when the principal quantum number n is equal to 15; namely, temporary disappearance of the signals from all lines as the field F varies from 1.5 to 3kVcm^{-1} (see Fig. 3.9 in [1]). As the field strength increases, the signals from the Stark sublevels reappear. There is also an alternative experimental verification of the theory, a verification not directly linked for the formulae (31) and (32). In the experiment reported in [1] the lithium atoms, being excited to Stark states, did not spontaneously decay, i.e., their lifetimes with respect to quantum tunneling in such fields were too long. The presence of the excited atoms can be assessed only by ionizing them by strong field pulses after a certain time lag and fixing the Li^+ ions that appear as result of this ionization. If the experimental method is changed to detecting the Li^+ ions immediately after the atoms have been excited, without ionizing the atoms with field pulses, the ions appear only in such fields F and at such energies E for which the signals corresponding to the presence of Stark states disappear. Another (also correct) explanation of the temporary disappearance of signals from the Stark levels of the lithium atom with $n = 15$ and $m = 0$ was given in [1]. According to this, in the field and energy ranges specified above the oscillator strengths responsible for the excitation of the specified levels sharply decrease. This in no way contradicts our explanation and, strictly speaking, only the two together provide a full picture of the behaviour of the atom in a field. The probability of an atom being directly ionized in a field increases in those fields and at those energies for which the probability of exciting quasidiscrete state drastically decreases.

4.3. THE EXPLANATION OF MICROWAVE IONIZATION OF HIGHLY EXCITED STATES OF THE HYDROGEN ATOM BY THE SPLITTING OF THE IONIZATION THRESHOLD IN THE PRESENCE OF THE FIELD

Let us apply the above theory to a time-dependent process. We start by briefly discussing the experiment described in [4,5,16,17]. A high-energy beam of excited hydrogen atoms interacted with radiation in a cavity. The atoms were "created" in strictly defined "one-dimensional" states with $n_1 = n - 1$ or $n_2 = n - 1$ and $m = 0$ (here we employ the terminology of [16]). The values of n varied from 32 to 90. Linearly polarized radiation did not change the magnetic quantum number $m = 0$. The temporal variation of the field obeyed the law $A(t)Fz\sin(\omega t + \varphi)$, where F is directed along the z axis, that is, in the direction of the beam, and $A(t)$ is a slowly varying function characterizing the moment when the field was switched on and off, the periods lasting about 60 oscillations.

We can assume that the quantity $A(t)F$ is time-independent, $A(t)F = \text{const}$, for about 300 oscillations in the cavity [5] and is determined to within 5%. We denote this product by F, and in what follows we will assume that the microwave field has its peak value in the cavity unless otherwise specified. In the first experiments, reported in [5,16], the frequency $\nu = \frac{\omega}{2\pi}$ was maintained at $9.92GHz$, while F varied from 1kVcm^{-1} to several Vcm^{-1}, depending on n. In recent years other frequencies have been used [17].

As a result of interacting with radiation the excited hydrogen atoms were found to decay. The percentage of decayed atoms in relation to the initial number of atoms was measured for every value of n as a function of F. Of all the data obtained so far, the literature most often gives the dependence on n and F of 10% and 90% "ionization", i.e., the percentage of decayed atoms.

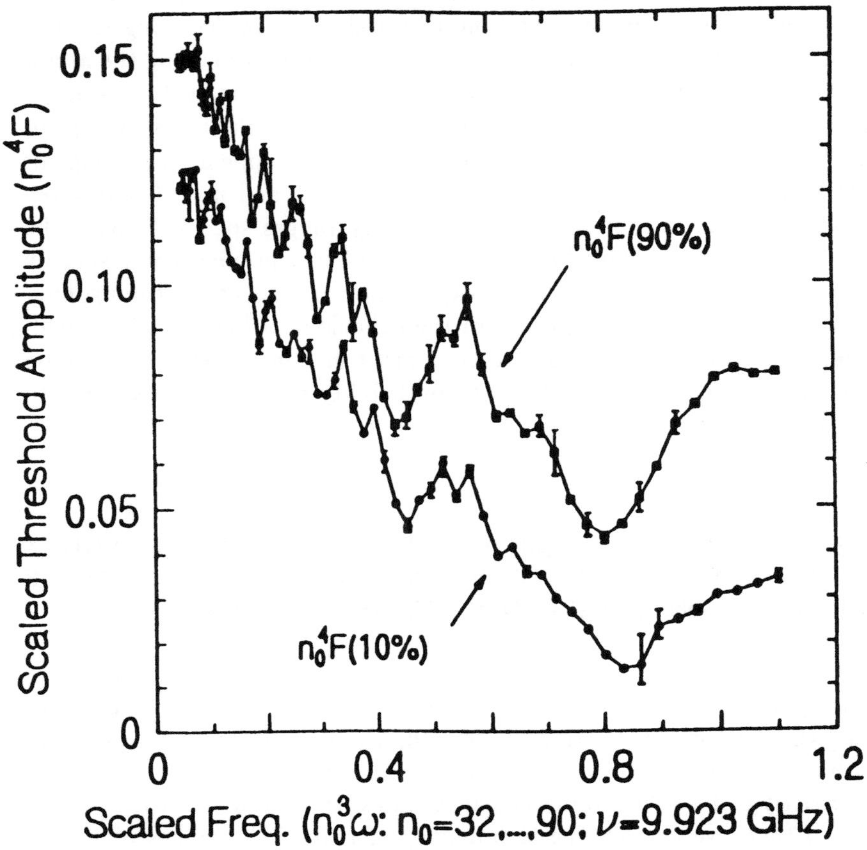

Figurre 2.

Let us examine the main feature of the experimental data more closely. The first experiments conducted at $n = 66$ showed that an atom with a binding energy of roughly 3.1meV effectively decays at a frequency of $9.92GHz$, at which the photon energy equals 41μeV and the reduced field strength is $n^4F \sim 0.09$a.u. (see [4]). For an excited atom to reach the ionization limit, for which the authors of [5,16] always took $E = 0$, it must absorb approximately 80 photons simultaneously as a result of a multiphoton process. The probability of this is, of course, negligible. There would seem to be the possibility of the atom decaying via quantum mechanical tunneling at the peak value of the oscillating field. But at $n^4F \sim 0.09$a.u. and lower the probability of this process is negligible, too, which follows from (72) of [14]. A thorough study, lasting many years and involving a broad range of principal quantum numbers n, revealed several factors whose presence had no apparent explanation. For instance, as n increased (starting at $n = 32$) and at a frequency of $9.92GHz$, the value of field F at which 10% and 90% ionization takes place falls, which appears natural since the energy with the binding energy of the electron in the atom decreases in the process. But examining the overall shape of the ionization curves with n varying from 32 to 90 [5], we discover at least six or seven sections where an increase in n by several units does not lead to a decrease in F. Such behaviour in ionization curves was named "staircase-like" in [5].

This effect manifests itself even more vividly on the ionization curves at $\nu = 36.02, 30.36$, and $26.43 GHz$ when n varies from 50 to 80 [17]. Alternation of locally stable and unstable states can be seen even more clearly at $\nu = 9.92 GHz$ if the ionization curves are plotted in the "natural" variables n^4F and $n^3\omega$. In this case the "steps" observed in the ionization curves plotted in the F and n variables transform into alternating maxima and minima [5,16] (see Fig. 2).

In view of their unconventional nature, the experimental data on microwave ionization produced a great number of theoretical papers. Reference to them can be found in [5,16]. We will not discuss the theoretical results arrived at in these papers. Suffice it to note that no simple and physically straightforward mechanism of the phenomena occurring in the microwave ionization of atoms was suggested. An attempt was made to guess and then theoretically substantiate a formula that yielded the sequence of local maxima on the ionization curves [5]. The formula considered was $n^3\omega = p/q$, with p and q integers ($p/q = 1$, 2/3, 2/5, 1/3, 1/4, 1/5, 1/6, etc.). This approach, in which the formula was initially "guessed", has been used by theoreticians in many problems and sometimes successfully. However, in the case at hand the suggested formula was on the whole unsatisfactory because only some of the values given by it succeed in explaining the local maxima.

Theoreticians in any case did not assume that microwave ionization can be understood solely by examining the behaviour of an atom in a constant field. Hence model time-dependent (classical and quantum) equations were solved and attempts to find a correlation between the obtained numerical and analytical results and the features of microwave ionization were made. Let us now turn to our study. As shown above, the spectrum of the hydrogen atom for $F \neq 0$ can be approximated by sets of quasidiscrete (Eq. (30)) and quasicontinuous (Eqs.(31) and (32)) states alternating in energy. Mathematically speaking, such an approximation of a basically continuous atomic spectrum is asymptotically meaningful as $F \to 0$ and hence can be used in practical applications of those fields and quantum states where the second terms in (30)-(32) prove to be much smaller than the first terms. In discussing the process of microwave ionization we assume that when the field F carries a level in state $n_2 = n - 1$ changes its energy according to (30). The change in the sign of F automatically transforms the level into one of the state $n_1 = n - 1$. By absorbing (emitting) a number of photons, the quasidiscrete state described by (30) with $n = n_1 + 1$ or $n = n_2 + 1$ and $m = 0$ may transform into a state of the continuous spectrum, (31) or (32), and hence become ionized. There is an infinite number of such possibilities, but only some are efficient. For instance, though most preferable energetically and requiring small values of F, the transition from state $E_{0n_20}(F)$ to state $E^+_{n_10}(F)$

is entirely ineffective because there is no overlap between the wave functions of the initial and final states (since n_1 and n_2 are very large). Hence we must find the transition energy

$$\Delta E(F) = E^{-}_{n+k,0}(F) - E_{0,n-1,0}(F) \tag{33}$$

where $n+k$ is the quantum number for the "ionization" curve, and k determines the channel "number". Assuming that $n \gg k$, we obtain

$$\Delta E = \frac{2k+1}{2n^3}\left[1 - \frac{6(k+1)}{2k+1}n^4 F\right]. \tag{34}$$

The transition between "discrete" and "ionization" states is most effective when

$$\Delta E = N\omega, \tag{35}$$

where N is integer equal to the number of absorbed photons. Naturally, transitions also occur when condition (35) is not met exactly, but their intensity proves to be much lower. Equating (34) and (35), we get

$$n^4 F = \frac{2k+1-2Nn^3\omega}{6(k+1)}. \tag{36}$$

Similarly, the transition between discrete states $E_{n_1 00}(F)$ and ionization states $E^{+}_{n+k}(F)$ satisfy the equation

$$n^4 F = \frac{2Nn^3\omega - 2k - 1}{6(k+1)}. \tag{37}$$

where N is a number of emitted photons.

It is seen from the equations (36) and (37) that the process of N-photon ionization through channel k by absorption is terminated, but similar process of ionization by emission is started at the following value of the scaled frequency

$$n^3\omega = \frac{2k+1}{2N}. \tag{38}$$

Of course, equation (38) (as well as equations (34), (36) and (37)) is not exact, if only because for $F = 0$ the ionization process is entirely impossible. It does provide, however, a qualitative description of the main features of the experimental data on microwave ionization of the hydrogen atom.

Obviously, the dominant processes leading to ionization are those for which N is small. But there is also a second, competing condition: the field F must be small. These two factors determine the experimentally obtained results.

Let us first consider the results qualitatively for relatively low values of n and $\nu = 9.92GHz$. The dominating process here is the one with $N = 1$ and $k = 0$. Using equation (36), we find that $n^4F = 0.15$ at $n = 32$ and $n^4F = 0.145$

at $n = 35$. This agrees with the experimental data on 90% ionization [5,16]. When the value of n corresponding to 90% ionization grows, n^4F decreases, but the equation (38) shows, the process of the ionization by absorption with $N = 1$ and $k = 0$ terminates at $n \sim 69$. But, in addition to one-photon ionization, two-photon ionization ($N = 2$), three-photon ionization ($N = 3$), etc. at $k = 0$ can also fairly effective if n is not too large. (As (37) shows, the two-photon process by absorption terminates at $n \sim 55$, the three-photon process at $n \sim 48$, the four-photon process at $n \sim 44$, etc.). The probability of such processes occurring is lower, but, nevertheless, they would seem to be responsible for 10% ionization [5,16]. For instance, (37) yields $n^4F = 0.133$ at $n = 32$ and $n^4F = 0.124$ at $n = 35$ for $N = 2$ and $k = 0$. These values are in satisfactory agreement with the experimental data on 10% ionization. Qualitatively it is easy to understand why a number of states in the vicinity of $n = 67 - 70$ (and to a great extent in a broader range of $66 - 72$) are ionized at roughly the same field strength F [5,16]. The value of n^4F for $k = 0$ and $N = 1$ (a one-photon process), for $k = 1$ and $N = 3$ (a three-photon process), and $k = 2$ and $N = 5$ (a five-photon process) reaches zero at $n \sim 69$. At $F = 0$ the ionization channel by absorption is "closed". Nevertheless, ionization may take place intensively at fairly low values of F, which is actually the case. This is why the given values of n the ionization curves plotted in F vs n variables exhibits a step, called an "island of stability" in [5,16].

For all the values of n ($32-90$) studied in [5], one-photon ionization through channels with $k \geq 1$ at the frequency $\nu = 9.92GHz$ requires too big value of F and, therefore, is ineffective. A number of other "stability islands" discovered at $9.92GHz$ correlate well with the termination of process at $N = 2 - 5$ and $k = 0-4$. The corresponding calculation and comparison with the experimental data can be carried out in elementary fashion, and of all the cases we note only one more, $N = 2$, $k = 0$, and $n = 54.9$, for the "island of stability" $n = 53 - 57$.

Let us recall one more experimental result obtained at $9.92GHz$ and explain it. Koch [5] gave the dependence of the ionization probability on n^4F for $n = 37$ and 38, and for $n = 47$ and 48. It was found in the first case ($n = 37$ and 38) the ionization curves exhibit a distinct peak at $n^4F \sim 0.12$, after which the ionization probability drops drastically as the field strength grows, but then again increases approaching a value of 100%. In the second case ($n = 47, 48$) the ionization probability monotonically increases with the field strength. Such a distinct difference in behaviour of the two pairs of curves can be understood from (36). A two-photon process at $n = 47$ and 48 proceeds most intensively at fields $n^4F \sim 0.05 - 0.06$ (for such values of n^4F (36) is satisfied exactly), a range where accurate measurements have yet to be done. At $n = 37$ and 38, two-photon ionization proceeds most intensively in a field n^4F of roughly 0.11 and three-photon ionization at fields $n^4F \sim 0.08 - 0.09$, which qualitatively explains the difference in the experimental data for $n = 37$ and 38 and for $n = 47$ and 48 [5] (see Fig. 3).

The values of n^4F obtained via (36) and (37) with properly-chosen values

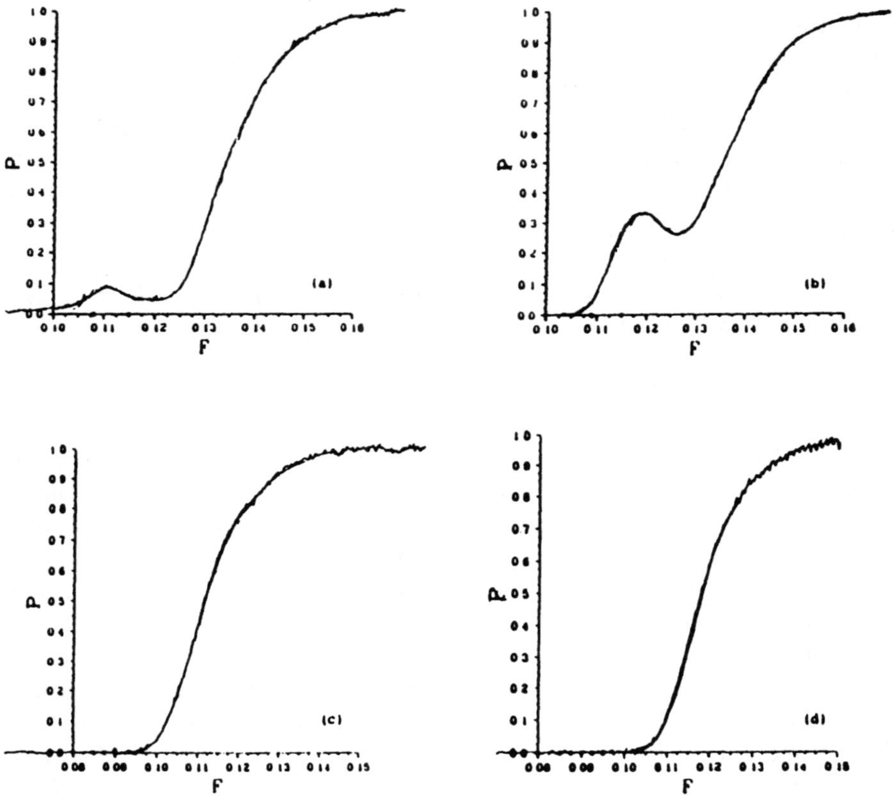

Figure 3. Frame (a): $n_0 = 37$ and $n_0^3\omega = 0.0764$. Frame (b): $n_0 = 38$ and $n_0^3\omega = 0.0828$. Frame (c): $n_0 = 47$ and $n_0^3\omega = 0.1567$. Frame (d): $n_0 = 48$ and $n_0^3\omega = 0.1669$.

of k and n usually proves to be somewhat lower than the one obtained through experiments. But it would be foolish to expect full agreement because not only one channel actually lead to "ionization". Besides, (36), (37) cannot answer the question of the percentage of atoms actually being ionized.

We made no attempt to improve our simple theory. Such improvement might appear to be just the exceeding of the accuracy. The reason for this is following. The experimental procedure, though ingenious and neat, cannot in any way claim to provide a strictly quantitative description of the process for several reasons. The number of oscillation in cavity cannot be predicted with 100% of accuracy. Experimental measurements do not yield completely reproducible results for all values of n, caused by uncontrollable and insignificant variations in experimental conditions.

But in no case by saying so we want to criticize these outstanding experimental studies, which, in our opinion, open the new area connected with the chaotic and stochastic behaviour in atomic physics.

We have shown that the experimental data on microwave ionization of highly excited states of hydrogen atom can be understood qualitatively by employing a simple model of a static electric field if one takes into account that for $E < 0$ the atomic spectrum contains states exhibiting the properties of a continuous spectrum in addition to essentially stationary states.

5. Another type of "Ionization" Splitting and its Verification in the Photoionization of Hydrogen Atom in the Presence of Uniform Electric Field

5.1. GENERAL CONSIDERATION

At $F = 0$ the starting point for ionization coincides with the end of the bound states: $E = 0$, $n_1 \to \infty$, n_2 and m are fixed; or $n_2 \to \infty$, n_1 and m are fixed.

At $F \neq 0$ the situation is different. We have considered above the splitting of the ionization boundary at $F \neq 0$. Similarly splits the end of bound states. At $F = 0$ the requirement $Z_1 = 0$ defines the boundary for the existence of the bound states: $E \to 0$, $n_2 \to \infty$, n_1 and m are fixed. At values of $Z_1 < 0$ an atom has only continuous states. Since the quasi-stationary states originate from bound states at $F = 0$, similar situation should take place obligatory at $F \neq 0$. The requirement $Z_1 = 0$ defines at $F \neq 0$ the boundaries for the existence of the quasi-stationary states $E^{bqs}_{n_1m}(F)$ for different quantum numbers n_1 and m. The values $E^{bqs}_{n_1m}(F)$ similarly to $E^{+}_{n_1m}(F)$ can be calculated exactly by solving the equation (4) at $Z_1 = 0$ numerically. Values $E^{+}_{n_1m}(F)$ can be both positive or negative,but values $E^{bqs}_{n_1m}(F)$ only positive.

$$E^{bqs}_{n_1m}(F) = k_{n_1m}F^{2/3}, \tag{39}$$

where k_{n_1m} are independent of F and can be calculated numerically. We present a few of them for $m = 0$ and 1.

We obtain $E^{bqs}_{n_1m}(F)$ in cm^{-1} by using k_{n_1m} from Tab. 3 and taking F in $kVcm^{-1}$.

TABLE 3. The values of $k_{n_1,0}$ and $k_{n_1,1}$

n_1	$k_{n_1,0}$	$k_{n_1,1}$	n_1	$k_{n_1,0}$	$k_{n_1,1}$
0	10.16	13.68	3	37.88	39.71
1	21.48	23.92	4	44.80	46.48
2	30.25	32.30	5	51.23	52.79

TABLE 4. The values of $E^+_{n_1,0}(F)$ (in cm^{-1}) at $F = 5.714\mathrm{kVcm}^{-1}$

n_1	$E^+_{n_1,0}(F)$	n_1	$E^+_{n_1,0}(F)$
24	7.3	29	126.2
25	33.8	30	146.6
26	58.8	31	166.1
27	82.4	32	184.9
28	104.9	33	202.9

5.2. COMPARISON OF THE THEORY WITH EXPERIMENTAL DATA OF ROTTKE AND WELGE [6] AND NUMERICAL CALCULATIONS

The calculation of the photoionization of a hydrogen atom performed by Alvarez *et.al.* [18] for $F = 5.714\mathrm{kVcm}^{-1}$ is, at least in principle, as well-founded as the exact calculation of the photoionization at $F = 0$, presented in the book of Bethe and Salpeter. First, we consider the trivial consequences of the above theory. At $F = 0$ the cross-section starts at $E = 0$ and is finite at the threshold. In the presence of a field the threshold for the "direct" ionization splits in quantum numbers n_1 ($m = 0$), as it is shown in Sec. 3. The starting points for the partial thresholds are characterized by values $E^+_{n_10}(F)$.

The values of $E^+_{n_10}(F)$ ($F = 5.714\mathrm{kVcm}^{-1}$) presented in Tab. 4 coincide with pinpoint accuracy with the positions of minima on the experimental spectrum (Fig. 4) for n_1 from 24 up to 28. They coincide also excellently with the calculated minima up to $n_1 = 33$ (not all of them are shown in the Fig. 4). We can clearly see that the qualitative picture of such splitting extends deeply to negative values of E. But pinpoint agreement between values of $E^+_{n_10}(F)$ and positions of minima on the cross-seçtion here can not be expected, since in deeply negative area "direct" ionization is essentially mixed with the decay of quasi-stationary states via tunneling. As it is shown above, the ionization threshold which starts at $E = 0$ at $F = 0$, at $F \neq 0$ experiences another type of splitting, namely to values $E^{bqs}_{n_10}(F)$.

We can compare the values presented in Tab. 5 with the experimental spec-

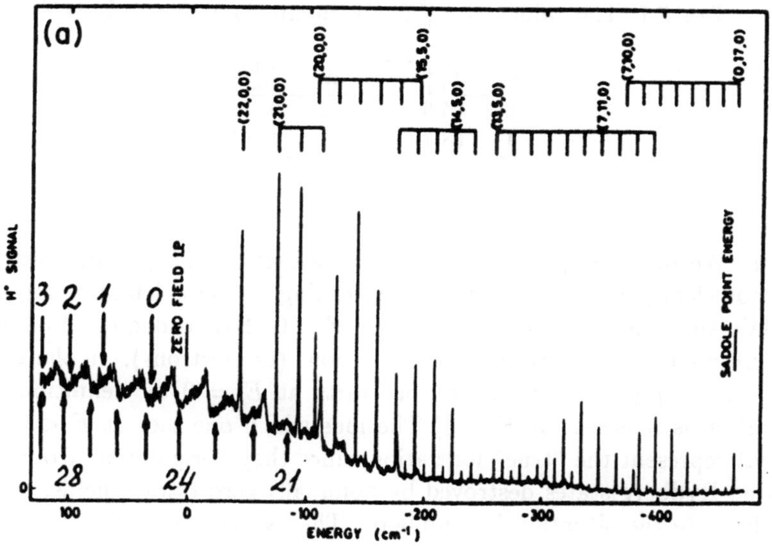

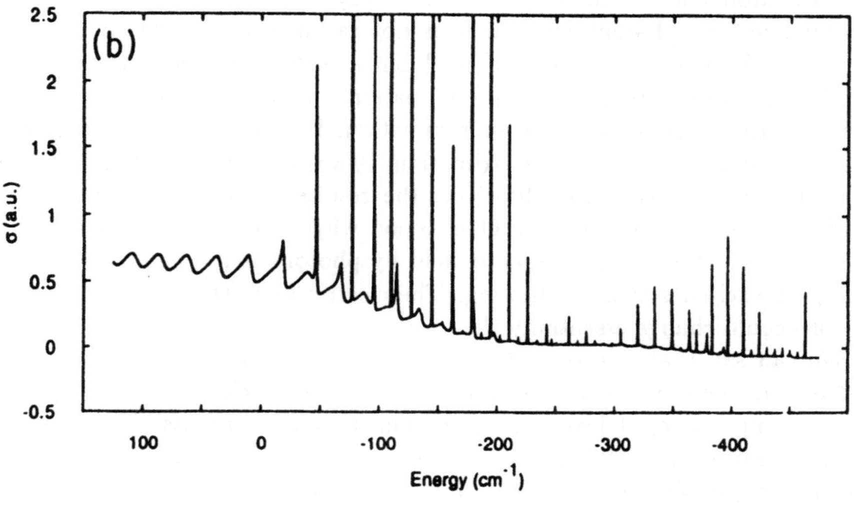

Figure 4. (a) Experimental photoionization spectrum at $F = 5.714\text{kVcm}^{-1}$ with π-polarization from $n = 2$ state from Rottke and Welge [6]. The arrows directed up and labeled by n_1 show positions of subthresholds $E^{+}_{n_1,0}(F)$ for the ionization. The arrows directed down and labeled by n_1 show positions of disappearances $E^{bqs}_{n_1,0}(F)$ of "quasistationary states".

(b) Theoretical *"ab initio"* photoionization spectrum that is purely superposition of resonance contributions [18].

TABLE 5. The values of $E^{bqs}_{n_1,0}(F)$ (in cm^{-1}) at $F = 5.714\mathrm{kVcm}^{-1}$

n_1	$E^{bqs}_{n_1,0}(F)$	n_1	$E^{bqs}_{n_1,0}(F)$
0	35.5	2	96.7
1	68.7	3	121.7

trum. We see in Fig. 4 only slight indications on the peculiarities, but they are not clearly resolved. But we see nothing in the calculated spectrum. To explain the situation, consider the case $F = 0$. Experimenters do not measure the cross-sections! They just count the ions (or electrons). In the case $F = 0$ one can simply say that ionization starts at $E = 0$ by definition. (Such a definition is senseless at $F \neq 0$) The ions which one sees at $E < 0$ at $F = 0$ do not represent the "true" ionization, since they "originated" from the highly excited Rydberg states destroyed by nominally zero (stray) fields. (It is useful to illustrate the situation by numbers. The "stray" field with $F \sim 0.5\mathrm{Vcm}^{-1}$ would ionize the states with n larger than 200, since such atoms effectively decay at $n^4F \sim 0.12 - 0.14$a.u. The energy of such states is $E \approx -2.7\mathrm{cm}^{-1}$). But such a "definition" even at $F = 0$ can not be considered as "rigorous" , since the atoms in excited states can be destroyed by the oscillating field which accompanies the photons and which is not taken into account in "rigorous" calculation both at $F = 0$ and $F \neq 0$. When we substitute the electro-magnetic field by photons we are doing an approximation. In most cases, especially for low excited states, it is an excellent approximation. But there are situations where we should abandon such an approximation. It was done by Damburg *et.al.* [19] in an attempt to explain qualitatively the results of microwave ionization of highly excited hydrogen atoms. One reason why we cannot consider that the substitution of the electro-magnetic field by photons is not enough "rigorous" for highly excited states is following. The electric field which accompanies the photons could change essentially the energies of highly excited states.

We can say that, although the situation of photoionization at $F = 0$ for $E < 0$ from the point of view of physics is absolutely clear, exact calculations here are impossible, at least for the reason, that it is impossible to take into account stray fields. The situation here is chaotic. It is always chaotic when discrete (or quasidiscrete) states are mixed with continuum. Such peculiarity of the photoionization spectrum at $F = 0$ was not discussed in the book of Bethe and Salpeter [9], but evidently the authors clearly understood it. In the energy scale presented in Fig. 37 of their book all the details of the spectrum which we are discussing here are squeezed into one single pinpoint and consequently can be disregarded. But what was unimportant for problems considered by them, becomes crucially important for our problem. We are studying here the physics in essentially different scales of energies and fields. For such reason we should take into account peculiarities of the problems which were unimportant for Bethe and Salpeter [9].

Quite similarly for the case $F \neq 0$ it is in principle impossible to calculate for all E the exact "true" cross-section of photoionization. Nevertheless, we can predict, how the peculiarity, which lies in the vicinity below $E = 0$ at $F = 0$ and which we are unable to calculate at $F = 0$, splits at $F \neq 0$. We might just calculate $E^{bqs}_{n_1 m}(F)$ and compare them with the peculiarities on the experimental spectrum.

We can present a visual though unrigorous explanation of the situation. At $F \neq 0$ the completely diffused relics of the states with very large values of n_2 and small n_1 are "pushed" by the field into region $E > 0$. Such "states" can, in principle, not be taken into account in any kind of "exact" calculations (quite similarly to the case $F = 0$, where we are unable to perform the "exact" calculations in the vicinity of $E = 0$). But such "states" are responsible for the chaotic regions at $E > 0$ and $F \neq 0$.

5.3. DISCUSSION ABOUT THE PHOTOIONIZATION SPECTRUM OBTAINED BY NAYFEH et al. [7] FOR $F = 2.5\text{kV cm}^{-1}$

Consider one difference between the experimental data of Rottke and Welge [5], and Nayfeh *et.al.* [7]. It would not only show one deficiency of the data presented by Nayfeh et.al. for $F = 2.5\text{kVcm}^{-1}$ and other F but would also indicate how this valuable information can be corrected. In both cases the authors were able to measure the value of F experimentally with accuracy $\sim 2 - 3\%$. Rottke and Welge extended their measurements into a region far below $E = 0$ (the threshold for ionization at $F = 0$). It allowed them to compare the positions of many sharp peaks on the photoionization curve with the values of Stark energies. Such a comparison performed by them is shown for the "best" value of F at Fig. 4, above the experimental spectrum. But Nayfeh *et.al.* [7], in their measurements of the photoionization, have not extended them far enough to negative E. Calculations of Stark energies in the regions close to $E = 0$ are not so trivial and, by comparing the positions of just a few essentially diffused peaks with the calculations, one can easily arrive at wrong conclusions about a more accurate value of F. The accuracy of $2 - 3\%$ in the value of F is not enough for such convincing agreement between the experimental data and calculation as it is shown in the Fig. 4: the positions of all peculiarities would be essentially shifted.

Basing on the theory presented above, which was, in addition, compared with the numerical calculations and experimental data for $F = 5.714\text{kVcm}^{-1}$, we can "improve" the value of $F = 2.5\text{kVcm}^{-1}$ "declared" by Nayfeh *et.al.* [7]. In Fig. 5 we show the experimental spectrum. In Tab. 4 we present the values of $E^{+}_{n_1 0}(F)$ for different n_1 calculated for $F_1 = 2.5\text{kVcm}^{-1}$ and $F_2 = 2.55\text{kVcm}^{-1}$.

TABLE 6. The values of $E^{+}_{n_1,0}(F)$ for $F_1 = 2.5\mathrm{kVcm}^{-1}$ and $F_2 = 2.55\mathrm{kVcm}^{-1}$

n_1	$E^{+}_{n_1,0}(F_1)$ cm^{-1}	$E^{+}_{n_1,0}(F_2)$ cm^{-1}	n_1	$E^{+}_{n_1,0}(F_1)$ cm^{-1}	$E^{+}_{n_1,0}(F_2)$ cm^{-1}
25	-73.3	-71.1	31	24.3	26.8
26	-54.4	-52.6	32	37.8	40.3
27	-36.8	-34.8	33	50.1	53.2
28	-20.2	-18.1	34	62.9	65.6
29	-4.5	-2.3	35	74.5	77.5
30	10.3	12.6			

TABLE 7. The values of $E^{bqs}_{n_1,0}(F)$ (in cm^{-1})

n_1	$E^{bqs}_{n_1,0}(F_2)$	n_1	$E^{bqs}_{n_1,0}(F_2)$
0	19.0	3	70.7
1	40.1	4	83.6
2	56.5	5	95.6

Comparison with the experimental spectrum at Fig. 5 shows that the values $E^{+}_{n_10}(F_2)$ coincide with the positions of minima quite satisfactorily, both at positive and negative E, especially if we take into account that the accuracy of the experimental measurements of E is $1.75\mathrm{cm}^{-1}$. We are not going to discuss the small roughness of the spectrum, including two small narrow minima in the vicinity of $E = 0$. The roughness of the spectrum can be explained by insufficient resolution, but only additional and much more accurate measurements can reliably resolve the problem.

But we clearly see in Fig. 5 some additional distinct peculiarities which can not be explained by $E^{+}_{n_10}(F_2)$ or by insufficient resolution. In Tab. 5 we present values $E^{bqs}_{n_10}(F_2)$.

Some values $E^{bqs}_{n_10}(F_2)$ may appear to be "disguised" by $E^{+}_{n_10}(F_2)$, and we are not able to discern them. But this is not a rule.

If we again would recollect that both values F and E are not defined in the experiment exactly, we can conclude that the agreement between the theory end experimental data is quite convincing.

We have explained by splitting $E^{bqs}_{n_10}(F_2)$ only the most prominent featúres of chaos in the spectrum. Strictly speaking, the spectrum possesses "chaotic" peculiarities at all energies. They originate from fundamental inability to determine exactly the widths of Stark states at large F as it was explained in Sec. 1. It is not the drawback of the theory which can be overcome. Such chaotic features disappear gradually when we are moving below the ionization threshold

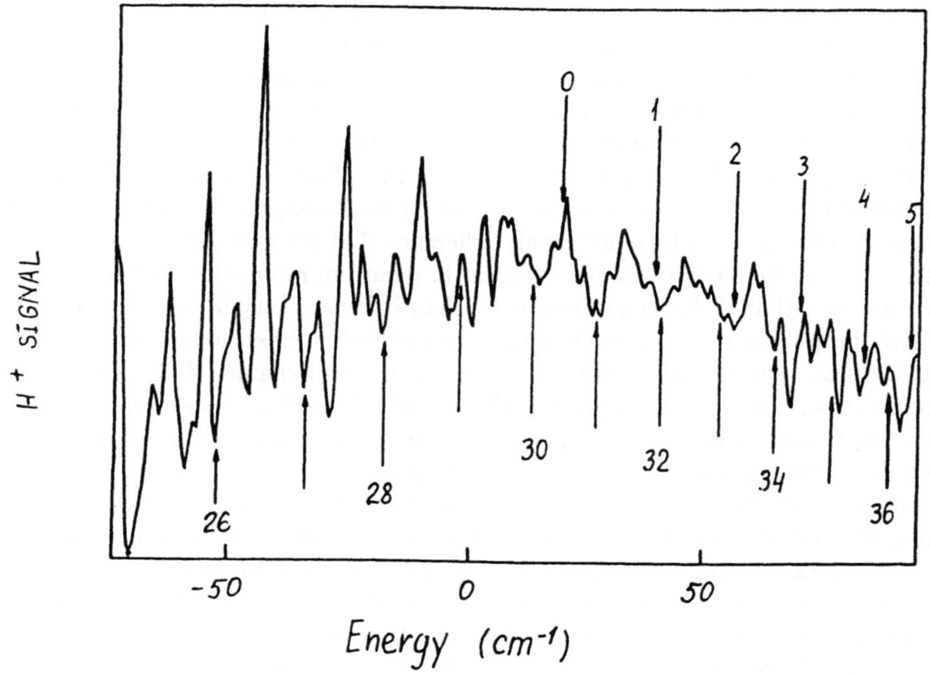

Figure 5. Experimental photoionization spectrum at $F = 2.55\text{kVcm}^{-1}$ with π-polarization from $n = 2$ state from Nayfeh *et.al.* [7]. The arrows directed up and labeled by n_1 show positions of subthresholds $E^+_{n_1,0}(F)$ for the ionization. The arrows directed down and labeled by n_1 show positions of disappearances $E^{bqs}_{n_1,0}(F)$ of "quasistationary states".

$E = 0$ at $F = 0$. In such case QS would resemble more and more bound states. The positions of small "chaotic" peculiarities of the spectrum can be associated with the positions of minima of amplitude for very "diffused" Stark states. But, experimental verification of such small peculiarities seems to be very difficult.

6. Conclusion

We have shown that the hydrogen atom in the presence of an electric field exhibits chaotic (stochastic) behaviour. The manifestation of it can be found in numerous experimental data. The main reason for such behaviour is the consequence of the different kinds of the splitting in the field the unique for $F = 0$ boundary at $E = 0$ which separates the discrete and continuous spectra.

At the end we shall say a few words about our understanding of chaos. The three-body problem in classical mechanics is known as chaotic. It does not prevent us, at least in principle, to calculate with any degree of accuracy energies and wave functions of bound spectrum of *He* by using nonrelativistic quantum mechanics. But it is clear that we are not able in this case (in the contrary to hydrogen) to ascribe strictly defined quantum numbers for electrons. By increasing the accuracy of calculation we should increase the amount of quantum numbers (though doing calculations practically we never worry about quantum numbers). In such a way quantum mechanics reflect the chaotic behaviour inherent to classical mechanics where we are not able to predict the trajectories of particles exactly. There is the only difference that the energies are quantized and Pauli principle is taken into account in quantum mechanics. Nevertheless, we are sure that if the energy is smaller than it is necessary for one particle to leave the system, the atom would remain bound both in classical and in quantum mechanics. Quite similar situation arises in both mechanics if the energy is larger, than it is required for the ionization. In both mechanics the system could remain in some cases for some period of time "quasibound". But one particle would eventually leave the system. In quantum mechanics the corresponding quantized "states" have got a name the quasistationary states. It is clear that we cannot determine for them "lifetimes" exactly. The introduction of the concept of "complex energies" which can be calculated exactly by Borel summation of asymptotically divergent series for the energy or by some other procedure destroys the fundamental link between quantum and classical mechanics. In such case the problem of quantum chaos becomes mysterious.

The reason why the chaotic behaviour in quantum mechanics caused a surprise was following. The irregular orbits of classical mechanics arise for nonlinear systems, whereas the Schrödinger equation is fundamentally linear. But according to the Copenhagen interpretation the link between quantum and classical mechanics lies on the different plane: "It is in principle impossible, however, to formulate the basic concepts of quantum mechanics without using classical mechanics" [8].

7. References

1. Kleppner, D., Littman, N.G. and Zimmerman, M.L. (1983) Rydberg atoms in strong fields, in R.F.Stebbings and F.B. Dunning (eds.), *Rydberg States of Atoms and Molecules*, Cambridge University Press, London, pp.73–116.

2. Nicholl, P., Bergeman, T. and Metcalf, H.J. (1988) Lifetime measurements of interference-narrowed sodium Stark resonances, *Phys. Rev.* **A 37**, 33O2–12.

3. Pinard, J. (1990) Atoms in static electric and magnetic field: the experimental aspect, in C.A. Nicolaides (ed.), *Atoms in Strong Fields*, Plenum Press, New York, pp.17–33.

4. Bayfield, E. and Koch, P.M. (1974) Multiphoton ionization of highly excited hydrogen atoms, *Phys. Rev. Lett.* **33**, 258–261.

5. Koch, P.M. (1990) Microwave ionization of excited hydrogen atoms: what we do and do not understand, in D.K. Campbell (ed.), *Chaos/XAOC: Proc. Soviet-American Conf.*, American Institute of Physics, New York, pp.441–475.

6. Rottke, H. and Welge, K.H. (1986) Photoionization of the hydrogen atom near the ionization limit in strong electric fields, *Phys. Rev.* **A 33**, 301–319.

7. Nayfeh, M.H., Humm, D. and Ng, K. (1990) Hydrogen in strong DC and low-frequency fields, in C.A. Nicolaides (ed.), *Atoms in Strong Fields*, Plenum Press, New York, pp.133–153.

8. Landau, L.D. and Lifshitz, E.M. (1965) *Quantum Mechanics*, Pergamon Press, Oxford.

9. Bethe, H.A. and Salpeter, E.E. (1957) *Quantum Mechanics of One- and Two-Electron Atoms*, Academic Press, New York!

10. Littman, M.G., Zimmerman, M.I. and Kleppner D. (1976) Electric field ionization rates of selected Stark states in sodium, in R. Marrus, M.H. Prior and H.A. Shugart (eds.), *Abstracts: Fifth International Conference on Atomic Physics*, Berkely, California, pp.236–237.

11. Reinhardt, W.P. (1982) Padé summation for the real and imaginary parts of atomic stark eigenvalues, *Int. Quantum Chem.* **XXI**, pp.133–146.

12. J.C.Le Guillou and J. Zinn-Justin (eds.) (1990) *Current Physics Sorces and Comments Vol. 7: Large-order Behaviour of Perturbation Theory*, North-Holland, Amsterdam.

13. Damburg, R.J., Propin, R.Kh., Graffi, S., Grecchi, V., Harrell, E.M., Zizek, J., Paldus, J. and Silverstone, H.J. (1984) $1/R$ expansion for H_2^+: analyticity, summability, asymptotics, and calculation of exponentially small terms, *Phys. Rev. Lett.* **52**, 1112–1115.

14. Damburg, R.J. and Kolosov, V.V. (1983) Theoretical studies of hydrogen Rydberg atoms in electric fields, in R.F. Stebbings and F.B. Dunning (eds.), *Rydberg States of Atoms and Molecules*, Cambridge University Press, London, pp.31–71.

15. Kolosov, V.V. (1991) The Stark resonances near zero energy, *J. Phys.* **B**:*At. Mol. Opt. Phys.* **24**, **L**185–**L**188.

16. Koch, P.M. (1988) Microvawe ionization of highly-excited hydrogen atoms: a driven quantal system in classically chaotic regime, in H.B. Gilbody, W.R. Newell, F.H. Read and A.C.H. Smith (eds.), *Electronic and Atomic Collisions* (Invited Papers of XV ICPEAC), North-Holland, Amsterdam, pp.501–515.

17. Koch, P.M. (1992) Experimental evidence for the influence of "scars" in hydrogen atoms driven by strong microwave fields, *Chaos* **2**, No. 1, 131–144.

18. Alvarez, G., Damburg, R. and Silverstone, H. (1991) Photoionization of atomic hydrogen in an electric field, *Phys. Rev.* **A 44**, 3060–3082.

19. Damburg, R., Baranovsky, B. and Silverstone, H. (1994) Direct "ionization" of the hydrogen atom by photons in the presence of an electric field, *JETP* **78**(6), 821–826.

MULTIELECTRON, MULTICHANNEL PROBLEMS AND THEIR SOLUTION TO ALL-ORDERS VIA THE DIAGONALIZATION OF STATE-SPECIFIC COMPLEX EIGENVALUE SCHRODINGER EQUATIONS

C.A.NICOLAIDES, TH.MERCOURIS AND S.I. THEMELIS
Theoretical and Physical Chemistry Institute,
National Hellenic Research Foundation,
48 Vas. Constantinou Ave., 116 35 Athens, Greece

Abstract

A variety of properties and processes in atomic, molecular and chemical physics can be conceptualized as being caused by nonstationary states satisfying state-specific complex eigenvalue Schrödinger equations (CESE). The imaginary part of the eigenvalue gives the rate of disintegration of the system. For ground or excited states dressed by an external dc- or ac-field, the energy shift is associated with linear and nonlinear polarizabilities. The CESE gives rise to all the conceptual and practical difficulties associated with the polyelectronic structures of excited states, as well as novel ones due to the presence of external fields and to the physical significance of the continuous spectrum. In a series of articles from this institute, it has been shown how advanced electronic structure theory and methods suitable for ground or excited states can be integrated in a practical way into selected elements of the rigorous theory of discrete states interacting with the continuous spectrum in order to solve the CESE nonperturbatively and efficiently and compute properties such as positions and partial widths of inner hole or multiply excited states, multiphoton ionization rates, multichannel predissociation lifetimes, nonlinear static and frequency-dependent polarizabilities, tunneling rates, magnetic field induced ionization of excited states and the Wannier problem of two-electron ionization near threshold. Here we outline this approach and present sample applications.

C. A. Tsipis et al. (eds.), New Methods in Quantum Theory, 135–147.

1. Difficulties due to electronic structure, to electron correlations and to the strength of interaction with external fields.

The traditional way of establishing quantum mechanical formalisms for a number of phenomena and processes in atomic, molecular and chemical physics, is to express the observables in terms of off-diagonal matrix elements involving the appropriate transition operator and stationary states. For example, the most common such expression is the golden rule for one-photon radiative transitions, whose derivation is part of almost every quantum mechanics text book. This stationary state, - transition operator approach indeed provides the framework for the establishment of formalisms appropriate to the phenomena of interest. On the other hand, the crucial problem in the quantum theory of atoms and molecules has been the quantitative computation for arbitrary systems and properties. As regards this goal, there are three major difficulties for systems with more than one electrons. The first has to do with the so-called bound **many-electron problem (MEP)**. That is, one has to face the fact that in order to be able to calculate the relevant off-diagonal matrix elements, reliable bound wavefunctions and energy differences which include the effects of electron correlation and of relativistic interactions must be computed and employed. Such wavefunctions are not only those of ground states (for which many advanced methods are applicable). They are also those of **excited states**, whose multitude of electronic structures and symmetries lead to the requirement of developing and using generally applicable methods of solving the corresponding Schrödinger equations.

The second difficulty has to do with the **multichannel scattering problem (MCSP)**. Most dynamical phenomena involve the continuous spectrum. For a polyelectronic atom or molecule, this spectrum contains, in general, more than one scattering channels, which interact with each other if symmetry and energy allow it. In the simplest of cases, (but still impossible to handle without applying advanced scattering theory), each channel represents one free particle continuum. Complexity increases when multielectron autoionizing states are present, or an external field is applied. Finally, things become even more difficult when the excitation process is such that more than one free electrons are created. For example, one weak intensity photon of say 80 eV energy interacting with the ground state of He, may create one electron continua ($1s^2\ {}^1S \xrightarrow{h\nu} 1s\varepsilon p, 2l\varepsilon l', 3l\varepsilon l'$, etc, ${}^1P^o$) as well as two electron continua ($1s^2\ {}^1S \xrightarrow{h\nu} \varepsilon l\varepsilon l'\ {}^1P^o$) with the same final state symmetry (${}^1P^o$). Furthermore, if ones considers the interaction of He $1s^2\ {}^1S$ with an intense laser beam, (between 10^{14}-10^{16} W/cm^2), with photon energy of say 5eV, multiphoton ionization occurs, where the intermediate as well as final real or virtual states have many more symmetries than just the ${}^1P^o$.

The third difficulty has to do with the challenging and often needed necessity of taking into account the physically relevant interactions to **all- orders in perturbation theory**. Two examples make the point: The first refers to the interelectronic Coulomb operator in relation to the reliable calculation of electron correlation . For both the MEP and the MCSP, it is insufficient, in general, to consider only first order matrix elements in the interaction. The second refers to the **strength** of the interaction caused by an external electromagnetic field. If it is not small compared to the internal Coulomb interaction of the active electrons, then higher-order (non-linear) couplings occur, contributing to both the nature and the magnitude of observable quantities.

In what follows, we outline essential aspects of theoretical and computational work carried out in this institute for the resolution of all of the three difficulties mentioned above, in dynamical problems such as autoionization, predissociation, field- induced multiphoton ionization, polarization and tunneling, etc.. The theory breaks down the many-electron and multichannel scattering problems, without or with an external field, in terms of their basic electronic structure, electron corrrelation and spectral features, thereby allowing the systematic as well as economical computation of many properties through the use of physically relevant **state-specific** function spaces. As regards general methods, although perturbation theory notions are used for heuristic reasons, (e.g. choice of important interacting states and correlation vectors), the final calculations are variational, establishing results to all-orders in the interactions within the full function space that is employed.

2. Outline of the theoretical approach

Observables for processes such as the ones mentioned in the introduction, which involve the continuous spectrum, can be conceptualized as being caused by the interaction of one or more discrete energy levels, treated as stationary states of an effective Hamiltonian, with the adjacent single or multichannel continuous spectrum. From a time-dependent point of view, such an interaction creates a **nonstationary** state and causes the decay of the initially localized state, Ψ_0, with rate Γ, (in a.u.), and the shift of the initial energy E_0, by a small amount Δ.

In the case of an isolated atom (molecule), the continuous spectrum which results from the free motion of an electron or a nucleus is an intrinsic characteristic of the Hamiltonian with all the interparticle interactions present. In the case of a "field-dressed" nonstationary state, the continuous spectrum becomes accessible because of the interaction with the external field (e.g. photoionization due to an ac-field or tunneling due to a dc-field).

The existing formal framework for treating the problem of the discrete- scattering state interaction can be divided into two categories. The first contains methods which use the stationary state-transition operator approach mentioned in the intro-

duction. The stationary states correspond to Hermitian matrices with real eigenvalues. The second contains methods which use the nonstationary state approach, where the Hamiltonian matrix is made nonHermitian by some technique, and the eigenvalue corresponding to the nonstationary state is complex, $z=E-\frac{i}{2}\Gamma$. In either case, the desideration is the one stressed above, i.e. how to develop and implement computationally practical formalisms for the solution, to all- orders in the perturbation parameter, of the MEP, of the MCSP and the field-free and field-induced processes. It is with this desideratum in mind that the **state-specific complex eigenvalue Schrödinger equation** (CESE) approach outlined below has been developed. Its hallmark is the breakdown of each case of interest into different function spaces, containing analytic as well as numerical orbitals, whose optimization is done by different techniques. The nonHermiticity of the problem is introduced by the use of a complex coordinate for the function representing the emitted particle.
The theory, methods and a number of applications can be found in [1-19]. A direct comparison with the complex coordinate rotation (CCR) method [20, 21] using as examples the He^- $2s2p^2$ ^4P autoionizing state and the hydrogen atom in a magnetic field, was discussed in [16]. The CCR method consists of the repeated construction from a large basis set of $\mathcal{L}^2$ functions and diagonalization of the rotated nonHermitian matrix $H(\varrho)$, $\varrho=re^{i\theta}$, until one or more complex eigenvalues exhibits stability with respect to the rotation angle θ and are identified as resonances. Another approach to the calculation of resonances in multichannel models using the complex coordinate regularization technique, is based on the solution of coupled scattering equations [22].
The present approach has its origin in the state-specific analysis of the many-electron autoionization problem [3-6], where the notion of a decaying state with a complex eigenvalue was combined with theory and methods for excited states in order to search for well-optimized function spaces. Accordingly, the total wavefunction is computed in terms of two general function spaces, representing respectively the localized part, Ψ_0, and the decayed part, X_{as}. Ψ_0 contains all the interelectronic and nuclear attraction interactions contributing to the stability of the state ("localized correlations") and X_{as} contains the final state interactions on resonance ("asymptotic correlations"). It can be shown [6, 13] that starting from Fano's formalism of discrete-continuous spectrum mixing for real energies, imposition of resonance asymptotic boundary conditions leads to the rigorous description of the nonstationary state in terms of a CESE

$$(H - z_0)\Psi = 0 \tag{1}$$

where

$$z_0 = E - \frac{i}{2}\Gamma \tag{2}$$

$$\Psi = \Psi_0 + X_{as} \tag{3}$$

$$E = E_0 + \Delta \tag{4}$$

$$E_0 = \langle\Psi_0|H|\Psi_0\rangle \tag{5}$$

Γ is the energy width (rate of decay) and Δ is the energy shift induced by the interaction.

The resonance wavefunction of eq. 1 is not square integrable, a fact representing the physics of outgoing flux. Its exact asymptotic form depends on the potential acting on the outgoing particle(s). Applying to eq.1 a perturbation of the asymptotic boundary conditions in the form of the coordinate transformation $r \rightarrow \varrho e^{i\theta}$, renders the resonance wavefunction normalizable and expressible in terms of square-integrable basis functions, while keeping the problem non Hermitian with the same complex eigenvalue [5, 13].

For one-dimensional problems, integration of eq. 1 using a complex coordinate can be done directly, numerically. For multidimensional problems arising with polyelectronic systems, Ψ_0 and X_{as} are represented by different function spaces which are optimized separately, the first on the real axis once, yielding a real energy, E_0, and the second partly on the real axis for its bound part and partly in the complex energy plane. The calculations of bound wavefunctions (either the Ψ_0 or the core of the final state) are done by paying attention to the electronic structure of the state of interest and to the self-consistent computation of the multiconfigurational zeroth order description. The second part of the computation depends only on the structure of the continuous spectrum and yields Δ and Γ.

For predissociation due to distinct operators such as radial or rotational or spin-orbit coupling, having obtained accurate CI wavefunction representations for the Ψ_0 vibronic states of interest, the X_{as} part is represented by suitable basis sets of complex coordinates [12]. For predissociation via tunnelling, as for example with the $^1\Sigma_g^+$ ground state of He_2^{++}, the conceptual separation of the calculation into two parts, one on the real axis and one in the complex energy plane, can be done via "*exterior complex scaling*" [11].

For Coulomb autoionization where the interaction operator is nonseparable, the state-specific calculation of Ψ_0 is based on the existence of a localized, self-consistent multiconfigurational zeroth order function satisfying the virial theorem and the physically correct orbital nodal structure. The remaining localized electron correlation is obtained variationally [1]. Both the zeroth order and the localized correlation part exclude the open channels by construction and by satisfying orthogonality constraints to electronic structure dependent core orbitals. This scheme can be used for the construction of diatomic diabatic states [23]. X_{as} incorporates the interchannel couplings and a simple expression yields the partial widths to all orders [8, 12, 19].

The same basic notions as above have been adapted to the general problem of computing properties induced by a strong external electromagnetic field [2, 9, 10, 13, 16-19]. The appropriate operators for an ac-field, a dc-field or a magnetic field are added to the field-free Hamiltonian **without** any of the coordinates being rotated.

The nonHermiticity is again introduced through the use of basis sets with a complex coordinate for the outgoing particle. Now, the functions Ψ_0 and X_{as} are generalized to represent multidimensional vector spaces, Q and P, where more than one physical and virtual N-electron states are included. For the case of the ac-field, the resulting quantities, e.g. the multiphoton ionization rates, the ATI peaks or the frequency-dependent linear and nonlinear polarizabilities, constitute averages over the field cycle. The total nonHermitian symmetric Hamiltonian matrix incorporates electron correlation and interchannel interactions to all orders. The nonorthonormality overlap matrix elements are accounted for explicitly. For the cases of static electric or magnetic fields, its diagonalization is done by standard subroutines. For the case of the ac-field, a self-consistent iterative scheme was developed [9]. In all cases, the sought after complex eigenvector is identified by following the root with the maximum overlap with the unperturbed state-specific wavefunction.

3. Examples

We now present as yet unpublished results from the application of the state-specific CESE theory to three problems of atom-field interactions.

3.1. dc-field induced tunneling rates in H^-

Using this approach it is possible to obtain ab initio dc-field induced tunneling rates in polyelectronic systems, [9, 18, 24 and refs. therein], except for very small field strengths, where the width (imaginary part of the complex eigenvalue) becomes extremely small for an expansion-based method to obtain via diagonalization of the nonHermitian matrix.

Since the theory for this process had earlier been developed for one-electron atoms or via semiclassical one-electron models, comparative calculations were recently carried out [18, 24]. Large discrepancies were found, especially for neutral atoms with equivalent electrons such as the He ^1S ground state. In this section we show how the tunneling rate of the H^- ^1S state is calculated and then we compare the results with available experimental values and semiclassical calculations.

We wish to solve eq. 1, where

$$H = H_A + \sum_i \vec{F} \cdot \vec{r}_i \tag{6}$$

and the solution z_0 is written as

$$z_0 = E_0 + \Delta(F) - \frac{i}{2}\Gamma(F) \tag{7}$$

The function space Q consists of the ground state correlated wavefunction, ^1S, which is computed according to the state-specific theory of electronic structure as

$$\Psi_o(^1S) = \Phi_{MCHF} + X_{loc} \qquad (8)$$

Φ_{MCHF} is the numerical multiconfigurational solution

$$\Phi_{MCHF} = 0.967(1s^2) + 0.217(2s^2) + 0.114(2p^2) \qquad (9)$$

and X_{loc} represents the most important remaining localized electron correlations, obtained variationally in terms of singly and doubly excited configurations with virtual orbitals up to l=4 and optimized Slater-type radials. The energy thus obtained is E_o=- 0.52748 a.u., sufficiently close to the exact value of -0.527751 a.u..

We note that, in this case, the Q space is one dimensional because there are no other singlet discrete states with which the ^1S ground state could mix via H. The only other localized singlet states are doubly excited resonances, but they are excluded since rough calculations show that their contribution is negligible.

The P space is multidimensional, containing $\mathcal{L}^2$ configurations with complex orbitals representing the continuous spectrum: P: { 1sεs, 1sεp, 1sεd, 1sεf, 1sεg, 1sεh }. Higher thresholds of the form 2sεl, 2pεl, etc, were not included since their contribution is negligible. The εl continua were discretized on a basis of complex STOs

$$\varepsilon l = \sum_n a_n \varphi_{nl}(r,\theta), \quad \varphi_{nl} = r^{n+l-1} e^{-\alpha r e^{-i\theta}} \qquad (10)$$

The final calculations employed 12s, 12p, 12d, 10f, 8g and 6h complex STOs. The spaces Q and P are not orthonormal between them. Hence, when Ψ of eq. 1 is projected onto Q and P, a generalized complex eigenvalue equation is obtained

$$\begin{Bmatrix} H_{QQ} & H_{QP} \\ H_{PQ} & H_{PP} \end{Bmatrix} \begin{pmatrix} C_Q \\ C_P \end{pmatrix} = z \begin{bmatrix} I & S_{QP} \\ S_{PQ} & I \end{bmatrix} \begin{pmatrix} C_Q \\ C_P \end{pmatrix} \qquad (11)$$

which is solved repeatedly by varying the parameters α and θ of the complex STOs, until the solution is found for the vector with the maximum overlap with Ψ_o when

$$\left.\frac{\partial z}{\partial \alpha}\right|_{z=z_o} = 0, \quad \left.\frac{\partial z}{\partial \theta}\right|_{z=z_o} = 0 \qquad (12)$$

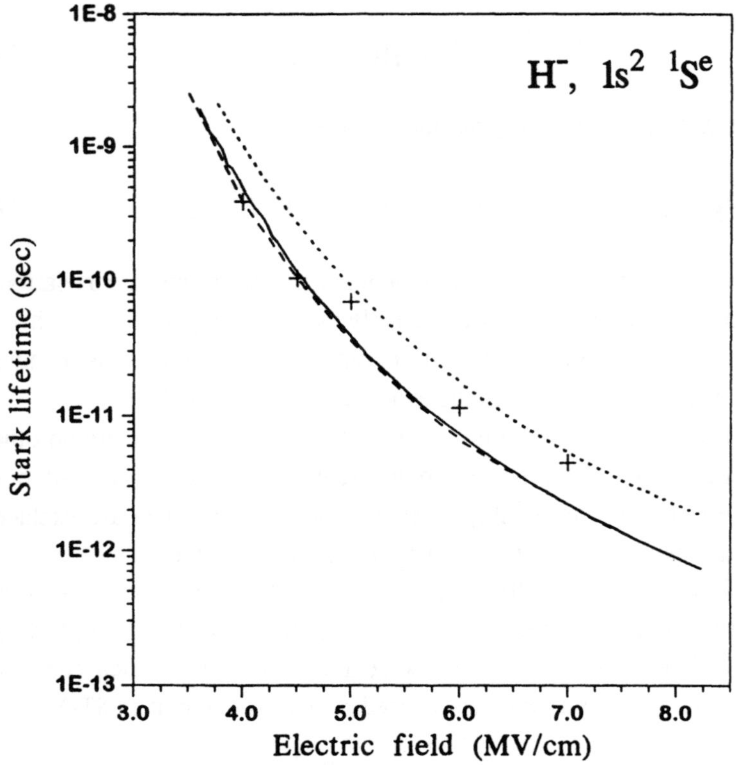

Figure 1. Lifetimes of H^- in a d.c. field: State-specific CESE theory (solid line), WKB expression (crosses)[25], semiclassical model (dotted line) [26], experimental values (dashed line)[27].

$F(10^{-2}$ a.u.)	Γ(a.u.)
0.070	1.27 [-8]
0.075	3.24 [-8]
0.080	6.92 [-8]
0.085	1.44 [-7]
0.090	2.81 [-7]
0.100	8.38 [-7]
0.120	4.17 [-6]
0.150	2.11 [-5]
0.200	1.09 [-4]
0.300	5.95 [-4]
0.400	1.46 [-3]
0.500	2.62 [-3]
0.700	5.51 [-3]
1.000	1.07 [-2]

TABLE 1. Field induced tunneling widths of H^-, calculated from the present theory. Numbers in square brackets denote powers of 10.

Our calculations covered the range of field strengths $0 \rightarrow 0.01$ a.u., corresponding to 0-50 MV/cm. Selected results are presented in Table 1. In figure 1, we plot on logarithmic scale our results, together with those obtained from the WKB expression [25] and a semiclassical model [26] and with the experimental values [27]. The agreement of the present quantum mechanical theory with experiment is excellent.

3.2. ac-field induced polarization and ionization of the hydrogen atom. The hydrogen atom in a ac-field has been treated by the CCR method [21]. For reasons of demonstration and comparison, here we treat the same problem for a long laser pulse of frequency $\omega=0.30$ a.u., (minimum number of photons for ionization to occur is two), and intensity $I=1.75 \times 10^{14}$ W/cm^2, which is one of the cases calculated by Chu and Cooper [21].

The square integrable wavefunction $\Psi(\varrho)$, ($\varrho=re^{i\theta}$), which is connected adiabatically to the unperturbed ground state of hydrogen, is formally written as

$$\Psi(\rho) = \sum_{i,n} a_{i,n}(\theta)|\varphi_i(\rho);n> + \sum_{j,n} b_{j,n}(\theta)|X_j(\rho);n> \tag{13}$$

The φ_i represent bound states, whose state-specific functions are known exactly, X_j represent open channels and $|n>$ denotes the photon states .

The time-independent Hamiltonian for linearly polarized light is

$$H = H_A + h\omega\, a_\omega^+ a_\omega - \frac{F_{ac}}{2} r\cos\varphi(a_\omega^+ + a_\omega) \tag{14}$$

where a_ω^+, a_ω are photon creation and annihilation operators determining the dipole operator angular momentum algebra. The sought after complex eigenvalue acquires the form

$$z_o = E_o + \Delta(\omega, F_{ac}) - \frac{i}{2}\Gamma(\omega, F_{ac}) \tag{15}$$

Expansion of the square integrable $X_j(\varrho)$ in a basis set of Slater-type orbitals (STOs) with real coordinates and use of the fact that for the bound functions the matrix elements are invariant under rotation of all the coordinates in the Hamiltonian and in the functions [5, 9], allows the construction of a Hamiltonian nonHermitian matrix where only the complex STOs of X_j are functions of $\varrho^*=re^{-i\theta}$. For the specific calculation, the bound state basis set consisted of the 1s, 2s, 3s, 2p, 3p and 3d hydrogen functions. The X_j consisted of nine s STOs, ten p STOs , eleven d STOs and twelve f and g STOs.

In general, use of the wavefunction form (13) gives rise to a many electron, many photon Hamiltonian matrix of the form of eq. 6 of ref. 9. The method for its solution is described by Mercouris and Nicolaides [9]. In the present case, the initial guess

for the complex eigenvalue was the energy E_0, i.e. the energy of the 1s state, E_0=-0.500 a.u.. The next guess for z_0 is obtained by a linear extrapolation technique and the final solution to the relevant matrix equation is determined following the criterion of maximum overlap with the immediately previous one. The matrix was truncated after the $\hat{A} + 5\omega \hat{I}$ and $\hat{A} - 5\omega \hat{I}$ blocks, (see eq. 6 of ref. 9), where $\hat{A}$ is the free atom Hamiltonian matrix.
The final result is (in a.u.)

$$z_0 = -0.50850 - i\ 0.1610 \times 10^{-2} \tag{16}$$

which is exactly the value obtained by Chu and Cooper [21, Table II].
The corresponding observable energy shift and multiphoton ionization rate are

$$-0.23 \text{ eV and } -\frac{2\,\mathrm{Im}(z_0)}{h} = 1.33 \times 10^{14}\ s^{-1}.$$

3.3. Single-and multiphoton two-electron ionization

As is well-known, absorption of one or more photons (in the case of strong laser light) can also result in the ejection of two electrons. For an ab initio calculation of the rate of such a process it is now necessary to be able to treat quantum mechanically the problem of pair correlation and of related intra-and inter-channel couplings in the continuous spectrum, as well as to compute the coupling to all orders with the initial state.
Aspects of the above difficult problem have been investigated over the past 40 years- since the work of Wannier by classical mechanics on the behavior of the cross-section at E=0- by semiclassical or quantum mechanical methods to some approximation. Until recently, a fully quantum mechanical many-electron, many-channel theory and computational method had been lacking. For although the wavefunction of the bound initial state (or intermediate states in the case of multiphoton steps) is computable via advanced theoretical methods, the implementation of conventional multichannel scattering theory to the problem of pair correlation in the continuum presents formidable difficulties, related to, among other things, strong singularities in the interaction matrix elements.
Here we point out that the state-specific CESE constitutes a rigorous framework for the solution of such problems. In fact it has already been applied to the calculation of the rate of multiphoton ionization of the $2p^2\ {}^3P$ bound state of H^- [28]. It has since been applied to the "Wannier problem" of near-threshold photoionization of the $1s^2\ {}^1S$ states of H^- and of He, and of the 1s2s2p $^4P^o$ state of He^- [29].
Briefly, the problem is formulated and solved as follows.
The total wavefunction is again written in the form (13), where now the X_j contain one- and two-electron open channels. Arguments analogous to the one-particle case

have led us to the conclusion that the two-electron dressed state resonance function must have an r_{12} dependence which ought to be representable to high accuracy by a square integrable expansion over products of back-rotated one-electron orbitals, such as STOs or Sturmians.

Let us use the example of the two-electron ^1S ground state in the case of one-photon absorption. The initially discrete state, Ψ_0, is mixed via the interaction with the field with the two-electron continuum, which can be expressed as an expansion over one electron $\mathcal{L}^2$ complex functions.

$$X_{as}^{++} = \sum_{s} \sum_{ij}^{M} a_{ij}^{(s)}(\theta)[\tilde{\chi}_i(\rho_1^*)\tilde{\chi}_j(\rho_2^*)]^{(s)} \qquad (17)$$

The $\mathcal{L}^2$ complex functions $\tilde{\chi}_i$ are complex Slater orbitals containing nonlinear parameters (α_1, α_2...). The index S runs over the different TEICs. Hence, the X_{as}^{++} contains terms reflecting the correlation of electronic structure dependent pairs of electrons, with a mixture of amplitudes representing equal or unequal sharing of the available total energy. Its optimization is achieved is by repeatedly diagonalizing the nonHermitian many-electron, one-photon matrix as a function of the parameters (θ, α_1,α_2...), until a stability region is found for the complex eigenvalue, z_0, for which the solution, Ψ, overlaps maximally with Ψ_0.

For H^- and He, good convergence is reached for M between 6 and 8. The (sp) two-electron continuum component dominates, with (pd) contributing more in H^- than in He. The contribution of the next combination, (df), is very small.

The quality of the expansion (17) depends on the type of $\tilde{\chi}_i$ and on their number M. As with every calculation, the latter is restricted by constraints of numerical accuracy and of computational economy. Constraints particular to the calculation of very small cross-sections close to E=0 are related to inaccuracies at double precision due to orthogonalization of basis sets and to inability of diagonalizers to obtain reliably extremely small complex eigenvalues.

References

1. Nicolaides, C.A., (1992), Theory and computation of nonstationary states of polyelectronic atoms and molecules, in D. Mukherjee (ed) "*Applied many-body methods in spectroscopy and electronic structure*", Plenum Pr. N.Y. , p. 233.
2. Nicolaides, C.A., Mercouris, Th., Komninos, Y. and Petsalakis, I.D., (1994), Many-electron, many-photon theory of nonstationary states, *Int. J.Qu. Chem.* **51**, 529.
3. Nicolaides, C. A., (1972), Theoretical approach to the calculation of energies and widths of resonant (autoionizing) states in many-electron atoms, *Phys. Rev.*

A6, 2078.

4. Nicolaides,C.A. and Beck, D.R., (1978), The variational calculation of energies and widths of resonances, *Phys. Lett.* **A65**, 11.
5. Nicolaides, C.A. and Beck, D.R., (1978), Time dependence, complex scaling and the calculation of resonances in many-electron systems, *Int. J.Qu.Chem.* **14**, 457.
6. Nicolaides, C.A. , Komninos Y. and Mercouris, Th., (1981), Theory and calculation of resonances using complex coordinates, *Int.J.Qu.Chem.* **S15**, 355 .
7. Nicolaides, C.A. , Mercouris Th. and Komninos,Y.K., (1984), Many-electron theory of autoionizing states using complex coordinates: The position and the partial and total widths of the Ne^+ 1s hole state, *Int.J.Qu.Chem.* **26**, 1017 .
8. Nicolaides, C.A. and Mercouris, Th., Partial widths and interchannel coupling in autoionizing states in terms of complex eigenvalues and complex coordinates, (1985), *Phys.Rev.* **A32**, 3247.
9. Mercouris , Th. and Nicolaides, C.A. ,(1990), Laser ionization of H^- in the presence of a DC field, *J.Phys.* **B23**, 2037.
10. Nicolaides, C.A. , Mercouris, Th. and Aspromallis, G. , (1990), Many-electron, many-photon theory of non-linear polarizabilities, *J.Opt. Soc.Am.* **B7**, 494.
11. Nicolaides, C.A., Gotsis, H.J., Chrysos, M. and Komninos, Y., (1990), Resonances and exterior complex scaling, *Chem.Phys.Lett.* **168**, 570.
12. Petsalakis,I.D., Mercouris, Th., Theodorakopoulos, G. and Nicolaides, C.A., (1991), Theory and ab-initio calculations of partial widths and interchannel coupling in predissociating diatomic states. Application to HeF,*Chem.Phys.Lett.* **182**, 561.
13. Nicolaides, C.A. and Themelis, S. I., (1992), Theory of the resonances of the LoSurdo-Stark effect, *Phys.Rev.* **A45**, 349.
14. Nicolaides, C.A. and Gotsis, H.J., (1992), On the calculation of complex energies of resonances, *J.Phys.* **B25**, L171.
15. Chrysos, M., Aspromallis, G., Komninos, Y. and Nicolaides,C.A., (1992), Partial widths of the He^- ^2S two-electron ionization ladder resonances, *Phys. Rev.* **A46**, 5789.
16. Bylicki, M. and Nicolaides, C.A.,(1993), Computation of resonances by two methods involving the use of complex coordinates, *Phys. Rev.* **A48**, 3589.
17. Bylicki, M., Themelis, S.I. and Nicolaides, C.A., (1994)., State specific theory and computation of a polyelectronic atomic state in a magnetic field. Application to doubly excited states of H^-, *J.Phys.* **B27**, 2741.
18. Themelis, S.I. and Nicolaides, C.A., (1995), Field induced tunneling rates and hyperpolarizabilities for low lying excited states of Li and Na, *Phys.Rev.* **A51**, 2801.
19. Themelis, S.I. and Nicolaides, C.A., (1995), Energies, widths and l-dependence

of the H^- ^3P and He^- ^4P TEIL states, *J.Phys.* **B28**, L379.

20. Junker, B.R. , (1982), Recent computational developments in the use of complex scaling in resonance phenomena, *Adv. At. Mol. Phys.* **18**, 207.
21. Chu, S. I. and Cooper, J. , (1985), Threshold shift and above-threshold multiphoton ionization of atomic hydrogen in intense laser fields, *Phys.Rev.* **A32**, 2769.
22. Atabek O. and Lefebvre, R. (1980), Multichannel Siegert quantization with complex rotated coordinates: Application to molecular resonances, *Phys. Rev.* **A22** 1817; Lefebvre, R., (1990), Diatomic predissociative widths and shifts from multichannel Siegert quantization and asymptotic analysis of resonance wavefunctions, *J.Chem. Phys.* **92** 2869.
23. Nicolaides, C.A., Bacalis, N.C. and Komninos,Y., (1992), Theory for the direct construction of diabatic states and application to the He_2^+ $^2\Sigma_g^+$ spectrum, *Chem.Phys.Lett.* **192**, 486.
24. Themelis, S.I. and Nicolaides, C.A., (1994), Dc-field tunneling of polyelectronic atoms and of negative ions: computations based on models and on ab initio theory, *Phys. Rev.* **A49**, 3089.
25. Lin, C.D. , (1986), Doubly excited states including new classification schemes, *Adv. At.Mol.Phys.* **22**, 77.
26. Guschina, N.A. and Nikulin,V.K. , (1975), Decay of atomic systems in a strong electric field. Negative ion and hydrogen atom, *Chem.Phys.* **10**, 23.
27. Bryant, H.C. quoted in [25], p. 127.
28. Mercouris, Th. and Nicolaides,C. A. ,(1993), Quantitative study of multiphoton multiple ionization: Second-harmonic Nd: YAG laser ionization of the doubly excited $2p^2$ ^3P bound state of H^-, *Phys. Rev.* **A48**, 628.
29. Nicolaides, C. A. , Haritos, C. and Mercouris, Th. , (1995) unpublished.

LARGE ORDERS OF $1/n$–EXPANSION IN QUANTUM MECHANICS AND ATOMIC PHYSICS

V.S.POPOV
Institute of Theoretical and Experimental Physics
B.Cheremushkinskaya, 117259, Moscow, Russia

The $1/n$–expansion, or dimensional scaling, is a new method in quantum mechanics and atomic physics, which has already found numerous applications. Interest has recently been attracted to the asymptotic form of higher orders of the $1/n$-expansion. Here we consider this question for the problem of two Coulomb centers and for the hydrogen atom in electric and magnetic fields. The energy eigenvalues, which are complex in the case of quasistationary states, can be expanded in powers of the "small parameter" $1/n$,

$$E_{nl} = E_r - i\Gamma/2 = E^{(0)} + \frac{E^{(1)}}{n} + ... + \frac{E^{(k)}}{n^k} + ..., \qquad (*)$$

where n is the principal quantum number and k is the order of $1/n$–expansion. The asymptotic expression for $E^{(k)}$ is of the Dyson form, see eq.(4) below. So, the $1/n$–expansion is divergent and can be summed to the energy eigenvalues E_{nl} by special summation methods. The asymptotical parameter a in eq.(4) is equal to

$$a = (\ 2\ ImS\ )^{-1}, \qquad S = \int_{\vec{r}_0}^{\vec{r}_2} \vec{p}\, d\vec{r} = \int_0^\infty \dot{\vec{r}}^{\,2} dt,$$

where the action S is calculated along a tunnelling trajectory which connects the minimum $\vec{r}_0$ of the effective potential $U(\vec{r})$ with the turning point $\vec{r}_2$, which lies on the constant–energy surface, i.e. $U(\vec{r}_2) = U(\vec{r}_0)$. A tunnelling trajectory satisfies the classical equation of motion but with an imaginary time, $t = i\tau$. In the case of the problem of two Coulomb centers the variables are separated in the elliptical coordinates $\xi, \eta = (r_1 \pm r_2)/R$, where R is the internuclear distance.

C. A. Tsipis et al. (eds.), New Methods in Quantum Theory, 149–182.

The calculation of the asymptotical parameter a can be performed analytically. For the hydrogen molecular ion the potential $V(\eta)$ is a two–well potential, if $R > R_* = 3^{3/2}/4$. The divergence of the series (*) is stem with the possibility of tunnelling of an electron from one minimum to the other ($\eta = \pm\eta_0$). Finally, we get the dependence $a = a(R)$ in parametric form:

$$a^{-1} = 4\left(\frac{\eta_0}{1-\eta_0^2} - \frac{1}{2}\,ln\,\frac{1+\eta_0}{1-\eta_0}\right), \quad \eta_0 = \sqrt{\frac{3\tau-1}{3\tau-\tau^2}}, \qquad (**)$$

$$R = 8\tau^{3/2}/(1-\tau)(1+\tau)^2, \qquad 1/3 < \tau < 1$$

If $0 < R < R_*$, the turning point move into the complex plane, hence $a(R) < 0$. The formulae similar to (**) were obtained, which are in agreement with the numerical results of López–Cabrera, Goodson et al., who calculated large orders of $1/D$–expansion and determined the singularity of the Borel transform closest to zero, $\delta_0 = 1/2a$. The series (*) for the H_2^+ ion is alternating in sign, if $R < R_*$, and of the sign–constant type for $R > R_*$.

For the problem of the hydrogen atom in electric and magnetic fields, variables in the Schrödinger equation are not separated. So, we have integrated equations of motion numerically. Some problems connected with the exponential instability of subbarrier trajectories are discussed. Two different calculation methods, namely, the "imaginary time" method and the solution of the Hamilton-Jacobi equation, were used, which determine the asymptotical parameter a as function of the reduced external fields, $F = n^4\mathcal{E}$ and $B = n^3\mathcal{H}$. The parameter $a(F, B) \to \infty$ as $F \to F_*(B)$. The nature of this singularity is also established.

If one of the external fields is zero (the Stark and Zeeman effects in a hydrogen atom), the parameter a can be found analytically.

1. Introduction

$1/n$–expansion, or dimensional scaling, is widely used in quantum mechanics and atomic physics (see, e.g., refs.[1–32]). The advantages of this method consist in its close relationship to classical mechanics, the possibility to use it in multidimensional quantum–mechanical problems without the separation of variables and in many–body problem. It should also be noted that for potentials commonly used in physics, already the first two or three terms of $1/n$–expansion determine energy eigenvalues E_{nl} with a fairly high accuracy even for small quantum numbers, $n \sim 1$.

We discuss below the version of the method [11–13] which can be used for both a discrete spectrum and quasistationary states (resonances). Some examples (the Yukawa, Hulthen and funnel potentials [11,14], a hydrogen atom in strong electric [12,15] and magnetic [7,17] fields, the hydrogen molecular ion [16,18] etc.) can be found in refs. [8,11–17]. A similar approach has been recently developed by Kais and Herschbach [24].

The energy eigenvalues (which are complex for quasistationary states, $E = E_r - i\ \Gamma/2$) are expanded in powers of the "small parameter" $1/n$,

$$\epsilon = \epsilon' - i\epsilon'' = \epsilon^{(0)} + \frac{\epsilon^{(1)}}{n} + \ldots + \frac{\epsilon^{(k)}}{n^k} + \ldots, \tag{1}$$

where n is the principal quantum number of the state, $\epsilon = 2n^2 E_{nl}$ is its "reduced" energy, $\epsilon'' = n^2 \Gamma_{nl}$, Γ_{nl} is the width of the nl–level and k is the order of $1/n$–expansion. The behaviour of the coefficients $\epsilon^{(k)}$ at $k \to \infty$ not only presents some theoretical interest[1)], but is of considerable importance in calculating the energy eigenvalues E_{nl} with high accuracy using the $1/n$–expansion (1).

Recently the behaviour of large orders of $1/n$–expansion has been studied by numerical [25,27] and analytical [26,28–30] methods. Here we discuss the results obtained, with the emphasis on the analytical methods and multidimensional problems of quantum mechanics. Below we confine ourselves to calculations of the energy eigenvalues E_{nl}. The calculation of wave functions with the help of $1/n$–expansion is considered in refs.[19,22,23].

2. The basic equations of the $1/n$–expansion method

For simplicity, let us begin with a one–dimensional case, e.g. a potential with spherical symmetry and a barrier (see Fig.1 in [28]). As is well known, the formal parameter $\hbar$ of the quasiclassical expa nsion turns out to be $1/n$ in final formulae. The widths of the highly excited ($n >> 1$) levels, within the exponential accuracy, are

$$\Gamma_n \approx const \cdot n^\sigma \; exp \left\{ -\frac{2}{\hbar} \int_{r_1}^{r_2} \sqrt{2[\, U(r) - E \,]}\, dr \right\} = e^{-nW} \cdot n^\sigma, \qquad (2)$$

where $W = 2\; ImS$ and the appropriatly "scaled" variables are intriduced.

For example, in the case of the Stark effect in a hydrogen atom

$$U(r) = \frac{Z}{r} - \mathcal{E}\, r, \qquad \hbar = m_e = e = 1$$

Proceeding to the scaled variables ϵ, ρ and F,

$$E = -\frac{Z^2 \epsilon}{2n^2}, \qquad r = Z^{-1} n^2 \rho, \qquad F = Z^{-3} n^4\, \mathcal{E}\,, \qquad (3)$$

one has ($l = 0$)

$$W = 2\; ImS = \int_{\rho_1}^{\rho_2} \sqrt{\epsilon - \frac{2}{\rho} - 2F\rho}\;\; d\rho, \qquad (3')$$

$$\rho_{1,2} = \frac{1}{4F}\,[\, \epsilon \mp \sqrt{\epsilon^2 - 16F}\,], \qquad (3'')$$

where $\rho_{1,2}$ are the turning points and F is the reduced electric field.[2)] Note that the quantities ϵ, F and W are ~ 1, as $n \to \infty$. A similar scaling can be done in other problems of quantum mechanics, when the states with $n >> 1$ are considered (see, e.g., refs.[11,12–16]).

Assuming the analyticity in the variable $\lambda = 1/n$ and using the dispersion relation,

$$\epsilon^{(k)} = 2n^2 E^{(k)} = \frac{1}{\pi} \int_0^\infty \Gamma_n \, \frac{d\lambda}{\lambda^{k+3}},$$

and eq.(2), one obtains the asymptotics

$$\epsilon^{(k)} \approx k!\, a^k\, k^\beta\, (c_0 + \frac{c_1}{k} + ...), \qquad k \to \infty, \tag{4}$$

where $\beta = \sigma + 1$,

$$a = (\ 2\ ImS\ )^{-1} = 2^{-3/2} \left(\int_{r_0}^{r_2} [\ U(r) - \epsilon^{(0)}\ ]^{1/2}\, dr \right), \tag{5}$$

$U(r) = n^2 V(n^2 r) + (2r^2)^{-1}$ is the effective potential including the centrigugal energy, $V(r)$ is the initial potential in the Schrödinger equation and $r_0 < r < r_2$ is the subbarrier region.[3)] Here we consider states with no radial nodes, $n_r = 0$, therefore the reduced energy $\epsilon^{(0)} = U(r_0)$ corresponds to the equilibrium point (minimum) of the effective potential $U(r)$ and the two turning points coincide, $r_0 = r_1$.

The next terms $\epsilon^{(1)}$, $\epsilon^{(2)}, ...$ of the $1/n$–expansion are calculated with the help of recurrence relations, convenient for a computer.

For the screened Coulomb potential, $V(r) = -r^{-1} f(\mu r)$, we have

$$U(r) = -\frac{1}{r}\, f(\nu r) + \frac{1}{2r^2}, \qquad \nu = n^2 \mu \tag{6}$$

(μ^{-1} is the radius of screening), and $1/n$–expansion is elaborated around the classical equilibrium point $x_0(\nu)$ in the potential $U(r)$. The quantities $x_0(\nu)$ and $\epsilon^{(0)}(\nu)$ are detemined by the equations [11]

$$\nu = xf - x^2 f', \qquad \epsilon^{(0)} = (xf')^2 - f^2\ |_{x=x_0(\nu)} \tag{7}$$

For example, the screening function $f(x) = exp(-x)$ corresponds to the Yukawa potential, frequently used in atomic and nuclear physics. In this case

$$\nu = x_0\ (1 + x_0)\ e^{-x_0}, \qquad \epsilon^{(0)} = (x_0^2 - 1)\ e^{-2x_0} \tag{8}$$

For sufficiently small values of ν the equilibtium point is real, $x_0(\nu) = \nu + \frac{1}{2}\nu^3 - \frac{1}{3}\nu^4 + ...$, as well as all the coefficients $\epsilon^{(k)}$ in (1). With the increasing ν the value $\nu = \nu_*$ is achieved, when the collision of two solutions (classical orbits) occurs, corresponding to the stable ($x_0(\nu)$) and unstable equilibrium points in $U(r)$. In the case of the Yukawa potential

$$x_* = \frac{1+\sqrt{5}}{2}, \quad \nu_* = x_*^3\, e^{-x_*} = 0.839962..., \quad \epsilon_* = x_*\, e^{-2x_*} = 0.0636... \quad (8')$$

It can be shown that [26,28]

$$a(\nu) \simeq A\,(1-\nu/\nu_*)^{-5/4}, \qquad \nu \to \nu_*, \tag{9}$$

where the constant A, as well as the value of ν_*, depends on the screening function, e.g. $A = 0.1116...$ in the case of the Yukawa potential. Thus, the coefficients $\epsilon^{(k)}(\nu)$ sharply grow[4)] at $\nu \to \nu_*$, and $1/n$–expansion is no longer applicable in the region $\nu \approx \nu_*$ (it was observed already in the first attempts to sum the series (1) for $\nu \sim \nu_*$ [11,12]). However, at $\nu > \nu_*$ the applicability of $1/n$–expansion is restored. The equilibrium point $x_0(\nu)$, as well as the coefficients $\epsilon^{(k)}(\nu)$ and the asymptotical parameter $a(\nu)$, become complex at $\nu > \nu_*$ and the summation of the first few terms in expansion (1) determine the width Γ_n of a quasistationary state with a reasonably good accuracy.

For an arbitrary screening function $f(x)$ the value ν_*, at which the collision of two classical solutions occurs, is determined from the equations

$$f - xf' - x^2 f'' = 0, \qquad \nu = x(\, f - xf'\,)\,|_{x=x_*} \tag{10}$$

and the coefficient A in (9) is equal to

$$A = 0.1997\, \left[\, 1 + \frac{x_* f'''(x_*)}{3f''(x_*)}\, \right]^{3/4} \tag{10'}$$

For example, if

$$f(x) = const\, \cdot x^\sigma\, exp(-x^\alpha), \qquad x = \mu r, \qquad \sigma > -1, \tag{11}$$

then

$$x_* = \left\{\, \frac{1}{2\alpha}\, [\, \alpha + 2\sigma + \sqrt{\alpha^2 + 4(\sigma+1)}\,] \right\}^{1/\alpha}, \tag{11'}$$

whence $\nu_* = 1.3442$, $A = 0.1997$ for $\alpha = \sigma = 1$ (the exponential potential), $\nu_* = 1.0827$, $A = 0.4167$ for $\alpha = 2$, $\sigma = 1$ (Gaussian), etc.[28,29].

In some cases the coefficients $\epsilon^{(k)}(\nu)$ in (1) oscillate for large values of k, and their asymptotics is of the form

$$\epsilon^{(k)} \propto \Gamma(k+\beta+1)\, Re\,(c_0\, a_c^k\,) \approx k!\, |\, a_c\,|^k\, k^\beta\, cos(k\varphi+\varphi_0), \qquad k \to \infty \tag{4'}$$

where $\varphi = arg\ a_c$ and a_c, c_0 are complex numbers. For instance, it takes place for the Yukawa potential [31], if $\nu = \nu^2\mu < \nu_{cr} = 0.73576...$, where ν_{cr} corresponds to $\epsilon^{(0)}(\nu) = 0$.

The confining potentials, e.g.

$$V(r) = -r^{-1} + g\, r^N, \qquad N > 0, \tag{12}$$

have only discrete spectrum at $g > 0$. In this case the collision of the classical solutions occurs at $g < 0$, so the equilibrium point $x_0(\nu)$ and the coefficients $\epsilon^{(k)}(\nu)$ remain real at all ν, $0 < \nu < \infty$ (here $\nu = n^2 g^{\frac{1}{N+1}}$ and the screening function in eq.(6) is $f(x) = 1 - x^{N+1}$). In particular, the case $N = 1$ corresponds to the funnel potential, frequently used in quantum chromodynamics.

The above behaviour of the parameter $a(\nu)$ is illustrated in Fig.1 for the Yukawa potential. The other examples, including Hulthen and funnel potentials, can be found in refs.[11,13,25].

3. $1/n$–expansion and the hydrogen molecular ion

The nonrelativistic problem of two Coulomb centers,

$$V(\vec{r}) = -(Z_1/r_1 + Z_2/r_2),$$
$$r_{1,2} = [\, \rho^2 + (z \pm R/2)^2\,]^{1/2}, \qquad \rho = (x^2 + y^2)^{1/2}, \tag{13}$$

is frequently encountered in different branches of physics, including the theory of molecules, μ–catalized nuclear fusion reactions, etc. Proceeding to the elliptic coordinaters $\xi = (r_1 + r_2)/R$, $\eta = (r_1 - r_2)/R$ and fulfilling the scaling transformation

$$E = \lambda\, \epsilon, \qquad R' = \lambda\, R, \qquad \vec{r}\,' = \lambda\, \vec{r}, \qquad \lambda = \frac{1}{m^2 - 1} \approx n^{-2}, \tag{14}$$

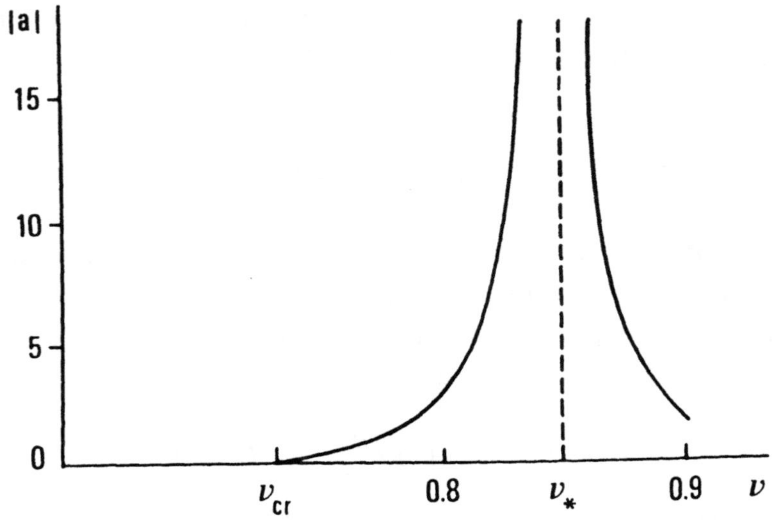

Figure 1. The parameter $| a(\nu) |$ of the asymptotics (4), (4') for the Yukawa potential ($\nu_* \approx 0.840$ corresponds to the collision of two classical solutions).

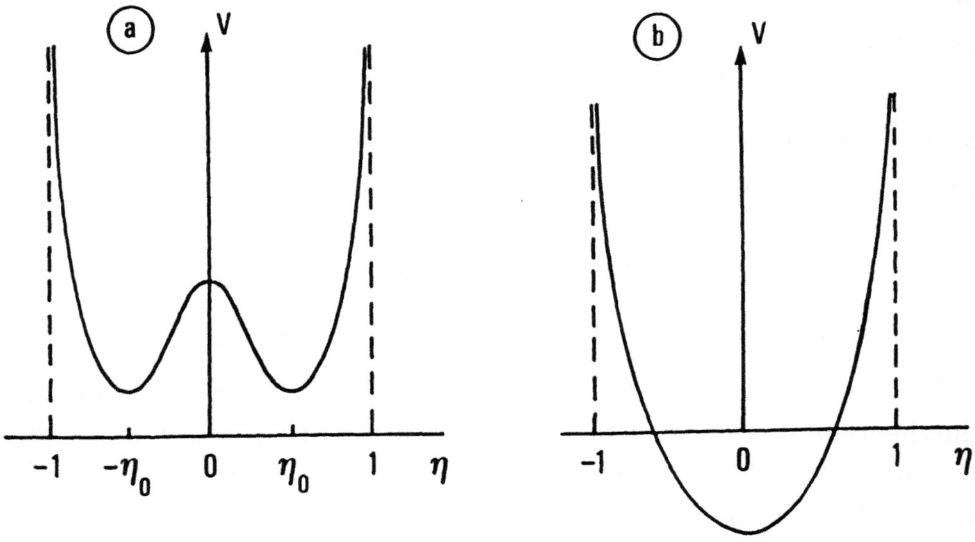

Figurre 2. The effective potential $V(\eta)$ in the problem of two centres with $Z_1 = Z_2$: (a) at $R' > R_*$, (b) at $R' < R_*$.

we arrive at the Shrödinger equations with the effective energy e and effective potentials $U(\xi)$ and $V(\eta)$,

$$e = \frac{1}{4}\,\epsilon\, R'^2 = \frac{E\,R^2}{4\,n^2}, \qquad U(\nu) = \frac{1}{2(\xi^2-1)} - \frac{\beta R'}{\xi^2-1},$$

$$V(\eta) = \frac{1}{2(1-\eta^2)^2} + \frac{\beta\,R'}{1-\eta^2} \tag{15}$$

(here $Z_1 = Z_2 = 1$, which corresponds to the hydrogen molecular ion H_2^+, $m = 0,\ \pm1,\ \pm2\ ...$ is the magnetic quantum number and R is the internuclear distance).

The potential $V(\eta)$ is shown in Fig.2. If $R' > R_* = 3^{3/2}\cdot 2^{-2}$, then

$$a^{-1} = 2\int_{-\eta_0}^{\eta_0} |\,p_\eta\,|\,d\,\eta = 4\left(\frac{\eta_0}{1-\eta_0^2} - \mathrm{Arth}\,\eta_0\right) > 0, \tag{16}$$

$$R' = \frac{8\tau^{3/2}}{(1-\tau)(1+\tau)^2}, \qquad \eta_0 = \left[\frac{3\tau-1}{\tau(3-\tau)}\right]^{1/2}, \tag{16'}$$

which coincides with eq.(28) in [26], see also Appendix. The value $R' = R_*$, or $\tau = 1/3$, corresponds to the collision of two classical electron orbits in the field of two Coulomb centres and is determined from the condition that the frequency of small oscillations near the equilibrium orbit turns into zero (compare with ref. [11]).

If $R' < R_*$, then $\eta_0 = 0$ and the turning points are complex, $\eta_1^2 = 1 + (2e)^{-1} < 0$. Therefore,

$$a^{-1} = 2\int_0^{\eta_1} [-p_\eta^2\,]^{1/2}\,d\eta = -2\,(\mathrm{Arth}\,w - w) < 0 \tag{17}$$

$(0 < w < 1)$. The dependence of the variable w on the reduced internuclear distance $R' = n^{-2}\,R$ is determined parametrically,

$$w = (2e+1)^{1/2} = (1-3\tau)^{1/2}\,(1-\tau)^{-1}, \tag{17'}$$

$$R' = \tau^{1/2}\,(1-\tau)^{-2}, \qquad 0 < \tau < 1/3$$

($\tau = 1/3$ corresponds to $w = 0$ and $R' = R_*$). So, the asymptotical parameter $a(R)$ in eq.(2) is obtained analytically for any R (in the parametric form).

The asymptotical parameter a has singularity as $R' \to R_*$. For instance, at $R' < R_*$ we have from eqs.(17), (17') that

$$R' = R_*(\,1 - \frac{3}{2}\,t + ...), \qquad w = \frac{3}{2}\,t^{1/2} + ...,$$
$$a = -\frac{3}{2}\,w^{-3} + ... = -\frac{4}{9}\,t^{-3/2} + ...$$

with $t = 1 - 3\tau \to 0$. It follows that at $R' \to R_*$

$$a(R) = A_\pm \mid h \mid^{-3/2} \;[\,1 + 0(h)\,], \qquad h = \frac{R' - R_*}{R_*}, \tag{18}$$

where $A_+ = 3^{-1/2}$ for $h > 0$ and $A_- = -(2/3)^{1/2}$ for $h < 0$. Calculation of the parameter a by the above equations gives the curve 1 in Fig.3. Some details of the calculations are discussed in Appendix.

If $Z_1 \neq Z_2$, the potential $V(\eta)$ is not symmetrical any more. The calculations are performed in a similar way as in the case $Z_1 = Z_2$, but the formulae become much more complicated. Just as for the discrete states in the Yukawa potential [11], the turning point r_2 moves into the complex plane, the parameter a becames complex and eq.(4') holds. Here $\mid a(R) \mid < \infty$ for any R (Fig.3), since the collision of classical orbits does not occur at $Z_1 \neq Z_2$, as opposed to the case of two equal charges. Note that the parameter $a(R) \to 0$ at $R \to 0$ and $R \to \infty$. In these limiting cases the problem of two Coulomb centres is reduced to the hydrogen–like atom.

Let us compare these results with that of ref.[27], where large–order dimensional perturbation theory for H_2^+ ion was developed and the singularity of the Borel transform closest to the origin, $\delta_0 = (2a)^{-1}$, was calculated numerically with the help of the quadratic Padé approximants. The dimensional scaling factor in [27] was choosen as f,

$$f = \frac{1}{6}\,D(D-1) \approx \frac{2}{3}n^2, \qquad \tilde{R} = \frac{R}{f} \approx \frac{3}{2}R' \tag{19}$$

and the expansion parameter $\delta = 1/D \approx 1/2n$ ($n = (D-1)/2 \to \infty$, D is space dimension). If the Borel transform

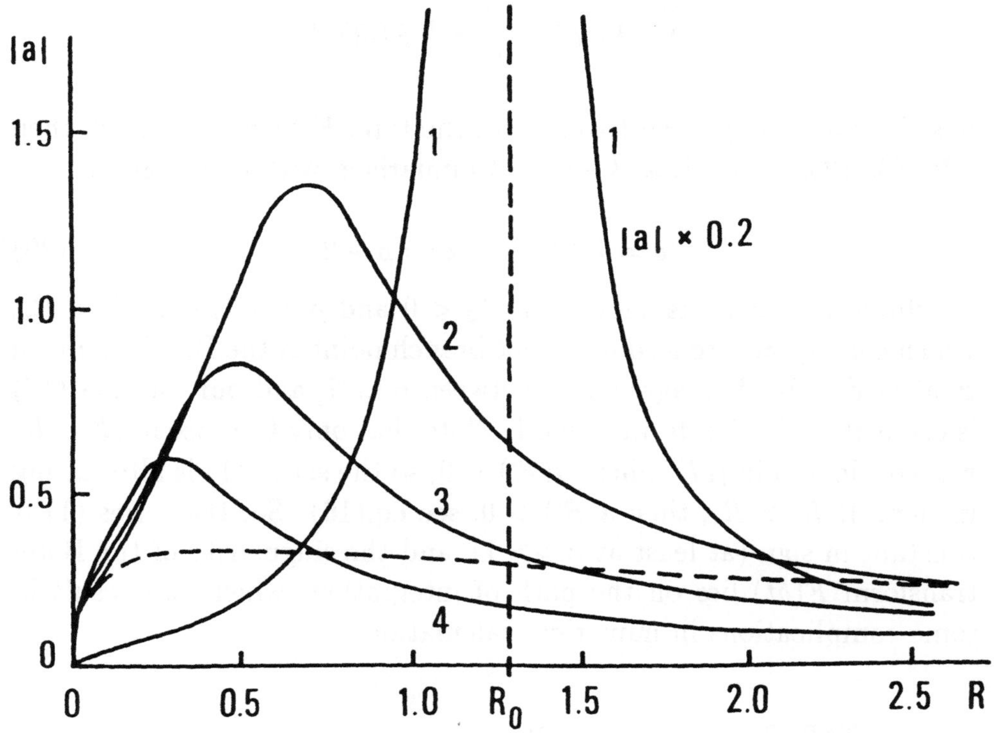

Figure 3. The asymptotical parameter a for the problem of two Coulomb centres. Curves 1–4 correspond to $Z_1/Z_2 = 1, 1.5, 2$ and 3. Note that for the curve 1 the ordinate scale has been changed, i.e., the values of $0.2 \mid a(R) \mid$ are plotted.

$$\sum_{k=0}^{\infty} E_k\, \delta^k = \int_0^{\infty} e^{-t}\, F(\delta t)\, dt$$

has the singularity closest to zero of the type $F(\delta) \propto (\delta_0 - \delta)^{-\rho}$, then $\epsilon^{(k)} \propto k!\, (2\delta_0)^{-k}\, k^{\rho-1}$, as $k \to \infty$. Comparison with eq.(4) gives

$$a = 1/2\delta_0, \qquad \sigma = \rho - 2 \tag{20}$$

Numerical analysis shows that $\delta_0 < 0$ and $\rho = -1/2$ at $R' < R_*$, which corresponds to a square–root branch point of the Borel tranform $F(\delta)$ at $\delta = \delta_0$. The agreement between ref.[27] and our formula (17) is exellent, as is seen from Table 1. Note that only the case of $R' < R_*$ was considered in [27], where $a(R') < 0$, so the series (1) is alternating in sign. If $R' > R_*$, then $a(R') > 0$, see eq.(16). So, the series (1) is constant in sign (at least at $k >> 1$), and the singularity of the Borel transform $F(\delta t)$ lies on the path of integration, which can result in some complications in numerical calculation.

TABLE 1. Values of the Borel parameter $\delta_0 = 1/2a$ for the hydrogen molecular ion H_2^+.

$\tilde{R}$	δ_0	$\tilde{R}$	δ_0
0.2	-1.720 950	1.4	-0.112 797
0.4	-1.062 376	1.6	-0.052 563
0.6	-0.705 527	1.8	-0.013 569
0.8	-0.474 801	2.0	0.003 681
	-0.474 795 (*)	3.0	0.294 164
1.0	-0.313 841 191	4.0	0.720 999
	-0.313 841 21 (*)	5.0	1.202 939
1.2	-0.197 516 621	10	3.943 259
	-0.197 516 618 (*)		

The definition of the scaled internuclear distance $\tilde{R}$ in ref.[27] differs from ours: $\tilde{R} = 1.5R'$, so $\tilde{R}_* = 1.94855...$ The values of δ_0, taken from [27], are marked with an asterisk.

4. The hydrogen atom in external fields

We consider the case of uniform and parallel fields $\mathcal{E}$ and $\mathcal{H}$ and $(0, 0, n-1)$ states with magnetic quantum number $m = n - 1$, which correspond to circular electron orbits (at $n >> 1$). The $1/n$–expansion is constructed around the classical orbit with radius $r_0 = r_0(F, B)$ determined from the equation [17]

$$r(1 - F^2\, r^4)^2\, (1 + \frac{1}{4}\, B^2\, r^3) = 1 \qquad (21)$$

(we use atomic units, $\hbar = e = m_e = 1$, and reduced variables $F = n^4\, \mathcal{E}$ and $B = n^3 \mathcal{H}$, convenient in the case of Rydberg, $n >> 1$, states).

To determine the asymptotical parameter a the "imaginary time" method was used, previously developed for calculation of tunneling probability through barriers, varying in time[5)]. The classical trajectory $\vec{r}(t)$, which joins the equilibrium point $\vec{r}_0$ of maximum of the effective potential

$$\tilde{U}(\vec{r}) = -U = \frac{1}{r} - \frac{1}{2\rho^2} + Fz - \frac{1}{8}\, B^2 \rho^2, \qquad \rho = (x^2 + y^2)^{1/2}, \quad (22)$$

with the turning point $\vec{r}_2 \in \Sigma$, was found by numerical integration (here Σ is an isoenergetical surface, i.e. $U(\vec{r}_2) = U(\vec{r}_0)$). The condition $\vec{r}_0 = \vec{r}_2 = 0$ selects the subbarier trajectory, which minimazes Im S [54]. Then the asymptotical parameter $a = a(F, B)$ in eq.(4) is equal to

$$a = (2\ \mathrm{Im}\ S)^{-1}, \qquad S = \int_{\vec{r}_0}^{\vec{r}_2} \vec{p}\, d\, \vec{r} = \int_0^{\infty} \dot{\vec{r}}^{\,2}\, d\, t, \qquad (23)$$

where $\dot{\vec{r}} = d\, \vec{r}/d\, t$ and $t = i\tau$ is the "imaginary time". Note that $\tilde{U}$ differs in sign from the potential U entering the Shrödinger equation, which exactly corresponds to description of subbarrier motion of a particle in terms of imaginary time. Since $\ddot{\vec{r}} = \nabla\, U = 0$ at $\vec{r} = \vec{r}_0$, the time of motion τ along the above trajectory is infinite. The lines of the constant potential $\tilde{U}$ (dashed curves) and a few classical trajectories (solid curves) are given in Fig.4. The integral S in eq.(23) was

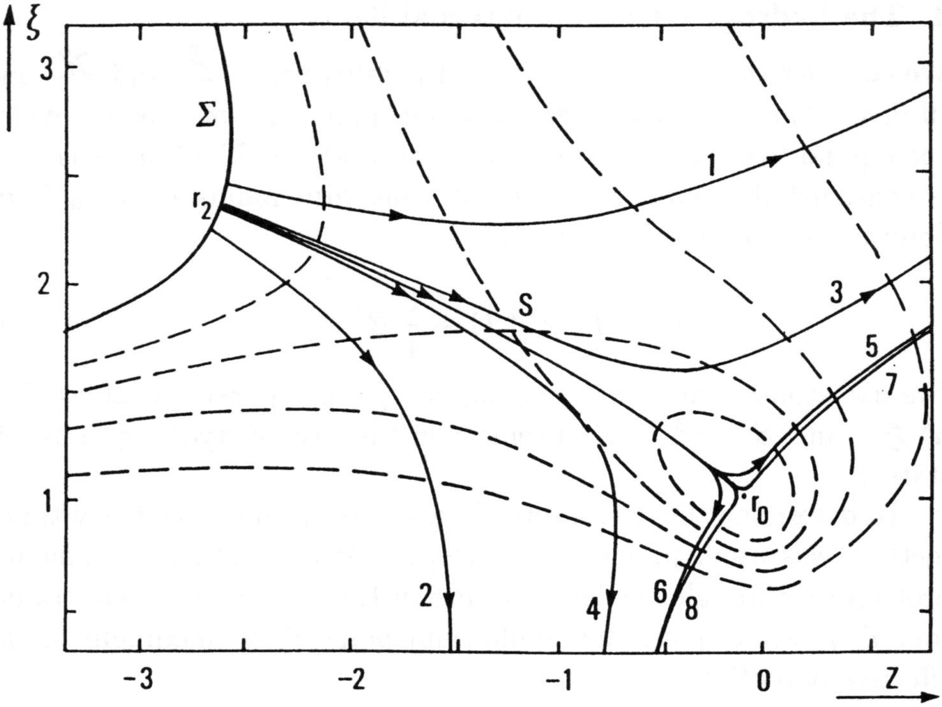

Figure 4. The subbarrier trajectories for the potential (22). Solid curves classical trajectories. with the values of parameters given in Table 2; dashed curves - constant - energy surfaces for the potential $\tilde{U}$; Σ – the isoenergetic surface, s – saddle point. Here $F = 0.15$ and $B = 0.2$ (in atomic units).

calculated from the initial point $\vec{r}_2 \epsilon \Sigma$ to the point $\vec{r}_m$ of the closest approach of the trajectory to the point of the unstable equilibrium $\vec{r}_0$. Since the potential $\tilde{U}(\vec{r})$ is approximated at $\vec{r} \approx \vec{r}_0$ by the potential of "upturned oscillator", arbitrary trajectory is unstable. The trajectories in Fig.4 correspond to initial deviations $\Delta\rho$ (at $t = t_0$), given in Table 2. Although it is impossible to get from a point $\vec{r}_2 \epsilon \Sigma$ precisely to $\vec{r}_0$ in numerical calculations, the value of Im S is readily determined. It can be shown that at small $\Delta\rho$

$$| 1 - \mathrm{Im} S/S_0 | \sim \left(\frac{\Delta\rho}{L}\right)^2, \qquad \tau \sim \frac{1}{\omega_1 + \omega_2} \ln \frac{L}{|\Delta\rho|}, \tag{24}$$

where $L = | \vec{r}_2 - \vec{r}_0 |$, ω_i are the frequencies of normal vibrations near $\vec{r} \approx \vec{r}_0$, and the action S_0 corresponds to the "instanton" trajectory, which connects the point $\vec{r}_0$ with the isoenergetic surface Σ.

The results of the calculations (at $F < F_*$) are shown in the left parts of Figs.5 and 6. Here F_* is the classical ionization threshold, e.g. $F_*(B) = 0.2081, 0.2207, 0.2532$ and 0.3449 at $B = 0, 0.25, 0.5$ and 1.0

TABLE 2. Convergence of the *Im S* values ($F = 0.15, B = 0.2$ and $F_* = 0.2163$)

	Imaginary time method				solution of the HJ equation	
No.	$\Delta\rho$	$r_{\min}$	τ	*Im S*	N	*Im S*
1	0.1	1.396	6.16	0.345	15	0.39924
2	−0.1	1.401	4.92	0.424	18	0.40072
5	10^{-4}	0.1347	11.35	0.40106	28	0.4011278
6	-10^{-4}	0.1335	11.24	0.40126	35	0.4011270
7	10^{-6}	0.0247	13.93	0.4011264	38	0.401126879
8	-10^{-6}	0.0246	13.91	0.4011278		
-	$\sim 10^{-8}$	$\sim 10^{-3}$	~ 20	0.401126879		

Nos. 1–8 in the Table correspond to the numbers of the curves in Fig.4. Here $r_{min} = | \vec{r}_m - \vec{r}_0 |$ is the distance of the closest approach of the classical trajectory $\vec{r}(t)$ to the point of (unstable) equilibrium $\vec{r}_0$, τ is the time of the subbarrier motion from the initial point $\vec{r}_2 \epsilon \Sigma$ to $\vec{r}_m$.

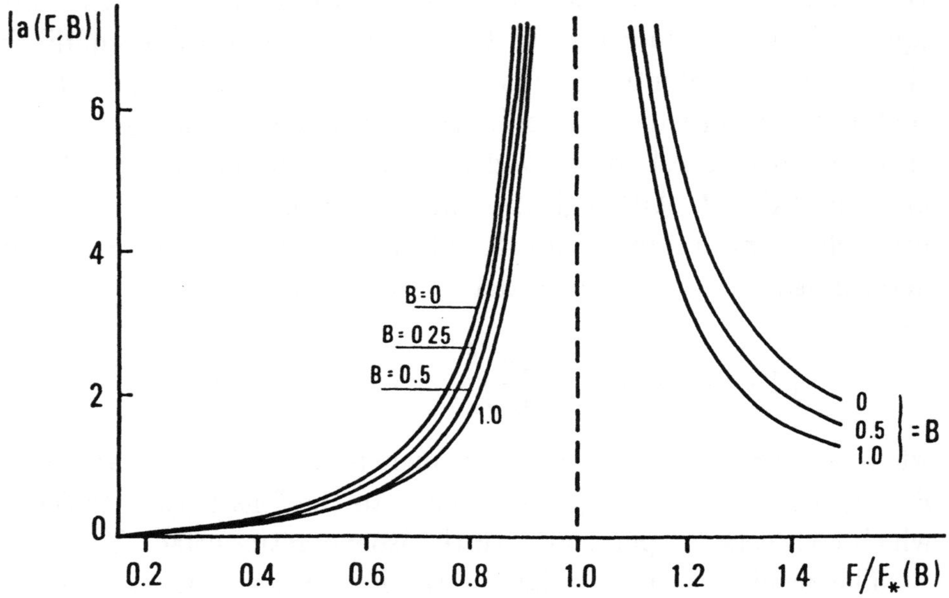

Figure 5. The asymptotical parameter $|\ a(F,B)\ |$ versus the ratio F/F_*, where $F_*(B)$ is the classical ionization threshold [13,17]. The curves are labelled with the values of the reduced magnetic field $B = n^3\mathcal{H}$.

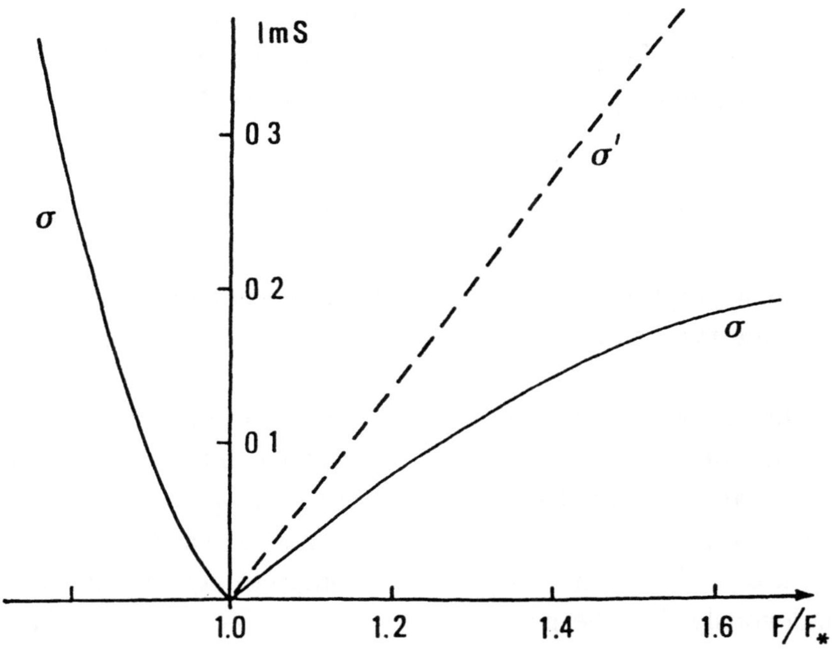

Figure 6. The action function, $Im\ S = \sigma + \sigma'$, calculated along the subbarrier trajectories for $B = 1.0$ and $F = 0.99F_* \approx 0.342$ (in atomic units). At $F > F_*$ the calculations were done using the HJ equation.

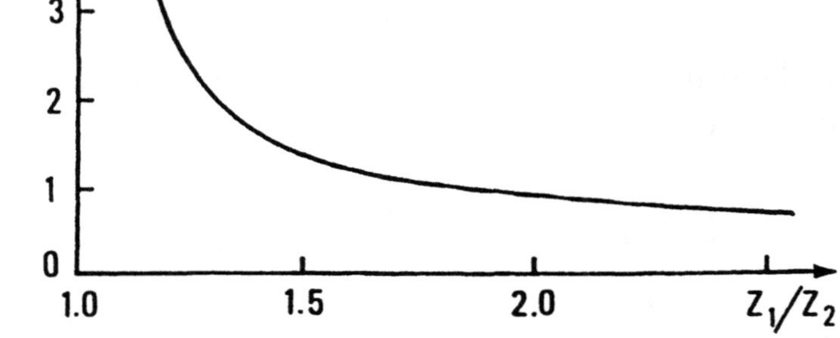

Figure 7. The maximal value a_{max} versus Z_1/Z_2 in the problem of two Coulomb centres Z_1, Z_2.

[12,17]. At $F \to F_*$ the points $\vec{r}_2$ and $\vec{r}_0$ converge, and the potential (22) takes the form $V = V_0 + \alpha q - \frac{1}{3}\beta q^3$, where $\alpha \propto F_* - F \to 0$ and $\beta \sim 1$. Taking eq.(3) into account, we find that

$$a^{-1} = 2 \int_{q_0}^{q_2} (-p^2)^{1/2}\, dq \propto \alpha^{5/4} \propto (F_* - F)^{-5/4} \tag{25}$$

So, the asymptotical parameter a has a power singularity at $F = F_*$ of the same type as in the one–dimensional case, compare eqs.(26) and (9).

On the other hand, in the two centres problem with $Z_1 = Z_2$ the collision of three, but not two, classical orbits occurs at $R = R_*$, which can be modelled by the potential $V(q) = V_0 - \frac{1}{2}\alpha q^2 + \frac{\beta}{4}q^4$ with $\alpha \propto R' - R_* \to 0$ and $\beta \sim 1$. It can be easily seen that in this case $a \propto \alpha^{-3/2} \propto (R' - R_*)^{-3/2}$, in complete agreement with eq.(19). Analogously, for the general case, $V(q) = V_0 - \frac{1}{2}\alpha q^2 + \beta q^N$ with N even, one obtains that [30]

$$a(\nu) \propto (\nu_* - \nu)^{-\frac{N+2}{2(N-2)}}, \qquad \nu \to \nu_* \tag{26}$$

The simplest case, $N = 4$, correspond to the singularity (19).

5. The Hamilton–Jacobi equation

Let us also give a brief description of the alternative calculation method for the action S along the subbarier trajectory, based on solution of the Hamilton–Jacobi, or eikonal equation (HJ)

$$(\nabla S)^2 = 2(U - U_0), \qquad U_0 = U(\vec{r}_0) \tag{27}$$

Choosing the origin at the equilibrium point $\vec{r}_0$ and axes x, y (normal coordinates), we get at $\vec{r} \approx \vec{r}_0$

$$U(x,y) = U_0 + \frac{1}{2}\,\omega_1^2\, x^2 + \frac{1}{2}\,\omega_2^2\, y^2 + \ldots.$$

$$S(x,y) = \int_0^x \dot{x}\, dx + \int_0^y \dot{y}\, dy = \frac{1}{2}(\omega_1\, x^2 + \omega_2\, y^2) + \ldots.$$

Substituting the expansions

$$U = \sum_{j,k=0}^{\infty} u_{jk}\, x^j y^k, \qquad S = \sum_{j,k} s_{jk}\, x^j y^k \tag{28}$$

into the HJ, we arrive at the recurrence relations, from which the coefficients s_{jk} can be calculated subsequently. For example,

$$s_{02} = \frac{1}{2}\,\omega_2, \qquad s_{11} = 0, \qquad s_{20} = \frac{1}{2}\,\omega_1,$$

$$s_{03} = (3\,\omega_2)^{-1}\,u_{03}, \quad s_{12} = (\omega_1 + 2\,\omega_2)^{-1}\,u_{12}, \quad s_{21} = (2\,\omega_1 + \omega_2)^{-1}\,u_{21},$$

$$s_{13} = (\omega_1 + 3\,\omega_2)^{-1}\,[u_{13} - 2\,s_{12}\,(s_{21} + 3\,s_{03})], \ldots$$

It is easy to obtain s_{jk} up to $n = j + k \sim 50$ with a computer, thus allowing one to use summation methods for divergent (or slowly convergent) series. So we proceed to power series in one variable λ,

$$f(\lambda) \equiv S(\lambda\, x,\, \lambda\, y) = \sum_{n=0}^{\infty} f_n\, \lambda^n, \qquad 0 < \lambda < 1,$$

$$f_n = \sum_{j=0}^{n} s_{j,n-j}\, x^j\, y^{n-j} \tag{29}$$

Then the value $f(1)$ is calculated by Padé (PA) or Hermite–Padé (HPA) approximants.

The subbarier trajectory ends at the turning point $\vec{r}_2$, where the line $S = S_0$ contacts the isoenergetic surface Σ, and $\partial_x S = \partial_y S = 0$. The derivatives $\partial_i S$ have a square root singularity at $\vec{r} = \vec{r}_2$, therefore, the calculation of $\vec{r}_2$ with high precision requires some effort[6]. Table 2 illustrates rapid convergence of the S values with growing order of approximation (N is the total number of the coefficients f_n, included into the calculation). It is evident from Table 2, that both methods used are in full agreement, thus providing the accuracy $\sim 10^{-8}$ in the S_0 values.

Note that the series (28) for $S(x, y)$ should be summed at the isoenergetic surface Σ, where the anharmonicity of the potential $U(x, y)$ is not small any more. The precise values of S_0 can be obtained only due to the availability of high–order coefficients s_{jk} with $n = j + k >> 1$ and the application of the effective methods for the summation of series.

6. The Stark and Zeeman effects in hydrogen

If $\mathcal{E} = 0$ or $\mathcal{H} = 0$, the asymptotical parameter a can be calculated analytically. Here we restrict ourselves to a qualitative discussion of these cases, referring to [30] for details.

For the $(0, 0, n-1)$ states which correspond to circular classical orbits, three–dimensional problem is analogous to one–dimensional potentials:

$$V(r) = -r^{-1} - \mathcal{E} r, \qquad -r^{-1} + \frac{1}{8}\mathcal{H}^2 r^2 \tag{30}$$

In the first case the potential has a barrier, and the "critical" field is given by the equations

$$x_* = 3^{-1/2}, \qquad \nu_* = x_* - x_*^3 = 2 \cdot 3^{-3/2}, \qquad F_* = \nu_*^2 = 4/27 \tag{31}$$

(here $\mu = \mathcal{E}^{1/2}$, $\nu = n^2 \mathcal{E}^{1/2}$ and the screening function $f(x) = 1 + x^2$, see sect.2). The value F_* is analogous to the critical electric fields for the Stark effect [12,15]. Note that the values of F_* for the real $3D$–problem exceed the estimate (31) by $1.5 \div 2$ times, but the character of singularity at $F \to F_*$ the same in both cases,

$$a(F) = A\, f^{-5/4}\, [\, 1 + 0(f^{1/2})\,], \qquad f = \frac{F_* - F}{F_*} \to 0, \tag{32}$$

where $A = 0.475$, $F_* = 0.148$ for a one–dimensional model (31) and $A = 0.572$, $F_* = 2^{12} \cdot 3^{-9} = 0.208$ for the $(0, 0, n-1)$ states of a hydrogen atom with $n >> 1$. The analytical formula for a is [26]

$$a(F) = \frac{1}{2} \left[z + \frac{z^3}{3(1 - z^2)} - Ar\, th\, z \right]^{-1}, \tag{33}$$

where $z = z(F)$ is determined parametrically:

$$z = \frac{(1 - 3\tau)^{1/2}}{1 - \tau}, \qquad \tau(1 - \tau^2)^4 = F, \qquad 0 < \tau < 1/3$$

A simple calculation leads from these equations to (32).

For the model (30) of the Zeeman effect we have

$$\nu = \frac{1}{2} n^2 \mathcal{H}^{2/3}, \qquad f(x) = 1 - x^3 \tag{34}$$

Therefore, the equation (10) for x_* has the only real solution, $x_* = -1/2$, which corresponds to

$$\nu_* = x_* + 2x_*^4 = -3/8 < 0, \qquad n^3 \mathcal{H}_* = \pm i\ 3^{3/2}\ /\ 8 \tag{35}$$

So, the collision of the two classical solutions corresponds to the case of purely imaginary magnetic field. The parameter $a = a(B)$ is complex and has no singularities at $0 < B < \infty$. In particular,

$$\mid a(B) \mid = \begin{cases} \pi^{-1} B + 0(B^2), & B \to 0 \\ \frac{1}{2\pi} \left\{ 1 + (2B)^{-1/2} - \frac{1}{16\pi^2} \cdot \frac{ln^2 B}{B} + ... \right\}, & B \to \infty \end{cases} \tag{36}$$

and $\mid a(B) \mid$ has a maximum at $B = B_m \approx 1.95$, where $\mid (B_m) \mid = 0.2133$ [30]. So, the asymptotics of $\epsilon^{(k)}$ in this case is of the type (4').

7. Concluding remarks

Thus, the coefficients of $1/n$–expansion $\epsilon^{(k)}$ grow as $k!$ at $k \to \infty$, so the series (1) is divergent (analogous to the "Dyson phenomenon" [33] for the usual perturbation series). In spite of its divergence, $1/n$–expansion is successfully applied in computations of energies [3–21] and wave functions [19,22,23] in quantum mechanics, atomic and molecular physics. As an illustration of the possibilities of the method, we give in Table 4 the positions and widths of the Stark resonances in strong ($F = \mathcal{E}\ /\ \mathcal{E}_a \sim 1$) electric fields, computed by different methods.[7)] It is seen, that these results are in a good agreement. Naturally, the accuracy of $1/n$–expansion quickly grows with n increasing. This method is the most useful for intense fields and the Rydberg, $n >> 1$, states, which at present are intensively investigated in atomic physics.

As was above demonstrated , in many cases asymptotical parameters (a, β, etc.) of the asymptotics (4) can be found analytically, otherwise the calculation of the subbarier trajectory by imaginary time method, or the HJ equation can be applied for the purpose. The use

TABLE 3. The asymptotical parameter $a(B)$ for the Zeeman effect in a hydrogen (the $(0,0,n-1)$ states).

B	$-Re\ a$	$Im\ a$	$\mid a(B) \mid$
0.01	1.44(-4)	-3.21(-3)	3.214(-3)
0.1	1.06(-2)	-3.19(-2)	3.365(-2)
0.5	0.134	-5.97(-2)	0.1468
0.7	0.175	-2.641(-2)	0.1765
0.8	0.1862	-7.919(-3)	0.1863
0.9	0.1935	9.879(-3)	0.1937
1.0	0.1975	2.637(-2)	0.1993
1.1	0.1992	4.132(-2)	0.2034
1.2	0.1991	5.469(-2)	0.2065
1.5	0.1931	8.64(-2)	0.2116
2.0	0.177	0.119	0.2133
2.5	0.161	0.138	0.2121
5	0.112	0.168	0.2023
10	7.65(-2)	0.175	0.1914
100	2.36(-2)	0.168	0.1698
∞	0	0.159	0.1592

$B = n^3\mathcal{H}$ is the reduced magnetic field; hereafter $a(-b)$ stands for $a \times 10^{-b}$, e.g.$1.06(-2) = 0.0106$, etc.

TABLE 4. Energies and halfwidths of the $(0,0,n-1)$ states in strong electric fields, calculated by different methods.

n	$\mathcal{E}$, *a.u.*	F	$-Re\ E_n$	$\Gamma_n/2$	calculation method
1	0.1	0.1	0.52742	7.27(-3)	HPA
			0.5274182	7.2691(-3)	ref.[61,62]
	0.5	0.5	0.6231	0.279	$1/n$
	1.0	1.0	0.624	0.647	$1/n$
			0.6243	0.6468	ref.[61]
	2.0	2.0	0.540	1.324	$1/n$
			0.539	1.3245	ref.[63]
2	0.05	0.8	0.15868	9.076(-2)	$1/n$
			0.158677	9.076(-2)	ref.[63]
	0.1	1.6	0.1540	0.20188	$1/n$
			0.15401	0.201879	ref.[63]
3	3.086(-3)	0.25	6.2467(-2)	4.047(-3)	$1/n$, HPA
	6.173(-3)	0.50	6.7994(-2)	1.761(-2)	–"–
10	5.0(-5)	0.5	6.0522(-3)	1.1430(-3)	$1/n$, HPA
	1.0(-4)	1.0	6.42589(-3)	3.36937(-3)	$1/n$
			6.426(-3)	3.369(-3)	HPA
20	1.250(-6)	0.2	1.3216(-3)	3.7135(-6)	$1/n$
			1.32157(-3)	3.713(-6)	HPA
	3.125(-6)	0.5	1.50886(-3)	2.6154(-4)	$1/n$, HPA

Here $F = n^4 \mathcal{E}$, $n_1 = n_2 = 0$ and $m = n-1$ are the parabolic quantum numbers of the state and atomic units are used, $\hbar = m_e = e = 1$ ($\mathcal{E}_a = m_e^2\, e^5/\hbar^4 = 5.142\ (9)\ V/cm$).

of the above parameters in the summation of series (1) can considerably enhance the accuracy of the calculations of the energy eigenvalues by $1/n$–expansion. Some interesting results of similar approach can be found in refs.[27,32].

In conclusion, mention should be made of a few points, which can not be discussed here in detail.

1) First, it is the calculation of wave functions. A few tems of $1/N$ expansion determine $\psi(x)$ in a classically allowed region with rather high accuracy [19], but they are not sufficient at all in a subbarrier region. Meanwhile, such quantities as $\psi^2(0)$ are of great interest in all those cases, when there are two interactions with considerably differing radii in the system (e.g., the strong and electromagnetic interaction). Combining $1/n$ expansion with the WKB method in a subbarrier region, formulae for $\psi(0)$ and $\psi(\infty)$ can be obtained [22,23] which are most accurate and convenient for numerical calculations.

2) $1/n$ expansion is in close relationship with the WKB method in quantum mechanics. The higher orders of the Bohr-Sommerfeld quantization rule for the energy eigenvalues were elaborated by many authors [49,50]. In particular, the powerlike potentials (i.e., $V(x) = x^N$, $N = 4, 6, 8, ...$) were discussed in ref.[49] and the anharmonic oscillator, $V(r) = \frac{1}{2}r^2 + \frac{1}{4}\lambda r^4$, in the recent paper [58] (up to the fifth WKB order). Nevertheless, the behaviour of the large orders of the WKB approximation for an arbitrary potential $V(r)$ does not seem to have been investigated so far.

3) At present $1/N$ expansion has been extended to the relativistic (the Klein–Gordon and Dirac) equations, see, e.g., the papers [59,60] and references therein. However, the large orders of $1/n$–expansion in the relativistic case have not been studied yet.

The author is grateful to Professor V.D.Mur, to Drs. A.V.Sergeev, A.V.Scheblykin and V.M.Weinberg, whose interest in the problem and permanent assistance made it possible to obtain the above results, as well as to O.P.Vilister–Tolstaya and M.N.Markina for their help in translating and preparing the manuscript.

Appendix

Here we discuss some details of the calculations for the problem of two Coulomb centres.

Proceeding to the elliptic coordinates ξ, η, φ, $\psi(\vec{r}) = [\,(\xi^2-1)(1-\eta^2)\,]^{1/2}\chi_1(\xi)\chi_2(\eta)\ exp(im\varphi)$, we arrive at the one-dimensional Shrödinger equations, e.g.,

$$\lambda^2 \frac{d^2\chi_1}{d\xi^2} + 2\,[\,e - U(\xi)\,]\,\chi_1 = 0, \tag{A.1}$$

with the effective energy e and effective potentials U, V:

$$e = \frac{1}{4}\epsilon R'^2 = \frac{ER^2}{4n^2}, \tag{A.2}$$

$$U(\xi) = \frac{1}{2(\xi^2-1)^2} - \frac{Z_+\xi+\beta}{\xi^2-1}R', \quad V(\eta) = \frac{1}{2(1-\eta^2)^2} + \frac{Z_-\eta-\beta}{1-\eta^2}R', \tag{A.3}$$

where $Z_\pm = \frac{1}{2}(Z_1 \pm Z_2)$ and β is the separation constant. At first we assume $Z_+ = 1$, $Z_- = 0$, which corresponds to H_2^+ ion, and, for brevity, omit primes at the variables R', r', etc.

The potential $U(\xi)$ has unique minimum at $\xi = \xi_0 > 1$, while $V(\eta)$ has different shapes for $R < R_*$ and $R > R_*$, see Fig.2. Let us consider two different cases.

a) $R > R_* = 3^{3/2}/4 = 1.299038...$

The potential $V(\eta)$ is a double-well potential, which has symmetric and antisymmetric states with slightly different energies, but identical $1/n$-expansions. The Borel sum of the series (1) has imaginary part $\Gamma/2$, which corresponds to the tunnelling of an electron from one vacuum to another, $\eta_0 \leftrightarrow -\eta_0$ [55], which is shown by arrows in Fig.2a (the calculation of higher orders of perturbation theory for this case is considered in ref.[56]).

If $n_\xi = n_\eta = 0$ and $n = |\,m\,| + 1 \to \infty$, the classical orbit is determined by the equations:

$$U(\xi_0) = V(\eta_0) = e, \quad U'(\xi_0) = V'(\eta_0) = 0, \tag{A.4}$$

which have a rather simple solution, expressed in terms of ξ_0:

$$R=\frac{8\,\xi_0^3}{(\xi_0^2-1)(\xi_0^2+1)^2},\qquad \eta_0=\left[\,(1-\frac{1}{3}\xi_0^2)/(1-\frac{1}{3\xi_0^2})\,\right]^{1/2},$$

$$\epsilon^{(0)}=-\frac{1}{32\,\xi_0^2}\left[\,(3-\frac{1}{\xi_0^2})\,(\xi_0^2+1)\,\right]^2, \tag{A.5}$$

$$e=-\frac{1}{2}\left(\frac{3\xi_0^2-1}{\xi_0^4-1}\right)^2,\qquad \beta=\frac{(3\xi_0^2-1)\,(\xi_0^2+1)}{8\,\xi_0^3}$$

Here $1<\xi_0\leq\sqrt{3},\ -\infty<e\leq-1/2$ and $\xi_0\to1$ corresponds to

$$R=\frac{1}{\xi_0-1}+1+...\to\infty,$$

$$\eta_0=1-\frac{4}{R}+...,\qquad \beta=\frac{1}{2}+\frac{1}{2R}+..., \tag{A.6}$$

$$\epsilon^{(0)}=-\frac{1}{2}-\frac{1}{R}+\frac{1}{2R^3}-\frac{1}{2R^4}-\frac{3}{8\,R^5}+\frac{1}{R^6}+\frac{143}{16\,R^7}-...,$$

while $\xi_0=\sqrt{3}$ corresponds to the case, when two classical orbits ($\eta=\pm\eta_0$) coincide and

$$R=R_*,\qquad \eta_0=0,\qquad e=-\frac{1}{2},\qquad \beta=\frac{1}{R_*}=0.7698...\,, \tag{A.7}$$

$\epsilon^{(0)}\equiv\epsilon_*=-32/27$. Since $V(\eta)-V(\eta_0)=-e\,[\,(\eta^2-\eta_0^2)/(1-\eta^2)\,]^2$, we arrive at eqs.(16), (16') with $\tau=\xi_0^{-2}$ and

$$w\equiv\eta_0=\left[\,\frac{3\tau-1}{3\tau-\tau^2}\,\right]^{1/2}=\begin{cases}c_0\,(\tau-1/3)^{1/2}, & \tau\to\frac{1}{3}\\[2ex] 1-\frac{1}{2}(1-\tau)+..., & \tau\to1\end{cases} \tag{A.8}$$

$c_0=(3/2)^{3/2}=1.837$ $(1/3<\tau<1)$. Note that

$$V(\xi)-V(0)=\begin{cases}-(\sqrt{-2e}-1)\,\eta^2+O(\eta^4), & e<-1/2\\[2ex] \frac{1}{2}\,\eta^4+..., & e=-1/2\end{cases} \tag{A.9}$$

So, the effective potential $V(\eta)$ changes its shape at $R = R_*$.

b) $\underline{0 < R < R_*}$. In this case $\sqrt{3} < \xi_0 < \infty$, $-1/2 < e < 0$ and

$$V(\eta) - V(0) = \frac{\eta^2}{(1-\eta^2)^2}\left(\frac{1}{2} + e - e\eta^2\right) =$$

$$(\frac{1}{2} + e)\eta^2 + (1+e)\eta^4 + ..., \qquad \eta \to 0 \tag{A.10}$$

So, the equilibrium point is $\eta_0 \equiv 0$ (see Fig.2b), the stable classical orbit is symmetrical and the turning points are complex: $\eta_1^2 = 1 + (2e)^{-1} < 0$, $\eta_1 = \pm i \mid \eta_1 \mid$. Therefore,

$$a^{-1} = 2^{3/2} \int_0^{\eta_1} [\, V(\eta) - V(0)\,]^{1/2} d\eta =$$

$$= 2^{3/2} \int_0^{\eta_1} \frac{\eta(\eta_1^2 - \eta^2)^{1/2}}{1-\eta^2}\, d\eta = -2\,(Arth\ w - w), \tag{A.11}$$

where

$$w = (2e+1)^{1/2} = \frac{(1-3\tau)^{1/2}}{1-\tau} = \begin{cases} 1 - \frac{1}{2}\tau - \frac{13}{8}\tau^2 + ..., & \tau \to 0 \\ c_0'\,(\frac{1}{3} - \tau)^{1/2}, & \tau \to 1/3 \end{cases} \tag{A.12}$$

$c_0' = 3^{3/2}/2 = 2.598$. Note that in this case $a(R) < 0$, due to the fact that the turning point η_1 is complex and lies at the imaginary axis.

If $\tau \to 1/3$, then $w \to 0$ and $\mid a \mid \to \infty$, see curve 1 in Fig.3. On the other hand, at small, as well at large values of R the asymptotical parameter tends to zero:

$$a = \frac{1}{2\ ln\ R} - \frac{1 - ln\ 2}{2(ln\ R)^2} + 0((ln\ R)^{-3}), \qquad R \to 0$$

$$a = \frac{1}{2R}\left[1 + \frac{ln\ 2R}{R} + 0\left(\left(\frac{ln\ R}{R}\right)^2\right)\right], \qquad R \to \infty \tag{A.13}$$

where R is the scaled internuclear distance (i.e., R' in eq.(14)). In these cases the asymptotics of the coefficients $\epsilon^{(k)}$ is determined by the complex parameter $a_c = 1/2\ S_\xi$ and is of the type (4').

Note that in addition to the turning points in the coordinate η, there are also some turning points in ξ. Since the potential $U(\xi)$ has a unique minimum $\xi = \xi_0$ $(1 < \xi_0 < \infty)$, these points can either be complex or lie in the uphysical region, $\xi < 1$. The calculation gives for $| a_c(R) |$ the dotted curve in Fig.3 (for details we refer to Appendix A in [30]).

c) For the circular Rydberg states $(0,0,m)$ with $| m | >> n_1, n_2 \sim 1$, the stable classical orbit is the circle perpendicular to z–axis (as in the case of the Stark efect [12]). Its elliptic coordinates, as well as other physical quantities, are easily expressed through an anxiliary variable τ,

$$
\begin{gathered}
\xi_0 = \tau^{-1/2}, \qquad 0 < \tau < \infty, \\
\eta_0 = \pm\ \tau^{-1/2}\ [\ (3\tau - 1)/(3 - \tau)\ ]^{1/2}\ \theta(\tau - 1/3), \qquad \text{(A.14)} \\
\rho_0 = \frac{1}{2}\ R\ [\ (\xi_0^2 - 1)(1 - \eta_0^2)\ ]^{1/2}, \qquad z_0 = \frac{1}{2}\ R\ \xi_0\eta_0, \\
\epsilon^{(0)} = -\frac{1}{32\tau}\ (3 + 2\tau - \tau^2)^2, \\
\epsilon^{(1)} = (n_1 + 1/2)\omega_1 + (n_2 + 1/2)\omega_2 - (n_1 + n_2 + 1)\rho_0^{-2}, \ldots \qquad \text{(A.15)}
\end{gathered}
$$

where ω_i are the frequences of normal vibrations around the classical orbit [12,16] ($\omega_2 = 0$ at $R = R_*$). Two possible signs for η_0 in (A.14) correspond to the two stable electron orbits, localized at either of the Coulomb charges, if $R > R_*$ (or $\tau > 1/3$).

These simple formulae make it possible to obtain $a(R)$ in the analytic form (eqs.(16) and (17)).

d) If $Z_1 \neq Z_2$, the tunneling can occur in either η or ξ. The asymptotics of $\epsilon^{(k)}$ has the form (4'), where the dominant term is that for which $| a |$ is maximal:

$$
| a(R) | = \frac{1}{2}\ max(|\ S_\xi\ |^{-1},\ |\ S_\eta\ |^{-1}) \qquad \text{(A.16)}
$$

The integrals for the action functions S_ξ, S_η can be evaluted analytically, but are rather cumbersome. For instance,

$$S_\xi = \int_{\xi_0}^{\xi_2} [\, U(\xi) - e_0 \,]^{1/2} d\xi =$$

$$= \frac{1}{2}\, Arth\; ap + \frac{bc}{q}\, \xi\; Arth(\frac{pq}{d}) - cp \pm i\frac{\pi}{2}, \tag{A.17}$$

and similar expression for S_η. Here

$$p = \sqrt{(\xi^2 - \eta^2)\,[\,\xi^2 - \eta^2 + 3(1 - \xi^2\eta^2)\,]}, \quad q = \sqrt{1 + \xi^2 + \eta^2 - 3\xi^2\eta^2},$$

$$a = \frac{2(\xi^2+1)(1-\xi^2)b}{h^2 - d^2 - (\xi^2-1)(1-\eta^2)p^2}, \qquad b = \xi^2 - 3(\xi^2-1)\eta^2 - \eta^4,$$

$$c = [\,(\xi^2-1)(1-\eta^2)\,]^{-1}, \qquad d = \xi\,[\,(1-\eta^2)^2 + q^2\,],$$

$$h = (\xi^2 - \eta^2)(1-\eta^2) + 2(\xi^2\eta^2 - 1)\eta^2$$

and $\xi \equiv \xi_0$, $\eta \equiv \eta_0$ are the coordinates of the equilibrium point.

Numerical calculations show that $|\, S_\xi \,|<|\, S_\eta \,|$ for sufficiently small values $R < R_c$, so in eq.(A.16) the first term dominates, while $|\, S_\xi \,|>|\, S_\eta \,|$ at $R > R_c$ and the second term dominates. Using these equations, the curves in Fig.3 were computed. As follows from eq.(A.17), the asymptotical parameter a is now complex and $|\, a(R) \,|< \infty$ for $0 < R < \infty$. The dependence of its maximal value a_{max} on the ratio of two Coulomb charges Z_1/Z_2 is presented in Fig.7. The details of the calculations for the case $Z_1 = Z_2$ can be found in refs.[29,30], while those for $Z_1 \neq Z_2$ are not published yet.

e) Finally, let us explain the origin of R_*. At $z \to 0$

$$U(\vec{r}) = -\frac{1}{r_1} - \frac{1}{r_2} + \frac{1}{2\rho^2} = -\frac{2}{r} + \frac{1}{2\rho^2} + (1 - 3\,cos^2\alpha)\,\frac{z^2}{r^3} + O(z^4),$$

where $Z_1 = Z_2 = 1$, $cos\alpha = R'/2r$, $r = (\rho^2 + z^2)^{1/2}$, α is the angle at Z vertex in a triangle (Z, Z, e) and ρ, z are the cylindrical coordinates of the electron. So, the symmetrical ($r_1 = r_2 = r$) orbit is stable, only if $1 - 3cos^2\alpha > 0$, or

$$\alpha > \alpha_* = arc\ cos\ 3^{-1/2} \quad (A.18)$$

On the other hand,

$$sin^4\alpha / cos\alpha = 1/R', \quad (A.19)$$

which follows from the equations of motion, $2\rho r^{-3} = v^2\rho^{-1}$, $\rho v = m \approx n$ ($\hbar = m_e = e = 1$, $n \to \infty$). Eqs.(A.18) and (A.19) immediately give

$$R' = n^2 R < R_* = 3^{3/2}/4 \quad (A.20)$$

Thus, the symmetrical electron orbit ($\eta_0 = 0$) becames unstable at $R' > R_*$.

The generalization of this mechanical approach to the case $Z_1 \neq Z_2$ is straightforward and can be found in refs.[16].

Footnotes

1) It is well known that the divergence of perturbation theory series (PT) in quantum mechanics and QFT is connected with the instability of the vacuum state, when coupling constant changes its sign (the famous "Dyson phenomenon", first established in QED [33] and later investigated for different quantum - mechanical problems, such as the anharmonic oscillator [34–38], Stark [39–43] and Zeeman [44] effects, etc.). Large orders of PT and its divergence for the Yukawa, funnel and other potentials were considered in refs. [45–48]. Large orders of the WKB approximation in quantum machanics were studied in refs.[49–50]. Certainly, the above references are not complete in any sense.

2) Namely, $F = \mathcal{E}/\mathcal{E}_a$, where $\mathcal{E}$ is an external electric field and $\mathcal{E}_a \sim Z/\bar{r}^2 \sim Z^3 n^{-4}$ is the characteristic atomic field in the electron orbit with principal quantum number n [15].

When $16F\ [\epsilon(F)]^{-2} = 1$, the turning points coincide, see eq.(3"). It corresponds to $F = F_*(n_1, n_2, m)$, see refs.[12,15]. At $F > F_*$ the quasiclassical solution moves into complex plane and begins to desribe above-the-barrier resonances [51].

3) The index σ in eqs.(2),(4) depends on the particular problem. A similar, though more complicated formula can be obtained [31] for the pre-exponential factor c_0.

4) The coefficients in (1) have singularities at $\nu = \nu_*$,

$$\epsilon^{(0)} = c_0 + c_1(\nu - \nu_*) + c_2(\nu_* - \nu)^{3/2} + ...,$$

$$\epsilon^{(1)} = c_0' + c_1'(\nu_* - \nu)^{1/4} + ..., \qquad \epsilon^{(2)} = c_0''(\nu_* - \nu)^{-1} + ...,$$

etc., with $\epsilon^{(k)}(\nu) \to \infty$ for $k \geq 2$. See, for instance, refs.[11,13].

5) The imaginary time was introduced to describe the subbarrier motion of a particle in the theory of multiphoton ionization of atoms and ions by a strong light wave [52,53], as well as for calculation of e^+e^- pair production from vacuum by a varying electric field [54]. The generalization of the method to the case of periodic fields was considered in [55]. As is well known, at present the imaginary time is often used in QFT.

6) For this purpose, we have used a special modification of the HPA, where the solution of the equations for $\vec{r}_2$ was obtained by a minimization procedure. The details of these calculations are not published yet.

7) The results of $1/n$-expansion and HPA (i.e., summation of the perturbation series with the help of Hermite–Pade approximants) are taken from refs.[12,13,15]. In these papers the values of E_r and Γ for different (n_1, n_2, m) states of a hydrogen atom in strong electric field can be found. Other methods are the use of the Borel–Pade transformation [62] and numerical solution of the Schrödinger equation via Sturmian expansions [63].

References

1. Yaffe L.G. (1982) Large-N limits as classical mechanics, *Reviews Mod. Phys.*, **54**, 407-435.
2. Chatterjee A. (1990) Large-N expansions in quantum mechanics, atomic physics and some $O(N)$ invariant systems, *Physics Reports*, **186**, 249-370.
3. Herschbach D.R., Avery J., and Goscinsky O., eds. (1993) *Dimensional Scaling in Chemical Physics*, Kluwer Academic Publishers, Dordrecht.
4. Dolgov A.D., and Popov V.S. (1979) The anharmonic oscillator and its dependence on space dimension, *Phys. Lett.*, **86B**, 185-188.
5. Dolgov A.D., Eletsky V.L., and Popov V.S. (1979) The anharmonic oscillator: perturbation theory, quasiclassics and $1/D$ expansion, *preprint ITEP* **No. 72**, pp.1-45 (in Russian).
6. Mlodinow L.D., and Papanicolaou N. (1980) SO(2,1) algebra and the large N-expansion in quantum mechanics, *Annals of Physics (N.Y.)*, **128**, 314-334.
7. Bender C.M., Mloinow L.D., and Papanicolaou N. (1982) Semiclassical perturbation theory for the hydrogen atom in a uniform magnetic field, *Phys. Rev.*, **A25**, 1305-1314.
8. Doren D.J., and Herschbach D.R. (1986) Spatial dimension as an expansion parameter in quantum mechanics, *Phys. Rev.*, **A34**, 2654-2664.
9. Doren D.J., and Herschbach D.R. (1986) Convergence properties and resummation of the $1/D$ expansion, *Phys. Rev.*, **A34**, 2665-2673.
10. Goodson D.Z., and Herschbach D.R. (1987) Recurrence calculation of dimensional expansion for two-electron atoms, *Phys. Rev. Lett.*, **58**, 1628-1631.
11. Popov V.S., Weinberg V.M., and Mur V.D. (1985) Divergent perturbation series and classical mechanics, *Pis'ma v ZhETF*, **41**, 439-442.
12. Popov V.S., Mur V.D., et al. (1987) $1/n$ expansion for the Stark effect in strong field, *Phys. Lett.*, **A124**, 77-88.
13. Mur V.D., Popov V.S., and Sergeev A.V. (1992) The $1/n$ expansion in quantum mechanics and quasistationary states, *J. Moscow Phys. Soc.*, **2**, 189-217.
14. Popov V.S. (1993) $1/n$ expansion for quasistationary states, in D.R. Herschbach et al. (eds.), *Dimensional Scaling in Chemical Physics*, Kluwer Academic Publishers, Dordrecht, pp. 179-195.
15. Popov V.S., Mur V.D., Sergeev A.V., et al. (1990) Strong-field Stark effect: perturbation theory and $1/n$ expansion, *Phys. Lett.*, A149, 418-424.
16. Mur V.D., Popov V.S., and Sergeev A.V. (1990) The $1/n$ expansion in quantum mechanics, *Zh. Eksp. Teor. Fiz.*, **97**, 32-46, *Preprint ITEP* **114-89,** pp. 1-44 (Russian).
17. Vainberg V.M., Popov V.S., and Sergeev A.V. (1990) $1/n$ expansion for the hydrogen atom in an external field, *Zh. Eksp. Teor. Fiz.*, **98**, 847-860.
18. Kais S., Morgan J.D., and Herschbach D.R. (1991) Electronic tunneling and exchange energy in the D-dimensional hydrogen-molecule ion, *J. Chem. Phys.*, **95**, 9028-9041.
19. Imbo T., Pagnamenta A., and Sukhatme U. (1984) Energy eigenvalues of spherically symmetric potentials using the shifted $1/N$ expansion, Phys. Rev., D29, 1669-1681.
20. Imbo T., Pagnamenta A., and Sukhatme U. (1984) Bound states of the Yukawa potential via the shifted $1/N$ expansion technique, *Phys. Lett.*, **105A**, 183-187.
21. Cristiansen H., Epele L.N., Fanchiotti H., et al. (1989) Improved shifted $1/N$ expansion, *Phys. Rev.*, **A40**, 1760-1764.
22. Mur V.D., and Popov V.S. (1987) $1/n$ expansion and wave functions, *Pis'ma v ZhETF*, **45**, 323-326.
23. Karnakov B.M., Mur V.D., and Popov V.S. (1994) $1/n$ expansion for wave functions, *Zh. Eksp. Teor. Fiz.*, **106**, 976-992.

24. Kais S., and Herschbach D.R. (1993) *J. Chem. Phys.*, **98**, 3990.
25. Popov V.S., and Scheblykin A.V. (1991) higher orders of $1/n$-expansion and its divergence, *Yad. Phys.*, **54**, 1582-1590.
26. Popov V.S., Sergeev A.V., and Scheblykin A.V. (1992) On the structure of large orders in $1/n$-expansion, *Zh. Exp. Teor. Fiz.*, **102**, 1453-1468.
27. Lopez-Cabrera M., Goodson D.Z., herschbach D.R., and Morgan J.D. (1992) Large-order perturbation theory for H_2^+, *Phys. Rev. Lett.*, **68**, 1992-1995.
28. Popov V.S., and Sergeev A.V. (1993) Large orders of $1/n$ expansion in quantum mechanics, *Phys. Lett.*, **A172,** !93-198.
29. Popov V.S., and Sergeev A.V. (1994) Large orders of $1/n$ expansion for multidimensional problems, *Phys. Lett.*, **A193**, 165-172.
30. Popov V.S., and Sergeev A.V. (1994) Asymptotic form of higher orders of the $1/n$ expansion for multidimensional problems, *Zh. Eksp. Teor. Fiz.*, **105**, 568-591.
31. Popov V.S., Sergeev A.V., Mur V.D., et al. (19940 on the asymptotic of high-order terms of the $1/n$ expansion, *Yad. Fiz.*, **57**, 1119-1128.
32. Goodson D.Z., Lopez-Cabrera M., herschbach D.R., and Morgan J.D. (1992) Large-order dimensional perturbation theory for two-electron atoms, *J. Chem. Phys.*, **97**, 8481-8496.
33. Dyson F.J. (1952) Divergence of perturbation theory in quantum electrodynamics, *Phys. Rev.*, **85**, 631-632.
34. Bender C.M., and Wu T.T. (1971) Large-order behavior of perturbation theory, *Phys. Rev. Lett.*, **27**, 461-465.
35. Bender C.M., and Wu T.T. (1973) Anharmonic oscillator. A study of perturbation theory in large order, *Phys. Rev.*, **D7**, 1620-1636.
36. Dolgov A.D., and Popov V.S. (1978) Higher orders and the structure of perturbation theory for an anharmonic oscillator, *Zh. Eksp. Teor. Fiz.*, **75**, 2010-2026; *Phys. Lett.*, **79B**, 403-405.
37. Silvestrone H.J., Harris J.G., Cizek J., and Paldus J. (1985) Asymptotics of high-order perturbation theory for the one-dimensional anharmonic oscillator by quasisemiclassical methods, *Phys. Rev.*, **A32**, 1965-1980.
38. Alvarez G. (1988) Coupling-constant behavior of the resonances in the cubic anharmonic oscillator, *Phys. Rev.*, **A37**, 4079-4083.
39. Silvestrone H.J. (1978) Perturbation theory of the Stark effect in hydrogen to arbitrary high order, *Phys. Rev.*, **A18**, 1853-1864.
40. Silvestrone H.J., Adams B.G., Cizek J., et al. (1979) Stark effect in hydrogen: dispersion relation, asymptotic formulas, and calculation of the ionization rate via high-order perturbation theory, *Phys. Rev. Lett.*, **43**, 1498-1501.
41. Alliluev S.P., Eletsky V.L., and Popov V.S. (1979) The high orders of perturbation theory for the Stark effect in hydrogen, *Phys. Lett.*, **73A**, 103-106.
42. Alliluev S.P., Eletsky V.L., Popov V.S., and Weinberg V.M. (1980) Stark effect in hydrogen: high orders of perturbation theory for excited states, *Phys. Lett.*, **78A**, 43-46.
43. Privman V. (1980) New method of perturbation-theory calculation of the Stark effect for ground state of hydrogen, *Phys. Rev.*, **A22**, 1833-1837.
44. Avron J.E., Adams B.G., Cizek J., et al. (1979) Bender-Wu formula, the $SO(4,2)$ dynamical group, and the Zeeman effect in hydrogen, *Phys. Rev. Lett.*, **43**, 691-693.
45. privman V. (1981) Ground state perturbation theory in one dimension: a new method with application to screened Coulomb and to charmonium potentials, *Phys. Lett.*, **81A**, 326-330.

46. Weinberg V.M., Eletsky V.L., and Popov V.S. (1981) Logarithmic perturbation theory for screened Coulomb potentials and charmonium potential, *Zh. Eksp. Teor. Fiz.*, **81**, 1567-1582.
47. Sergeev A.V., and Sherstyouk A.I. (1982) Higher orders and structure of perturbation theory series for screened Coulomb potentials, *Zh. Eksp. Teor. Fiz.*, **82**, 1070-1078.
48. Vrscay E.R. (1986) Hydrogen atom with a Yukawa potential: perturbation theory and continued-fractions-Pade approximants at large order, *Phys. Rev.*, **A33**, 1433-1436.
49. Bender C.M., olaussen K., and Wang P.S. (1977) Numerological analysis of the WKB approximation in large order, *Phys. Rev.*, **D16**, 1740-1748.
50. Kesarwani R.N., and Varshni Y.P. (1980) Five-term WKBJ approximation, *J. Math. Phys.*, **21**, 90-92.
51. Mur V.D., and Popov V.S. (1993) The WKB method for resonances, *Pis'ma v ZhETF* **57**, 406-409: *Zh. Eksp. Teor. Fiz.*, **104**, 2293-2313.
52. Perelomov A.M., Popov V.S., and terentiev M.V. (1966) Ionization of atoms in a varying electric field,*Zh. Eksp. Teor. Fiz.*, **51**, 309-326.
53. Popov V.S., Kuznetsov V.P., and perelomov A.M. (1967) Quasiclassical approximation for nonstationary problems, *Zh. Eksp. Teor. Fiz.*, **53**, 331-347.
54. Popov V.S. (1971) Pair production in a varying external field: Quasiclassical approximation, *Zh. Eksp. Teor. Fiz.*, **61**, 1334-1351.
55. Popov V.S. (1974) Method of imaginary time for periodic fields, *Yad. Fiz.*, **19**, 1140-1156.
56. Cizek J., Damburg R.J., Graffi S., et al. (1986) $1/R$ expansion for $H_2{}^{+}$: Calculation of exponentially small terms and asymptotics, *Phys. Rev.*, **A33**, 12-54.
57. Zinn-Justin J. (1981) Expansion around instantons in quantum mechanics, *J. Math. Phys.*, **22**, 511-520.
58. Lakshmanan M., Kaliappan P., Larsson K., Karlsson F., and Fröman P.O. (1994) Phase-integral approach to quantum two- and three-dimensional isotropic anharmonic oscillators, *Phys. Rev.*, **A49**, 3296-3309.
59. Panja M.M., and Dutt R. (1988) Shifted large N-expansion for the energy levels of relativistic particles, *Phys. Rev.*, **A38**, 3937-3943.
60. Goodson D.Z., Morgan J.D., and Herschbach D.R. (1991) Dimensional singularity analysis of relativistic equations, *Phys. Rev.*, **A43**, 4617-4624.
61. Benassi L., and Grecchi V. (1980) Resonances in the Stark effect and strongly asymptotic approximants, *J. Phys. B: At. Mol. Phys.*, **13**, 911-930.
62. Franceschini V., Grecchi V., and Silverstone H.J. (1985) Complex energies from real perturbation series for the Lo Surdo-Stark effect in hydrogen by Borel-Pade approximants, *Phys. Rev.*, **A32**, 1338-1340.
63. Telnov D.A. (1989) DC Stark effect in a hydrogen atom via Sturmian expansions, *J. Phys. B: At. Mol. Opt. Phys.*, **22**, L399-L404.

SIZE-EXTENSIVE BRILLOUIN-WIGNER PERTURBATION THEORY. SIZE-EXTENSIVE BRILLOUIN-WIGNER COUPLED CLUSTER THEORY

IVAN HUBAČ
Division of Chemical Physics, Faculty of Mathematics and Physics, Comenius University, 842 15 Bratislava, Slovakia

1. Introduction

In present day quantum chemistry the methods which scale properly with the number of particles (i.e. they are size-extensive) play an important role in calculating the correlation energy as well as properties of molecular systems. Among these methods, many-body Rayleigh-Schrödinger perturbation theory (MB-RSPT) [1, 2, 3, 4] and coupled cluster (CC) theory [1, 5, 6, 7, 8] are well established methods. Both these methods in their diagrammatic formulation obey the linked cluster theorem and the size-extensivity is achieved by the fact that the energy is given only through the so called connected diagrams. On the other hand, the limited configuration interaction (LCI) method [9, 10], which is variational method and which represents an upper bound to the exact energy, does not scale properly with the number of particles. This method is widely used for practical applications due to its simplicity. Since the size-extensivity was found to be very important (this property is crucial when studying extended systems) there are attempts to correct LCI method for the size-nonextensivity [11, 12, 13, 14, 15, 16, 17, 18, 19, 20].

In this article we will discuss the size-extensivity problem of LCI through the Brillouin-Wigner perturbation theory (BWPT). This theory (BWPT) was not used in quantum chemistry because it is not size-extensive and is believed that it is slowly convergent. We show in this article that BWPT plays an important role in understanding the size-extensivity problem in LCI method and also that it makes the connection between LCI method and CC theory. We shall formulate the Brillouin-Wigner form of CC the-

C. A. Tsipis et al. (eds.), New Methods in Quantum Theory, 183–202.

ory which will be size-extensive. Using the Lippmann-Schwinger equation we show how on one side we get from BWPT the LCI method (size-nonextensive) and on the other side introducing the Bloch equation in the Brillouin-Wigner form we get the Brillouin-Wigner coupled cluster (BWCC) method (size-extensive).

2. Brillouin-Wigner Perturbation Theory and the Limited Configuration Interaction Method.

As is usual in perturbation theory we assume that in the Schrödinger equation

$$H\,\Psi_i = \mathcal{E}_i\,\Psi_i \tag{1}$$

we are able to split the Hamiltonian H into two parts, namely

$$H = H_0 + V \tag{2}$$

where H_0 is the unperturbed Hamiltonian and V is the perturbation. Further we assume that we know the solution of the unperturbed eigenvalue problem

$$H_0\,\Phi_i = E_i\,\Phi_i \tag{3}$$

Let Φ_i be the configuration state functions constructed from the Hartree-Fock molecular orbitals. Then, according to the BWPT expansion [3, 21, 22], the exact wave function for the ground state Ψ_0 is given by the expression

$$|\Psi_0\rangle = (1 + BV + BVBV + \dots)\,|\Phi_0\rangle \tag{4}$$

and the exact energy of the ground state $\mathcal{E}_0$ is given as

$$\mathcal{E}_0 = \langle\Phi_0|H_0|\Phi_0\rangle + \langle\Phi_0|V + VBV + VBVBV + \dots\,|\Phi_0\rangle \tag{5}$$

where B is the Brillouin-Wigner (BW) type of propagator

$$B = \sum_{i\neq 0} \frac{|\Phi_i\rangle\langle\Phi_i|}{\mathcal{E}_0 - E_i} \tag{6}$$

Let us introduce, in formal analogy with scattering theory, the operator T_B as in [23]

$$T_B = V + VBV + VBVBV + \dots \tag{7}$$

which also may be written in the form

$$T_B = V + VB\,T_B \tag{8}$$

This equation is known in scattering theory as the Lippmann-Schwinger equation [21, 24]. It is interesting to note that in many-electron theory this equation was used to generate the Brillouin-Wigner perturbation series and it was not used as a recurrent expression which can be brought into matrix form and applied to practical calculations.

Using Eq. (8) we can now rewrite (5) as

$$\mathcal{E}_0 = \langle\Phi_0|H_0|\Phi_0\rangle + \langle\Phi_0|T_B|\Phi_0\rangle \tag{9}$$

and the exact wave function (4) as

$$|\Psi_0\rangle = (1 + BT_B)\,|\Psi_0\rangle \tag{10}$$

The Eq. (8) can be used for practical calculations. When we substitute from Eqs. (6) and (8) into (9) we get

$$\mathcal{E}_0 = \langle\Phi_0|H_0|\Phi_0\rangle + \langle\Phi_0|V|\Phi_0\rangle + \sum_{i\neq 0} \frac{\langle\Phi_0|V|\Phi_i\rangle\langle\Phi_i|T_B|\Phi_0\rangle}{\mathcal{E}_0 - E_i} \tag{11}$$

Let us now introduce the following notation:

$$V_{ij} = \langle\Phi_i|V|\Phi_j\rangle \tag{12}$$

$$T_{ij} = \langle\Phi_i|T_B|\Phi_j\rangle \tag{13}$$

Then we may write Eq. (11) as

$$\mathcal{E}_0 = \langle\Phi_0|H_0|\Phi_0\rangle + V_{00} + \sum_{i\neq 0} \frac{V_{0i}\,T_{i0}}{\mathcal{E}_0 - E_i} \tag{14}$$

This equation cannot be used to calculate the exact energy $\mathcal{E}_0$ because we do not know the matrix elements T_{i0}. However, using Eq. (8) we may write

$$\langle\Phi_i|T_B|\Phi_0\rangle = \langle\Phi_i|V|\Phi_0\rangle + \sum_{j\neq 0} \frac{\langle\Phi_i|V|\Phi_j\rangle\langle\Phi_j|T_B|\Phi_0\rangle}{\mathcal{E}_0 - E_j} \tag{15}$$

or

$$T_{i0} = V_{i0} + \sum_{j\neq 0} \frac{V_{ij}\,T_{j0}}{\mathcal{E}_0 - E_j} \tag{16}$$

Note that the Lippmann-Schwinger equation (8) has been used twice, namely for the exact energy (11) and for the T_{i0} matrix elements (15). Let us define the U matrix in the form

$$U_{jk} = \frac{V_{jk}}{\mathcal{E}_0 - E_k} \tag{17}$$

Then we can rewrite Eq. (16) into the form

$$\sum_{j\neq 0}(\delta_{ij} - U_{ij})T_{j0} = V_{i0} \tag{18}$$

which is a system of linear equations for T_{i0} elements. This system of equations together with Eq. (14) in the form

$$\mathcal{E}_0 = \langle\Phi_0|H|\Phi_0\rangle + \sum_{i\neq 0} U_{0i}T_{i0} \tag{19}$$

represent a system of coupled equations for $\mathcal{E}_0$ and can be solved iteratively in the following way: set $\mathcal{E}_0 = E_0$, get the U matrix elements from Eq. (17), solve the system of linear equations (18) for T_{i0} and calculate a new $\mathcal{E}_0$ from Eq. (19). This procedure is repeated till selfconsistency. The approach represented by Eqs. (18) and (19) is equivalent to the CI calculation for the particular level of excitation. For example, if we restrict the V matrix elements to single, double, triple and quadruple excitations, we get the CI-SDTQ energies. In Table 1 we present the results of BWPT calculations according to Eqs. (18) and (19) for LiH and BeH_2 molecules. Thus we have shown that the Lippmann-Schwinger equation and BWPT brings us to the size-nonextensive results of the limited CI theory.

3. Brillouin-Wigner Perturbation Theory and Coupled Cluster Method.

Now we turn our attention to size-extensive theories in order to analyze the source of size-nonextensivity in BWPT [23]. The most general approach to the calculation of correlation energy is the method based on the effective Hamiltonian and the Bloch equation [3, 26, 27]. The basic idea lies in the introduction of the effective Hamiltonian H_{eff}, which acts within the d-dimensional model space and has the property that when diagonalized in this space it gives the part of the spectrum of our exact Hamiltonian. If we separate the complete configurational space into two disjoint subspaces, namely the model space P and its orthogonal complement Q, the projection operators associated with these subspaces will have the form

$$P = \sum_{i\in P}|\Phi_i\rangle\langle\Phi_i| \tag{20}$$

$$Q = \sum_{i\in Q}|\Phi_i\rangle\langle\Phi_i| \tag{21}$$

$$P + Q = 1 \tag{22}$$

TABLE 1. BWPT calculations for LiH and BeH_2 in DZv basis set.[a]

order of BWPT	energy (a.u.)		
	LiH[b]	BeH_2[c]	
		up to 4-excit.	up to 6-excit.
2	-7.9979303	-15.7808272	-15.7808272
3	-7.9996000	-15.7819585	-15.7819585
4	-7.9997856	-15.7821103	-15.7821103
5	-7.9998374	-15.7821266	-15.7821267
6	-7.9998571	-15.7821305	-15.7821307
7	-7.9998647	-15.7821316	-15.7821317
8	-7.9998675	-15.7821318	-15.7821319
9	-7.9998685	-15.7821319	-15.7821320
10	-7.9998689		
11	-7.9998691		
12	-7.9998691		
13	-7.9998691		
14	-7.9998692		
15	-7.9998692		
FCI	-7.9998692	-15.7821319[d]	-15.7821320
SCF	-7.9810242	-15.7602797	-15.7602797

[a] DZv basis set taken from Ref. [25]
[b] internuclear distance R(Li–H) = 3.1049 a.u.
[c] linear H–Be–H arrangement, R(Be–H) = 2.5223 a.u., two highest virtual orbitals were frozen
[d] CI-SDTQ

According to Lindgren [3] the construction of H_{eff} is possible through the wave operator Ω

$$H_{eff} = PH\Omega P \tag{23}$$

which is defined through the equation

$$\Psi_i = \Omega\,(P\Psi_i)\,; \qquad i = 1, 2, \ldots\ d \tag{24}$$

and which fulfills the so-called Bloch equation

$$[\Omega, H_0]P = V\Omega P - \Omega PV\Omega P \tag{25}$$

that may be viewed as an equivalent to the Schrödinger equation for wave functions (1). If we split our Hamiltonian into the unperturbed part H_0

and the perturbation V, the effective Hamiltonian (23) can be expressed in the form

$$H_{eff} = PH_0P + PV\Omega P \tag{26}$$

which implies that in the case of the one-dimensional (non-degenerate) model space the ground state energy is given as

$$\mathcal{E}_0 = \langle\Phi_0|H_0|\Phi_0\rangle + \langle\Phi_0|V\Omega|\Phi_0\rangle \tag{27}$$

The Bloch equation (25) for the degenerate case may be rewritten in the form [3]

$$(E_0 - H_0)\Omega P = QV\Omega P - \chi PV\Omega P \tag{28}$$

where χ is defined as

$$\Omega = 1 + \chi \tag{29}$$

and for the non-degenerate (one-dimensional) case we have

$$\Omega\,|\Phi_0\rangle = |\Phi_0\rangle + \sum_{i\neq 0} \frac{|\Phi_i\rangle\langle\Phi_i|V\Omega|\Phi_0\rangle}{E_0 - H_0} - \frac{\chi|\Phi_0\rangle\langle\Phi_0|V\Omega|\Phi_0\rangle}{E_0 - H_0} \tag{30}$$

The realization of the wave operator Ω can be done in various ways, e.g., using the expansion

$$\Omega = \sum_{i=0}^{\infty} \Omega^{(i)} \tag{31}$$

and we can obtain the perturbation theory form of the Bloch equation and the perturbation theory formulation for the effective Hamiltonian method. Note that in the Bloch equation (30) we have the Rayleigh-Schrödinger (RS) resolvent

$$R = \sum_{i\neq 0} \frac{|\Phi_i\rangle\langle\Phi_i|}{E_0 - H_0} \tag{32}$$

A great simplification in Eq. (30) is achieved by the realization of the linked cluster theorem [3] which says that in Eq. (30) we calculate only linked connected (LC) diagrams, and therefore, Eq. (30) reads

$$\Omega\,|\Phi_0\rangle = |\Phi_0\rangle + \left\{\sum_{i\neq 0} \frac{|\Phi_i\rangle\langle\Phi_i|V\Omega|\Phi_0\rangle}{E_0 - H_0}\right\}_{LC} \tag{33}$$

Now, let us introduce the RSPT expansion for the exact energy. Using the linked cluster theorem we have

$$\mathcal{E}_0 = \langle\Phi_0|H_0|\Phi_0\rangle + \langle\Phi_0|V + VRV + VRVRV + VRVRVRV + \ldots\; |\Phi_0\rangle_{LC} \tag{34}$$

If we use the Lippmann-Schwinger equation (8) with the resolvent of the RS type (32) we can introduce the operator T_R

$$T_R = V + VRT_R|_{LC} \tag{35}$$

and the exact energy can be expressed in the form

$$\mathcal{E}_0 = \langle\Phi_0|H_0|\Phi_0\rangle + \langle\Phi_0|T_R|\Phi_0\rangle_{LC} \tag{36}$$

where by $|_{LC}$ we mean that when calculating matrix elements we consider only contributions from connected diagrams. Comparing Eqs. (9), (27) and (36) we have three different equations for the same energy $\mathcal{E}_0$

$$\mathcal{E}_0 = \langle\Phi_0|H_0|\Phi_0\rangle + \langle\Phi_0|V\Omega|\Phi_0\rangle \tag{37}$$

$$\mathcal{E}_0 = \langle\Phi_0|H_0|\Phi_0\rangle + \langle\Phi_0|T_R|\Phi_0\rangle_{LC} \tag{38}$$

$$\mathcal{E}_0 = \langle\Phi_0|H_0|\Phi_0\rangle + \langle\Phi_0|T_B|\Phi_0\rangle \tag{39}$$

Another possibility how to realize the wave operator in Eq. (37) is to use the exponential ansatz [5]

$$\Omega = e^S \tag{40}$$

and up to unitary transformation we have

$$\langle\Phi_0|T_R|\Phi_0\rangle_{LC} = \langle\Phi_0|Ve^S|\Phi_0\rangle_{LC} \tag{41}$$

$$\langle\Phi_0|T_B|\Phi_0\rangle = \langle\Phi_0|Ve^S|\Phi_0\rangle \tag{42}$$

or

$$T_R = Ve^S|_{LC} \tag{43}$$

$$T_B = Ve^S \tag{44}$$

Using Eqs. (8), (35) and (43–44) we have

$$Ve^S|_{LC} = V + VRVe^S|_{LC} \tag{45}$$

and

$$Ve^S = V + VBVe^S \tag{46}$$

Note that when substituting Eq. (40) into Eq. (37) we can develop the coupled cluster method. Instead of doing so we utilize the exponential ansatz (40) to develop the Lippmann-Schwinger equation. Substituting Eqs. (45) and (46) into (38) and (39) we get

$$\mathcal{E}_0 = \langle\Phi_0|H_0|\Phi_0\rangle + \langle\Phi_0|V|\Phi_0\rangle + \langle\Phi_0|VRVe^S|\Phi_0\rangle_{LC} \tag{47}$$

and

$$\mathcal{E}_0 = \langle \Phi_0 | H_0 | \Phi_0 \rangle + \langle \Phi_0 | V | \Phi_0 \rangle + \langle \Phi_0 | VBVe^S | \Phi_0 \rangle \tag{48}$$

If we substitute for resolvents B and R from (6) and (32) we get

$$\mathcal{E}_0 = \langle \Phi_0 | H_0 | \Phi_0 \rangle + \langle \Phi_0 | V | \Phi_0 \rangle + \sum_{i \neq 0} \frac{\langle \Phi_0 | V | \Phi_i \rangle \langle \Phi_i | Ve^S | \Phi_0 \rangle}{E_0 - E_i} \Bigg|_{LC} \tag{49}$$

and

$$\mathcal{E}_0 = \langle \Phi_0 | H_0 | \Phi_0 \rangle + \langle \Phi_0 | V | \Phi_0 \rangle + \sum_{i \neq 0} \frac{\langle \Phi_0 | V | \Phi_i \rangle \langle \Phi_i | Ve^S | \Phi_0 \rangle}{\mathcal{E}_0 - E_i} \tag{50}$$

Both Eqs. (49) and (50) can be solved iteratively but in both cases we have to know equations for matrix elements $\langle \Phi_i | Ve^S | \Phi_0 \rangle$. These can be obtained from Eqs. (45) and (46) which can be simplified as (multiplying both equations by V^{-1} from the left)

$$e^S |_{LC} = 1 + RVe^S \Big|_{LC} \tag{51}$$

$$e^S = 1 + BVe^S \tag{52}$$

or in matrix elements

$$e^S | \Phi_0 \rangle |_{LC} = | \Phi_0 \rangle + \sum_{i \neq 0} \frac{| \Phi_i \rangle \langle \Phi_i | Ve^S | \Phi_0 \rangle}{E_0 - E_i} \Bigg|_{LC} \tag{53}$$

$$e^S | \Phi_0 \rangle = | \Phi_0 \rangle + \sum_{i \neq 0} \frac{| \Phi_i \rangle \langle \Phi_i | Ve^S | \Phi_0 \rangle}{\mathcal{E}_0 - E_i} \tag{54}$$

Equation (53) is identical to the Bloch equation (33) and Eq. (54) is the analogue of the Bloch equation (33) with the BW type of denominator. The Lippmann-Schwinger equation (54) also gives us a new expression, namely, the Bloch equation in the BW form. Equations (49), (50) and (53), (54) can be used for practical calculations.

Let us further adopt a simple approximation in the exponential ansatz (40), such that

$$S = S_2 \tag{55}$$

and simultaneously

$$e^{S_2} = 1 + S_2 + \frac{1}{2} S_2^2 \tag{56}$$

This expansion is complete since the higher terms do not contribute (as will be seen below). Substituting Eq. (56) into Eqs. (49) and (53), we get

$$\mathcal{E}_0 = \langle\Phi_0|H_0|\Phi_0\rangle + \langle\Phi_0|V|\Phi_0\rangle + \sum_{i\neq 0} \frac{\langle\Phi_0|V|\Phi_i\rangle\langle\Phi_i|V|\Phi_0\rangle}{E_0 - E_i}\Bigg|_{LC}$$

$$+\sum_{i\neq 0} \frac{\langle\Phi_0|V|\Phi_i\rangle\langle\Phi_i|VS_2|\Phi_0\rangle}{E_0 - E_i}\Bigg|_{LC} + \frac{1}{2}\sum_{i\neq 0} \frac{\langle\Phi_0|V|\Phi_i\rangle\langle\Phi_i|VS_2^2|\Phi_0\rangle}{E_0 - E_i}\Bigg|_{LC} \tag{57}$$

and

$$\{1 + S_2 + \frac{1}{2}S_2^2\}|\Phi_0\rangle\,|_{LC} = |\Phi_0\rangle + \sum_{i\neq 0} \frac{|\Phi_i\rangle\langle\Phi_i|V|\Phi_0\rangle}{E_0 - E_i}\Bigg|_{LC}$$

$$+\sum_{i\neq 0} \frac{|\Phi_i\rangle\langle\Phi_i|VS_2|\Phi_0\rangle}{E_0 - E_i}\Bigg|_{LC} + \frac{1}{2}\sum_{i\neq 0} \frac{|\Phi_i\rangle\langle\Phi_i|VS_2^2|\Phi_0\rangle}{E_0 - E_i}\Bigg|_{LC} \tag{58}$$

Substituting Eq. (56) into Eqs. (50) and (54) we get

$$\mathcal{E}_0 = \langle\Phi_0|H_0|\Phi_0\rangle + \langle\Phi_0|V|\Phi_0\rangle + \sum_{i\neq 0} \frac{\langle\Phi_0|V|\Phi_i\rangle\langle\Phi_i|V|\Phi_0\rangle}{\mathcal{E}_0 - E_i}$$

$$+\sum_{i\neq 0} \frac{\langle\Phi_0|V|\Phi_i\rangle\langle\Phi_i|VS_2|\Phi_0\rangle}{\mathcal{E}_0 - E_i} + \frac{1}{2}\sum_{i\neq 0} \frac{\langle\Phi_0|V|\Phi_i\rangle\langle\Phi_i|VS_2^2|\Phi_0\rangle}{\mathcal{E}_0 - E_i} \tag{59}$$

and

$$\{1 + S_2 + \frac{1}{2}S_2^2\}|\Phi_0\rangle = |\Phi_0\rangle + \sum_{i\neq 0} \frac{|\Phi_i\rangle\langle\Phi_i|V|\Phi_0\rangle}{\mathcal{E}_0 - E_i}$$

$$+\sum_{i\neq 0} \frac{|\Phi_i\rangle\langle\Phi_i|VS_2|\Phi_0\rangle}{\mathcal{E}_0 - E_i} + \frac{1}{2}\sum_{i\neq 0} \frac{|\Phi_i\rangle\langle\Phi_i|VS_2^2|\Phi_0\rangle}{\mathcal{E}_0 - E_i} \tag{60}$$

The equations for cluster amplitudes can be obtained by projection of Eqs. (58) and (60) onto the corresponding unperturbed state functions.

Detailed analysis of Eqs. (57) and (58) shows that these are coupled clusters with doubles (CCD) equations. The analysis of Eqs. (59) and (60) shows that even these equations are equivalent to CCD equations but they are in the BW form. To distinguish the BW form of CC equations from standard CC equations we will call Eqs. (59) and (60) BWCCD equations. Similarly, when adopting approximation

$$S = S_1 + S_2 \tag{61}$$

we will speak about BWCCSD equations. The results for both sets of Eqs. (57), (58) and (59), (60) are equivalent and therefore Eqs. (59) and (60) are size-consistent.

The equivalence of both approaches mentioned above can also be shown in another way. For practical reasons, below we will use operators in their normal product form (denoted by subscript N), e.g.

$$H_N = H - \langle\Phi_0|H|\Phi_0\rangle \tag{62}$$

Let us employ the exponential ansatz for the wave operator (40) in the case of the nondegenerate ground state $|\Phi_0\rangle$

$$|\Psi_0\rangle = e^S|\Phi_0\rangle \tag{63}$$

Then, in accordance with the Baker-Campbell-Hausdorff formula and the linked cluster theorem we could develop the standard CC theory; for more details see e.g., [8]. The expression for the ground state correlation energy $\Delta\mathcal{E}_0$

$$\Delta\mathcal{E}_0 = \mathcal{E}_0 - E_0 \tag{64}$$

can be written as

$$\Delta\mathcal{E}_0 = \langle\Phi_0|V_N\, e^S|\Phi_0\rangle_{LC} \tag{65}$$

and the equations for cluster amplitudes can be derived from the condition

$$\langle\Phi_i|H_N\, e^S|\Phi_0\rangle_{LC} = 0 \tag{66}$$

so

$$\langle\Phi_i|e^S|\Phi_0\rangle_{LC} = \frac{1}{E_0 - E_i}\langle\Phi_i|V_N\, e^S|\Phi_0\rangle_{LC} \tag{67}$$

where $|\Phi_i\rangle$ represent excited unperturbed wave functions. From Eqs. (1), (2) and (63) we also have

$$(H_{oN} + V_N)\, e^S|\Phi_0\rangle = \Delta\mathcal{E}_0\, e^S|\Phi_0\rangle \tag{68}$$

Projecting the previous equation onto the unperturbed state functions $|\Phi_i\rangle$ we have

$$\langle\Phi_i|V_N\, e^S|\Phi_0\rangle = \langle\Phi_i|(\Delta\mathcal{E}_0 - H_{oN})\, e^S|\Phi_0\rangle \tag{69}$$

and when we exploit Eq. (3) with H_0 in its normal product form

$$H_{oN}\,|\Phi_i\rangle = (E_i - E_0)\,|\Phi_i\rangle \tag{70}$$

we obtain the final equation for amplitudes

$$\langle\Phi_i|e^S|\Phi_0\rangle = \frac{\langle\Phi_i|V_N\, e^S|\Phi_0\rangle}{\Delta\mathcal{E}_0 + E_0 - E_i} = \frac{\langle\Phi_i|V_N\, e^S|\Phi_0\rangle}{\mathcal{E}_0 - E_i} \tag{71}$$

Analogously, if we project Eq. (68) onto the ground state wave function $|\Phi_0\rangle$ with the use of Eq. (70), we obtain the expression for the ground state energy in the form

$$\Delta\mathcal{E}_0 = \langle\Phi_0|V_N\, e^S|\Phi_0\rangle \tag{72}$$

Using the resolution of unity we can write

$$\Delta\mathcal{E}_0 = \langle\Phi_0|V_N|\Phi_0\rangle + \sum_{i\neq 0}\langle\Phi_0|V_N|\Phi_i\rangle\langle\Phi_i|e^S|\Phi_0\rangle \tag{73}$$

If we realize that the first term is zero and if we substitute for $\langle\Phi_i|e^S|\Phi_0\rangle$ from Eq. (71), we finally get

$$\Delta\mathcal{E}_0 = \sum_{i\neq 0}\frac{\langle\Phi_0|V_N|\Phi_i\rangle\langle\Phi_i|e^S|\Phi_0\rangle}{\mathcal{E}_0 - E_i} \tag{74}$$

It is obvious that Eqs. (65) and (67) are equivalent to Eqs. (57) and (58) and Eqs. (74) and (71) are equivalent to Eqs. (59) and (60) (of course, in appropriate normal product form) when projecting Eqs. (58) and (60) onto the corresponding unperturbed wave functions. Since Eqs. (67) and (65) and Eqs. (71) and (74) represent two different ways of solving the same problem without the use of any approximation, their results have to be equivalent.

The implementation of the BWCCSD theory was discussed in [23]. As was demonstrated, it is rather simple. The additional contributions of BWCCSD theory with respect to CCSD theory are proportional to N^4. Since the most time consuming processes in one iteration step of the CCSD method depends on the number of basis functions as N^6, it is evident that the additional consumption of the computational time required for the realization of one iteration in the BWCCSD method is obviously negligible.

We give some numerical results to illustrate the methods introduced in this work. We calculate BWCCD and BWCCSD energies for three small closed shell systems, namely, for Be, H_2O and F^-; and we also present some single-reference open-shell BWCCSD results for two doublet states (Na and CN) and one triplet state (O_2). The corresponding results are listed in Table 2. As a reference wave function we have used the restricted HF (RHF) solution for closed-shell systems and restricted open-shell HF (ROHF) solution according to Roothaan [30] for open-shell systems. It is worth noting that the BWCCD method for open-shell systems is not very suitable due to the much more important role of single excitations.

As we can see from Table 2, the results for BWCCD and BWCCSD methods are actually identical to those for the CCD and CCSD methods in all cases, so one can conclude that the numerical equivalence of CC and

TABLE 2. Self-consistent field (SCF) energies and correlation energies for some selected small closed-shell and open-shell systems (in a.u.).

system	SCF	CCD	CCSD	BWCCD	BWCCSD
Be [a]	-14.572984	-0.051871	-0.052486	-0.051871	-0.052486
2 Be [a,b]	-29.145969	-0.103743	-0.104972	-0.103743	-0.104972
H_2O [c,d]	-76.064564	-0.245442	-0.247283	-0.245442	-0.247283
F^- [a]	-99.459102	-0.306461	-0.308943	-0.306461	-0.308943
CN ($^2\Sigma^+$) [e,f,g]	-92.208831	—	-0.321720	—	0.321720
Na (^2S) [g,h]	-161.846414	—	-0.083908	—	-0.083908
O_2 ($^3\Sigma_g^-$) [e,g,i]	-149.650063	—	-0.403366	—	-0.403366

[a] ANO (5.4.3.2) spherical basis set [31]
[b] R = 1000.0 a.u. (quasi-infinite distance)
[c] ANO (5.3.2;3.2) spherical basis set [31]
[d] R(O–H) = 1.790 a.u., ϕ=105.5 [32]
[e] POL (5.3.2) Cartesian basis set [33]
[f] R = 2.2144 a.u. [34]
[g] corresponding SCF and CCSD values were taken from Ref. [29]
[h] POL (7.5.2) Cartesian basis set [35]
[i] R = 2.207 a.u. [32]

BWCC has been confirmed. In order to prove the size-consistency of the BWCCD and BWCCSD methods, we have chosen a simple example with two Be atoms in quasi-infinite distance (1000 a.u.). In this sense it is not surprising that the BWCCD and BWCCSD energies for 2 Be supersystem are exactly two times larger than those for the single Be atom (see Table 2) even though it is clear that this feature directly follows from the equivalence with corresponding CC results.

4. Size-Extensivity and the Limited Configuration Interaction Method.

In up-to-date literature there are many attempts to develop size-extensive LCI methods [15, 20]. In the previous part of this article we have shown that BWPT together with the Lippmann-Schwinger equation can be used to generate the size-nonextensive LCI method. In order to do so, we used the Lippmann-Schwinger equation twice; we used it to get the energy expression (11) as well as for amplitudes (15). We also showed that using BWPT we can generate also the size-extensive theory (BWCC). In order to do so, we used the Lippmann-Schwinger equation only once, namely to get the energy expression (48), but the amplitudes were calculated by the Bloch equation (54). For the wave operator we used the exponential ansatz (40).

In this part of the article we shall use the Lippmann-Schwinger equation for the energy and the Bloch equation for amplitudes to try to develop a size-extensive CI method assuming that the Lippmann-Schwinger operator T_B (8) is in general an N-particle operator. We show that using this concept we arrive at an exponential ansatz for the wave operator. Thus, asking the LCI method to be size-extensive we arrive directly at the CC theory.

According to the Lippmann-Schwinger equation (8) the energy $\mathcal{E}_0$ is given by Eq. (11) and the T_{i0} amplitudes can be calculated from the Bloch equation which we get multiplying Eq. (8) from the left by V^{-1} assuming that the inverse operator V^{-1} exists and is nonsingular, i.e.

$$V^{-1}T_B = 1 + BT_B \tag{75}$$

If we define a new wave operator M in the following manner

$$M = V^{-1}T_B \tag{76}$$

we have

$$T_B = VM \tag{77}$$

and the Bloch equation (75) now reads

$$M = 1 + BT_B \tag{78}$$

In this article we will not use an exponential ansatz for the wave operator M, however, we will assume that it is in general an N-particle operator. Substituting Eq. (77) into Eq. (78) we get

$$M = 1 + BVM \tag{79}$$

and now we can write the expression for the energy in the form

$$\mathcal{E}_0 = \langle\Phi_0|H_0|\Phi_0\rangle + \langle\Phi_0|VM|\Phi_0\rangle \tag{80}$$

Since we do not know the matrix elements of VM operator, we employ the resolution of an identity operator

$$\mathcal{E}_0 = \langle\Phi_0|H_0|\Phi_0\rangle + \sum_j \langle\Phi_0|V|\Phi_j\rangle\langle\Phi_j|M|\Phi_0\rangle \tag{81}$$

The operator M is in general an N-particle operator

$$M = \sum_{i=0}^{N} M^{(i)} \tag{82}$$

Analyzing Eq. (81) we find that we need to know at most the $M^{(2)}$ operator, so we have (we assumed the use of Hartree-Fock orbitals)

$$\mathcal{E}_0 = \langle\Phi_0|H_0|\Phi_0\rangle + \sum_j \langle\Phi_0|V|\Phi_j^{(2)}\rangle\langle\Phi_j^{(2)}|M^{(2)}|\Phi_0\rangle \tag{83}$$

where the configuration state functions $|\Phi_j^{(2)}\rangle$ are doubly excited configurations. To make our derivation more transparent, we omitted the effect of single excitations, i.e. we do not include the effect of the $M^{(1)}$ operator. If we want the derivation to be complete, the $M^{(1)}$ operator should be included from the very beginning.

Matrix elements $\langle\Phi_j^{(2)}|M^{(2)}|\Phi_0\rangle$ can be obtained from the Bloch equation (79)

$$\langle\Phi_j^{(2)}|M^{(2)}|\Phi_0\rangle = \frac{\langle\Phi_j^{(2)}|VM|\Phi_0\rangle}{\mathcal{E}_0 - E_j} \tag{84}$$

Using the resolution of identity we can write

$$\langle\Phi_j^{(2)}|M^{(2)}|\Phi_0\rangle = \sum_k \frac{\langle\Phi_j^{(2)}|V|\Phi_k\rangle\langle\Phi_k|M|\Phi_0\rangle}{\mathcal{E}_0 - E_j} \tag{85}$$

Analyzing Eq. (85) we find that we have to know at most the $M^{(4)}$ matrix elements

$$\langle\Phi_j^{(2)}|M^{(2)}|\Phi_0\rangle = \sum_{n=0}^{4}\sum_k \frac{\langle\Phi_j^{(2)}|V|\Phi_k\rangle\langle\Phi_k|M^{(n)}|\Phi_0\rangle}{\mathcal{E}_0 - E_j} \tag{86}$$

or

$$\begin{aligned}\langle\Phi_j^{(2)}|M^{(2)}|\Phi_0\rangle = \sum_k \Bigg\{ & \frac{\langle\Phi_j^{(2)}|V|\Phi_k\rangle\langle\Phi_k|M^{(0)}|\Phi_0\rangle}{\mathcal{E}_0 - E_j} \\ + & \frac{\langle\Phi_j^{(2)}|V|\Phi_k\rangle\langle\Phi_k|M^{(1)}|\Phi_0\rangle}{\mathcal{E}_0 - E_j} + \frac{\langle\Phi_j^{(2)}|V|\Phi_k\rangle\langle\Phi_k|M^{(2)}|\Phi_0\rangle}{\mathcal{E}_0 - E_j} \\ + & \frac{\langle\Phi_j^{(2)}|V|\Phi_k\rangle\langle\Phi_k|M^{(3)}|\Phi_0\rangle}{\mathcal{E}_0 - E_j} + \frac{\langle\Phi_j^{(2)}|V|\Phi_k\rangle\langle\Phi_k|M^{(4)}|\Phi_0\rangle}{\mathcal{E}_0 - E_j}\Bigg\}\end{aligned} \tag{87}$$

We see that in order to obtain the $M^{(2)}$ matrix elements we need to know the $M^{(4)}$ matrix elements. To obtain the $M^{(4)}$ matrix elements we shall need

$M^{(6)}$ matrix elements, etc. In order to solve Eq. (87), we do the following. Starting from Eq. (82) we put

$$M^{(1)} = m_1 \tag{88}$$

$$M^{(2)} = m_2 + \frac{1}{2!} m_1 m_1 \tag{89}$$

where

$$m_2 = M^{(2)} - \frac{1}{2!} m_1 m_1 \tag{90}$$

$$M^{(3)} = m_3 + \frac{1}{3!} m_1 m_1 m_1 + \frac{1}{2} m_1 m_2 \tag{91}$$

where

$$m_3 = M^{(3)} - \frac{1}{3!} m_1 m_1 m_1 - \frac{1}{2} m_1 m_2 \tag{92}$$

etc. In an analogous manner we can write for $M^{(4)}$

$$M^{(4)} = m_4 + \frac{1}{4!} m_1 m_1 m_1 m_1 + \frac{1}{3!} m_1 m_1 m_2 + \frac{1}{2} m_2 m_2 + \frac{1}{2} m_1 m_3 \tag{93}$$

In this respect we can also write Eq. (82) as

$$M = e^m \tag{94}$$

where

$$m = \sum_{i=0}^{N} m^{(i)} \tag{95}$$

As we have said at the beginning, we omit the effect of m_1 operator (in order to make our derivation simpler). If we adopt the approximation that

$$M^{(2)} = m_2 \tag{96}$$

and

$$M^{(4)} = \frac{1}{2} m_2 m_2 \tag{97}$$

we arrive at the following equations: Eq. (83) now reads

$$\mathcal{E}_0 = \langle \Phi_0 | H_0 | \Phi_0 \rangle + \sum_j \langle \Phi_0 | V | \Phi_j \rangle \langle \Phi_j | m_2 | \Phi_0 \rangle \tag{98}$$

and Eq. (87) for the m_2 matrix elements reads

$$\begin{aligned} \langle \Phi_j^{(2)} | m_2 | \Phi_0 \rangle \;&=\; \sum_k \frac{\langle \Phi_j^{(2)} | V | \Phi_k \rangle \langle \Phi_k | m_2 | \Phi_0 \rangle}{\mathcal{E}_0 - E_j} \\ &+\; \frac{1}{2} \sum_k \frac{\langle \Phi_j^{(2)} | V | \Phi_k \rangle \langle \Phi_k | m_2 m_2 | \Phi_0 \rangle}{\mathcal{E}_0 - E_j} \end{aligned} \tag{99}$$

Eqs. (98) and (99) are equivalent to Eqs. (50) and (54).

The problem of different representation of the wave operator was also studied by Shork and Fulde [36] and they showed that the wave operator must be of the exponential ansatz form. Thus, asking LCI to be size-extensive we arrive at BWCC theory.

5. Formulation of Multireference BWCC theory

An extension of the nondegenerate (single-reference) CC theory to multireference (MR) situations is essential for a proper description of general open-shell systems or even nearly degenerate closed-shell systems. An appropriate treatment of general open-shell and quasidegenerate situations requires the multireference formalism. While the CC theory for nondegenerate closed-shell ground states is well established and widely used method, its extension to MR situation proved to be very difficult to accomplish.

The existing MR CC theories fall into two basic categories (i) Fock space or valence universal methods, and (ii) Hilbert space or state universal methods [37]. They both rely on the same MR formalism based on the effective or model Hamiltonian and the Bloch equation. However, they differ in the cluster ansatz for the open-shell wave operator. We are not going to discuss here both approaches. We present a simple MR formulation of BWCC theory. The formulation is based on BWPT, Lippmann-Schwinger equation and the Bloch equation. We shall use the similar ideas as in Section 2. Our derivation will be similar as given by Brandow [27, 38].

We start from the Schrödinger equation (1) and we expect the validity of Eqs. (2) and (3). If we split our Hilbert space by projector operators P and Q defined by Eqs. (20), (21) and (22) we can write for the exact wave function Ψ

$$\Psi = (P+Q)\Psi = \sum_{i\in P} |\Phi_i\rangle\langle\Phi_i|\Psi\rangle + \sum_{i\in Q} |\Phi_i\rangle\langle\Phi_i|\Psi\rangle \tag{100}$$

or

$$\Psi = \sum_{i\in P} a_i|\Phi_i\rangle \; + \sum_{i\in Q} a_i|\Phi_i\rangle \tag{101}$$

Multiplying the Schrödinger equation (1) by $\langle\Phi_i|$ from the left and using Eq. (2) we get

$$\langle\Phi_i|H_0 + V|\Psi\rangle = \mathcal{E}\,\langle\Phi_i|\Psi\rangle \tag{102}$$

or

$$E_i\,\langle\Phi_i|\Psi\rangle + \langle\Phi_i|V|\Psi\rangle = \mathcal{E}\,\langle\Phi_i|\Psi\rangle \tag{103}$$

The previous Eq. (103) can be rewritten as

$$\langle\Phi_i|V|\Psi\rangle = (\mathcal{E} - E_i)\,\langle\Phi_i|\Psi\rangle \tag{104}$$

Using Eq. (101) and restricting ourselves to the model space P (dimension of P is the dimension of quasi-degeneracy) we get

$$\langle \Phi_i | V | \Psi \rangle = (\mathcal{E} - E_i)\, a_i \,; \qquad i \in P \tag{105}$$

Analogously with Eq. (24) we introduce the wave operator Ω

$$\Psi = \Omega\,(P\Psi) \tag{106}$$

Then we can write Eq. (105) as

$$(\mathcal{E} - E_i)\, a_i = \langle \Phi_i | V\Omega P | \Psi \rangle \,; \qquad i \in P \tag{107}$$

or

$$(\mathcal{E} - E_i)\, a_i = \sum_{j\in P} \langle \Phi_i | V\Omega | \Phi_j \rangle\, a_j \,; \qquad i \in P \tag{108}$$

Eq. (108) can also be written as [27, 38]

$$\{H_0 + P(V\Omega)P - \mathcal{E}\}A = 0 \tag{109}$$

This is the fundamental equation. For example, in the case of the two-dimensional model space

$$P = |\Phi_0\rangle\langle\Phi_0| + |\Phi_1\rangle\langle\Phi_1| \tag{110}$$

we have

$$\begin{vmatrix} E_0 + \langle \Phi_0 | V\Omega | \Phi_0 \rangle - \mathcal{E} & \langle \Phi_0 | V\Omega | \Phi_1 \rangle \\ \langle \Phi_1 | V\Omega | \Phi_0 \rangle & E_1 + \langle \Phi_1 | V\Omega | \Phi_1 \rangle - \mathcal{E} \end{vmatrix} = 0 \tag{111}$$

Now we can employ the Lippmann-Schwinger equation and also the Bloch equation in the Brillouin-Wigner form. According to Eq. (46) we have for the Lippmann-Schwinger equation

$$V\Omega = V + VBV\Omega \tag{112}$$

and

$$\Omega = 1 + BV\Omega \tag{113}$$

for the Bloch equation. Substituting Eqs. (112) and (113) into Eq. (111) gives us possibility to calculate the energies $\mathcal{E}_i$. The selection of the wave operator Ω for $\Omega|\Phi_0\rangle$ and $\Omega|\Phi_1\rangle$ can be done either as in the Fock space approach or as in the Hilbert space approach.

6. Discussion

The exact energy dependence of individual orders of the BWPT expansion and the size-inconsistency of the method were the reasons why this method was not further studied and used for practical calculations. The many-body formulation of BWPT was done by Brandow [27, 38]. He used BWPT to develop quasidegenerate RSPT introducing the concept of folded diagrams [27]. However, the size-inconsistency of BWPT was not studied in detail. In this article, analyzing the connection of BWPT and the Lippmann-Schwinger equation with the size-consistent effective Hamiltonian method and CC theory and using the exponential ansatz for the exact nondegenerate wave function, we developed BWCC theory.

The connection of RSPT with CC theory is well known; see e.g., [39]. We can see that the fully analogous relation is also between BWPT and BWCC theory. While in the RSPT case the individual orders of perturbation expansion as well as the individual iterations of the CC procedure are size-consistent, in the BWPT and BWCC cases they are not. However, as was shown in this work, converged BWCC results are not only size-consistent but also identical to corresponding CC results. It can be said that the cluster expansion forms of both different perturbational approaches is identical. It is therefore interesting to note that the complete expansion in both BWPT and RSPT theories is not the only common result of both approaches, but as presented in this paper, also the limited versions like BWCCD and (BWCCSD) are common. It is worth noting that the BWCCD (BWCCSD) theory contains disconnected diagrams. The practical implementation and the numerical complexity is approximately the same as that for CCD (CCSD) theory.

One could ask, what is the reason for developing the size-consistent BWPT theory? We note that such a theory may play an important role in the development of size-consistent CI theories. Another problem is that the present stage of developing multireference (MR) RSPT and MR CC theories is still far from being satisfactory. One of the crucial problems in existing MR RSPT an MR CC methods is the presence of the so-called intruder states. Intruder states may cause the divergence problems, which is due to vanishing denominators in the RS-type propagators. The formulation of MR BWPT of MR BWCC theories is much simpler than that of MR RSPT and MR CC theories. Moreover, the presence of the BW-type propagators can avoid the intruder state problem.

Acknowledgements

I would like to thank Professor C. A. Tsipis from the Aristotle University of Thessaloniki as well as the organizers of the 1995 NATO Advanced Study Institute in Halkidiki for their kind invitation. I also want to thank Professor A. Mavridis from the University of Athens for his very kind hospitality and discussions during author's stay in his laboratory. I am also grateful to Professor P. Čársky, Drs. P. Babinec, P. Mach, J. Mášik, P. Neogrády, and J. Urban for discussion and their help in preparing the manuscript. This work was supported by Grant No. 1/1134/94 of the Slovak Grant Agency, NATO Science Fellowships Programme for the year 1994, and also by COST Action D3 project.

References

1. Paldus, J. and Čížek, J. (1975) *Adv. Quantum Chem.* **9**, 105.
2. Hubač, I. and Čársky, P. (1978) *Topics Curr. Chem.* **75**, 99.
3. Lindgren, I. and Morrison, J. (1982) *Atomic Many-Body Theory*, Springer-Verlag, Berlin.
4. Wilson, S. (1984) *Electron Correlation in Molecules*, Clarendon Press, Oxford.
5. Čížek, J. (1966) *J. Chem. Phys.* **45**, 4256.
6. Čížek, J. (1969) *Adv. Chem. Phys.* **14**, 35.
7. Bartlett, R. J. (1981) *Ann. Rev. Phys. Chem.* **32**, 359.
8. Paldus, J. (1992) in S. Wilson and G. H. F. Diercksen, (eds.), *Methods in Computational Molecular Physics*, NATO ASI Series, Plenum, New York, pp. 99–194.
9. Shavitt, I. (1977) in H. F. Schaefer III, (ed.), *Methods of Electronic Structure*, Plenum Press, New York.
10. Karwowski, J. (1992) in S. Wilson and G. H. F. Diercksen, (eds.), *Methods in Computational Molecular Physics*, NATO ASI Series, Plenum, New York, pp. 65–98.
11. Davidson, E. R. (1974) in R. Daudel and B. Pullman, (eds.), *The World od Quantum Chemistry*, Reidel, Dordrecht.
12. Langhoff, S. R. and Davidson, E. R. (1974) *Int. J. Quantum Chem.* **8**, 61.
13. Duch, W. and Diercksen, G. H. F. (1994) *J. Chem. Phys.* **101**, 3018.
14. Ahlrichs, R., Scharf, P., and Ehrhardt, C. (1985) *J. Chem. Phys.* **82**, 890.
15. Pople, J. A., Head-Gordon, M., and Raghavachari, K. (1987) *J. Chem. Phys.* **87**, 5968.
16. Povill, A. R., Caballol, R., Rubio J., and Malrieu, J. P. (1993) *Chem. Phys. Lett.* **209**, 126.
17. Szalay, P. G. and Bartlett, R. J. (1993) *Chem. Phys. Lett.* **214**, 481.
18. He, Z. and Cremer, D. (1991) *Int. J. Quantum Chem.* **25**, 43.
19. He, Z. and Cremer, D. (1993) *Theor. Chim. Acta* **85**, 305.
20. Malrieu, J. P., Daudey, J. P., and Caballol, R. (1994) *J. Chem. Phys.* **101**, 8908.
21. Löwdin, P. O. (1968) *Int. J. Quantum Chem.* **2**, 867.
22. Čársky, P., Hrouda, V., Sychrovský, V., Hubač, I., Babinec, P., Mach, P., Urban, J., and Mášik, J. (1995) submitted to *Coll. Czech. Chem. Commun.*
23. Hubač, I. and Neogrády, P. (1994) *Phys. Rev. A* **50**, 4558.
24. Lippmann B. A. and Schwinger, J. (1950) *Phys. Rev.* **79**, 469.
25. Dunning, T. H. and Hay, P. G. (1977) in F. Schaefer III, (ed.), *Methods of Electronic Structure*, Plenum Press, New York, p. 23.
26. Lindgren, I. (1974) *J. Phys. B* **7**, 2441.
27. Brandow, B. H. (1967) *Rev. Mod. Phys.* **39**, 771.
28. Stanton, J. F., Gauss, J., Watts, J. D., and Bartlett, R. J. (1991) *J. Chem. Phys.* **94**, 4334.
29. Neogrády, P., Urban, M., and Hubač, I. (1994) *J. Chem. Phys.* **100**, 3706.
30. Roothaan, C. C. J. (1960) *Rev. Mod. Phys.* **32**, 179.
31. Widmark, P. O., Persson, B. J., and Roos, B. O. (1991) *Theor. Chim. Acta* **79**, 419.
32. De Frees, F. J., Krishnan, R., Schlegel, H. B., and Pople, J. A. (1982) *J. Amer. Chem. Soc.* **104**, 5576.
33. Sadlej, A. J. (1990) *Coll. Czech. Chem. Commun.* **53**, 1995.
34. Urban, M., Watts, J. D., and Bartlett, R. J. (1994) *Int. J. Quantum Chem.* **52**, 211.
35. Sadlej, A. J. and Urban, M. (1991) *J. Mol. Structure (Theochem)* **80**, 234.
36. Shork, T. and Fulde, P. (1992) *J. Chem. Phys.* **97**, 9195.
37. Paldus, J. (1994) in G. L. Malli, (ed.), *Relativistic and Correlation Effects in Molecules and Solids*, NATO ASI Series, Plenum, New York, pp. 207–282.
38. Brandow, B. H. (1979) *Int. J. Quantum Chem.* **15**, 207.
39. Kucharski, S. A. and Bartlett, R. J. (1986) *Adv. Quantum Chem.* **18**, 281.

LEAST MOTION, HYPERSPHERICAL COORDINATES, AND POLYTOPES

JAN LINDERBERG
Department of Chemistry
Aarhus University
DK-8000 Aarhus C, Denmark

ABSTRACT

A picture of chemical processes as being characterised by means of directions in a multidimensional space has emerged in recent years. It serves both as a conceptual tool and as a means for effective numerical calculations. The emphasis of this presentation is placed on the definition of conformational directions and the related kinematic rotations. Examples are offered for bimolecular and unimolecular rearrangements.

THE HISTORICAL PERSPECTIVE.

Quantum mechanics provided the theoretical background for a description of the chemical bond through the work of Heitler and London [1] while Born and Oppenheimer [2] gave the formal justification for the practice of classification of molecular energy levels as characterised by rotational, vibrational, and electronic quantum numbers. The separate treatment of electronic and nuclear degrees of freedom gave rise to an industry for the production of *electronic energy surfaces* from which conclusions regarding properties of chemical compounds were inferred. Much of the information pertained to the relative stability of families of substances and modern electronic structure codes can give accurate values for heats of reaction, bond distances and angles.

Formation and breaking of chemical bonds offers a great challenge to the Born-Oppenheimer separation since the rearrangement of the nuclear geometry is associated

C. A. Tsipis et al. (eds.), New Methods in Quantum Theory, 203–223.

with changes in the electronic structure in a way that might violate the criteria for reasonable usage of the normal picture. Eyring created, in spite of these difficulties, a working theory for the rates of chemical reactions [3] which was based on the notion that the system moved along a reaction path on a many-dimensional energy surface. The surface for the collinear arrangement of three hydrogen atoms [4] gave the conceptual framework for a general theory that remains significant to this day.

Gronwall [5] and Bartlett [6] carried out calculations for three particle systems in a hyperspherical representation for six dimensional spaces and Wannier [7] used this for a conceptual development of relevance to highly excited states of Helium like ions. These developments did not have a great impact on the evolution of theoretical atomic physics for some time, but the group around Fano were able to resurrect and further advance these methods with a variety of applications [8]. It was Kuppermann [9] who initiated the hyperspherical treatment in the field of reaction dynamics and demonstrated its feasibility. Conceptual advantages of the hyperspherical representation were demonstrated by Mead [10] in addressing the intricacies of conical intersections and phase relations in the trihydrogen problem. Manz [11] suggested the possibility of vibrational bonding on the basis of a hyperspherical analysis of a collinear heavy-light-heavy molecular system, but no confirmation has yet been found to support this concept.

Transformations from standard cartesian coordinates in a laboratory frame to a set of internal coordinates in a rotating, center-of-mass frame will be described in the next section. Arrangement channels may be labeled as distinct directions in a kinematic rotation space and three- and four-particle systems are used, in the third section, to make these notions less abstract. Molecular systems with more than four atoms are considered in the fourth section with a view towards the classification of passages between different conformations as proceeding in a planar subspace. The possible rearrangements in the barbaralyl cation are examined and detailed in the fifth section where the particular intricacies of a highly degenerate situation are exposed. Conclusions regarding the principle of least motion and its relevance for the hyperspherical representation are offered in the final section.

EXTERNAL AND INTERNAL COORDINATES.

Laboratory coordinates for a system of N particles and their masses are denoted as a set of vectors in R^3 and associated scalars : $\{(\mathbf{R}_1,m_1),(\mathbf{R}_2,m_2),\ldots,(\mathbf{R}_N,m_N)\}$.The center-of-mass position is irrelevant for the discussion of the internal motions of the system and it is then relevant to introduce a transformation to variables which are independent of the choice of origin. We follow the procedure given by Radau in 1868 [12] and define a diagonal matrix **D** of dimension N where the elements are

$$\mathbf{D}_{jk} = \delta_{jk}\mu_j; \mu_j = \sqrt{\frac{m_j}{M}}; M = \sum_1^N m_j .$$

An elementary reflection matrix **K** of dimension N is defined with elements

$$\mathbf{K}_{jk} = \delta_{jk} + \frac{(\mu_j - \delta_{jN})(\mu_k - \delta_{kN})}{\mu_N - 1}$$

New variables are defined as

$$\mathbf{r}_j = \sum_1^N \mathbf{R}_s \mathbf{D}_{st} \mathbf{K}_{tj}$$

and it follows that $\mathbf{r}_N$ is the center-of-mass vector. There are $n = N\text{-}1$ internal vectors, each associated with the formal mass M to be dealt with in the following. They maintain the symmetry features of the original set when $(\mathbf{R}_N, m_N)$ refers to a unique choice of a particle to be formally eliminated and provides maximum remaining equivalence otherwise. Further orthogonal transformations may be applied within the set of n internal vectors in order to obtain coordinates adapted to particular preferences.

Different orientations of the axes of the reference frame are related through rotations which are parametrized by means of an axis and an angle such that

$$\begin{aligned}\mathbf{R}'_s &= \mathbf{R}_s \cos\phi + \mathbf{n}(\mathbf{n}\cdot\mathbf{R}_s)(1-\cos\phi) + (\mathbf{n}\times\mathbf{R}_s)\sin\phi \\ &= \mathbf{R}_s\left(\cos^2\tfrac{1}{2}\phi - |\mathbf{a}|^2\right) + 2\mathbf{a}(\mathbf{a}\cdot\mathbf{R}_s) + 2\cos\tfrac{1}{2}\phi(\mathbf{a}\times\mathbf{R}_s)\end{aligned}$$

The new vector **a** is parallel to the unit axis vector **n** and has the length $\sin\frac{1}{2}\phi$.It forms, together with $\cos\frac{1}{2}\phi$, a four vector α of unit length. Its components are the Euler-Rodrigues parameters for the rotation group and we have that

$$\mathbf{R}_s = \begin{pmatrix}\mathbf{e}_1 & \mathbf{e}_2 & \mathbf{e}_3\end{pmatrix}\begin{pmatrix}R_{1s}\\ R_{2s}\\ R_{3s}\end{pmatrix}$$

$$= \begin{pmatrix}\mathbf{e}_1 & \mathbf{e}_2 & \mathbf{e}_3\end{pmatrix}\begin{pmatrix}\alpha_0^2+\alpha_1^2-\alpha_2^2-\alpha_3^2 & 2\alpha_1\alpha_2-2\alpha_0\alpha_3 & 2\alpha_1\alpha_3+2\alpha_0\alpha_2\\ 2\alpha_1\alpha_2+2\alpha_0\alpha_3 & \alpha_0^2-\alpha_1^2+\alpha_2^2-\alpha_3^2 & 2\alpha_2\alpha_3-2\alpha_0\alpha_1\\ 2\alpha_3\alpha_1-2\alpha_0\alpha_2 & 2\alpha_3\alpha_2+2\alpha_0\alpha_1 & \alpha_0^2-\alpha_1^2-\alpha_2^2+\alpha_3^2\end{pmatrix}\begin{pmatrix}R'_{1s}\\ R'_{2s}\\ R'_{3s}\end{pmatrix}$$

It is then useful to represent the coordinates in a reference system and to refer to the inertial system through the auxilliary variables α.

There are three conditions that may be imposed on the internal coordinates on account of the three degrees of freedom in the Euler-Rodrigues parameter set. An internal system may be chosen so as to diagonalize the tensor of moments of inertia, but the precise choice is of little significance for the analysis of reactive transformations, where the classification of the modes of motion is inadequate due to the strong coupling between modes in different channels. It should be realized that when reactants and products are well separated there is a dominant contribution to the internal coordinates from the vector that describes the distance between the species. The typical binary reaction, $A+B \Rightarrow C + D$, has linear initial and final conformations at asymptotically large separations, and it is then conventional to choose this axis as the internal third axis. Particular internal coordinates are associated with limiting situations where the reaction is a one-particle exchange, such as in the triatomic hydrogen system where one atom replaces one of those in the diatomic. Thus we choose a representation where the cluster of *N-1* particles is placed at the origin and the one labeled s is far away. The internal coordinates are then given as

$$\mathbf{r}_j^s = \mathbf{R}_s \mathbf{D}_{ss} \mathbf{K}_{sj}$$

and we see that there is a factorization in terms of a length, given by the position $\mathbf{R}_S$ and the mass factor $\mathbf{D}_{SS}$, and a direction in the *N-1* dimensional kinematic space as the vector $\{\mathbf{K}_{sj} | j = 1, N-1\}$.

A useful concept for the discussion of different arrangements of a system of mass point is the hyperspherical one. The internal coordinates, $\{\mathbf{r}_j | j = 1, N-1\}$, define a hyperspherical radius as the translational and rotational invariant

$$q = \sqrt{\sum_{j=1}^{N-1} |\mathbf{r}_j|^2} = \sqrt{\sum_{j=1}^{N-1} |\mathbf{r}'_j|^2}$$

It follows that the hyperradius for the previously defined case is

$$q_s = \sqrt{\sum_{j=1}^{N-1} |\mathbf{r}_j^s|^2} = |\mathbf{R}_s| \mu_s \sqrt{1-\mu_s^2}\,.$$

The length concept provides a measure of the distance, but in order to allow for the isotropy of the space it is necessary to account for possible overall rotations when comparing two conformations. Thus we define the distance between two conformations, each given by a set of internal coordinates, as the least value of a form:

$$d_{st} = \min_{\phi,\mathbf{n}} \sqrt{\sum_{j=1}^{N-1} \left| \mathbf{r}_j^s - \mathbf{r}_j^t \cos\phi - \mathbf{n}\left(\mathbf{n}\cdot\mathbf{r}_j^t\right)(1-\cos\phi) - \left(\mathbf{n}\times\mathbf{r}_j^t\right)\sin\phi \right|^2}\,.$$

It follows that a kinematic angle may be defined through the expression

$$\omega_{st} = \arccos\left(\frac{q_s^2 + q_t^2 - d_{st}^2}{2q_s q_t}\right)$$

and that we have a vehicle for the description of processes in terms of motions in the *3N-6* dimensional space where the different conformations. This aspect has been stressed by Fano.

A particular result of the previous analysis is the concept of an angular distance between different arrangements. It follows that an elementary process where one particle is exchanged by another in a system is represented by an angular displacement in the internal kinematic space of the magnitude

$$\omega_{st} = \arccos\sqrt{\frac{\mu_s^2\mu_t^2}{\left(1-\mu_s^2\right)\left(1-\mu_t^2\right)}} = \arccos\sqrt{\frac{m_s m_t}{(M-m_s)(M-m_t)}}$$

and we recognize the familiar 60° angle for the case of three identical hydrogen atoms as used by Eyring, Gershinowitz, and Sun [4]. Four particle systems have linear asymptotic conformations with four kinds of 1+3 arrangements and three different 2+2 arrangements. The latter ones have asymptotic coordinates of the form

$$\mathbf{r}_j^{ss'} = \mathbf{R}_s\left(\mathbf{D}_{ss}\mathbf{K}_{sj} + \mathbf{D}_{s's'}\mathbf{K}_{s'j}\right)$$

with the hyperradius

$$q_{ss'} = \sqrt{\sum_{j=1}^{N-1}\left|\mathbf{r}_j^{ss'}\right|^2} = \left|\mathbf{R}_s\right|\sqrt{\left(\mu_s^2+\mu_{s'}^2\right)\left(1+\delta_{ss'}-\mu_s^2-\mu_{s'}^2\right)}.$$

Angular displacements are readily calculated as

$$\omega_{ss',tt'} = \arccos\frac{\left|\mu_s^2(\delta_{st}+\delta_{st'})+\mu_{s'}^2(\delta_{s't}+\delta_{s't'})-\left(\mu_s^2+\mu_{s'}^2\right)\left(\mu_t^2+\mu_{t'}^2\right)\right|}{\sqrt{\left(\mu_s^2+\mu_{s'}^2\right)\left(1+\delta_{ss'}-\mu_s^2-\mu_{s'}^2\right)\left(\mu_t^2+\mu_{t'}^2\right)\left(1+\delta_{tt'}-\mu_t^2-\mu_{t'}^2\right)}}$$

and it is seen that $\omega_{12,34} = 0$ since the situation when the pair (12) is far away from the pair (34) at the origin is kinematically equivalent to the reciprocal one when there are only four particles.

An example is provided by the frequently studied water-hydrogen atom and hydroxyl-hydrogen molecule system. There are seven distinct axis for asymptotically linear conformations and their relative kinematic directions are given in degrees as:

Channel	O	H_a	H_b	H_c	OH_a	OH_b	OH_c
O	0	57	57	57	37.6	37.6	37.6
H_a	57	0	86.8	86.8	85.4	46.6	46.6
H_b	57	86.8	0	86.8	46.6	85.4	46.6
H_c	57	86.8	86.8	0	46.6	46.6	85.4
OH_a	37.6	85.4	46.6	46.6	0	63.8	63.8
OH_b	37.6	46.6	85.4	46.6	63.8	0	63.8
OH_c	37.6	46.6	46.6	85.4	63.8	63.8	0

The three equivalent axes for the separations $H\text{-}H_2O$ form a set of nearly orthogonal directions while the set for separations $H_2\text{-}HO$ are at about 64° from one another. Each of the latter are close to the bisector of a pair of the former. We find that the process

$$H_2 + HO \Leftrightarrow H + H_2O$$

represents an angular displacement of nearly 47°.

Our development indicates that a useful formulation for the discussion of molecular processes derives from a formulation in terms of internal coordinates separated into a size variable, the hyperradius, and directional coordinates in a kinematic space where total orientational coordinates are eliminated. These concepts will be further developed from the quantum mechanical variational principle in the Wigner-Eisenbud R-matrix theory.

DIGRESSION ON SCATTERING THEORY.

Reaction rate theory has numerous equivalent formulations. The result is eventually put in terms of elements of the unitary S-matrix which expresses the transformation between a set of solutions of the time-independent Schrödinger equation with incoming boundary conditions and another set with outgoing boundary conditions. It was due to Wigner and Eisenbud [13] that a theory appeared where the asymptotic boundary conditions were transferred to the boundary of the effective interaction region and expressed in terms of an R-matrix. Applications of R-matrix methods are numerous in the theory of nuclear reactions, electron scattering, and in molecular reaction dynamics. It is particularly well suited to the hyperspherical description where the interactive region can be enclosed in a hypersphere and the various entrance and exit channels correspond to distinct narrow "windows" on the surface of the hypersphere.

The variational expression is a functional containing the trial wave function, $\Psi(x)$, for the system inside the surface and an auxiliary function, $\Phi(x)$, on the boundary. Appropriate units are chosen so as remove the explicit appearance of Planck's constant and the total mass of the system. The set of internal coordinates is represented as a vector,

$$x = \left\{ r_{1,1}, r_{2,1}, r_{3,1}, \ldots, r_{1,N-1}, r_{2,N-1}, r_{3,N-1} \right\}$$

the interior, interactive region Ω and its boundary Γ are defined as

$$\Omega : \left\{ x \middle| |x| < Q \right\}; \Gamma : \left\{ x \middle| |x| = Q \right\}.$$

Thus we define the functional

$$J(\Psi,\Phi)=\int_{\Omega}dx\left\{[E-W(x)]|\Psi(x)|^2-|\nabla\Psi(x)|^2\right\}$$
$$+\int_{\Gamma}dx\left[\Psi^*(x)\Phi(x)+\Phi^*(x)\Psi(x)\right]$$
$$-\int_{\Gamma}dx\Phi^*(x)\int_{\Gamma}dx'\mathbf{R}(x,x')\Phi(x')$$

and conclude that stationarity of $J(\Psi,\Phi)$ implies that the Schrödinger equation applies,

$$(W(x)-E)\Psi(x)+\nabla^2\Psi(x)=0; x\in\Omega,$$

and that the normal derivative at the boundary satisfies the relations

$$x\cdot\nabla\Psi(x)=Q\Phi(x); x\in\Gamma,$$
$$\Psi(x)=\int_{\Gamma}dx'\mathbf{R}(x,x')\Phi(x').$$

The kernel $\mathbf{R}(x,x')$ is the continuous representation of the R-matrix and determines the coupling among the wave functions in the asymptotic region, $|x|>Q$, where their form is provided by the separable problem for the initial and final situations.

The potential function $W(x)$ is a translational and rotational invariant and while the former property is explicitly utilized in the variational formulation the latter remains to consider. It will also be assumed that the potential is invariant under inversion and that, as pronounced by Weyl, the $W(x)$ can be expressed as a function only of the various scalar products of the internal coordinate vectors. Equivalently we assume that the interparticle distances are the essential variables in W. These assumptions cannot be applied to the wave function except for a state with zero total angular momentum. A general wave function may be expanded in terms of the irreducible representations of the rotation group and it is relevant here to indicate the steps of this analysis.

Euler-Rodrigues parameters will be used for the elements of the rotation group and the coordinate vector x with components given in a fixed reference frame is expressed in terms of a vector y where the components refer to an internal reference frame:

$$x=\tilde{\mathrm{G}}(\alpha)y=\begin{bmatrix}\mathrm{G}(\alpha) & \mathbf{0} & \cdots & \mathbf{0}\\ \mathbf{0} & \mathrm{G}(\alpha) & \cdots & \mathbf{0}\\ \vdots & \vdots & \ddots & \vdots\\ \mathbf{0} & \mathbf{0} & \cdots & \mathrm{G}(\alpha)\end{bmatrix}y$$

The previously defined matrix

$$\mathbf{G}(\alpha) \equiv \begin{pmatrix} \alpha_0^2+\alpha_1^2-\alpha_2^2-\alpha_3^2 & 2\alpha_1\alpha_2-2\alpha_0\alpha_3 & 2\alpha_1\alpha_3+2\alpha_0\alpha_2 \\ 2\alpha_1\alpha_2+2\alpha_0\alpha_3 & \alpha_0^2-\alpha_1^2+\alpha_2^2-\alpha_3^2 & 2\alpha_2\alpha_3-2\alpha_0\alpha_1 \\ 2\alpha_3\alpha_1-2\alpha_0\alpha_2 & 2\alpha_3\alpha_2+2\alpha_0\alpha_1 & \alpha_0^2-\alpha_1^2-\alpha_2^2+\alpha_3^2 \end{pmatrix}$$

occurs as blocks along the diagonal and the coordinates of y may be subjected to three conditions. Interparticle distances are independent of the Euler-Rodrigues parameters and it is then conventional to seek a wave function representation in terms of products of α-dependent functions and rotational invariants:

$$\Psi(x) = \sum_j \Psi_j(y)\mathrm{A}_j(\alpha),$$

and we proceed to calculate the gradient.

The gradient of the new variables in the wave function are obtained by implicit differentiation of the defining relation and the observation

$$\nabla\mathbf{G}(\alpha) = \begin{bmatrix} 0 & u_3 & -u_2 \\ -u_3 & 0 & u_1 \\ u_2 & -u_1 & 0 \end{bmatrix}\mathbf{G}(\alpha) = \mathbf{G}(\alpha)\begin{bmatrix} 0 & v_3 & -v_2 \\ -v_3 & 0 & v_1 \\ v_2 & -v_1 & 0 \end{bmatrix}$$

where the vectors $\{u_j, v_k\}$ in the *3N-3* space are defined as

$$2\begin{bmatrix} \alpha_0 & \alpha_1 & \alpha_2 & \alpha_3 \\ \alpha_1 & -\alpha_0 & \alpha_3 & -\alpha_2 \\ \alpha_2 & -\alpha_3 & -\alpha_0 & \alpha_1 \\ \alpha_3 & \alpha_2 & -\alpha_1 & -\alpha_0 \end{bmatrix}\begin{bmatrix} \nabla\alpha_0 \\ \nabla\alpha_1 \\ \nabla\alpha_2 \\ \nabla\alpha_3 \end{bmatrix} = \begin{bmatrix} 0 \\ u_1 \\ u_2 \\ u_3 \end{bmatrix}$$

and

$$2\begin{bmatrix} \alpha_0 & \alpha_1 & \alpha_2 & \alpha_3 \\ \alpha_1 & -\alpha_0 & -\alpha_3 & \alpha_2 \\ \alpha_2 & \alpha_3 & -\alpha_0 & -\alpha_1 \\ \alpha_3 & -\alpha_2 & \alpha_1 & -\alpha_0 \end{bmatrix}\begin{bmatrix} \nabla\alpha_0 \\ \nabla\alpha_1 \\ \nabla\alpha_2 \\ \nabla\alpha_3 \end{bmatrix} = \begin{bmatrix} 0 \\ v_1 \\ v_2 \\ v_3 \end{bmatrix}$$

respectively. The unit length property of the four-vector α is accounted for as seen by the zeroes on the right hand sides. One arrives then at the formula

$$\nabla \mathrm{G}(\alpha)=\sum_{j=0}^{3}\frac{\partial \mathrm{G}}{\partial \alpha_j}\nabla \alpha_j=\sum_{k=1}^{3} iJ_k \mathrm{G} u_k=\sum_{k=1}^{3} iP_k \mathrm{G} v_k$$

where the new differential operators are

$$\frac{-i}{2}\begin{bmatrix}\alpha_0 & \alpha_1 & \alpha_2 & \alpha_3\\ \alpha_1 & -\alpha_0 & \alpha_3 & -\alpha_2\\ \alpha_2 & -\alpha_3 & -\alpha_0 & \alpha_1\\ \alpha_3 & \alpha_2 & -\alpha_1 & -\alpha_0\end{bmatrix}\begin{bmatrix}\partial/\partial\alpha_0\\ \partial/\partial\alpha_1\\ \partial/\partial\alpha_2\\ \partial/\partial\alpha_3\end{bmatrix}=\begin{bmatrix}J_0\\ J_1\\ J_2\\ J_3\end{bmatrix}$$

and

$$\frac{-i}{2}\begin{bmatrix}\alpha_0 & \alpha_1 & \alpha_2 & \alpha_3\\ \alpha_1 & -\alpha_0 & -\alpha_3 & \alpha_2\\ \alpha_2 & \alpha_3 & -\alpha_0 & -\alpha_1\\ \alpha_3 & -\alpha_2 & \alpha_1 & -\alpha_0\end{bmatrix}\begin{bmatrix}\partial/\partial\alpha_0\\ \partial/\partial\alpha_1\\ \partial/\partial\alpha_2\\ \partial/\partial\alpha_3\end{bmatrix}=\begin{bmatrix}J_0\\ P_1\\ P_2\\ P_3\end{bmatrix}$$

These operators are the representatives of the angular momentum operators in the fixed reference frame, (J_1,J_2,J_3), and in the rotated frame, (P_1,P_2,P_3). A coordinate expression for this is

$$\begin{aligned}\nabla x &= \nabla\left[\tilde{\mathrm{G}}(\alpha)y\right]\\ &= i\left(u_1 j_1 x+u_2 j_2 x+u_3 j_3 x\right)+\tilde{\mathrm{G}}(\alpha)\nabla y\\ &= \tilde{\mathrm{G}}(\alpha)\left[i\left(v_1 p_1 y+v_2 p_2 y+v_3 p_3 y\right)+\nabla y\right]\end{aligned}$$

which implies that the operators (j_1,j_2,j_3) are the familiar angular momentum operators in the coordinate representation while (p_1,p_2,p_3) have the same form but apply to the new coordinates y. At this point we introduce the constraints on the new coordinates which simplify the following analysis. The rotated frame is chosen such that the tensor of moments of inertia is diagonal and thus we define

$$F_1(y) \equiv \sum_{k=1}^{3N-3}(p_2y)_k^*(p_3y)_k; F_2(y) \equiv \sum_{k=1}^{3N-3}(p_3y)_k^*(p_1y)_k; F_3(y) \equiv \sum_{k=1}^{3N-3}(p_1y)_k^*(p_2y)_k$$

and require that they all vanish:

$$F_1(y) = F_2(y) = F_3(y) = 0.$$

The diagonal elements are the principal moments of inertia, in the present units, with the expressions

$$I_j(y) = \sum_{k=1}^{3N-3}(p_jy)_k^*(p_jy)_k, j = 1,2,3.$$

It also holds that

$$0 = \nabla F_1(y) = \sum_{k=1}^{3N-3}(p_2p_3y + p_3p_2y)_k \nabla y_k$$

since the p_j are hermitian. Further calculations give results such as

$$\sum_{k=1}^{3N-3}(p_2p_3y + p_3p_2y)_k(p_1y)_k = i[I_3(y) - I_2(y)]$$

and its cyclic equivalents while other expressions with three p's vanish.

The original coordinates x are independent and orthogonal and the orthogonality of the tranformation $\tilde{G}(\alpha)$ allows us to conclude that a set of orthonormal vectors exist as the forms

$$\hat{f}_k = iv_1(p_1y)_k + iv_2(p_2y)_k + iv_3(p_3y)_k + \nabla y_k$$

from which it is concluded that

$$v_1 = \sum_{k=1}^{3N-3}\hat{f}_k \frac{\partial F_1}{\partial y_k}\frac{1}{I_2 - I_3}$$

with its cyclic equivalents.

The gradient of the wave function

$$\nabla\Psi(x)=\nabla\Psi(\alpha,y)=\sum_{k=1}^{3}iP_k\Psi(\alpha,y)v_k+\sum_{k=1}^{3N-3}\frac{\partial\Psi(\alpha,y)}{\partial y_k}\nabla y_k$$

obtains, with a substitution for ∇y_k,as the expression

$$\nabla\Psi(x)=\nabla\Psi(\alpha,y)=\sum_{k=1}^{3}iP_k\Psi(\alpha,y)v_k+\sum_{k=1}^{3N-3}\frac{\partial\Psi(\alpha,y)}{\partial y_k}\hat{f}_k$$

since the wave function depends on the y's through rotational invariants only. Thus

$$\sum_{k=1}^{3N-3}\frac{\partial\Psi(\alpha,y)}{\partial y_k}(p_jy)_k=0, j=1,2,3.$$

This feature will be further exploited in the choice of the invariants to be the primary variables. One of these is the hyperradius, to be denoted q, with the definition

$$q\equiv|x|\equiv|y|=\sqrt{\sum_{k=1}^{3N-3}y_k^2}.$$

It has the properties that

$$\sum_{k=1}^{3N-3}\frac{\partial q}{\partial y_k}\frac{\partial F_j(y)}{\partial y_k}=\frac{2}{q}F_j(y)=0;\ \sum_{k=1}^{3N-3}\frac{\partial q}{\partial y_k}(p_jy)_k=0.$$

Additional invariants will not be explicitly introduced at this point, they will typically be interparticle distances divided by the hyperradius. Here they are represented by a set $\{\alpha_k | k=4,3N-4\}$ with the properties

$$\sum_{k=1}^{3N-3}\frac{\partial q}{\partial y_k}\frac{\partial\alpha_{k'}}{\partial y_k}=0;\ \sum_{k=1}^{3N-3}\frac{\partial q}{\partial y_k}(p_jy)_k=0.$$

According to the relations above we can construct a set of vectors,

$$\hat{h}_j = \sum_{k=1}^{3N-3} \hat{f}_k (p_j y)_k, j = 1,2,3;$$

$$\hat{h}_{k'} = \sum_{k=1}^{3N-3} \hat{f}_k \frac{\partial \alpha_{k'}}{\partial y_k}, k' = 4,3N-4$$

$$\hat{h}_{3n-3} = \sum_{k=1}^{3N-3} \hat{f}_k \frac{\partial q}{\partial y_k}.$$

where the first three and the last are orthogonal to all others. Orthogonality among the remaining ones requires that

$$\sum_{k=1}^{3N-3} \frac{\partial \alpha_{k'}}{\partial y_k} \frac{\partial \alpha_{k''}}{\partial y_k} = 0, k' \neq k'',$$

which may be convenient. The situations to be considered later are sometimes discussed more readily in a nonorthogonal basis.

The vectors $\left\{\hat{h}_{k'} | k' = 4,3N-4\right\}$ will not be orthogonal to the set $\left\{v_j | j = 1,2,3\right\}$ and this will result in the Coriolis' coupling between rotational and internal motion. We define the relevant coefficients for this coupling through the expansions

$$v_j = \frac{-i}{I_j} \hat{h}_j + \sum_{k'=4}^{3N-4} \beta_{jk'} \hat{h}_{k'}$$

and obtain for the wave function gradient the form

$$\nabla \Psi(\alpha) = \sum_{j=1}^{3} P_j \Psi(\alpha) \frac{\hat{h}_j}{I_j} + \sum_{k'=4}^{3N-4} \left[\frac{\partial \Psi(\alpha)}{\partial \alpha_{k'}} + \sum_{j=1}^{3} i P_j \Psi(\alpha) \beta_{jk'} \right] \hat{h}_{k'} + \frac{\partial \Psi(\alpha)}{\partial q} \hat{h}_{3N-3}$$

Thus we have the kinetic energy term in the functional as the expression

$$|\nabla \Psi(\alpha)|^2 = \sum_{j=1}^{3} \frac{|P_j \Psi(\alpha)|^2}{I_j} + \left| \frac{\partial \Psi(\alpha)}{\partial q} \right|^2$$
$$+ \sum_{k',k''}^{4,3N-4} \left[\frac{\partial \Psi(\alpha)}{\partial \alpha_{k'}} + \sum_{j=1}^{3} i P_j \Psi(\alpha) \beta_{jk'} \right]^* \hat{h}_{k'}^* \cdot \hat{h}_{k''} \left[\frac{\partial \Psi(\alpha)}{\partial \alpha_{k''}} + \sum_{j=1}^{3} i P_j \Psi(\alpha) \beta_{jk''} \right]$$

which displays the rigid rotor terms and the hyperradial contribution as uncoupled to the others except for the Coriolis' terms. Dimensional considerations show that the scalar products in the last term are proportional to the inverse square of the hyperradius. A careful choice of the internal dimensionless coordinates restricts the Coriolis' terms to three of them. The Jacobian of the coordinate transformation is proportional to the hyperradius to the power *3N-4* and reflects that the Euler-Rodrigues parameters represent a point on the unit sphere in four dimensions.

Our analysis pertains to the general situation when the three moments of inertia are different and larger than zero. The regions of the space where this is not so needs to be given special attention. A linear conformation is characterized by a vanishing moment of inertia and equality between the two others. This is associated with a factor in the Jacobian which goes to zero as $\sqrt{I_1(y)I_2(y)I_3(y)}$ and limits the variable domain of the set $\{\alpha_k | k = 4, 3N-4\}$. A planar conformation is implied by the condition

$$I_3(y) = I_1(y) + I_2(y) = q^2$$

which is particularly relevant for three particle systems where all conformations are planar.

The hyperspherical coordinates are such that we can associate different cluster formations with limited regions on the hypersphere specified by a few generalized coordinates. A prototypical reaction,

$$A + BC \Rightarrow AB + C,$$

requires that the bond between the species B and C is broken and that a new bond is formed between A and B. An illustration of this is offered by a three atom system on a Born-Oppenheimer surface. It is readily calculated that the hyperradius can be expressed in terms of the interatomic distances as

$$q^2 = \frac{m_1 m_2 R_{12}^2 + m_2 m_3 R_{23}^2 + m_3 m_1 R_{31}^2}{(m_1 + m_2 + m_3)^2}$$

and that internal hyperangles can be defined as

$$\alpha_4 = \frac{R_{12} - R_{23} + R_{31}}{2(R_{12} + R_{23} + R_{31})},$$
$$\alpha_5 = \frac{R_{12} + R_{23} - R_{31}}{2(R_{12} + R_{23} + R_{31})},$$
$$\alpha_6 = \frac{-R_{12} + R_{23} + R_{31}}{2(R_{12} + R_{23} + R_{31})} = 1 - \alpha_4 - \alpha_5;$$

where only two are independent. The first of these, α_4, is close to unity when particle 1 is distant from the others, while α_5 and α_6 relate to particles 2 or 3 being distant from the other two. These three coordinates have the property of baricentric coordinates for a triangle and for each hyperradius we can map the potential W onto the triangle. We find that there will be potential troughs near the corners of the triangles, parallel to the opposite sides, when the hyperradius is large and the corresponding diatomics are bound. A symbolic picture of this is offered in the figure:

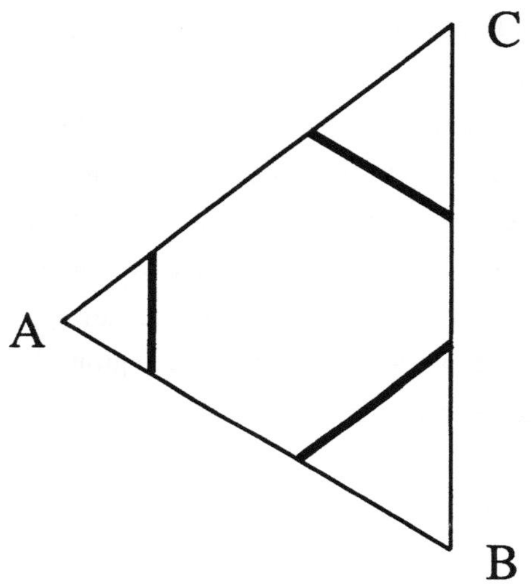

Heavy lines indicate the bottoms of potential troughs. The edges of the triangle represent colinear structures and its center the equilateral triangular conformation. These coordinates give rather awkward expressions for the kinetic energy and have their main usefulness for illustrative purposes. Davidson discussed analytical features of Born-Oppenheimer potentials in a related format.[14]

Four atom systems require five internal hyperangles and an illustration involving all becomes is quite complex. The set of six interparticle distances forms a basis for a reducible representation of the particle permutation group as the direct sum of the totally symmetric, the two dimensional, and one of the three dimensional irreducible ones. A basis for the three dimensional irreducible representation is constructed from the differences between the lengths of opposite edges in the tetrahedron:

$$\left\{R_{12}-R_{34},R_{13}-R_{24},R_{14}-R_{23}\right\}$$

The ranges of these variables are intricately connected and cannot be readily disentangled as a simplex in the fashion which was used for the three-particle system. Characterization of different asymptotic situations in the kinematic space is more easily referred back to the vectors arising in the coordinate transformations. Since there are three internal vectors in a four particle system we have that the kinematic rotations are elements of the three dimensional rotation group and may be parametrized accordingly [15]. A shortest path in the Euler-Rodrigues parameter space can be defined so as effectuate the required kinematic rotation between two asymptotic situations. Such a path and the hyperradius define a subspace of configuration space where the principal changes in the molecular conformation takes place.

The hyperspherical formulation provides the channel concept with a quantitative content as a direction in a high-dimensional space and a corresponding angular displacement from other arrangement channels. Wigner and Eisenbud's R-matrix formulation with a hyperspherical boundary shows that there are windows in the boundary corresponding to open channels in the scattering. Wave functions may be joined through such windows to the asymptotic region Γ and the elements of the scattering matrix are deduced accordingly.

REARRANGEMENT REACTIONS.

Reaction coordinates for rearrangements in complex systems have been sought as paths on many dimensional energy hypersurfaces through application of concepts from differential geometry [16] and applied in variational transition state theory [17]. These procedures are often costly and require extensive calculations and searches in the nuclear conformation space. The class of reactions where the initial and final conformations are identical in geometry and differ through the labeling of the atoms present situations where the two (or more) equilibrium coordinate sets give the same hyperradius. We may define a coordinate system such that the arc length on the hypersphere between the two points is the shortest. Given a transition state geometry it is possible to determine its arc length to the two equilibria. An example of this is offered by the two hydron exchange in the formamidine dimer. The geometries determined by Truong [18] place the reactant and the product conformations on a hypersphere with radius about 2 Bohr radii and 6.48° apart. A concerted transition state was found on a hypersphere with the radius 1.9 Bohr radii and with an angular distance from each of the others of 4.94°. Isotopic substitution changes these angles to 7.04° and 5.14° when one of the "reactive" protons is exchanged with a deuteron and to 7.56° and 5.33° when both are. The reaction path becomes longer for these substitutions and the system is heavier, thus there will be smaller reaction rate as expected. It is clear that the passage from the reactant through the transition state to the product is longer than the direct path and that there is at least one "knee" to be considered so that a model surface requires two coordinates.

Oxalamidine offers an example, similar to the formamidine dimer, where two protons can be transferred. Truong [18] found here that the concerted transition state had a higher energy than points on a two-step path with an intermediate stationary geometry connected with transition states leading to the alternative equilibria. The six structures have hyperradii around 1.6 Bohr radii and their relative angular positions define a six dimensional subspace and cannot be faithfully represented in a plane. A least energy transition path traverses at least a four dimensional subspace defined by the simplex formed by five independent conformations. Angular distances are listed below for the cases of pure protonic species and those with one or two deuterons in the active sites.

Arc length	Protonic	Singly deuterated	Doubly deuterated
R - TS'	7.73°	8.60°	8.52°
TS" - P	7.73°	7.66°	8.52°
TS' - I	7.02°	7.09°	7.06°
I - TS"	7.02°	6.99°	7.06°
R - TS_c	7.51°	7.89°	8.24°
TS_c - P	7.51°	7.89°	8.24°
TS_c - I	4.72°	5.19°	5.61°
R - P	8.04°	9.45°	10.61°

It appears that the least energy reaction path nears an arc length of 30° while the direct route only is 8° for the protonic form. Thus we may question the principle of least motion [19] if only the height of the energy barrier is considered.

Complex geometrical considerations are associated with the study of the rearrangements in the highly degenerate system $C_9H_9^+$, the barbaralyl cation [20]. The equilibrium geometry has C_s symmetry and the nine CH groups occupy the vertices of a polyhedron with two triangular and three hexagonal faces.

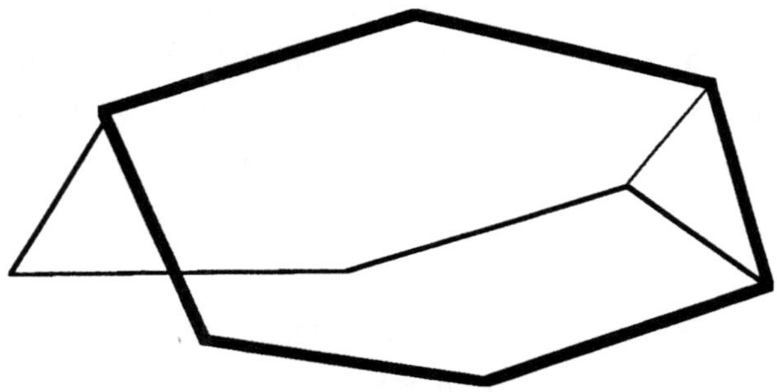

Sketch of the nine vertices of the barbaralyl cation structure.

The two triangular faces are not equivalent while two of the hexagonal ones are. There are 9! = 362 880 equivalent assignments of the groups to the locations and it becomes a major effort to enumerate these and their respective angular distances from one another. Experimental studies have concluded that a rapid rearrangement takes place, at low temperatures, between six equilibrium structures with identical values for the hyperradius. They are formed by a six membered subgroup of the symmetric group S_9 and is realized as the group generated by a twofold "rotation" of the labels around an axis through one of the dual hexagons and the center position of the edge opposing it. There are two, noncommuting, operations of this type and the group is isomorphic to D_3, which has led to speculations that the rearrangements took place through a transition state with D_{3h} symmetry [21]. A direct analysis of the angular distances from accurate electronic structure calculations [22] shows that the changes effectuated by the generators discussed above give the shortest arc lengths. Each such arc is associated with a transition state within a short distance and with C_2 symmetry.

Higher temperatures allows the barbaralyl cation to escape the subspace of the six structures and to access all possible ones. Cremer *et.al.* [22] found that the energetically favored path from a previously defined C_2 structure to an equivalent one *via* a higher order transition state of C_{2v} symmetry. This is formally a cyclic permutation of labels around one hexagon and this operation generates the entire S_9 when combined with the elements of the subgroup. It is possible to picture the set of 9! equilibria as a set of

60 480 clusters of six, each connected, from short edge to short edge, with six neighboring clusters. A selection of the relevant arc lengths [23] are given in the sketches below.

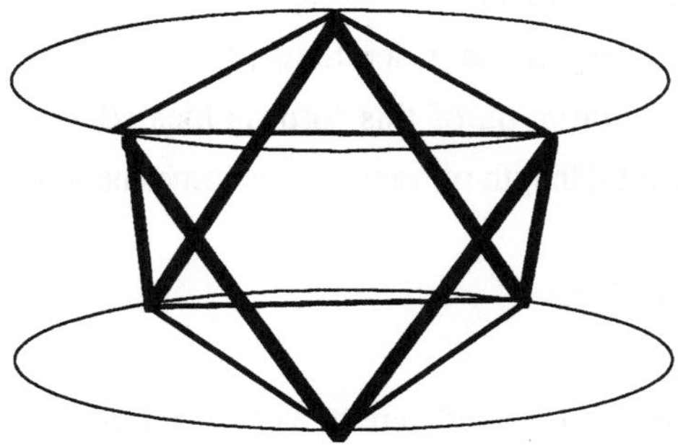

Sketch of a set of six points representing equilibrium conformations of the barbaralyl cation. The heavy lines indicate the shortest arc lengths on the hypersphere and they are 10.41°. The sides of the equilateral triangles on top and bottom are 13.25° while the diagonal of the octahedron has the length 12.74°. (The circles indicate only the planes through next nearest neighbors.)

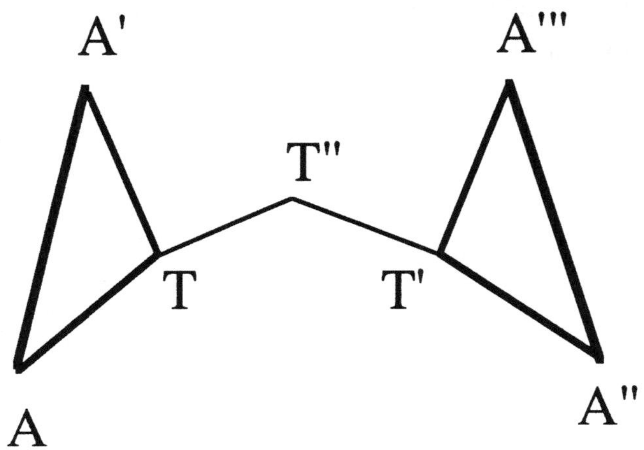

Sketch of the transition paths between equivalent C_S-structures, the A's, of the barbaralyl cation. The arc lengths AA' and A"A"' are those of nearest neighbors, 10.41°.

Those of AT, A'T, A"T', and A'"T' equal 6.43°, and between T and T", or T' and T", there is 4.22°. A direct path TT' requires a displacement of 8.16°. The many dimensional nature of the picture is evident when the direct distance AA" (or A'A'") is found to be 14.30°, while AA'" (or A'A") equals 15.23°.

Path length measures in terms of angular displacement can give quantitative content the *principle of least motion* [19] and for the barbaralyl cation it holds that the favorable rearrangements occur between near neighbors. Passage through a D_{3h} -structure is disfavored by energetic considerations [22] as well as by the path length it requires. An optimal, local minimum, of this form is located 7.92° from each of its six associated forms and the total length of transition becomes nearly 16°.

CONCLUSIONS.

Examples have been offered here of a characterization of the various reactive channels in bimolecular scattering and of initial and final conformations in unimolecular reactions as related to locations on a hypersphere. It is argued, in agreement with the *principle of least motion* , that the angular displacement is the smallest for the most favored transition. This is particularly relevant for rearrangements which take place with little change in the hyperradius.

ACKNOWLEDGEMENT.

The author is grateful to Professor Thanh Truong for providing data for the formamidine dimer and oxalamidine.

REFERENCES

1. Heitler, W. and London, F. (1927) Z. *Physik* **44**, 455.
2. Born, M. and Oppenheimer (1927) J.R. *Ann. Phys. (Germany)* **84**, 457.
3. Eyring, H. (1935) *J. Chem. Phys.* **3**, 107.
4. Eyring, H., Gershinowitz, H., and Sun, C.E. (1935) *J. Chem. Phys.* **3**, 786.
5. Gronwall, T.H. (1937) *Phys. Rev.* **51**, 695.
6. Bartlett Jr, J.H. (1937) *Phys. Rev.* **51**, 661.
7. Wannier, G.H. (1953) *Phys. Rev.* **90**, 817.
8. Macek, J. (1967) *Phys. Rev.* **160**, 170.
   Fano, U. (1981) *Phys. Rev.* **A24**, 2402.
   Fano, U. and Rau, A.R.P. (1986) *Atomic Collisions and Spectra*, Academic press, Orlando.
9. Kuppermann, A., Kaye, J.A., Dwyer, J.P. (1980) *Chem. Phys. Letters* **74**, 257.
10. Mead, C.A. (1980) *J. Chem. Phys.* **72**, 3839.
11. Manz, J. and Römelt, J. (1985) *Nachr. Chem. Tech. Lab.* **33**, 210.
12. Smith, F.T. (1980) *Phys. Rev. Lett.* **45**, 1157.
13. Wigner, F.P. and Eisenbud, L. (1947) *Phys. Rev.* **72**, 29.
14. Davidson, E.R.(1977) *J. Am. Chem. Soc.* **99**, 397.
15. Öhrn, Y. and Linderberg, J. (1983) *Mol. Phys.* **49**, 53.
16. Tachibana, A. and Fukui, K. (1978)*Theoret. Chim. Acta (Berlin)* **49**, 321.
17. Truhlar, D.G., Hase, W.L., and Hynes, J. T. (1983) *J. Phys. Chem.* **87**, 2664, 5523.
18. Truong, T.N. *private communication..*
19. Rice, F.O. and Teller, E. (1938) *J. Chem. Phys.* **6**, 489.
20. Ahlberg, P., Jonsäll, G., and Engdahl, C. (1983) *Adv. Phys. Org. Chem.* **19**, 223.
21. Hoffmann, R., Stohrer, W.-D., and Goldstein, W.-D. (1972) *Bull. Chem. Soc. Jpn.* **45**, 2513.
22. Cremer, D., Svensson, P., Kraka, E., and Ahlberg, P. (1993) *J. Am. Chem. Soc.* **115**, 7445, 7457.
23. Larsson, P. E. and Linderberg, J. *to be published.*

EVOLUTION OF ATOM-MOLECULAR EIGENSTATES

U. FANO and JOHN L. BOHN
Department of Physics and James Franck Institute
University of Chicago
5640 S. Ellis Avenue
Chicago, IL 60637 USA

KEYWORDS / ABSTRACT: compact / fragmented / Jost matrices / (hyper)-spherical harmonics expansion / eigenphases / eigenchannels / R-matrix

A prototype is outlined, aimed at replacing current procedures of quantum chemistry, which generalizes to multiparticle systems the treatment of a single particle in a potential well. Diagonalization of R-matrices documents each eigenchannel's evolution from compaction to fragmentation; initial results for doubly excited helium illustrate the wave functions' evolution.

Stationary states of atoms, molecules and molecular complexes reflect mainly the transition from dominance of centrifugal forces when compressed (c) to that of Coulomb interactions when fragmented (f). Jost matrices [1, 2] ($J^{\pm}_{cf}$ with indices c and f = 1, 2, ...) interlink the contrasting features of the limits (c, f), providing all observable cross sections and spectral data. We aim at describing the evolution of energy eigenstates from their c limit to their f limit as a function of a hyperradius parameter R, thereby constructing the relevant Jost matrices.

Displaying this evolution contrasts with the usual procedures of atomic theory and quantum chemistry, whose computations yield results without tracing their origins. We thus view here quantum chemistry processes in the frame of the partial wave expansion of collision theory, rather than in the frame of Hartree-Fock and Born-Oppenheimer procedures.

C. A. Tsipis et al. (eds.), New Methods in Quantum Theory, 225–232.

This shift of reference frames appears rich with promising implications. Even static properties of multiparticle systems (energy levels, transition amplitudes, etc.) fit in our frame naturally.

1. Approach

The simple example of a single particle moving in a spherically symmetric potential-well affords conservation of angular momentum, thus separating radial from angular motions. The latter have spherical harmonic eigenfunctions $Y_{lm}(\theta,\varphi)$; to each of them belongs a radial wave function $f_l(r)$ that propagates from the well's center to infinity, being characterized outside the well by a phaseshift δ_l at positive energies, and by discrete negative energy levels at which $f_l(r)$ fails to diverge as $r \to \infty$.

Full separation of radial and angular motions fails in the absence of spherical symmetry under rotations about the well's center. Eigenfunctions $F_{l_0 m_0}(r,\theta,\varphi) \propto Y_{l_0 m_0}(\theta,\varphi)$ as $r \to 0$ turn then into superpositions

$$F_{l_0 m_0}(r,\theta,\varphi) = \Sigma_{lm} F_{l_0 m_0, lm}(r) Y_{lm}(\theta,\varphi). \tag{1}$$

As r increases, separability holds again beyond the potential-well's rim at radial distances $r > r_0$, where it is demonstrated by constructing superpositions

$$\Sigma_{l_0 m_0} A^{(\rho)}_{l_0 m_0} F_{l_0 m_0}(r,\theta,\varphi) = \sin(kr + \phi_\rho) \Sigma_{lm} b^{\rho}_{lm} Y_{lm}(\theta,\varphi) \tag{2}$$

whose normal (i.e. radial) logarithmic derivative remains uniform over each spherical surface at $r > r_0$. The eigenvector sets $\left\{A^{(\rho)}_{l_0 m_0}\right\}$ of the relevant homogeneous linear problem form "Jost matrices" with rows (ρ) and columns $(l_0 m_0)$ [1, 2]. These matrices embody the effect of all asymmetries within the well that intermix $(l_0 m_0)$ and (lm) "channels"; they thus provide all the information required to solve scattering problems at positive energies and to determine energy levels and other features at negative energies.

Calogero and collaborators [3] introduced an illustration of the *channel mixing* process by considering sequences of potential wells *truncated* at successively larger radial distances r_0. This physico-mathematical procedure has been dubbed an "invariant embedding".

The logical structure thus developed to describe the motion of a single particle within and about a potential-well $V(r,\theta,\varphi)$ extends readily to describe the motions of multi-particle systems under the influence of all the particle interactions. The abstract ($3n$)-dimensional space of n-particle coordinates replaces now the physical space. Choosing the particles' center of mass as the center of their mock-potential-well reduces the dimensions to $3(n-1)$. The *inertial radius*.

$$R = [\Sigma_{i=1}^{n} m_i r_i^2 \,/\, M]^{1/2}, \qquad M = \Sigma_i m_i \tag{3}$$

replaces the potential-well's r by a "hyper-radial" polar coordinate that becomes separable in the $R \to \infty$ limit.

1.1 EXTENSION Studies of multiparticle systems in hyperspherical coordinates utilized initially an *adiabatic* approximation, serviceable only over a limited range of applications [2, 4]. The alternative Calogero procedure of direct integration of close-coupled ordinary equations in R has been pursued in our group since 1988, with preliminary results reported at the Copenhagen 1991 meeting [5]. The main qualitative result, already apparent at that time, reveals how sets of $(\lambda\vec{\mu})$ channels-analogues of the eigenchannels (2) - focus early, as R_0 increases, toward the fragmentations to be achieved as $R_0 \to \infty$.

The $(\lambda\vec{\mu})$ labels of these channels, corresponding to the labels of the $Y_{lm}(\theta,\varphi)$ functions, characterize hyperspherical harmonic functions $Y_{\lambda\vec{\mu}}(\hat{R})$ with a number of indices $\vec{\mu}$ equal to that of $\hat{R}$'s independent components. The set of $(\lambda\vec{\mu})$ spans a space corresponding to the phase space of classical mechanics. Within this space eigenfunctions $\psi(\vec{R})$ of time-independent Schrödinger equations develop as R increases, much as classical trajectories of a phase space do in the course of time [6]. Full mastery of the hyperspherical harmonics' rich manifolds remains a rather distant objective; progress toward this objective limits current applications [7]. A broader aspect of the same effort, the "Geometry of Multiparticle Systems", forms the subject of a forthcoming paper [8].

2. Analytical Treatment.

As the spherical polar coordinates facilitate the partial-wave treatment of a single particle in a potential well, so do polar hyperspherical coordinates (3), $\vec{R} \equiv (R, \hat{R})$, for aggregates of n particles with pair Coulomb interactions. The Laplacian $\Delta_{\vec{R}}$ resolves into two terms, $\Delta_{\vec{R}} = \partial^2 / \partial R^2 + \Delta_{\hat{R}} / R^2$, as it does for a single particle; the sum over all pairs' interactions takes the simple form $Z(\hat{R}) / R$, with the "effective charge" Z depending only on the mutual orientations $m_i^{1/2} \hat{r}_i$ of the particles about their center of mass.

The aggregate's energy eigenfunctions,

$$\Psi(\vec{R}) = \Sigma_{\lambda\vec{\mu}} F_{\lambda\vec{\mu}}(R) \mathrm{Y}_{\lambda\vec{\mu}}(\hat{R}) \tag{4}$$

normalized at $R \to 0$ as

$$F_{\lambda_0\vec{\mu}_0,\lambda\vec{\mu}} \propto R^{\lambda_0+(3n-4)/2} \delta_{\lambda_0\vec{\mu}_0,\lambda\vec{\mu}} \quad \text{at } R \to 0, \tag{4'}$$

obey then the Schrödinger equation system

$$\frac{d^2}{dR^2} F_{\lambda_0\vec{\mu}_0,\lambda\vec{\mu}}(R) + \Sigma_{\lambda'\vec{\mu}'} (\lambda\vec{\mu} | k^2 | \lambda'\vec{\mu}') F_{\lambda_0\vec{\mu}_0,\lambda'\vec{\mu}'}(R) = 0 \tag{5}$$

with the squared wavenumber matrix

$$(\lambda\vec{\mu} | k^2 | \lambda'\vec{\mu}') = \left[2ME - \frac{[\lambda + (3N-5)/2]^2 - 1/4}{R^2} \right] \delta_{\lambda\vec{\mu},\lambda'\vec{\mu}'} + \\ + \frac{(\lambda\vec{\mu} | Z(\hat{R}) | \lambda'\vec{\mu}')}{R}. \tag{5'}$$

Main features of the governing Eq. (5) are:

1) The set of eigenfunction indices $\{\lambda\vec{\mu}\}$, seemingly infinite, reduces effectively to a finite dimension at *any* finite value of R, owing to the repulsive action of the second (centrifugal) term of Eq. (5'), $\propto \lambda^2 / R^2$ at large λ. This effect is embodied in the exponent λ_0 of the boundary condition (4')

2) The basic mechanism embodied in Eq. (5) lies in the shifting balance of the second and third terms on the right of (5'), the centrifugal one dominating at shorter ranges R, the third one at larger R. Analogues of the long-range eigenchannels (2) - i.e. *fragmentation channels* - stabilize accordingly astride eigenvectors of the matrix $(\lambda\vec{\mu}|Z(R)|\lambda'\vec{\mu}')$.

3) Eigenfunctions $\psi(R;\hat{R})$, analogues to single particles' (2) at alternative finite ranges R_0, are represented by *standing waves*, implying that each of their incoming components $a_{\lambda\vec{\mu}}(R_0)\exp(-iKR_0)$ is reflected back by the interactions over the range $0<R<R_0$ with *unchanged coefficients* $|a_{\lambda\vec{\mu}}|$ i.e., is backscattered *elastically*.

These particular energy eigenfunctions are identified as the eigenvectors of the "*short-range* R_0" R-matrix - equivalent to a scattering matrix - thus embodying the effect of *all* particle interactions $Z(\hat{R})/R$ *truncated* at R_0. Their pilot evaluation, for ^1S states of He in ref. [9], labels each of them with the index ρ. Each of them displays an eigenphase

$$\phi_\rho(R_0) = \cot\mathrm{an}^{-1}\left[d\ln\psi^{\rho}_{\lambda\vec{\mu}}(R)/dR\right]_{R_0} \tag{6}$$

equal for all $\lambda\vec{\mu}$ *components*. Ref. [9] has calculated these eigenphases, and their corresponding eigenvectors $a^{(\rho)}_{\lambda\vec{\mu}}(R) \equiv (\rho|\lambda\vec{\mu})$ by solving numerically the pair of equations [(14) in Ref. [9]],

$$d\phi_\rho / dR = \cos^2\varphi_\rho(R) + \Sigma_{\rho'}\bar{k}^2_{\rho\rho'}\sin^2\phi_{\rho'} \tag{7a}$$

$$\frac{d(\rho|\lambda\vec{\mu})}{dR} = \Sigma_{\rho'}\frac{\sin^2\phi_\rho\bar{k}^2_{\rho\rho'}\sin^2\phi_{\rho'}}{\sin(\phi_\rho-\phi_{\rho'})}(\rho'|\lambda\vec{\mu}). \tag{7b}$$

with the transform of (5').

$$\bar{k}^2_{\rho\rho'}(R) = \Sigma_{\lambda\vec{\mu},\lambda'\vec{\mu}'}(\rho|\lambda\vec{\mu})(\lambda\vec{\mu}|k^2|\lambda'\vec{\mu}')(\lambda'\vec{\mu}'|\rho') \tag{8}$$

This procedure proved rather economic, extending in R until all relevant effects of interaction had become manifest.

A major feature of the Eqs. (7) emerges at near-degeneraciess of eigenphase pairs $(\phi_\rho, \phi_{\rho'})$ through near-zero values of the denominator in Eq. (7b). The sharp peak of the coupling matrix in this equation prevents any actual crossing of $\phi_\rho(R)$-curve pairs in Fig. 1. There result "avoided crossings", of the Landau-Zener type, having major or lesser effect depending on the magnitude of the numerator in (7b)'s fraction. Most of the avoided crossing in the lower parts of Fig. 1's graphs are not perceptible, implying their "diabatic" behavior, in which pairs of eigenvectors $[(\rho|\lambda\vec{\mu}),(\rho'|\lambda\vec{\mu})]$ and of eigenvalue slopes $[d\rho / dR, d\rho' / dR]$ appear merely to interchange. The major effects of channel mixing stem, however, from the easily perceivable avoided crossings, as indicated by Eq. (7b). Even in these cases, however, the curves attain a nearly perfect interchange of slopes over a wider range of R. The detailed mapping of the channel-mixing process, displayed in Fig. 1 and its analogues, is unprecedented (even though discernible in analogous maps of ref. [7b]). Whether the qualitative aspects of channel mixing can be assessed by fitting Landau-Zener parameters to each avoided crossing remains to be determined.

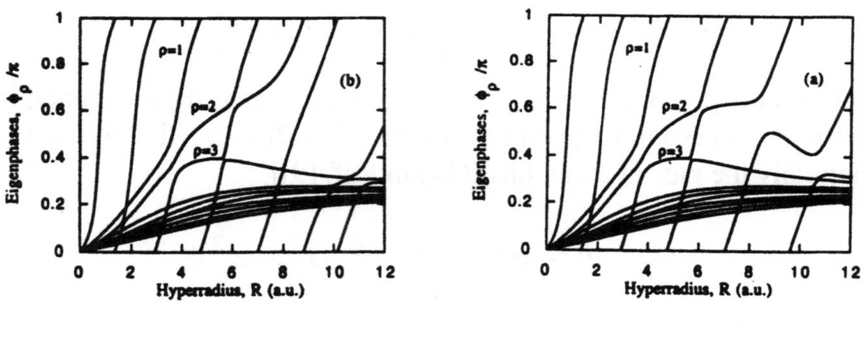

Figure 1. Eigenphases (modulo π) φ_ρ / π in an 18-channel calculation at the pair of total energies (a) E = -0.78 a.u. and (b) E = -0.76 a.u. The $_2(+1,0)_2{}^+$ resonance lies between these energies, as discussed in the text.

Unexplored also remained until recently the extraction, from solutions of the Eqs. (5), of the eigenphases $\phi_\rho(R_0 \to \infty)$ and of the corresponding eigenvectors, required to predict experimental observables.

3. Discussion

The main feature of the approach outlined here appears to lie in its *bypassing* preliminary calculations of particle orbitals, potential curves or surfaces, and other ingredients of independent-particle approaches. Such elements should emerge instead by complementing solutions of Eq. (5) with appropriate harmonics $Y_{\lambda\vec{\mu}}(\hat{R})$.

Treatments of spin-orbit interactions, of relativity, hyperfine and related effects should proceed by complementing the matrix $<\lambda\vec{\mu}|Z(\hat{R})|\lambda'\vec{\mu}'>$ in Eq. (5') and refining its application, as relevant, possibly expanding Z into multipole components. Such procedures would specifically bypass the current reliance on matrices in single-orbital bases. These extended potential applications of the present approach have been identified for some time but remain unexplored owing to manpower limitations.

For example, the initial applications of ref. [9]'s techniques to the simplest molecule of H_2 and to multi-electron atoms (whose matrix of $Z(\hat{R})$ had been calculated in ref. [4]), are currently being pursued slowly by undergraduate students.

Besides pursuing these explorations lies the need to provide a quorum of eigenfunctions (4), demonstrating the internal consistency of the present approach and its ability to achieve a wide range of applications. Past evidence with analogous developments e.g., with the introduction of Quantum Defect procedures, suggests the requirement for a few dozen man-years.

References

1. Newton, R.G. (1988) *Scattering Theory of Waves and Particles*, McGraw Hill, New York.

2. Fano, U. (1981) *Unified treatment of collisions*, Phys. Rev. A, **24**, 2402-2415 ; (1983) ibid. **27**, 1208.

3. Calogero, F. (1967) *Variable Phase Approach to Potential Scattering*, Chap. 19, Academic, New York.

4. Cavagnero, M. (1984) *Electron correlations in atomic shells: Systematics of high angular-momentum admixtures*, Phys. Rev. A, **30**, 1169-1174; (1986) II: *Antisymmetric basis functions*, ibid. **33**, 2877- 2886; (1987)III: *Algebraic form of the Hamiltonian matrix*, ibid **36**, 523-538.

5. Fano, U. (1993) *Dimensional Scaling in Chemical Physics*, in D.R. Herschback et al (eds.), Kluwer, Dordrecht, pp. 461-470.

6. Bohn, J.L. and Fano, U. (1994) *Multichannel quantum mechanics as a Hamiltonian phase flow*, Phys. Rev. A, **50**, 2893-2898.

7. Fano, U. and Sidky, E.Y, (1992) *Semi-analytic theory of Rydberg electron diamagnetism*, Phys. Rev. A, **45**, 4776-4791; Sidky, E.Y, (1993) *Semi-analytic theory of Rydberg electron diamagnetism*, II: ibid., **47**, 2812-2817 ; Bohn, J.L. (1994) *Phase amplitude method applied to doubly excited states of He*(^1S), ibid., **49**, 3761-3769.

8. Fano, U., Green, D. and Bohn, J.L., (1995) *Geometry of multi-particle systems*, to be published.

9. Bohn, J.L., (1995) *Total eigenphase description of multiparticle quantum systems*, Phys. Rev. A, **51**, 1110-1118.

10. Bohn, J.L., (1995) *Development of ridge resonances in helium* (University of Chicago thesis), and to be published.

HYPERANGULAR MOMENTUM: APPLICATIONS TO ATOMIC AND MOLECULAR SCIENCE

V. AQUILANTI, S. CAVALLI, C. COLETTI, D. DE FAZIO AND G. GROSSI
Dipartimento di Chimica
Università di Perugia
I-06123 Perugia, Italy

1. Introduction

These notes deal with the extension of the concept of angular momentum to spaces mathematically larger than the physical one. Such a generalization leads to the introduction of hyperspherical harmonics as orthonormal expansion basis for a variety of problems. The harmonics are eigenfunctions of generalized angular momenta: for them, an algebra can be developed, similar to the one describing coupling and recoupling of ordinary angular momenta; the possibilities which open up (for spaces greater than three) of alternative angular parametrizations of hyperspheres, corresponding to alternative reduction schemes of the rotation group O(n) into its subgroups, are easily visualized through the tree method and enrich the algebra by orthonormal transformations which connect the representations. Any chosen representation carries quantum numbers which appear as labels for the harmonics - specifically, the Casimir's eigenvalues for the group and its subgroups: they are conveniently interpreted as eigenvalues of hyperangular momenta or their projections. The search for alternative coupling schemes and frame transformations aims at finding evidence for approximate symmetries and quasiseparable variables.

As eigenfunctions of the Laplace operators on the hyperspheres, the hyperharmonics can be used to expand potential energy functions and allow to classify - and establish relations among - alternative hyperspherical coordinate systems for three or few body problems. Specifically, asymmetric (or Fock's or Delves') coordinates are related to symmetric (or Smith's) coordinates by rotations on a four dimensional sphere: such a rotation, as well as kinematic rotations among Jacobi vectors for different rearrangement channels, can be explicitly described by orthogonal transformations

C. A. Tsipis et al. (eds.), New Methods in Quantum Theory, 233–250.

belonging to the algebra. Sec. 2 outlines the basic concepts and formulas, summarizing and updating a previous paper [1].

Fock's projection leading to the connection between hydrogenic orbitals and four dimensional harmonics - and to the association of the principal quantum number to a four dimensional angular momentum - can be generalized (sec. 3) to higher dimensions to study atomic and molecular structure from the point of view of the broken symmetry of hyperspheres. The frame transformation from polar to parabolic coordinates of hydrogenic orbitals, corresponding in momentum space to the connection between alternative four dimensional harmonics, is exploited for one-electron fixed nuclei problems, extending the Shibuya-Wulfman treatment for H_2^+. This case and the construction of alternative wavefunctions for the Coulomb three body problems, such as bielectronic orbitals for helium, are presented in sec. 4.1 and 4.2 respectively: the use of hyperharmonics in momentum space - or their counterparts, generalized Sturmians in configuration space - allows all integrals to be written in closed form as matrix elements corresponding to coupling, recoupling or transformation coefficients.

The connection between all these coefficients, which are proper generalizations of Clebsch-Gordan and Racah's coefficients, and hyperharmonics is such that the former can be obtained as the high (hyper)-angular momentum limit of the latter (in the physical space, this corresponds to the semiclassical limit). The specular view is that the coefficients, which are discrete polynomials of the Hahn and Racah's type (essentially finite sums of the ${}_3F_2(1)$ and ${}_4F_3(1)$ type, respectively) can be taken as the discrete analogues of hyperspherical harmonics, when angular ranges are assumed to be discretized in grids, whose points are labeled by quantum numbers corresponding to hyperangular momenta and their projections. This *hyperquantization* technique, which leads to useful representations for molecular collisions, including reactive scattering, is briefly presented in sec. 5.

2. Hyperspherical harmonics as hyperangular momentum wavefunctions

2.1. GENERAL

The purpose of this paper is to present some results on the hyperspherical coordinates for the representation of the many-body problem. Several reviews are now available, [2, 3, 4] and are useful introduction to various aspects of this approach. Although the hyperradius, ρ, the variable which is basic to this method, appears already in Jacobi's investigations on the three-body problem it is in quantum mechanics that its use for the helium problem was first exploited [5]. Since then, the hyperspherical method became one of the main tools in the quantum mechanical treatment of the interaction between few particles. Extensive applications to nuclear, atomic, and molecular processes have been made, and are being made. They involve both studies of bound states, and the treatment of scattering problems.

Substantial advancements both for the semianalytical, qualitative understanding and for quantitative treatment of bound states and reactions has been obtained by the introduction of the hyperradius as a nearly separable variable [6]. The other variables, which, besides the hyperradius are needed for the complete formulation of the problem, can be parametrized as (hyper-)angles spanning the surface of hyperspheres. This can be done in many ways [2, 7], many possibilities exist for exploiting harmonics analysis on the surface of the hyperspheres, and in general the full power of the theory of representations for the higher rotation groups can be made to enter into play. The assumption of the validity of some of the symmetries which these group imply (and the variety of choices for hyperangles) will be shown in the following to correspond to the choice of coupling schemes, which suggest approximate decoupling of the dynamic equation: the appearance of approximate quantum numbers labeling the elements of the bases may serve as powerful starting points for the numerical approach to the solution of the problem [8].

The key papers for modern developments were by Smith, who introduced a generalization to hyperspaces of the concept of angular momentum [9] and gave a full treatment of the motion of three particles on a plane [10]. The early motivation of the work by Smith was to deal with the rearrangement problem which arises in the treatment of chemical reactions [11]: this led him to introduce, in the treatment of several particles constrained to move on a line, the concept of kinematic rotations, which appeared as a generalization of the concept of the reaction skewing angle, already well known in the chemical kinetics. The usefulness of this concept in the general case is now apparent, as demonstrated in previous papers [1, 12]. Section 2 provides an introduction to the tree method for the building up of hyperspherical coordinates and harmonics.

2.2. HYPERSPHERICAL COORDINATES AND THE TREE REPRESENTATION

The method of trees [3, 13, 14] is very useful to represent in a graphical way the relationship between Cartesian coordinates and angular parametrizations on the hyperspheres. A d-dimensional sphere will be parametrized either by d Cartesian coordinates or by a (hyper)radius ρ and d-1 (hyper)angles. To the Cartesian coordinates, we will put into correspondence d leaves, which will be connected by branches which join at d-1 nodes, representing the angles. As a convention, we will consider as representing the cosine (sine) of the hyperangle the branch converging to the node from the left-hand side (right-hand side). Starting then from a leaf and descending to the root of the tree, through the various nodes, we have a relationship between coordinates and hyperangles: Figure 1 shows different ways of doing so by reference to the examples for the three-dimensional sphere $S^2 \subset R^3$ and the four-dimensional hypersphere $S^3 \subset R^4$.

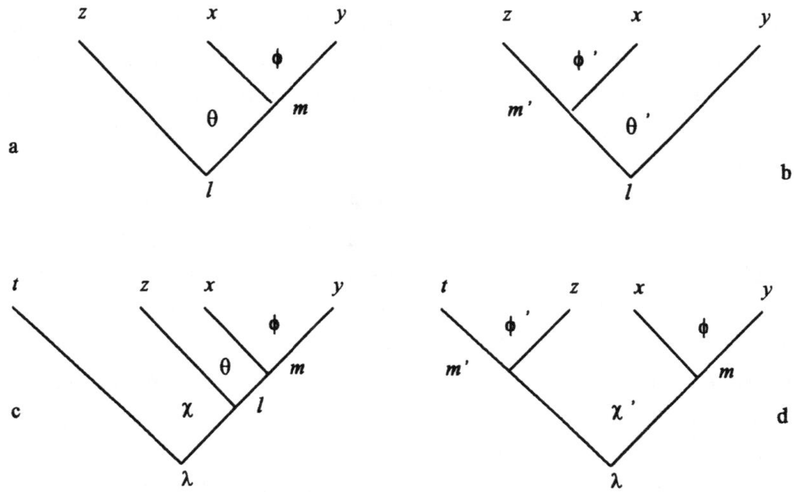

Figure 1. Trees a and b represent for sphere S^2 the correspondence $x = \rho\sin\theta\cos\phi = \rho\cos\theta'\sin\phi'$, $y = \rho\sin\theta\sin\phi = \rho\sin\theta'$, $z = \rho\cos\theta = \rho\cos\theta'\cos\phi'$. They also represent harmonics of $O(3)$ labeled by l, m and l, m', respectively. Similarly, trees c and d represent two alternative parametrizations for S^3 and the labels for the corresponding harmonics of $O(4)$: tree c corresponds to the standard subgroup reduction chain $O(4) \supset O(3) \supset O(2)$ and the tree d to the chain $O(4) \supset O(2) \times O(2)$.

When for a hyperspherical parametrization a graphical representation is possible by the tree method, then the coordinates can be shown to form an orthogonal system, *i.e.* metric tensor is diagonal and the Laplacian on the hypersphere does not contain products of derivatives corresponding to different hyperangles, the corresponding Laplace equations are separable and the hyperspherical harmonics will be constructed in closed form. The elements of the metric tensor are given explicitly in a previous paper [1], from them we can write the angular Laplacian, which acts on the d space of hyperspherical harmonics:

$$\Lambda(\Omega_{\mathrm{d}-1})Y_{\lambda\{\mu\}}(\Omega_{\mathrm{d}-1}) = -\lambda(\lambda + \mathrm{d} - 2)Y_{\lambda\{\mu\}}(\Omega_{\mathrm{d}-1}) \qquad (1)$$

with eigenvalues $-\lambda(\lambda + \mathrm{d} - 2)$. The number λ labels the root of any tree with d leaves. For the physical case d=3, eq. (1) is immediately interpreted as the defining one for angular momentum wavefunctions, and therefore λ plays the role of the *hyperangular momentum* quantum number [9]. The remaining set of d-2 quantum numbers $\{\mu\}$ is associated with the other nodes in the tree.

2.3. HYPERSPHERICAL HARMONICS BY THE METHOD OF TREES

Trees not only relate Cartesian coordinates and (hyper)angles: they also can be taken to represent hyperspherical harmonics $Y_{\lambda\{\mu\}}$, *i.e.*, solutions to eq. (1)and therefore hyperangular momentum wavefunctions. In general, when the angular coordinates chosen to parametrize the Cartesian coordinates form an orthogonal system, then the Laplace equation (1) can be solved by the method of separation of angles. The eigenfunctions $Y_{\lambda\{\mu\}}$ are then given by products of orthogonal functions, which are represented by the fundamental forks shown in Figure 2 which form a given tree. The relevant formulas are given in the literature [3, 1, 13, 15]: here, we collect some of them which will be needed later.

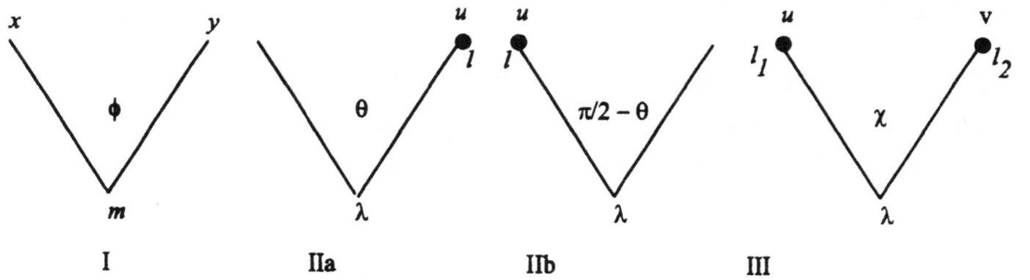

Figure 2. Elementary binary forks. Ranges spanned by angles are $0 \le \phi < 2\pi$, $0 \le \theta \le \pi$, $0 \le \chi \le \pi/2$. In I, m takes the values $m = 0, \pm1, \pm2, \cdots$. In II and III, λ is zero or a positive integer; l, l_1, l_2 assume the values $0, 1, \cdots, \lambda$ (and also the negative values if at the roots of forks of type I). In III, there is the further restriction that $\lambda - l_1 - l_2$ must be even and positive, or zero.

Binary forks can be classified in four types (ternary forks have been introduced and briefly discussed in [1]). The simplest fork is a node with two leaves, its orthonormal system is given by

$$|\phi; m\rangle = (2\pi)^{-1/2} e^{im\phi} \tag{2}$$

and the Cartesian coordinates, which span the circle given by $x^2 + y^2 = \rho$, are

$$x = \rho\cos\phi \qquad y = \rho\sin\phi \tag{3}$$

The fork of the Type IIa is given by a leaf on the left-hand side, and a node on the right-hand side. When a node is a root with u leaves, the corresponding function is a Gegenbauer, or ultraspherical polynomial [16]:

$$|\theta, \lambda; l, u\rangle = \mathcal{N}^u_{\lambda l} \sin^l\theta C^{l+(u-1)/2}_{\lambda-l}(\cos\theta) \tag{4}$$

where the normalization factor is:

$$\mathcal{N}_{\lambda l}^{u} = \frac{(2l+u-2)!}{2^{l+(u-1)/2}(l+u-1)!}\left[\frac{(2\lambda+u-1)(\lambda-l)!}{(\lambda+l+u-2)!}\right]^{1/2}$$

It has to be noted that the Gegenbauer polynomial can be written as associated Legendre polynomials [14, 16]. If the number of leaves is even, the corresponding associated Legendre polynomial will be labeled by integer quantum numbers, if it is odd it will be labeled by half-odd quantum numbers. Forks of type IIb are related to the previous ones (Fig. 2) by a reflection on the root: their functions can be obtained by eq.(4) interchanging sines and cosines.

With these forks it is possible to build up (see eq. (1)) the standard hyperspherical harmonics $Y_{\lambda\{\mu\}}$, which are functions of the $\mathrm{d}-1$ hyperangles $\Omega_{\mathrm{d}-1}$ which parametrize the sphere $S^{\mathrm{d}-1}$, and are labeled by the grand orbital angular momentum quantum number λ and d-2 integers $\mu_{\mathrm{d}-2}, \cdots, \mu_1$; with the further restriction that $\lambda \geq \mu_{\mathrm{d}-2} \geq \cdots \geq \mu_1$.

Such a standard tree is shown in Figure 3: according to eq.(4) the corresponding hyperspherical harmonics are given by a product of Gegenbauer polynomials. To exhibit the connection with the familiar eigenfunctions of

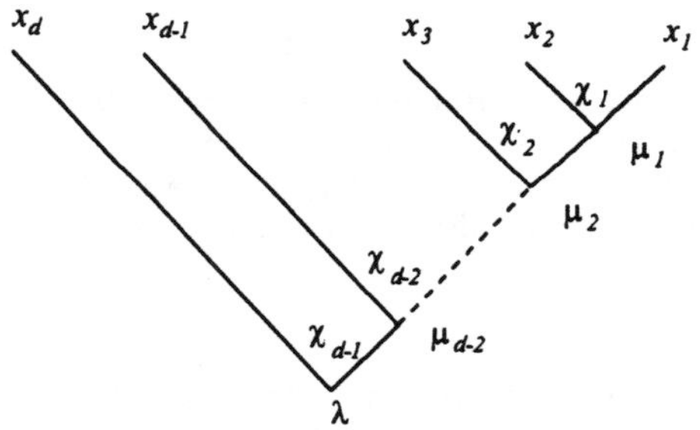

Figure 3. Tree for the standard hyperspherical parametrization of $S^{\mathrm{d}-1}$ corresponding to the subgroup reduction chain $O(\mathrm{d}) \supset O(\mathrm{d}-1) \supset \cdots \supset O(2)$.

angular momentum in quantum physics, it is customary to use the symbols l and m for μ_1 and μ_2. For example, in order to obtain the harmonics, corresponding to the tree in Fig. 1a, we have to combine two forks, one of type I, and one of type IIa, obtaining

$$Y_{l,m}(\theta, \phi) = |\theta; l; m, 2\rangle|\phi; m\rangle \tag{5}$$

which is recognized as the usual spherical harmonic in R^3. For the example of the sphere S^3 (Fig. 1c) we have

$$Y_{\lambda,l,m}(\chi, \theta, \phi) = |\chi; \lambda; l, 3\rangle Y_{l,m}(\theta, \phi) \tag{6}$$

Finally, let us consider now a fork of type III. The corresponding harmonics can be written as a Jacobi polynomial

$$|\chi;\lambda;l_1,u;l_2,v\rangle = \mathcal{N}^{u,v}_{\lambda,l_1,l_2}\sin^{l_2}\chi\cos^{l_1}\chi P^{l_2+(v-2)/2,l_1+(u-2)/2}_{(\lambda-l_1-l_2)/2}(\cos\chi) \qquad (7)$$

where u and v are the numbers of leaves above l_1 and l_2, respectively. The normalization factor $\mathcal{N}^{u,v}_{\lambda,l_1,l_2}$ is:

$$\mathcal{N}^{u,v}_{\lambda l_1 l_2} = \left[\frac{(2\lambda+u+v-2)[(\lambda-l_1-l_2)/2]![(\lambda+l_1+l_2+u+v-4)/2]!}{[(\lambda+l_1-l_2+u-2)/2]![(\lambda-l_1+l_2+v-2)/2]!}\right]^{\frac{1}{2}}$$

For the particular important case, $u = v = 2$, it is convenient to identify the Jacobi polynomial in eq. (7) with a reduced Wigner rotation matrix [17], obtaining

$$|\chi;\lambda;l_1,2;l_2,2\rangle = (-1)^{l_1}\left[2(\lambda+1)\right]^{1/2} d^{\lambda/2}_{(l_1+l_2)/2,(l_1-l_2)/2}(2\chi) \qquad (8)$$

Therefore, the tree in Fig. 1d corresponds to the harmonic

$$\overline{Y}_{l,m,m'}(\phi,\chi',\psi') = (-)^{m'}\left(\frac{\lambda+1}{2\pi^2}\right)^{\frac{1}{2}} D^{\lambda/2}_{\frac{m'+m}{2},\frac{m'-m}{2}}(-\psi'-\psi,2\chi',\psi-\psi') \qquad (9)$$

The bar over the symbol for the harmonic (9) indicates that we are not dealing with a standard tree of the type c in Fig. 1 (or as in Fig.3). The connection of harmonic in eq. (9) with that in eq. (6) will be established in the next section.

2.4. FRAME TRANSFORMATIONS, COUPLINGS AND RECOUPLINGS OF HYPERANGULAR MOMENTUM

Several types of operations on harmonics can be visualized and performed by the aid of the tree method and interpreted as generalizations of operations in angular momentum algebra. A basic operation is the multiplication of harmonics, which is exemplified by the well known Clebsch-Gordan series for $O(3)$. The extension to higher rotation groups[3] makes it possible, at least in principle, to expand the product of two harmonics of the same angles in terms of a linear combination of the same harmonics. The coefficient are an extension of the Clebsch-Gordan coefficients and the operation has the meaning of coupling of hyperangular momenta.

A more complicated operation on harmonics is addition: addition theorems combine harmonics which are functions of different sets of angles [1]. Both multiplication and addition formulas are often given in literature for the standard tree representation [18] of coordinates and harmonics (Fig.

3). However, the corresponding formulas for non standard trees have to be obtained through the use of the so-called timber coefficients, first discussed by Kil'dyushov [15].

The possibility of changes in coupling schemes is described by particular rotations, and can be written as matrix elements (Timber coefficients). These coefficients have been classified and computed in ref. [15], and the tree method visualizes them as graftings. The simplest ones, which we shall need in the following, are for the trees illustrated in Fig. 1. The relationship between harmonics corresponding to the S^2 trees a and b in fig. 1, can be established using properties of the familiar $O(3)$ harmonics,

$$\overline{Y}_{lm'}(\theta'\phi') = \sum_{m=-l}^{l} Y_{lm}(\theta\phi)D^{l}_{mm'}(\frac{\pi}{2},\frac{\pi}{2},\pi) \tag{10}$$

where the bar over the symbol for the harmonics indicates that we are considering the nonstandard tree b in Fig. 1. It can be verified that

$$\overline{Y}_{lm'}(\theta'\phi') = Y_{lm'}(\pi/2-\theta',\phi') \tag{11}$$

For S^3 (trees c and d in fig. 1), we have an interesting formula from the theory of $O(4)$ group [19]:

$$\begin{aligned} Y_{\lambda lm}(\chi,\theta,\phi) &= \sum_{m'=-l}^{l} (-)^{\frac{\lambda-m+m'}{2}} \langle\frac{\lambda}{2},\frac{m-m'}{2};\frac{\lambda}{2},\frac{m+m'}{2}|lm\rangle \\ &\times \ \overline{Y}_{\lambda mm'}(\phi',\chi',\phi) \end{aligned} \tag{12}$$

whereby the alternative harmonics given by eqs. (6) and (9) are related by an orthogonal transformation matrix, whose elements are just the familiar su_2 Clebsch-Gordan coefficients. We offer an alternative view of the coefficient in (12): by Regge's symmetry [20], one has identically

$$\langle\frac{\lambda}{2},\frac{m-m'}{2};\frac{\lambda}{2},\frac{m+m'}{2}|lm\rangle = \langle\frac{\lambda+m}{2},\frac{-m'}{2};\frac{\lambda-m}{2},\frac{m'}{2}|l0\rangle \tag{13}$$

this leads to the physical interpretation of the transformation as the introduction of a new reference frame, where the quantization axis is such that the orbital angular momentum has zero projection and of $\frac{m'}{2}$ as a *helicity* quantum number. An operation of this kind finds its analogue in a *space fixed-body fixed* transformation in molecular collisions [21, 22], in Hund's case $(e) \rightarrow (c)$ in molecular spectroscopy and atomic scattering, or in the passage between symmetric and asymmetric coordinates in the hyperspherical treatment of the three-body problem [23, 24] (this will be exploited in sec. 4.2, fig. 4). In sec. 3 we will see that the same transformation also encompasses the polar-to-parabolic transformation in the hydrogen atom and its multidimensional generalization.

These are the simplest transformations between harmonics: in general, they are described by coefficients which are generalizations of Racah's recoupling coefficients in ordinary angular momentum algebra, and mathematically belong to the class of ${}_4F_3$ functions of unit argument.

More on trees can be traced in the cited literature. They can be particularly useful to exploit symmetries which lead to simplifications. An example of such important *pruning* [1] is useful for the J=0 three-body problem (see sec. 4.2), a case also basic for the mapping of potential energy surfaces [1, 12].

3. Hyperspherical harmonics as wavefunctions for the multidimensional hydrogen atom

In 1935, Fock [25] proved the relationship between the hydrogen atom wavefunction in momentum space and four dimensional spherical harmonics. In position space, closed form solutions can be obtained either in spherical or in parabolic coordinates. In this section we are interested in the problem of a hydrogen atom in a d-dimensional space, and in particular in the transformation between wavefunctions of the H-atom in polar and parabolic coordinates in view of their use as expansion bases for atomic and molecular problems, see below. We demonstrate that these two complete orthonormal bases are related by orthogonal transformations whose elements are as in (10) for d=2. It has already been noted that (12) is the transformation for d=3. Generalized vector coupling coefficients will be shown elsewhere to occur for d>3. To illustrate this procedure we work out the details for the two dimensional hydrogen atom in position and momentum space respectively, and derive the transformation coefficients. The solutions to the Schrödinger equation for an electron in a two-dimensional attractive Coulomb field in polar coordinates (ρ, ϕ) [26], see eq.(3) for their definition, take the form

$$\Psi_{nm}(\rho,\phi) = \mathcal{N}_{nm}(2p_0\rho)^{|m|}e^{-p_0\rho}L^{2|m|}_{n-|m|-1}(2p_0\rho)e^{\imath m\phi} \tag{14}$$

and the corresponding energy levels are

$$E = -\frac{2}{(2n-1)^2} \tag{15}$$

In the above equations we have considered the mass of the electron, its charge and $\hbar$ equal to 1. The principal quantum number n takes the values $1, 2, \ldots$ and $m = 0, \pm1, \cdots, \pm(n-1)$; $L^{2|m|}_{n-|m|-1}$ is a Laguerre polynomial.

The eigenfunction in eq.(14) is the product of a radial part and an angular part: a Laguerre polynomial and a harmonic of the circle respectively, see previous section, $p_0 = \sqrt{-2E}$ and the normalization factor $\mathcal{N}_{nm}$ is:

$$\mathcal{N}_{nm} = \left[\frac{2^3(n- \mid m \mid -1)!}{\pi(2n-1)^3(n+ \mid m \mid -1)!^3}\right]^{\frac{1}{2}}$$

A slightly different normalization applies if this set is interpreted as Sturmian, *i.e.* hydrogenlike wave functions of constant p_0.

In the parabolic coordinate system the problem of two-dimensional hydrogen atom has been treated in [27]. Parabolic coordinates, λ_1^2 and λ_2^2, are defined through the equations:

$$x = \frac{1}{2}(\lambda_1^2 - \lambda_2^2) \qquad y = \lambda_1\lambda_2 \tag{16}$$

The expression for wave functions is:

$$\Psi_{n_1 n_2}(\lambda_1, \lambda_2) = \mathcal{N}_{n_1 n_2} e^{-p_0 \frac{\lambda_1^2+\lambda_2^2}{2}} H_{n_1}(\sqrt{p_0\lambda_1^2}) H_{n_2}(\sqrt{p_0\lambda_2^2}) \tag{17}$$

and the energy levels have the form:

$$E = -\frac{2}{(n_1+n_2+1)^2} \tag{18}$$

n_1 and n_2 are the parabolic quantum numbers and H_{n_1} and H_{n_2} are Hermite polynomials. The normalization factor $\mathcal{N}_{n_1 n_2}$ is:

$$\mathcal{N}_{n_1 n_2} = \left(\frac{2^3}{\pi(n_1+n_2+1)^3 \, 2^{n_1+n_2} \, n_1! n_2!}\right)^{\frac{1}{2}}$$

By comparison of equations (15) and (18) the relation of n_1 and n_2 with the principal quantum number n can be worked out:

$$n_1 + n_2 + 1 = 2n - 1 \tag{19}$$

As emphasized in [27], only values of n_1 and n_2 with the same parity give the correct energy levels. The acceptable wave functions are those given in eq.(17) with the constraint that n_1 and n_2 are both even or both odd, *i.e.* wavefunctions must satisfy $\Psi_{n_1 n_2}(\lambda_1, \lambda_2) = \Psi_{n_1 n_2}(-\lambda_1, -\lambda_2)$ as (λ_1, λ_2) and $(-\lambda_1, -\lambda_2)$ represent the same point in the plane. In order to make the two-valuedness of the wave functions more evident it is convenient to use parabolic coordinates, see eqs.(16) and (17), which are squares of those used in three and d dimensions. Also, in order to facilitate the comparison with the well known 3-dimensional case, we remind that Hermite polynomials can be considered as particular cases of Laguerre polynomials [28].

Let us consider now the transformation between the solutions Ψ_{nm} and $\Psi_{n_1 n_2}$:

$$\Psi_{n_1 n_2}(\lambda_1, \lambda_2) = \sum_m T_{nm}^{n_1 n_2} \Psi_{nm}(\rho, \phi) \tag{20}$$

where $T_{nm}^{n_1 n_2}$ are the transformation coefficients which are elements of a unitary matrix. Below we prove that these coefficients are the Wigner d-functions already encountered in the previous section. Letting ρ go to infinity in eq.(20), from eqs.(14) and (17), we find:

$$\sum_m T_{nm}^{n_1 n_2} \frac{(-)^{n-|m|-1} e^{\imath m\phi}}{[(n- \mid m \mid -1)!(n+ \mid m \mid -1)!]^{\frac{1}{2}}} = \left[\frac{2^{n_1+n_2}}{n_1! n_2!}\right]^{\frac{1}{2}} \cos^{n_1}\frac{\phi}{2} \sin^{n_2}\frac{\phi}{2}$$

where use has been made of the relation between parabolic and polar coordinates: $\lambda_1^2 = 2\rho\cos^2(\phi/2)$, $\lambda_2^2 = 2\rho\sin^2(\phi/2)$ and of the expression of Laguerre and Hermite polynomials for $\rho \to \infty$ [28]. Using the property of orthogonality of the functions $e^{\imath m\phi}$ we obtain an integral representation for the coefficients of the transformation and we find that they allow the representation through the Wigner d-functions:

$$T_{nm}^{n_1 n_2} = (-)^{\frac{n_1-n_2}{2}} e^{-\imath\pi\frac{n_1}{2}} d^{\frac{n_1+n_2}{2}}_{\frac{n_1-n_2}{2},m}(\frac{\pi}{2}) \tag{21}$$

We have demonstrated that when the motion of the electron around the nucleus is constrained in a plane the transformation coefficients between the wave functions in polar and parabolic coordinates are Wigner d-functions. The appearance of these functions proves that the symmetry group for the two-dimensional hydrogen atom is $O(3)$ as stressed in [27].

Let us now consider the wave functions of the two dimensional Coulomb problem in momentum space. The momentum eigenfunctions are solutions of the related Schrödinger equation obtained by Fourier-transforming the position space Schrödinger equation. In [29], applying Fock's projection to the 2-dimensional hydrogen atom in momentum space, it is shown that the eigenfunctions are ordinary spherical harmonics:

$$\chi_{n,m}(p, \varphi) = \imath^m (-)^{n-1} \frac{\sqrt{8} p_0^2}{(p_0^2 + p^2)^{\frac{3}{2}}} Y_{n-1,m}(\vartheta, \varphi) \tag{22}$$

In order to obtain momentum space wavefunctions $\chi_{n_1 n_2}(p_1, p_2)$ corresponding to position space ones in parabolic coordinates, eq.(17), we use the transformation given in eq.(21) above:

$$\chi_{n_1 n_2}(p_1, p_2) = e^{\imath\pi\frac{n_1}{2}} \frac{\sqrt{8} p_0^2}{(p_0^2 + p^2)^{\frac{3}{2}}} \sum_m D^{n-1}_{m, \frac{n_1-n_2}{2}}(\frac{\pi}{2}, \frac{\pi}{2}, \pi) Y_{n-1,m}(\vartheta, \varphi) \tag{23}$$

where $D^{n-1}_{m,\frac{n_1-n_2}{2}}$ is a rotation matrix. Using the known relation between spherical harmonics in two different angular coordinates, (see eqs.(10) and (11) of previous section) we obtain:

$$\chi_{n_1 n_2}(p_1,p_2) = e^{i\pi\frac{n_1}{2}} \frac{\sqrt{8}p_0^2}{(p_0^2+p^2)^{\frac{3}{2}}} Y_{n-1,\frac{n_1-n_2}{2}}(\frac{\pi}{2}-\vartheta',\varphi') \tag{24}$$

where the angles ϑ' and φ' are related to ϑ and φ through the equations reported in figure 1. Note that in this case ϑ and φ are used instead of θ and ϕ to denote that we are considering the momentum space. These relations illustrate the two alternative angular parametrizations of the rotation group $O(3)$ [1], see fig. 1, which is the symmetry group of the bidimensional hydrogen atom [27]. Here z, x, y define a sphere of radius 1 and their definition coincides with Fock stereographic projection [29].

These results - that wavefunctions in parabolic and polar coordinates of the H-atom in a plane are related through a Wigner rotation matrix and that these transformation coefficients are timber coefficients [15] connecting spherical harmonics parametrized with alternative angular variables - can be generalized to spaces of higher dimensions [30] where the transformation is formally a vector coupling coefficient of type (12) (see fig. 1).

4. Applications

The previous sections suggest we have orthonormal bases of two kinds - the (hyper)harmonics and the (hyper)hydrogenic position wavefunctions: since Fock's projection relates the harmonics to hydrogenic momentum wavefunctions, as shown by Wulfman and Shibuya [31, 32, 33] for the physical d=3 case, both can be used to expand the d-dimensional plane wave and to Fourier transform between position and momentum space. Current applications from our laboratory follow.

4.1. HYPERHARMONIC TREATMENT OF A MULTICENTER PROBLEM: H_2^+

In Wulfman and Shibuya treatment of one electron multicenter problem the basic ingredient is the explicit calculation of the elements of a matrix, usually indicated with S, representing the overlap between the basis eigenfunction relative to a certain nucleus and the eigenfunction basis set relative to another nucleus placed at an arbitrary distance. Monoelectronic energies and orbitals can be obtained solving a secular problem involving essentially the S matrix [19].

Being the S matrix elements strictly linked to the tetradimensional harmonics structure, coinciding with the molecular orbitals for the tridimensional hydrogen atom in momentum space (as we have seen in section 3), we have considered an alternative representation obtainable by using the orthogonal transformation (12), that is the suitable timber coefficient. We have shown that in this case, corresponding to a configuration space eigenfunction written in parabolic coordinates, the S matrix can be calculated by group theory using an angular momentum algebra resulting simpler than the conventional formulation as it doesn't require 6-j or 9-j coupling coefficients, but just 3-j coefficients. The utility of this representation has been proved by a series of calculations on the molecular ion H_2^+, by comparing the results obtained with reduced bases, assuming in the various cases the conservation of different quantum numbers. The hypothesis of the approximate conservation of the parabolic quantum number n_1 leads to an approximate eigenfunction of only three orbitals describing the electronic energy at the equilibrium state within 5%. Such a quantum number has the physical meaning of a helicity quantum number and the investigation on the possibility that its conservation could be useful, even besides this context, represents an interesting perspective for the research in this field. We have also examined the effectiveness of the two kinds of representations concerning the eigenvalues calculated in the two ways as the internuclear distance varies and have noted a better approximation at short range for energies calculated according to the representation in polar coordinates, while for those calculated in parabolic coordinates the convergence is better at long range. In particular for internuclear distances close to the potential energy minimum the parabolic representation, as formally obtained as an alternative coupling scheme, provides a better approximation than the polar one. Perspectives for further developments include extension of the S matrix to the molecular multielectron case, employing the hyperspherical harmonic atomic basis to be discussed next for the example of helium.

4.2. HYPERSPHERICAL HARMONICS AS FEW BODY COULOMB WAVEFUNCTIONS

In configuration space, the many body Coulomb problem, describing atomic and molecular structure, is isomorphic to that for a multidimensional hydrogen atom with an anisotropic charge [34]. This encourages multidimensional hydrogenic Sturmian expansions [18, 35], such as eq.14 for d=2 in sec. 3, consisting of a (hyper)-radial part and of a hyperspherical harmonic. As shown before, the corresponding eigenfunctions used in momentum space are hyperspherical harmonics belonging to a space of one dimension higher. The calculation of matrix elements for the secular equation for the spectrum involves radial and angular integrals. Guided by the previous outlined duality of multidimensional hydrogenic Sturmians and hyperspherical harmonics, we have shown that not only angular integrals, but also radial ones can be obtained as (generalized) vector coupling coefficients.

Therefore, the numerical solution of the secular equation for different representations for three-body problems has enabled us to make interesting considerations on the various quantum numbers we used in correspondence to the various kind of harmonics.

Alternative subgroup reductions, corresponding to alternative hyperspherical harmonics expansions, can now be dealt with by the generalized angular momentum algebra outlined in sec. 2.

The three-body problem requires a mapping of the S^6 hypersphere embedded in a d=7 Euclidean space. The corresponding harmonics and quantum number labels are shown in fig 4.

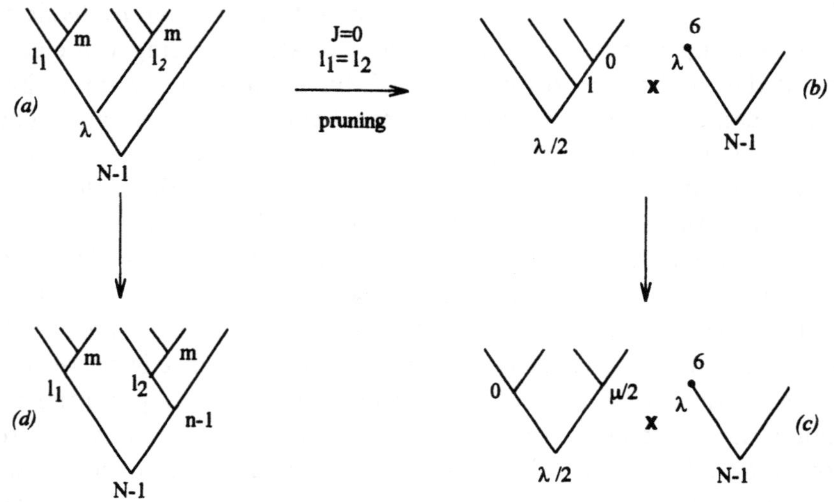

Figure 4. Hyperspherical harmonic basis for the three body Coulomb problem in d=7 momentum space. Tree *a* represents the parametrization of the S^6 sphere. Pruning (see ref. 1) for J=0 leads to tree *b*, and the appearance of a $O(4) \supset O(3) \supset O(2)$ subgroup chain. Tree *c* is obtained by timber vector coupling leading to a $O(4) \supset O(2) \times O(2)$ subgroup chain. Tree *d*, obtained from tree *a* by timber vector coupling involves a $O(7) \supset O(4) \times O(3)$ subgroup chain, where $O(4)$ is the tridimensional hydrogen parametrization in asymmetric coordinates.

To test the effectiveness of the different parametrizations in some specific cases we have applied the above considerations to some numerical examples explicitly for the helium atom and for H_2^+ beyond the fixed nuclei scheme as seen in sec. 4.1. Concerning the helium we can reduce the basis on the λ quantum number (and consider all the correspondent quantum numbers μ and l for both cases, respectively, see trees in (b) and (c) in fig. 4) and for both bases we get, for the same dimensions and for a suitable truncation, a better convergence than a for closed-shell approach.

By cutting also on the quantum numbers μ and l we have seen that convergence gets worse. In both cases a truncation leads to similar results, even if slightly better for a truncation on μ, so we can deduce that, for He, μ is a slightly better quantum number in comparison with l.

With respect to H_2^+ the convergence is very slow, and a truncation only on λ does not lead to any improvement. Reducing on μ isn't worth either, while the convergence becomes far more rapid reducing abruptly on l, as also found before [8]. The passage between tree a and tree d in fig. 4 allows to verify the effectiveness of the reductions on λ and on n. For the H_2^+ molecular ion the truncation on n, even if not drastic, produces better results than a truncation on λ, so n is certainly a better quantum number than λ. Dealing with the helium case, here a reduction on n is strictly connected to a reduction on l, which, as we already know, doesn't lead to good results. However for the same values of l, n results to be a better quantum number with respect to λ.

5. Further application: discrete analogs of hyperspherical harmonics for scattering problems

In the application of quantum mechanics to few-body problems, the solutions of Schrödinger equation are often sought by expanding the unknown eigenfunctions in a suitable basis reducing the problem to algebra. Extensions of the quantum mechanical theory of angular momentum to spaces with dimensions greater than the physical dimension are an interesting source of tools for dealing with atomic and molecular collision processes including rearrangement. Reference is made to our previous work [21, 22, 23, 24, 36, 37, 38].

In this section we review some aspects of the numerical implementation of a technique to solve the Schrödinger equation for scattering problems, including rearrangement. Our technique exploits the discrete analogs of orthonormal bases usually defined on continuous variables. These bases are orthonormal polynomial sets (see section 2), and our representation requires the discrete counterparts of orthogonal polynomials, *i.e.* polynomials orthogonal on lattice points.

From the *mathematical* viewpoint, the harmonics belong to the class of Jacobi polynomials, which are special cases of the hypergeometric function, ${}_2F_1$. Their discrete analogs belong to higher hypergeometric families: in our applications, we will use Hahn polynomials - ${}_3F_2$ hypergeometric functions of unit argument [39, 40]. The latter discrete polynomials can be identified, in particular cases, with the Clebsch-Gordan coefficients or vector coupling coefficients in the quantum theory of angular momentum. The limit relations connecting Jacobi polynomials to their discrete counterparts, the Hahn polynomials, hold in the limit of large quantum numbers and are usually employed in the semiclassical limit to evaluate vector coupling coefficients. This fact provided the original motivation [36] of our work and leads to a *physical* picture of the technique, whereby the discretization is interpreted as a quantization of an artificial hyperangular momentum. Therefore in a previous paper [24] we introduced a natural definition of this procedure termed the *hyperquantization* algorithm. The characteristic feature of our discretization technique is just the introduction of an artificial (hyper)-vector which plays the role of a (hyper)-angular momentum whose length is given by the number of lattice points. It divides the range of the angular variable into slices and its projection on the quantization axis selects the quadrature points.

In this representation, exploiting orthogonality properties of the discrete analogs of the spherical and hyperspherical harmonics [24, 37], the coupling is transferred to the kinetic energy operator. The interaction potential matrix is diagonal, and its elements are simply the potential evaluated at the grid points.

The use of such an orthonormal discrete basis set leads to a computationally convenient algebraic formulation and is also of interest for finding useful decoupling schemes for the three-body dynamical problem. These schemes involve the full or partial conservation of quantum numbers corresponding to angular momenta, their projections and their generalization on the hypersphere. The algorithm can be extended to spaces of any dimension; parity and identical particle permutation symmetry are easily introduced into the discretization procedure. The symmetry properties of the angular momentum algebra coefficients that we exploit gives the matrix interesting symmetry properties. In particular every block results skewsymmetric and in the particular case of total angular momentum equal to zero the global matrix itself results skewsymmetric.

For numerical implementation, the present technique finds its natural setting in the hyperspherical approach to few-body problems. It must be noted however that in our implementation we chose the asymmetric parametrization (see sec. 2) rather than the symmetric one [41], as defined in [1]. Our motivation is that no analytic harmonics are known for the latter parametrization for nonzero total angular momentum, and we are presently extending our procedure to those cases. We list some characteristic features of the numerical implementation of the Hyperquantization algorithm, starting from matrix elements. As in finite difference or discrete variable representation techniques, no integrals are involved. Furthermore, it is not necessary to compute the discrete analogs of the Jacobi polynomials, since this part of the problem can be performed analytically. Numerically, only the Clebsch-Gordan coefficients need be computed. This step in the numerical implementation is crucial because the accuracy of the method and the rate of convergence of the eigenvalues and the eigenvectors greatly depends on the evaluation of those coefficients. To compute these coefficients, we have developed a routine based on recursion relations. This routine is very efficient and allows to save cpu time in the construction of the kinetic energy matrix, typically a few cpu minutes. The Hamiltonian matrix is block-tridiagonal, this property permits the use of a very accurate and efficient routine to diagonalize the hamiltonian matrix. So to optimally exploit this property, we diagonalize the matrix using the STLM package [42] and LANZ package [43] based on the Lanczos algorithm which is very well adapted to sparse matrices or diagonally dominant ones.

Further simplifications can be obtained by neglecting kinetic energy couplings between points far apart on the hypersphere. The diagonal (local) potential energy representations allows drastic reduction in the matrix size by neglecting hyperspherical regions where the potential substantially exceeds the total available energy.

In general, the asymmetric parametrization is dependent on the kinematic rotation [11]: we found it interesting to run test cases with different choices of the kinematic rotations. Through the use of Raynal-Revail coefficients [44] it is possible to project the eigenvectors found on grids based on different Jacobi vectors, thus allowing kinematic angles changes on the coordinates as the hyperradius varies.

We are applying the methodology to prototypical chemical reactions extending earlier results presented elsewhere [37]. The technique is also of interest for the quantum mechanical representations which describe *steric* effects [45].

References

1. Aquilanti V., Cavalli S., J. Chem. Phys., **85** (1986), 1355 Aquilanti V., Cavalli S., Grossi G., J. Chem. Phys. **85** (1986), 1362
2. Louck J. D., Galbraith H. W., Rev. Mod. Phys., **44** (1972), 540
3. Smirnov Yu. F., Shitikova K. V., Sov. J. Part. Nucl., **8** (1976), 344
4. Ballot L., Fabre de la Ripelle M., Ann. Phys., **127** (1980), 62
5. Fock V., Kgl. Norske Videnskab. Selkabs Forh. **31** (1958), 138, 145
6. See, for example, U. Fano, Phys. Rev. A **24**, 2402 (1981); Rep. Progr. Phys. **46**, 97 (1983)
7. Klar H. in *Physics of Electronic and Atomic Collisions*; ed. S. Datz, North Holland, Amsterdam, 1982; Aquilanti V., Grossi G., Laganá A., J. Chem. Phys., **76** (1982), 1588
8. Aquilanti V., Grossi G., Laganá A., Pelikan E., and Klar H., Lett. Nuovo Cimento **41**, 541 (1984)
9. Smith F. T., Phys. Rev., **120** (1960), 1058
10. Smith F. T., J. Math. Phys., **3** (1962), 735
11. Smith F.T., J. Chem. Phys., **31** (1959), 1352
12. Grossi G., J. Chem. Phys., **81** (1984), 3355; Aquilanti V., Laganá A., Levine R.D., Chem. Phys. Lett., **158** (1989), 87; Aquilanti V., Cavalli S., Chem. Phys. Lett., **141** (1987), 309; Aquilanti V., Cavalli S., Grossi G., Rosi M., Pellizzari V., Sgamellotti A., Tarantelli F., Chem. Phys. Lett., **16** (1989), 179; Aquilanti V., Cavalli S., Grossi

G., Anderson R. W., J. Chem. Soc. Faraday Trans., **86** (1990), 1681
13. Vilenkin N.Ya., Kuznetsov G.I., Smorodinskii Ya.A., Sov.J.Nucl.Phys. **2**(1966),645
14. Vilenkin N. J., *Fonctions Spéciales et Théorie de la Représentation des Groupes*; Dunod: Paris, 1969
15. Kil'dyushov M. S., Sov. J. Nucl. Phys. , **15** (1972), 113
16. Abramowitz M., Stegun I. A., *Handbook of Mathematical Function* ; Dover, New York, 1964. Chap. 22
17. Rose M. E, *Elementary Theory of Angular Momentum*; Wiley: New York, 1957; Fano U., Racah G., *Irreducible Tensorial Sets*, Academic Press Inc: New York, 1959; Edmonds A. R., *Angular Momentum in Quantum Mechanics*; Princeton University Press: Princeton, 1960; Brink D. M. and Satchler G. R.,*Angular Momentum*; Oxford University Press: Oxford, 1968; Zare R.N., *Angular Momentum*; John Wiley & Sons, New York, 1988
18. Avery, J., *Hyperspherical Harmonics, Application in Quantum Theory*; Kluwer Academic Publishers: Dordrecht, The Netherlands, 1989; Avery J., Herschbach D. R., Int. J. Quantum Chem., **41** (1992), 673
19. Judd, B. R., *Angular Momentum Theory for Diatomic Molecules*; Academic Press: New York, 1975
20. Regge T., Nuovo Cimento, **10** (1958), 544
21. Aquilanti V., Beneventi L., Grossi G., Vecchiocattivi F., J. Chem. Phys. **89** (1988), 751
22. Aquilanti V., Grossi G., J. Chem. Phys. **73** (1980), 1165
23. Aquilanti V., Cavalli S., Grossi G., Teor. Chim. Acta **79** (1991), 283
24. Aquilanti V., Cavalli S., *Few Body Systems*; **suppl. 6** (1992), 573
25. Fock, V., Z. Phys., **98** (1935), 145
26. Zaslow B., Zandler M.E., Am. J. Phys., **35** (1967),1118
27. Cisneros A., McIntosh H.V., J. Math. Phys., **10** (1969), 277
28. Morse P. M. and Feshbach H., *Methods of Theoretical Physics*; McGraw-Hill: New York, 1953
29. Shibuya I., Wulfman C.E., Am. J. Phys., **33**, (1965) 570
30. Aquilanti V., Cavalli S., De Fazio D., J. Phys. Chem., in press 1995
31. Shibuya I., Wulfman C.E., Proc. Roy. Soc., **A286** (1965) 376
32. Dunlap B. I., Chem. Phys. Lett., **30** (1975), 39
33. Novosadov B. K., Opt. Spectrosc., **41** (1976), 490
34. Klar H., J. Phys. B: Atom. Molec. Phys., **7** (1974), L436
35. Whitten R. C., Sims J. S., Phys. Rev. A, **9** (1974), 1586
36. Aquilanti V., Grossi G., Lett. Nuovo Cimento, **42** (1985), 157
37. Aquilanti V., Cavalli S., Monnerville M., in *Grid Methods in atomic and molecular quantum calculations*, Cerjan C. Ed.: Kluwer Academic Publishers, 1993; 25-48
38. Aquilanti V., Cavalli S., Grossi G. in *Advances in Molecular Vibrations and Collision Dynamics*, **2A** (1994), 147
39. A.F. Nikiforov, Suslov S.K., Uvarov V.B., *Classical Orthogonal Polynomials of a Discrete Variable*; Springer-Verlag: Berlin (1991)
40. Karlin S., Mc Gregor J.L., Scripta Math. **26** (1961), 33
41. Whitten R. C., Smith F. T., J. Math. Phys., **9** (1968), 1103
42. Ericsson T., Ruhe A., Math. Comp., **35** (1980), 1251
43. The routine Lanz is available in the Netlib
44. Raynal J., Revai J., Nuovo Cimento **A 68** (1970), 612; Raynal J., Nucl. Phys. **A 259** (1976), 272
45. Anderson R. W., Aquilanti V., Cavalli S., Grossi G., J. Phys. Chem., **95** (1991), 8184; Anderson R. W., Aquilanti V., Cavalli S., Grossi G., J. Phys. Chem., **97** (1993), 2443

QUANTUM AND SEMI-CLASSICAL METHODS IN REACTIVE SCATTERING

GERT D. BILLING

Department of Chemistry, H. C. Ørsted Institute
University of Copenhagen Ø
Denmark

Abstract

The wavepacket method for treating 3-atomic reactive scattering problems is outlined and approximate semi-classical methods for larger reactive scattering problems are discussed.

1. Introduction

The fundamental problem in exact, i.e. quantum mechanical treatments of reactive scattering problems is that whenever chemical bonds are broken the system can access a large number of continuum states, states which for bound state problems are of minor importance. Thus methodology based on quantum state expansion techniques will inevitably be expensive to work with in practical calculations. Even the treatment of nonreactive scattering problems involving heavy molecules will pose problems if conventional methods are used. It is the purpose of the present paper to outline various approximate methods which can be used for complex chemical reactions - as well as to discuss the exact treatment of three-particle reactive systems using wave-packet technique in hyperspherical coordinates.
Many chemical reactions involve non-adiabatic electronic transitions, spin-orbit couplings, geometric phase effects - problems which involve the effect of more than one adiabatic electronic potential energy surface. Thus it is necessary to develope the theoretical tools which can handle multi-surface problems. Such methods are e.g. the semi-classical two and three dimensional wave-packet methods described in section 3 below. Recent investigations [1] indicate that the more traditional trajectory surface hopping (TSH) technique [2] may not be adequate for treating multi-surface problems quantitatively.

C. A. Tsipis et al. (eds.), New Methods in Quantum Theory, 251–264.

The advantage of using hyperspherical coordinates for a semi-classical treatment of reactive scattering problems is that all reaction channels are treated evenhandedly, i.e. the classical approximation can be introduced in e.g. the motion of the hyperradius and/or the overall rotation of the plane formed by the three atoms. These approximations reduce the quantum problem to a three or eventually a two-dimensional one. The introduction of a wave-packet treatment of the quantum degree's of freedom makes the method easy to apply, i.e. no basis set considerations are needed. The wavefunction is expanded on a two or a three-dimensional grid and the time-dependent Schrödinger equation solved using conventional and well-established time propagation methods. The use of time-dependent methods converts the problem to an initial value problem, makes it possible to treat an energy range rather than a single energy and to incorporate approximate classical description of part of the systems. However the propagation of the wavepackets for systems with large attractive wells does sometimes pose a problem - since part of the wavepacket has a tendency to stick in the well for a long time.
For an exact treatment of three-particle scattering problems we have considered the formulation in terms of a coupled 3D-wavepacket solution method. The reason being that the expansion in the rotational basis set for the Euler-angles involve a finite and often a small number of basis-functions ($J/2$), where J is the total angular momentum. For the remaining degree's of freedom it is advantageous to use a grid expansion. In such an expansion the effect of dissociative channels is automatically incorporated and it is furthermore for heavy mass systems possible to introduce a more coarse grained grid - if resolution on e.g. rotational states is not of interest. The method is easy and "inexpensive" to extend to multi-surface, i.e. electronically non-adiabatic problems. Thus the CPU-time increases only about linearly with the number of potential energy surfaces.

2. Theory

2.1. THREE-ATOMIC REACTIVE SCATTERING

For an approximate treatment of three-atomic reactive scattering problems it is as mentioned convenient to introduce hyperspherical coordinates. The reason being that the three arrangement channels are treated evenhandedly, i.e. we can in a more natural manner introduce approximate descriptions of e.g. the rotational motion and/or the motion along the hyper-radius. But also for the exact quantum mechanical treatment of the collision problem the use of hyper-spherical coordinates have some advantages. Thus the reactive scattering problem is transformed to that of an inelastic collision just with three distinct channels. Also the treatment of the geometric phase problem is easier in hyper-spherical coordinates [4].
In hyperspherical coordinates the hamiltonian for the three-body problem is [5]:

$$\hat{H} = -\frac{\hbar^2}{2\mu}\frac{\partial^2}{\partial\rho^2} + \frac{2}{\mu\rho^2}\hat{L}^2(\theta,\phi) +$$

$$\frac{1}{\mu\rho^2}[\frac{\hat{J}^2}{cos^2\theta}+\frac{\hat{J}_z^2-4cos\theta\hat{J}_z\hat{P}_\theta}{2sin^2\theta}+\frac{sin\theta}{cos^2\theta}\frac{1}{2}(\hat{J}_+^2+\hat{J}_-^2)]$$
$$+V(\rho,\theta,\phi)+\Delta V(\rho,\theta) \quad (1)$$

where

$$\mu=\sqrt{m_A m_B m_C/(m_A+m_B+m_C)} \quad (2)$$

$$\hat{J}_z=-i\hbar\frac{\partial}{\partial\gamma} \quad (3)$$

$$\hat{P}_\phi=-i\hbar\frac{\partial}{\partial\phi} \quad (4)$$

$$\hat{L}^2(\theta,\phi)=-\hbar^2[\frac{\partial^2}{\partial^2\theta}+\frac{1}{sin^2\theta}\frac{\partial^2}{\partial\phi^2}] \quad (5)$$

$$\Delta V(\rho,\theta)=-\frac{\hbar^2}{2\mu\rho^2}[\frac{1}{4}+\frac{4}{sin^2 2\theta}] \quad (6)$$

$\hat{J}_+$ and $\hat{J}_-$ are raising and lowering operators, which couple the K states. The initial wavefunction is taken to be a product of a Gaussian wave-packet in the translational coordinate R and a vibrational/rotational eigenfunction of the diatomic molecule, i.e.

$$\psi_{initial}=\phi_{vjm}(r,\eta,\xi)\Phi(R,\Phi',\Theta',t) \quad (7)$$

Here r is the bond-distance and η, ξ angles which specify the orientation of the diatomic molecule in space. R is the distance for the atom to the center of mass of the diatomic molecule. The angles Φ' and Θ' give the orientation of the $\mathbf{R}$ -vector. Instead of using the rotational eigenfunctions $Y_{jm}(\eta,\xi)$ it is possible to introduce the socalled Arthurs-Dalgarno states labelled by the orbital angular momentum l instead of m, i.e. to use:

$$\mathcal{Y}_{jl}^{JM}=(-1)^{j-l-M}\sqrt{(2J+1)(2l+1)/4\pi}$$
$$\times\sum_m\begin{pmatrix} j & l & J \\ m & 0 & -m \end{pmatrix}Y_{jm}(\eta,\xi)D^J_{-M,-m}(\Phi',\Theta',0) \quad (8)$$

where J is the total angular momentum and M its projection on a space-fixed axis. Thus the wavefunction is initialized as:

$$\psi_{initial}=\frac{1}{rR}\phi_{vj}(r)\mathcal{Y}_{jl}^{JM}\Phi_{WP}(R,t) \quad (9)$$

where Φ_{WP} is a Gaussian wavepacket centered around R_0 and P_0.
By using the relation between the Jacoby and the hyperspherical coodinates [3] we can express the initial wavefunction as:

$$\psi_{initial}=\sum_{K=-J}^{K=J}\Phi_K(\rho,\theta,\phi,t)\sqrt{\frac{2J+1}{8\pi^2}}D^{J*}_{MK}(\alpha,\beta,\gamma) \quad (10)$$

TABLE 1. Expansion coefficients w_K for the (v,j)=0,0 and $l = J$=10. Due to symmetry around $\phi = 2\pi$ only the even K states are involved and the expansion coefficients for the negative K components are identical to those of the positive. The classical weight obtained by initializing in cartesian coordinates and then transforming to hyperspherical coordinates are also shown. 10000 trajectories were used.

K	w_K	w_K^{clas}
10	0.1762	0.156
8	0.0927	0.113
6	0.0736	0.077
4	0.0655	0.066
2	0.0617	0.061
0	0.0606	0.056

where D^J_{MK} are rotation matrix elements. A given A-D state is projected on the K states $-J \leq K \leq +J$ (see table 1).

These states are coupled together during the propagation of the solution to the time dependent Schrödinger equation (TDSE). Thus by inserting this expansion in the TDSE we obtain a set of coupled equations in the expansion-coefficients $\Phi_K(\rho,\theta,\phi,t)$

$$i\hbar\frac{\partial\Phi_K}{\partial t} = \{\frac{1}{2\mu}[\hat{P}_\rho^2 + \frac{4\hat{L}^2(\theta,\phi)}{\rho^2}] + \frac{\hbar K(\hbar K - 4cos\theta\hat{P}_\phi)}{2\mu\rho^2 sin^2\theta} + \frac{\hbar^2 J(J+1)}{\mu\rho^2 cos^2\theta} + V(\rho,\theta,\phi) + \Delta V(\rho,\theta)\}\Phi_K + \frac{sin\theta}{\mu\rho^2 cos^2\theta}(M_{K,K+2}\Phi_{K+2} + M_{K,K-2}\Phi_{K-2}) \qquad (11)$$

where $\hat{P}_\rho = (\hbar/i)(\partial/\partial\rho)$, $\hat{P}_\phi = (\hbar/i)(\partial/\partial\phi)$ and

$$M_{K,K\pm 2} = \frac{\hbar^2}{2}\sqrt{(J\mp K)(J\pm K+1)(J\mp K-1)(J\pm K+2)} \qquad (12)$$

Since the coupling bewteen the various K-channels is simple (multiplication with a tri-diagonal matrix) the coupled wavepacket approach is ideal for parallel computing facilities, where the evaluation of the diagonal working on a K wavepacket is handled on one node.. Thus we need $J/2$ such nodes -the information from which is used when propagating the system one time-step using e.g. the Lanczos iterative reduction method [6]. Each Lanczos iteration involves the evaluation of the coupling matrix on a wavefunction. Typically 10-20 iterations are necessary per time-step - one timestep being 10^{-16} sec. One propagation involves typically between 1000 and 3000 time-steps.

The hyperspherical coordinates are not convenient to use asymptotically. It is therefore necessary to switch to Jacoby or mixed hyperspherical/Jacoby coordinates when the value of the hyper-radius exceeds a certain value ρ^*. The value of ρ^* should be chosen large enough for the reactive coupling to have siezed. Once the projection on these coordinates is done the remaining propagation can then be carried out channel by channel. The possibility of using a mixed Jacobi-hyperspherical coordinate approach is based upon the mapping

$$\rho\theta\phi \rightarrow \rho x \eta \tag{13}$$

and an expansion of the Hamiltonian in a power serie:

$$\hat{H} = \hat{H}_a + \frac{1}{\rho^2}\hat{H}_b + \frac{1}{\rho^4}\hat{H}_c + ... + W \tag{14}$$

where H_a, H_b etc. depend upon the variables η, x, γ and the momentum operators $\hat{P}_\rho, \hat{P}_\gamma, \hat{P}_x$ and $\hat{P}_\eta$ [9]. W is the difference between the actual potential and the Morse potential

$$W = V(x,\eta,\rho) - V_{Morse}(r(x)) \tag{15}$$

where $V_{Morse}(r(x))$ is included in $\hat{H}_a$. The mixed Jacobi-hyperspherical coordinate approach is sometimes convenient to use when the wavepacket covers a large ρ region. The advantage is furthermore that the physical allowed space does not shrink with increasing values of the hyper-radius. A disadvantage is that the coupling between the fragment vibrational states vanishes slowly, as $1/\rho$.

In order to obtain state-resolved transition probabilities and cross-sections the wavepacket is after the propagation projected onto the appropriate asymptotic basis functions, i.e. eigenfunctions of the asymptotic hamiltonian. The projection gives the scattering amplitudes in each channel, i.e. $u^J_{v'j'l'}(R,t)$. When the wavepacket has passes the projection point in R-space we can obtain the energy resolved scattering amplitudes by a time- energy Fourier transfer:

$$b^J_{v'j'l'}(E;R) = \frac{1}{\sqrt{2\pi}} \int dt u^J_{v'j'l'}(R;t) exp(iEt/\hbar) \tag{16}$$

These amplitudes are projected onto incoming and outgoing waves to get $A^{out}_{v'j'l'}$ and $A^{in}_{v'j'l'}$. We notice that the projection on the incoming waves should be zero. This projection is only performed in order to secure that no reflection has occured. The probability for a given transition is then given as:

$$P_{v'j'l'} = \frac{F_{v'j'l'}}{F_{vjl}} \tag{17}$$

i.e. as the ratio between the outgoing and incoming fluxes [3]:

$$F_{v'j'l'} = \frac{1}{\mu'} k_{v'j'} \mid A^{out}_{v'j'l'} \mid^2 \tag{18}$$

$$F_{vjl} = \frac{1}{\mu} k_{vj} (\frac{\mu}{\hbar k})^2 \sqrt{2/\pi} \sigma exp(-2\sigma^2(k-k_0)^2) \tag{19}$$

where σ is the width of the initial wavepacket, μ the reduced mass for the entrance and μ' for the exit channel. Before this calculation is carried out it is usually necessary to propagate the the time-independent coupled equations for the amplitudes

$b_{vjl}(R)$ from the projection point R and a few Å out in each rearrangment channel.

Since the method is time-dependent we can by introducing a Gaussian wavepacket in the translational coordinate cover a large energy range with just a single propagation of the solution. Nevertheless the computation is heavy and can only be carried out on computers with Gflop operational performance. For this reason we have developed semi-classical methods in which part of the problem is treated by classical mechanics. Thus the rotational motion defined by the Euler-angles is a "slow" motion which often can be described classically. Another candidate for a classical treatment is the motion along the hyper-radius ρ. With these approximations the quantum problem is reduced from a 4D to a 3D or a 2D respectively. Two or three dimensional wavepackets can be handled on an ordinary work-station. If only the Euler-angle γ is treated classically then the method becomes exact for J=0 and larger J-values are as easy the deal with as the J=0 problem [8].

3. Semi-classical treatment of chemical reactions

Introducing a classical description of the Euler angle γ and the conjugate momentum P_γ we obtain the following hamiltonian:

$$\hat{H} = -\frac{\hbar^2}{2\mu}\frac{\partial^2}{\partial\rho^2} + \hat{H}_0(\rho,\theta,\phi) + \hat{H}_1(\rho,\theta,\phi,t;J) \tag{20}$$

where

$$\hat{H}_0 = \frac{2}{\mu\rho^2}\hat{L}^2(\theta,\phi) + V(\rho,\theta,\phi) + \Delta V(\rho,\theta) \tag{21}$$

$$\hat{H}_1 = \frac{P_\gamma(P_\gamma - 4cos\theta\hat{P}_\phi)}{2\mu\rho^2 sin^2\theta} + \frac{P_J^2 - P_\gamma^2}{\mu\rho^2 cos^2\theta}(1 + sin\theta cos2\gamma) \tag{22}$$

The time-dependent Schrödinger-equation

$$i\hbar\frac{\partial\Psi}{\partial t} = \hat{H}\Psi \tag{23}$$

is solved using a 3D-grid representation of the wavefunction and the Lanczos time-propagator. The time dependence arising from the classical variables is obtained by solving the equations of motion:

$$\dot{\gamma} = \frac{\partial H_{eff}(P_\gamma,\gamma,t)}{\partial P_\gamma} \tag{24}$$

$$\dot{P}_\gamma = -\frac{\partial H_{eff}(P_\gamma,\gamma,t)}{\partial\gamma} \tag{25}$$

where the effective hamiltonian is obtained as the expectation value of the hamiltonian, i.e.

$$H_{eff}(P_\gamma,\gamma,t) = <\Psi \mid \hat{H} \mid \Psi> \tag{26}$$

The initial values of the classical variables are:

$$\gamma \text{ between } 0 \text{ and } 2\pi$$
$$P_\gamma \text{ between } -P_j \text{ and } +P_J$$

We notice however that the distribution is not uniform in the above range (see table 1). Thus the values of P_γ and γ have to be compatible with the constraint that $\mid \nu \mid$ (see below) should be smaller that j.

Also the motion along the hyper-radius is a candidate for a classical treatment. Thus the first term in eq.(20) is replaced by $P_\rho^2/2\mu$. Hereby the quantum problem is reduced to a two-dimensional one in the θ and ϕ angles. In order to investigate what this corresponds to it is interesting to consider the asymptotic hamiltonian. The asymptotic mapping can be introduced by introducing the variables x, η, such that:

$$\theta = \frac{\pi}{2} - \frac{x}{\rho} sin\eta \tag{27}$$

$$\phi = \phi_0 + \frac{x}{\rho} cos\eta \tag{28}$$

where ϕ_0 is a channel dependent angle [7] and the variable x is related to the diatomic internuclear distance:

$$r = \frac{d_i x}{2} \tag{29}$$

The constant d_i depends on the masses and the channel "i". In the limit $\rho \to \infty$ we see that the physical allowed region in hyper-spherical coordinates shrink to a point. Inserting the above representation of θ and ϕ in eq.(20) we get:

$$\begin{aligned} \hat{H} \to \frac{P_\rho^2}{2\mu} - \frac{2\hbar^2}{\mu}\left[\frac{\partial^2}{\partial x^2} + \frac{1}{x}\frac{\partial}{\partial x} - \frac{1}{4x^2}\right] + V(x) \\ -\frac{2\hbar^2}{\mu x^2}\left[\frac{\partial^2}{\partial \eta^2} + \frac{1}{4sin^2\eta} + \frac{1}{4}\right] + \frac{P_J^2 - P_\gamma^2}{\mu x^2 sin^2\eta}(1 + cos(2\gamma)) \end{aligned} \tag{30}$$

with the eigenfunctions:

$$\phi_{vj}(x, \eta; \nu) = \frac{1}{\sqrt{x}}\sqrt{sin\eta}\, g_{vj}(x) P_j^\nu(cos\eta) \tag{31}$$

It can be shown [3] that the angle η is the angle between the distance form the atom to the center of mass of the diatomic molecule and the r-axis. The quantity ν is given by

$$\nu = \frac{cos\gamma}{\hbar}\sqrt{P_J^2 - P_\gamma^2} \tag{32}$$

Thus we see that a quantum mechanical treatment of the variabels θ and ϕ asymptotically corresponds to a quantum treatment of the vibrational and the rotational motion, whereas the rotational projection is treated classically.

The mixed 2D quantum/classical treatment of the three-particle collision problem is especially convenient for multi-surface problems using a grid-representation of the

wave function. Thus the wavefunction (31) is put on the θ, ϕ grid using the relations (28). The classical equations of motion for ρ, P_ρ, γ and P_γ are solved simultaneously with the wavepacket propagation. The effective potential which governs the classical motion is obtained from the expectation value of the Hamiltonian.
Considering the classical hamiltonian in the asymptotic limit $\rho \rightarrow \infty$ we get

$$H_{cl} \rightarrow \frac{1}{2\mu}P_\rho^2 + V(x) + \frac{P_\gamma^2}{2\mu\rho^2} + \frac{2}{\mu\rho^2}(P_\theta^2 + P_\phi^2) + \frac{(P_\alpha^2 - P_\gamma^2)}{\mu x^2}\frac{(1 + cos2\gamma)}{sin^2\eta} \tag{33}$$

Introducing the F_3-generator $F_3(P_\theta, P_\phi, x, \eta)$ where:

$$F_3 = -\theta P_\theta - \phi P_\phi \tag{34}$$

where

$$\theta = -\frac{\partial F_3}{\partial P_\theta} \tag{35}$$

$$\phi = -\frac{\partial F_3}{\partial P_\phi} \tag{36}$$

and the new momenta:

$$P_x = -\frac{\partial F_3}{\partial x} \tag{37}$$

$$P_\eta = -\frac{\partial F_3}{\partial \eta} \tag{38}$$

we have:

$$P_x = \frac{1}{\rho}(sin\eta P_\theta + cos\eta P_\phi) \tag{39}$$

$$P_\eta = \frac{x}{\rho}(cos\eta P_\theta - sin\eta P_\phi) \tag{40}$$

Inserting this in the asymptotic hamiltonian we get:

$$H_{cl} = \frac{1}{2\mu}P_\rho^2 + \frac{P_\gamma^2}{2\mu\rho^2} + \frac{2}{\mu}(P_x^2 + \frac{P_\eta^2}{x^2}) + \frac{P_\alpha^2 - P_\gamma^2}{\mu x^2}\frac{(1 + cos2\gamma)}{sin^2\eta} + V(x) \tag{41}$$

Thus we can identify the first term as the kinetic energy along the hyper- radius, using that $x = 2r/d_1$ and $d_1^2 = \mu/m$ we have the vibrational energy of the fragment considered as:

$$\frac{2}{\mu}P_x^2 + V(x) = \frac{P_r^2}{2m} + V(r) \tag{42}$$

and introducing $P_\xi = \sqrt{P_\alpha^2 - P_\gamma^2 cos\gamma}$, i.e. the momentum conjugate to the angle ξ, we have the diatomic rotational energy given as:

$$\frac{1}{2mr^2}(P_\eta^2 + \frac{P_\xi^2}{sin^2\eta}) \tag{43}$$

This analyses also explains why a given diatomic rotational state is represented by all the K states of a given J-manifold. Thus with restrictions on the rotational energy we see that e.g. P_ξ ~0 requires $\gamma \sim \frac{\pi}{2}$.

4. Three-center reactions in large systems

Many chemical reactions are although they involve a large number of atoms characterized as being three-center reactions. It is therefore convenient to derive a hamiltonian, which allows for a free unconstrained motion of the three center/atoms. The motion of these atoms characterise the reaction. The remaining atoms follow paths in which they are rigidly bound to the moving atoms in some reference configuration - typically defined by the equilibrium geometry. These atoms are however allowed to vibrate around their equilibrium position. The effect of these degree's of freedom on the motion of the three atoms is of interest for reactions in liquids or of polyatomic systems. Furthermore an overall rotation of the complete system is allowed for.
The three atoms in question form a triangle - the shape and size of which can be defined by the three hypervariables: the hyperradius ρ and the two angles θ, ϕ. But sofar we shall just formulate the problem in terms of three parameters s_1, s_2 and s_3. The reference vectors $\mathbf{a}_i(s_j)$ $i = 1, ...N$ will depend parametrically on these variables. $\mathbf{a}_i$ are position vectors measured from the center of mass of the system. The usual Eckart condition

$$\sum_{i=1}^{N} m_i(\mathbf{a}_i \times \mathbf{v}_i) = \mathbf{0} \tag{44}$$

where the summation runs over all atoms.
The position vector in a space fixed coordinate system is

$$\mathbf{r}_{0i} = \mathbf{R}_p + \mathbf{a}_i(s_j) + \rho_i \tag{45}$$

where ρ_i is a displacement vector and the position vector in the body fixed coordinate system

$$\mathbf{r}_i = \mathbf{a}_i(s_j) + \rho_i \tag{46}$$

Thus the center of mass condition is

$$\sum_{i=1}^{N} m_i \mathbf{v}_i = \mathbf{0} \tag{47}$$

where $\mathbf{v}_i = \frac{d}{dt}\mathbf{r}_i$.
The kinetic energy relative to the center of mass is:

$$2T = \sum_i m_i(\mathbf{v}_i + \omega \times \mathbf{r}_i)^2$$

$$= 2(T_t + T_r + T_c) = \sum_i m_i \mathbf{v}_i^2 + \sum_i m_i(\omega \times \mathbf{r}_i) \cdot (\omega \times \mathbf{r}_i) + 2\sum_i m_i \mathbf{v}_i \cdot (\omega \times \mathbf{r}_i) \quad (48)$$

the last term can using the Eckart condition be rewritten as:

$$2T_c = 2\omega \cdot \sum_i m_i \mathbf{v}_i \times \rho_i \quad (49)$$

Considering now the potential energy we expand it around the reference vectors $\mathbf{a}_i$ (i=1,...N), i.e.

$$V(\mathbf{r}_1...\mathbf{r}_N) = V(\mathbf{a}_1...\mathbf{a}_N) + \sum_{i=1}^{N} \frac{\partial V}{\partial \mathbf{r}_i} |_{\mathbf{r}_i=\mathbf{a}_i} \cdot \rho_i + \frac{1}{2}\sum_i \sum_j \rho_j^T \cdot \frac{\partial^2 V}{\partial \mathbf{r}_i \partial \mathbf{r}_j} \cdot \rho_i + ... \quad (50)$$

If we now introduce the constraint that the first derivative should be a linear combination of the vectors $\frac{\partial \mathbf{a}_i}{\partial s_j}$, i.e.

$$\frac{\partial V}{\partial \mathbf{r}_i} |_{\mathbf{r}_i=\mathbf{a}_i} = m_i \sum_j c_j \frac{\partial \mathbf{a}_i}{\partial s_j} \quad (51)$$

and the constraints that:

$$\sum_i m_i \frac{\partial \mathbf{a}_i}{\partial s_j} \cdot \rho_i = 0 \ (j = 1, 2, 3) \quad (52)$$

we see that the linear terms vanishes. Hence by introducing the normal mode coordinates such that

$$\rho_i = \frac{1}{\sqrt{m_i}} \sum_k \ell_{ik}(s_j) Q_k \quad (53)$$

we get:

$$V(\mathbf{r}_1, ...\mathbf{r}_N) = V(s_1 s_2 s_3) + \frac{1}{2} \sum_{k=1}^{3N-9} \omega_k (s_1 s_2 s_3)^2 Q_k^2 + .. \quad (54)$$

The parameters s_j are now introduced such that the vectors $\sqrt{m_i}\frac{\partial \mathbf{a}_i}{\partial s_j}$ are normalized, i.e. such that

$$\sum_i m_i \mid \frac{\partial \mathbf{a}_i}{\partial s_j} \mid^2 = 1 \quad (55)$$

We introduce the notation

$$\ell_{ik} = \sqrt{m_i} \frac{\partial \mathbf{a}_j}{\partial s_j} \quad (56)$$

where $k = 3N - 9 + j$ (j=1,2,3) and use that

$$\sum_i \ell_{ik} \cdot \ell_{ik'} = \delta_{kk'} \quad (57)$$

$k, k' = 1, \ldots 3N-9$ we can derive the following expression for the kinetic energy:

$$2T = \sum_j \dot{s}_j^2 + \sum_k \dot{Q}_k^2 + 2\sum_j \dot{s}_j \sum_{kk'} Q_k \dot{Q}_{k'} B^j_{kk'} + \sum_{jj'} \sum_{kk'} C^{jj'}_{kk'} \dot{s}_j \dot{s}_{j'} Q_k Q_{k'} \tag{58}$$

where

$$B^j_{kk'} = \sum_i \sqrt{m_i} \ell_{ik'} \cdot \frac{\partial \ell_{ik}}{\partial s_j} \tag{59}$$

$$C^{jj'}_{kk'} = \sum_i \frac{\partial \ell_{ik}}{\partial s_j} \cdot \frac{\partial \ell_{ik'}}{\partial s_{j'}} \tag{60}$$

Considering now a term in the Coriolis coupling (49) we have:

$$\sum_i m_i (\rho_i \times \mathbf{v}_i)_x = \sum_k X_k \dot{Q}_k + \sum_j \dot{s}_j \left(\sum_k Q_k \zeta^{(x)}_{k,3N-9+j} + \sum_l Q_l h^{(x)}_{kl;j}\right) \tag{61}$$

$$X_k = \sum_l Q_l \zeta^{(x)}_{kl} \tag{62}$$

with similar definitions of the (y) and (z) components. We have in the above equation used the following definitions:

$$\zeta^{(x)}_{kl} = \sum_i (l^{(y)}_{ik} l^{(z)}_{il} - l^{(z)}_{ik} l^{(y)}_{il}) \tag{63}$$

$$\zeta^{(x)}_{k,3N-9+j} = \sum_i (l^{(y)}_{ik} l^{(z)}_{i,3N-9+j} - l^{(z)}_{ik} l^{(y)}_{i,3N-9+j}) \tag{64}$$

$$h^{(x)}_{kl;j} = \sum_i (l^{(y)}_{ik} \frac{\partial l^{(z)}_{il}}{\partial s_j} - l^{(z)}_{ik} \frac{\partial l^{(y)}_{il}}{\partial s_j}) \tag{65}$$

Thus the Coriolis term can be expressed as:

$$\begin{aligned} 2T_c = {} & 2\omega_x[\sum_k X_k \dot{Q}_k + \sum_j \dot{s}_j (\sum_k Q_k \zeta^{(x)}_{k,3N-9+j} + \sum_l Q_l h^{(x)}_{kl;j})] \\ & +2\omega_y[\sum_k Y_k \dot{Q}_k + \sum_j \dot{s}_j (\sum_k Q_k \zeta^{(y)}_{k,3N-9+j} + \sum_l Q_l h^{(y)}_{kl;j})] \\ & +2\omega_z[\sum_k Z_k \dot{Q}_k + \sum_j \dot{s}_j (\sum_k Q_k \zeta^{(z)}_{k,3N-9+j} + \sum_l Q_l h^{(z)}_{kl;j})] \end{aligned} \tag{66}$$

The rotational energy is expressed as:

$$2T_r = A\omega_x^2 + B\omega_y^2 + C\omega_z^2 - 2D\omega_x\omega_y - 2E\omega_y\omega_z - 2F\omega_x\omega_z \tag{67}$$

where

$$A = \sum_i m_i[(a_{iy} + \rho_{iy})^2 + (a_{iz} + \rho_{iz})^2] \tag{68}$$

$$D = \sum_i m_i(a_{iy} + \rho_{iy})(a_{iz} + \rho_{iz}) \tag{69}$$

and similar equations for B, C, E and F. We can now introduce the momenta, p_{s_j} (j=1,2,3), p_k and P_x, P_y, P_z as:

$$p_{s_j} = \frac{\partial T}{\partial \dot{s}_j} \tag{70}$$

$$p_k = \frac{\partial T}{\partial \dot{Q}_k} \tag{71}$$

$$P_x = \frac{\partial T}{\partial \omega_x} \tag{72}$$

The momenta can be introduced in the hamiltonian, i.e. for the classical hamiltonian we get:

$$H = \frac{1}{2}(P_x P_y P_z \{p_k\} p_{s_1} p_{s_2} p_{s_3}) \mathbf{M}^{-1} (P_x P_y P_z \{p_k\} p_{s_1} p_{s_2} p_{s_3})^T + V(s_1 s_2 s_3) + \frac{1}{2}\sum_k \omega_k(s_1 s_2 s_3) Q_k^2 \tag{73}$$

where the symmetric matrix $\mathbf{M}$ is defined by:

$$(P_x P_y P_z \{p_k\} p_{s_1} p_{s_2} p_{s_3})^T = \mathbf{M}(\omega_x \omega_y \omega_z \{\dot{Q}_k\} \dot{s}_1 \dot{s}_2 \dot{s}_3)^T \tag{74}$$

The evaluation of the inversion of the $\mathbf{M}$ matrix is carried out to second order in the normal mode coordinates and momenta Q_k and p_k. The final expression is rather lengthy and will therefore not be given here. A significant simplification is obtained for a system so large that the overall rotational motion can be neglected [10]. Here we get:

$$H = \frac{1}{2}\sum_k (p_k^2 + \omega_k(s_1, s_2, s_3) Q_k^2) + V_0(s_1 s_2 s_3) + \frac{1}{2}\sum_i p_{si}^2 (1 - 2\sum_k Q_k B_{ki}^{(i)}) - \sum_{i>j}\sum_k p_{si} p_{sj} Q_k (B_{ki}^{(j)} + B_{kj}^{(i)}) \tag{75}$$

where only the linear coupling terms are included.

5. Discussion

The quantum and semiclassical methods outlined here are capable of treating accurately three particle reactive scattering problems and larger system approximately.

For intermediate sized, i.e. 4-atomic systems it is with present computer facilities possible to obtain the total reaction probability by invoking the absorbing potential surfaces and solving the time-dependent Schrödinger equation with these surfaces placed where bondbreaking lead to the reaction of interest. The total flux over a dividing line before the barrier is recorded and used to compute reaction cross sections and rates [11]. The method which treats two to three degree's of freedom by quantum mechanics and the remaining by classical mechanics has turned out to be accurate for total reaction probabilities. Thus the comparison with 6D quantum calculations [12] was very satisfactory [11]. We notice that the full 7D quantum calculation has not yet been carried out.

For larger systems it is then possible to use the methodology outlined above. The method projects out a reaction-volume in a manner similar to the one used in the reaction path method [13]. The dynamics in the projected space is followed exactly using either classical, semi-classical or quantum dynamics. The remaining space is treated within the harmonic approximation. The advantage of this approach is that it can be generalized to large systems involving e.g. chemical reactions in solution. Also the space in which an accurate description of the interaction potential is needed is reduced drastically. If an accurate quantum treatment is preferred, then the space which should be treated quantally is reduced to just three dimensions and this is certainly possible using the wave-packet methodology described above.

References

1. N. Markovic and G. D. Billing, Chem. Phys. 191(1995)247.
2. A. Bjerre and E. E. Nikitin, Chem. Phys. Lett. 1(1967)179; R. K. Preston and J. C. Tully, J. Chem. Phys. 54(1971)4297.
3. G. D. Billing and N. Markovic, J. Chem. Phys. 99(1993)2674; N. Markovic and G. D. Billing, J. Chem. Phys. 100(1994)1085.
4. Y.-S. Wu and A. Kuppermann, Chem. Phys. Lett. 195(1992)178.
5. B. R. Johnson, J. Chem. Phys. 79(1983)1906; 79(1983)1916.
6. See e.g. H. Tal-Ezer and J. C. Light, J. Chem. Phys. 81(1984) 3967.
7. J. T. Muckerman, R. D. Gilbert and G. D. Billing, J. Chem. Phys. 88(1988)4779.
8. N. Markovic and G. D. Billing, J. Chem. Phys. 97(1992)8201.
9. G. D. Billing, N. Markovic and N. Balakrishnan, (in press)
10. G. D. Billing, Chem. Phys. 146(1990)63.
11. N. Balakrishnan and G. D. Billing, J. Chem. Phys. 101(1994) 1785.
12. U. Manthe, T. Seideman and W. H. Miller, J. Chem. Phys. 99(1993)10078.

13. See e.g. W. H. Miller, N. C. Handy and J. E. Adams, J. Chem. Phys. 72(1980)99; G. D. Billing, Chem. Phys. 89(1984)199; Chem. Phys. 135(1990)63; 161(1992)245; 159(1992)109.

MULTIDIMENSIONAL ANGULAR FUNCTION METHODS IN THEORETICAL AND APPLIED PHYSICS

A. A. SADOVOY
Russian Federal Nuclear Center,
All-Russian Research Institute of Experimental Physics,
Arsamas-16, Russia

YU. A. SIMONOV
Institute of Experimental and Theoretical Physics,
Moscow, Russia

The general regularity of using multidimensional angle functions methods for solution of the diverse problems in the theory of the atomic nucleus and elementary particles , atomic and molecular physics have been considered. The investigation results of the multiparticle of the hyperspherical functions method and in the multielectron atomic theory with multidimensional angle Coulomb functions method have been presented.

Modern technical progress requires more and more information on material properties under various, including extreme, conditions. At the same time requirements to accuracy of these data are much increased, because they determine calculation validity of many unique technical devices and physical installations, allow to predict effect on environment more reliably. Therefore, more and more attention is paid in recent decades to studying material properties under various condition. The majority of objects, investigated in physics, are multiparticle systems, and their theoretical description require to take into account a large number of freedom degrees. Therefore, no real problem can be precisely resolved and various approximate models are developed in theoretical physics. They base on concepts of average fields, independent particles or the strongly correlated motions in multiparticle systems such as a liquid drop. In recent decades methods using multidimensional angular functions /1/-/6/ are finding ever-widening application in the solution of multiparticle problems in molecular and atomic physics, the theory of the atomic nucleus and elementary particles. An attractive feature of these methods consists in possibility to consider collective and one-particle degrees of freedom simultaneously at a strict mathematical approach to solution of multiparticle problems. Recently found and analysed common laws in spectra of elementary particles, atomic nuclei, atomic and molecular systems serve an experimental basis for methods of the multidimensional angular functions. The multidimensional angular function methods in research of multiparticle systems of various physical

C. A. Tsipis et al. (eds.), New Methods in Quantum Theory, 265–280.

nature allow to identify dominant movement by introduction of appropriate collective variables. A specific definition of the latter makes a subject of the detailed theoretical analysis of specific character of a physical system under research . As a rule, collective variables are used as dynamic variables. Dependence of the wave function on them is found from the solution of multiparticle Schroedinger or Dirac equations at arbitrary interaction among particles. Dependence of the multiparticle wave function on other variables is usually chosen from natural or kinematical conditions of a particular problem. For example, similar conditions may be as follows: space symmetry of a problem, translational invariance; Pauli principle, requirement of isotopic spin or color symmetry, etc. Essentially it permits to seek for the multiparticle wave function in the form of a series in multidimensional angular functions, which form complete orthonormalized basis. The amplitude of wave function decomposition in multidimensional angular functions depends on collective variables. Using of projective methods in the multidimensional angular function methods allows to reduce the solution of multiparticle Schroedinger or Dirac equations to the solution of a system of ordinary differential equations of the second or first order which depend on collective variables. All that essentially simplifies calculation algorithms of solution of multiparticle quantum-mechanical problems and in many cases allows to find the multiparticle wave function analytically. It provides capacious storage of the large information about multiparticle systems behavior and rational use of this information at calculation of various properties. Among other features of multidimensional angular function methods (MAF) we shall note only their universality and variational properties. The MAF methods universality is that computed data, obtained at solution of problems with smaller number of particles, may be used at solving problems with a large number of particles. The matrix elements of the potential and kinetic energy operators on MAF, needed for calculations of neutral atom properties in the method of multidimensional angular coulomb functions (MACF) may be used to calculate properties of ions of any degree of ionization for a given isoelectron series. A similar picture also takes place in relativistic version of the MACF method. Universal factors are introduced and partially calculated /3/-/4/ in the hyperspherical function method in the nuclear theoretical physics. They express matrix elements of nucleon-nucleon potentials on MAF through unitary integrals. Variational properties of the MAF methods with consideration of completeness of used MAF basis allow to control an accuracy of conducted numerical calculations. It is essential at realization of processional calculations, for example, in atomic or molecular physics, where to achievement relative accuracy of ~ 10^{-4} many members of wave function decomposition are taken into account and it leads to calculations with poorly stipulated matrices of high ranks. In summary we note that in various methods of multidimensional angular functions from one up to six collective variables were introduced and review of all these methods can make the subject of no one monograph. Below we shall dwell upon some

MAF methods by which results of scientific and technical interest were obtained.

I. For multi-nucleon and multi-quark systems in the nucleus and elementary particle theory various versions of the huperspheric function method (HSFM) /1/-/6/ which most consistently accounts the short-term nature of inter-particle interaction have found a wide utility. An advantage of the HSFM as compared to other atomic nucleus models consists in absence of any fitting parameters, the method allows account both two - and three -particle interaction. Various approximate models of the atomic nucleus theory (the shell model, models of deformed nuclei, some nuclear reaction models) are obtained as limit cases in the HSFM method, which allows to interrelate phenomenological parameters of these models and study their dependence on nucleon-nucleon forces. The solution of the multi-particle Schroedinger equation in the HSFM method is sought for in the form

$$\Psi(\vec{r}_1,\vec{r}_2,\ldots\vec{r}_A)=\sum_{k\nu}\frac{\chi_{k\nu}(\rho)}{\rho^{\frac{3A-3}{2}}}U_{k\nu}(\Omega)\ .$$

As a collective variable, the hyperspherical coordinate system hyperradius is used:

$$\rho^2=\sum_{i=1}^{A}\vec{r}_i^{\,2}-AR^2=\sum_{i=1}^{A-1}\vec{\xi}_i^{\,2}\quad ,$$

and the basis functions are of the form:

$$\Phi_{njljz}(\vec{\rho}_i)=C_{nl}\left(\vec{\rho}_i^{\,2}\right)^{n+l/2}Y_{lm}(\Omega_i)\,\chi_{1/2\sigma}(i)\,\xi_{1/2\tau}(i)\quad .$$

The multi-particle Schroedinger equation eventually reduces to a system of second-order ordinary differential equations:

$$\left\{\frac{d^2}{d\rho^2}-\frac{Z_k(z_k+1)}{\rho^2}-\frac{2m}{\hbar^2}\left[E+W_{k'\upsilon'}^{k\upsilon}(\rho)\right]\right\}\chi_{k\upsilon}(\rho)=$$

$$=\frac{2m}{\hbar^2}\sum_{k'\upsilon'\neq k\upsilon}W_{k'\upsilon'}^{k\upsilon}(\rho)\cdot\chi_{k'\upsilon'}(\rho)\quad ,$$

where

$$W_{k\upsilon}^{k'\upsilon'}(\rho)=\int U_{k\upsilon}^{+}(\Omega)\,\hat{V}(\vec{r}_1\ldots\vec{r}_A)\,U_{k'\upsilon'}(\Omega)\,d\Omega_{3A-3}$$

$$Z_k=K+\frac{3A-6}{2}$$

When computing properties of specific nuclear systems it is convenient to use the Sturmian representation of the Green function for an arbitrary number of particles interacting according to the oscillatory law:

$$\left\{-\frac{\hbar^2}{2m}\sum_{i=1}^{A-1}\frac{\partial^2}{\partial\vec{\xi}_i^2}+\frac{m\omega^2}{2}\sum_{i\succ j=1}^{A}\left(\vec{r}_i-\vec{r}_j\right)^2-E\right\}G_E\left(\left\{\vec{\xi}_i\right\},\left\{\vec{\xi}'_i\right\}\right)=$$

$$=\prod_{i=1}^{A-1}\delta\left(\vec{\xi}_i-\vec{\xi}'_i\right)\prod_{j=1}^{A}\delta_{\mu_j\mu'_j}\,\delta_{\tau_j\tau'_j}.$$

This Green function has analytic representation of the form /1/:

$$G\left(\vec{\xi},\mu,\tau;\vec{\xi}',\mu',\tau'\right)=\left(\rho\rho'\right)^{\frac{-3A-4}{2}}\sum_j U_j^+(\Omega)U_j(\Omega')\times$$

$$\times\sum_{n=0}^{\infty}\frac{\Psi_{kjn}(\rho)\cdot\Psi_{kjn}(\rho')}{\frac{m\omega^2 A}{2}-\frac{2m}{\hbar^2}\left(\frac{E}{4n+2Z_k+3}\right)^2},$$

where $U_j(\Omega)$ - MAF of the HSFM method, $Z_k=k+(3A-3)/2$, $\Psi_{nlm}(\vec{\rho})$ Sturmian functions for the oscillatory potential.

Such Sturmian series does not contain a continuum and each its member has a regular asymptotics in infinity.

Recently some papers were published demonstrating successful HSFM application to solution of diverse problems of discrete and continuous spectrum in nuclear physics. In connection with the interest to properties of exotic, including neutron-excessive, super-heavy and purely neutronic nuclei it is deemed possible to study HSFM in more detail. One of first theoretical researches are performed in work these nuclei on a various component nucleon-nucleon potentials and simultaneously have allowed to coordinate calculation of their properties to properties of stable nuclei. Despite dependence many calculative results in nuclear physics from used models of nuclear forces, nevertheless, for some phenomenological nucleon-nucleon potentials calculations indicate on nucleon stability of superheavy nuclei $^{280}112$, $^{408}168$. Naturally, the results of these work could be specifies by using of modern achievement of multidimensional angular functions.

II. The long- range nature of Coulomb interaction and availability of the heavy particle - nucleus are considered in the method of multidimensional angle Coulomb functions the following value is introduced:

$$\rho=\sum_{i=1}^{A}|\vec{r}_i|$$

Parametrization of 3A-dimensional space, formed by direct product of three-dimensional spaces of all A - electrons, is conducted so, that the volume element of the angular variables equals /10/:

$$d\Omega_{3A}=\prod_{i=1}^{A}d\Omega_i\prod_{j=1}^{A-1}\left(1-\sin\theta_j\right)^2\left(\sin\theta_j\right)^{3j-1}d\left(\sin\theta_j\right)$$

In the MACF method multidimensional angular functions having a definite value of the total angular moment, its projection, parity are built using Clebsh-Gordon coefficients from multidimensional polynomials $P_{k\upsilon}\left(\vec{r}_1,\vec{r}_2,\ldots\vec{r}_A\right)$. After orthogonalization they are of the form

$$U_{k_i}(\Omega)=\sum_j\Delta_{ij}\sqrt{\frac{\tilde{A}\left(2k_j+3A\right)}{A!}}\frac{1}{\rho^{k_j}}P_{k_j}\left(\Psi(\vec{r}_i)\right),$$

where Δ_{ij} - orthogonalization constant, $k=\sum_{\omega}(l+n)$, polynoms P_{k_j} constitute the Slater determinants constructed from basic functions of the form

$$\Phi_{nlm\mu}(\vec{r}_i)=\frac{(-1)^n}{\tilde{A}\,(n+1)}r_i^n\,Y_{lm}(\vec{r}_i)\,\chi_{1/2\mu}(i)$$

and therefore they are uniform.

At presenting the multielectronic wave function as a series in MAF

$$\Psi(\vec{r}_1,\ldots\vec{r}_A)=\sum_{k\upsilon}\frac{\chi_{k\upsilon}(\rho)}{\rho^{\frac{3A-1}{2}}}U_{k\upsilon}(\Omega_{3A})$$

it was possible to reduce the nonrelativistic Schroedinger equation

$$\left\{-\frac{\hbar^2}{2m}\sum_{i=1}^{A}\frac{\partial^2}{\partial\vec{r}_i^{\,2}}-\sum_{i=1}^{A}\frac{Ze^2}{r_i}+\sum_{i\succ j}^{A}\frac{e^2}{|\vec{r}_i-\vec{r}_j|}-E\right\}\Psi(\vec{r}_1,\vec{r}_2,\ldots\vec{r}_A)=0$$

after projecting to a system of ordinary differential equations for amplitudes of wave function decomposition

$$\chi''(x)+\frac{C}{x}\chi'(x)+\left(\frac{B}{x^2}+\frac{\rho F}{x}-\frac{1}{4}I\right)\chi(x)=0$$

where a matrix representation form is used, I - unity matrix, $x=\alpha\rho$, $\alpha=\sqrt{\frac{8E}{A}}$. Matrices C and B are expressed through matrix elements of the kinetic and potential energy operator respectively. The vector $\chi(x)$ for a discrete spectrum satisfies to the following boundary conditions:

$$\lim_{x\to 0}\chi(x)=\lim_{x\to\infty}\chi(x)=0$$

To obtain an analytical solution of differential equations of MAF method simple Sturmian decomposition and Sturmian decomposition with separated asymptotics are developed now. Illustrate the method of simple Sturmian decomposition. When using Sturmian representations of the Green functions (GF) one can transfer from the differential equation system (2) to the integral equation system:

$$\chi_i(x)=-\sum_{j\neq i}^{N}\int G_i(x,y)\left[\frac{C_{ij}}{y}\frac{d}{dy}+\frac{b_{ij}}{y^2}+\frac{\rho f_{ij}}{y}\right]\chi_j(y)\,dy\ .$$

In this equation the Green function is specified in the following way:

$$\left(\frac{d^2}{dx^2}+\frac{b_{ii}}{x^2}+\frac{\rho f_{ii}}{x}-\frac{1}{4}\right)G_i(x,y)=\delta(x-y)\ .$$

One can obtain the following Sturmian representation for it [1]:

$$G_i(x,y)=\sum_{k=0}^{\infty}\frac{\Psi_{k\beta_i}(x)\,\Psi_{k\beta_i}(y)}{\rho f_{ii}-k-\beta_i}\ ,$$

if one uses the Sturmian functions

$$\Psi_{n\beta}(x)=\sqrt{\frac{\tilde{A}(n+1)}{\tilde{A}(n+2\beta)}}\ e^{-x/2}\,x^{\beta}\,L_n^{2\beta-1}(x)\ ,$$

which are solutions to the problem

$$\left(\frac{d^2}{dx^2}+\frac{b_{ii}}{x^2}-\frac{1}{4}\right)\Psi_{k\beta_i}(x)=\frac{\lambda_k}{x}\Psi_{k\beta_i}(x)$$

and which vanish in infinity and at coordinate origin under the condition

$$\lambda_k=-(k+\beta_i)\,,\qquad \beta_i(\beta_i-1)-b_{ii}=0\quad .$$

It allowed to find solution of the system (x) in the form

$$\chi_i(x)=\sum_{n=0}^{\infty} J_{in}\,\Psi_{n\beta_i}(x)\quad ,$$

where coefficients J_{in} and eigenvalues ρ are found from solution of the uniform algebraic system

$$\left(\rho f_{ii}-k-\beta_i\right)J_{ik}=-\sum_{j\neq i}^{N}\sum_{n=0}^{\infty} M_{ij}^{kn}(\rho)\,J_{jn}\quad ,$$

where matrix M is expressed through the matrix elements of the operators $\frac{1}{y}\frac{d}{dy}$; y^{-2} , y^{-1} and by Sturmian functions [1].

It is important to note, that it results in the fact that the MAF method without use of any variational parameters yields an analytical representation for multielectronic wave function, which, in addition, in contrast to a wave function of the Hartree-Fock method, has correct asymptotic in infinity and automatically takes into account configurations mixing. To illustrate the quality of the wave function of the MACF method the moments of atomic radii of some helium-like ions are presented in Table 1. They are compared with calculations of Smith and Thakker group /11/, using up to hundred variational parameters.

TABLE 1. Comparison of < R^n > for helium-like ions.

Z	< R >	-2	-1	2	3	4
3	Smith	14.927	2.6879	0.4463	0.4413	0.5296
	MACF	14.923	2.6870	0.4486	0.4466	0.5447
9	Smith	15.241	8.6876	0.4057-1	0.1188-1	0.4186-2
	MACF	15.243	8.6874	0.4061-1	0.1190-1	0.4195-2
12	Smith	275.16	11.6876	0.2230-2	0.4850-2	0.1258-2
	MACF	275.18	11.6874	0.2232-2	0.4839-2	0.1262-2

The MACF method predicts existence of radialy excited levels in atoms and ions, corresponding to collective fluctuations of electronic density. Similar results qualitatively agree with calculations by a method of density functional, however, in connection with exact account of Pauli principle in MACF method additional levels are predicted, which is illustrated by the data of table 2.

TABLE 2. Spectrum of 1/2 levels of lithium-like ions.

Ion	Levels	MACF (at.un.)	Data /11/ (at.un.)
Be I I	1	14.245	14.296
	2	13.443	12.485
	3	12.323	10.037
Be I I I	1	23.321	23.389
	2	21.829	20.280
	3	20.127	16.239
C I V	1	34.648	34.739
	2	32.364	29.989
	3	29.816	23.941

The increasing role of configuration mixing with the number of electrons illustrates the contribution to w.f. of normalizing various MAF corresponding to the following one-electron $(1s \to 3s)$, $(1s \to 4s)$, $(1s \to 5s)$ and two-electron excitations $\begin{pmatrix} 1s \to 3s \\ 2s \to 3s \end{pmatrix}$, $\begin{pmatrix} 1s \to 4s \\ 2s \to 3s \end{pmatrix}$, $\begin{pmatrix} 2s \to 4s \\ 2s \to 4s \end{pmatrix}$ from the main configuration of beryllium-like ions $\left(1s^2 2s^2\right)$.

TABLE 3. Contribution of various MAF to w.f. normalization of beryllium-like ions.

Ion	E_0 at.un.	1	2	3	4	5	6	7
B I I I	24.162	.739	.206	.023	.006	.009	.001	.016
	22.965	.506	.262	.111	.011	.024	.06	.078
C I I I	36.326	.774	.184	.018	.005	.007	.001	.011
	34.498	.505	.265	.117	.011	.023	.005	.074
N I V	50.966	.795	.170	.016	.004	.006	.001	.009
	48.339	.506	.267	.119	.009	.022	.005	.071

The software available for the authors also allowed to study O^+ states of carbon-like ions using the MACF method as a seven-body problem. In so doing MAF were taken into account corresponding to one-electron $(1s \to 3s)$, $(1p \to 3p)$, $(1s \to 4s)$, $(1s \to 5s)$ and two-electron $\begin{pmatrix} 1s \to 3s \\ 2s \to 3s \end{pmatrix}$, $\begin{pmatrix} 1s \to 4s \\ 2s \to 3s \end{pmatrix}$,

$\begin{pmatrix} 1s \to 3s \\ 1s \to 4s \end{pmatrix}$, $\begin{pmatrix} 3s \to 6s \\ 1s \to 6s \end{pmatrix}$ excitations from the basic configuration $1s^2 2s^2 1p_{1/2}^2$

. Binding energies of the basic and first radially excited O^+ levels of carbon-like ions are given below.

TABLE 4. Energy of O⁺ states of carbon isoelectron series (at units).

Ion	E_0 MACF	E_1 MACF	E_c H-F with corrections
CI	37.25	35.50	37.64
NI	53.47	50.55	53.82
OTY	72.68	69.24	73.01
S I V	94.88	90.35	95.014
Ne V	120.07	114.30	120.42
NaVi	148.25	141.08	148.63
MqVII	179.42	170.68	179.84
ALVIII	213.60	203.12	214.05
Si IX	250.76	238.38	251.27
PX	290.92	276.48	291.48
SXI	334.07	317.40	334.70
SL XII	380.12	361.15	380.92
Ar XIII	429.37	407.93	430.13

The data calculated using the MACF and Hartree-Fock method with corrections [12] agree within some tenths of per cent. It is noted that there is decrease in contribution of the main configuration of $1s^2 2s^2 1p_{1/2}^2\ {}^1S_0$ with ion charge (60% for CI and 83% for CAXV). Figs. 1 and 2 show amplitudes χ_i for decomposition of multi-electron wave functions in MAF for the basic and first radially excited O^+ of CI levels. The basic level amplitudes have no zeros and all amplitudes of the radially excited level have one zero each which is a numerical justification of multi-electron w.f. oscillatory properties in the MACF method. Figs. 3 and 4 give electron densities calculated using the MACF method and the Hartree-Fock method ($\int_0^\infty n(r)\,dr = A$ normalization).

The authors attribute the difference in electron density distribution on the ion boundary to exactness of multi-electron wave function asymptotics description where Hartree-Fock method wave functions are known to be not exact [14].

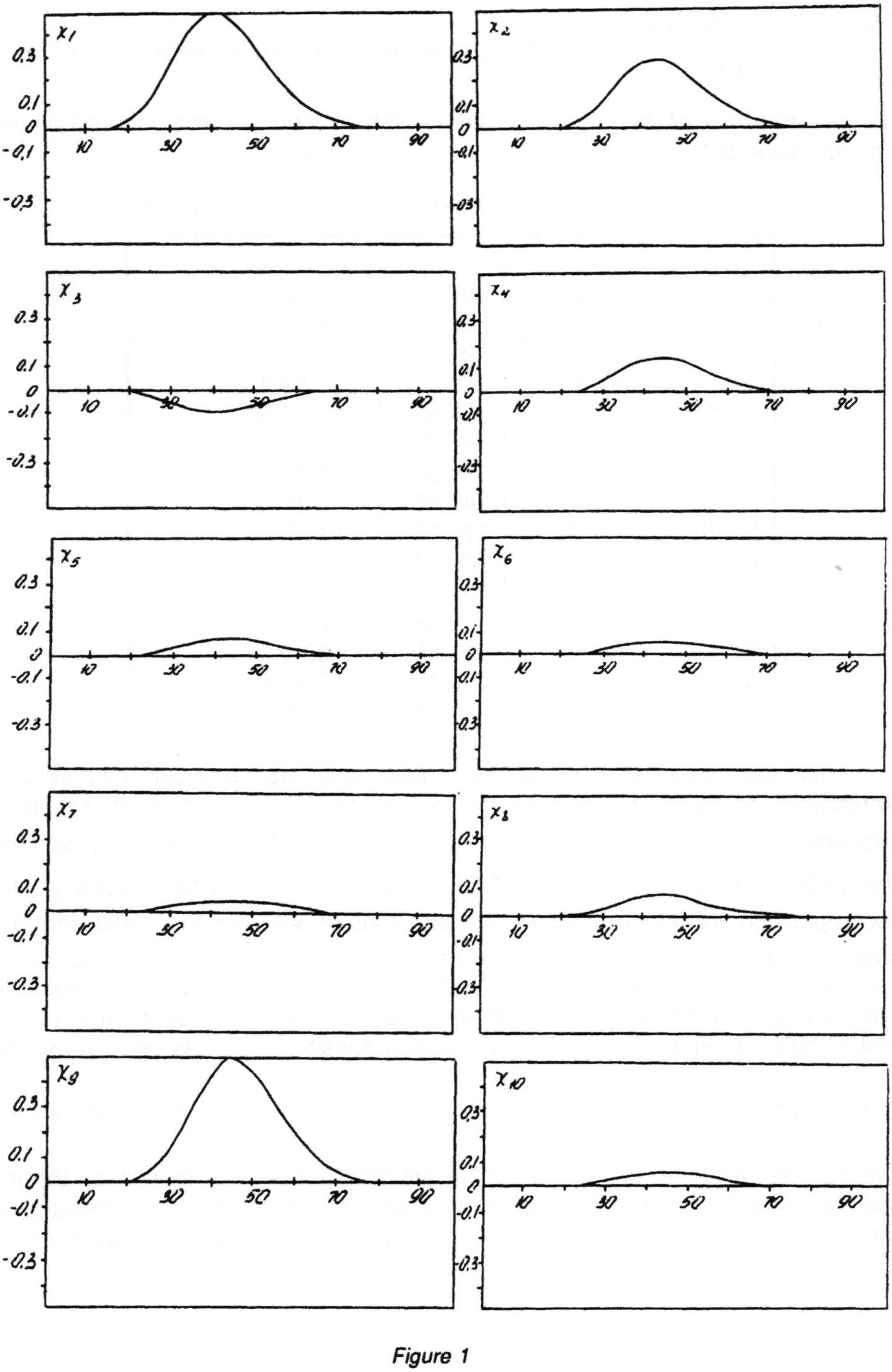

Figure 1

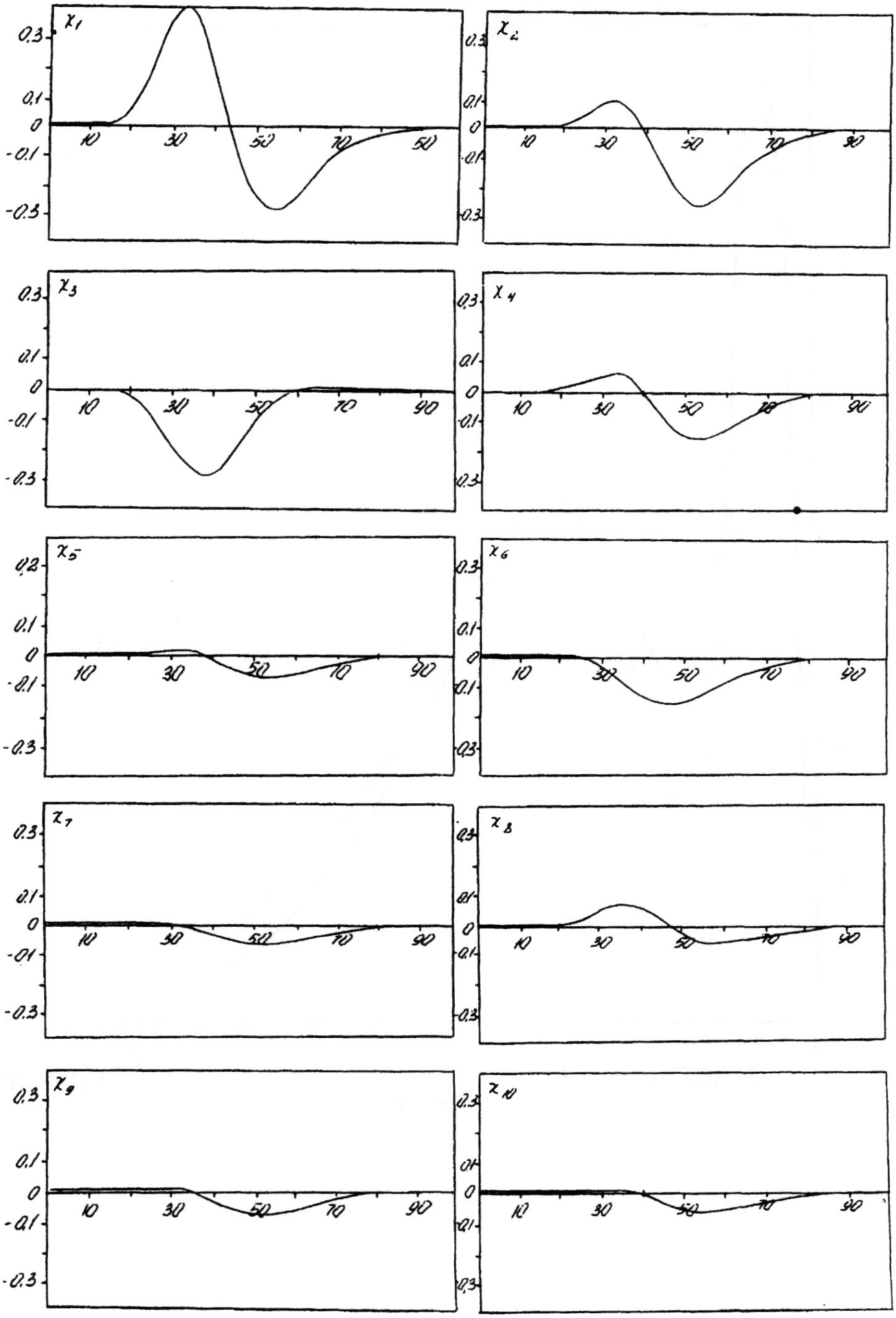

Figure 2

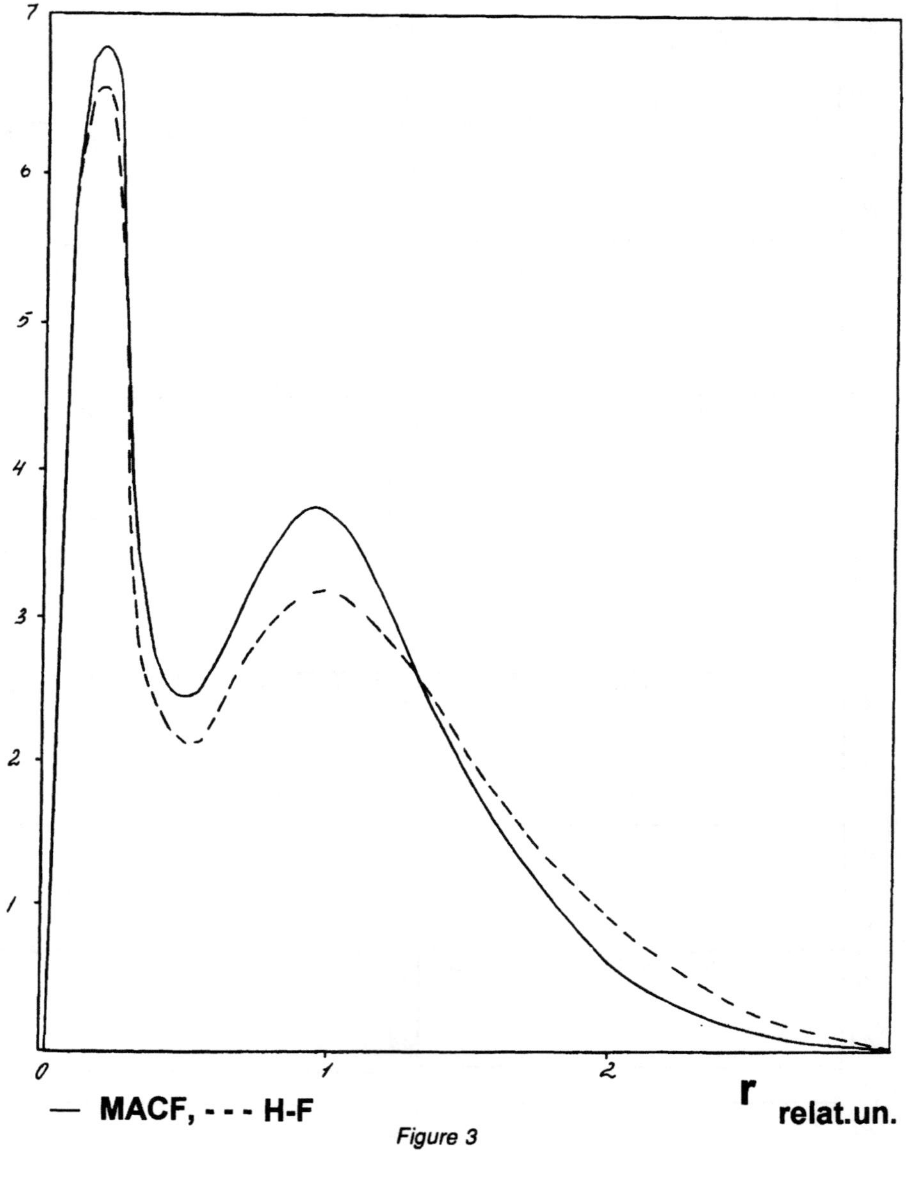

Figure 3

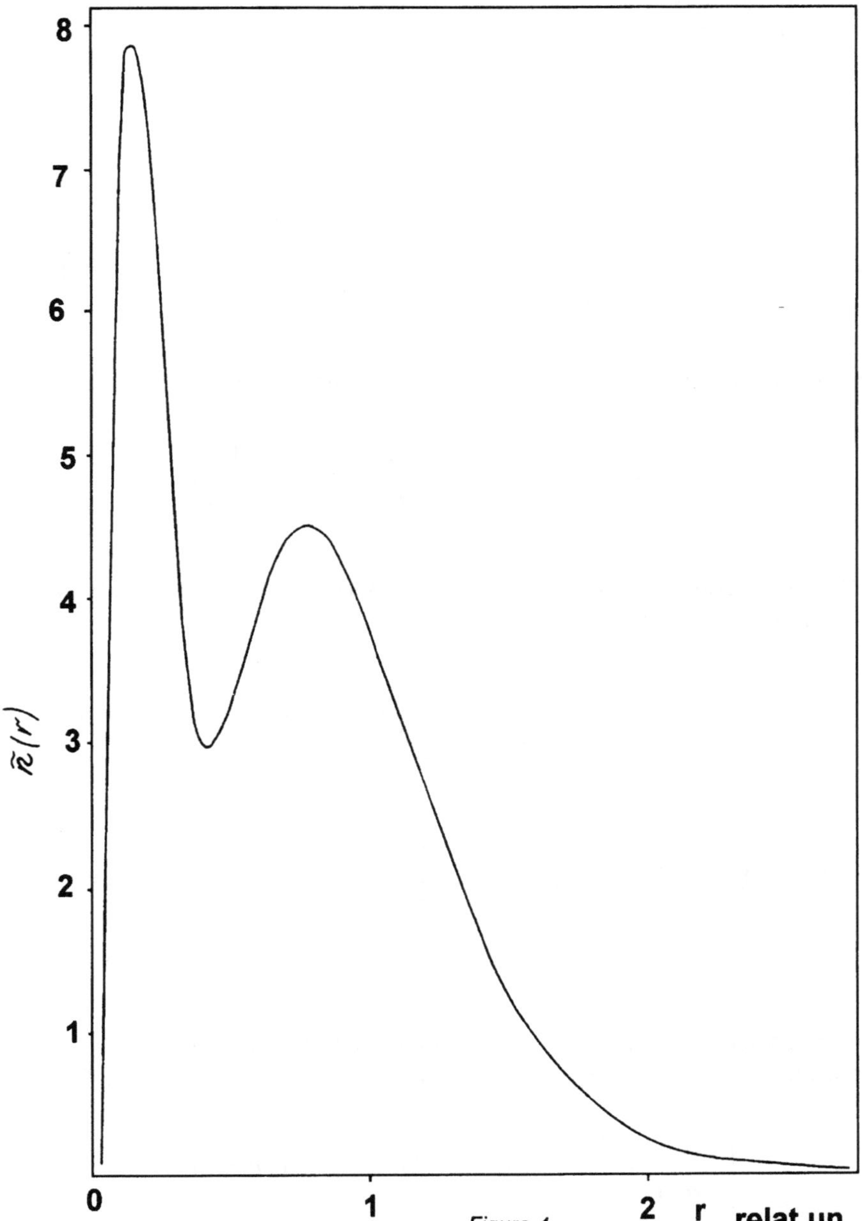

Figure 4

The data on distribution of electron density in multielectron atoms and ions, elastic and inelastic electron scattering on ions in the Born-Coulomb approximation and a number of other characteristics of multielectron atoms and ions with an arbitrary degree of ionization are also published. According to estimates the most interesting results are possible for transition probabilities, in particular, for forbidden transitions.

III. In relativistic version of the MACF method /1/ the solution of a multielectronic Dirac equation

$$\left\{\sum_{i=1}^{A}\left(C\vec{\alpha}_i\,\vec{p}_i + mC^2\beta_i' - \frac{Ze^2}{r_i}\right) + \sum_{i\succ j=1}^{A}\frac{e^2}{\left|\vec{r}_i - \vec{r}_j\right|} - E\right\}\Psi(\vec{r}_1,\vec{r}_2,\ldots\vec{r}_A) = 0$$

is sought for as a number of four-component multidimensional angular functions

$$\Psi(\vec{r}_1,\vec{r}_2,\ldots\vec{r}_A) = \sum_{k\upsilon}\rho^{-\frac{3A-1}{2}}\begin{pmatrix} M_{k\upsilon}(\rho)\,U_{k\upsilon}(\Omega) \\ N_{k\upsilon}(\rho)\,W_{k\upsilon}(\Omega)\end{pmatrix},$$

which represent determinants, constructed by way of their consecutive filling by basic functions of the form

$$\varphi_{njlj_z}(\vec{r}_i) = \frac{(-1)^n r_i^n}{\tilde{A}(n+1)}\sum_{m\mu}\left(lm\frac{1}{2}\mu / j\,j_z\right) Y_{lm}(\Omega_i)\,\chi_{1/2\,\mu}(i),$$

$$\chi_{njl'j_z}(\vec{r}_i) = \frac{(-1)^n r_i^n}{\tilde{A}(n+1)}\sum_{m\mu}\left(l'm\frac{1}{2}\mu / j\,j_z\right) Y_{l'\mu}(\Omega_i)\,\chi_{1/2\mu}(i),$$

$$l' = 2j - l\ .$$

After projection using the four component MAF for amplitudes of wave function decomposition we have the following system of hyperradial equations:

$$\begin{cases}\dfrac{dN}{d\rho} - \dfrac{B}{\rho}N - \alpha\dfrac{A}{\rho}M - \varepsilon_o M(\rho) = 0\\[2mm] \dfrac{dM}{d\rho} + \alpha\dfrac{\overline{A}}{\rho}N + \dfrac{\overline{B}}{\rho}M + (\varepsilon_o - 2Z)N(\rho) = 0\end{cases}$$

in which matrices B and $\overline{\mathrm{B}}$ are expressed through matrix elements of the Dirac operator, and the matrices A and $\overline{\mathrm{A}}$ through matrix elements of potential energy in MAF. One method of solving this system consists in the presentation of the solution in the form of a series of Sturmian functions and in concept this method is close to the method described in equation an analytic representation for the wave function was also obtained. It significantly simplifies carrying out systematic calculations to study properties of various Coulomb systems. In so doing asymptotics of the wave function has to satisfy some requirements to avoid the "variational collapse" [15] According to estimations, most interesting results are expected for high ionization ions with Z>>1, and similar computations are currently underway.

IV. The analysis submitted and published on methods of angular functions materials shows that these methods allow to receive unique calculated data. The main difficulties hindering wide distribution of these methods, are connected to construction of orthogonalized MAF in many dimensions, computations of matrix elements from various operators. In the light stated international cooperation is obviously possible, on development of computational apparatus and realization of massive calculations using installed base of various countries for studying properties:

- ions of any degree of ionization of helium-, lithium-, beryllium-, boronlike series;
- ions of high degree of ionization of heavy elements, for which relativistic effects are essential;
- multielectronic atoms and ions in external electrical and external fields;
- interatomic potentials, taking into account Pauli principle, for development of the microscopic approach to construction of material equations of state;
- exotic nuclei, including neutron-excessive and superheavy.

It is natural, that the presented proposals are preliminary in character and can be adjusted by interested organizations with consideration of their technical and scientific interests.

References

1. Sadovoy A. A. (1994) *Methods of multidimensional wave functions in theoretic and applied physics*, RFNC-VNIIEF, Arzamas-16.
2. Dzhibuti R. T., Shitikova K. V. (1993) *Method of hyperspheric functions in atomic and nuclear physics*, Energoatomizdat Publishers, Moscow.
3. Filippov G. E., Ovcharenko V. Z. , Smirnov Ju. F. (1981) *Microscopic theory of atomic nuclei collective excitations*. Naukova Dumka Publishers, Kiev.
4. Simonov Yu. A. (1968) Nuclear functions of arbitrary nucleon number. *Jadernaya Fizika* **7**,1210-1221.
5. Hutson J. M., Jaon S. (1989) On the coupled-channel calculation of bound states for triatomic systems using hyperspherical coordinates. *J. Chem. Phys.* **91**, 4197-4203.
6. Badalyan A. M. (1987) United description of ground state mesons and baryons in a potential model. *Preprint ITEP*.
7. Badalyan A. M., Calogero F., Simonov Y. A. (1970) The hyperspherical expansion approach to nuclear bound states. *Nuovo Cimento* **A-68**, 572-584.
8. Baz A. I., Averyanov I. K., Golubev A. I., Sadovoy A. A. (1974) Tables of universal factors of hyperspheric function method for central and coulomb forces. Kurcbatov institute of Atomic Energy *Preprint N2355*.
9. Baz A. I., Gol'dansky V. I., Goldberg V. Z., Zel'dovich Ja. V. (1972) *Light and intermediate nuclei near nucleon stability boundaries*, Nauka Publishers, Moscow.
10. Sadovoy A. A. (1978) On multi-electronic atom theory. *Dokl. AN USSR,* **238**, 77-81.
11. Thakkar A. J., Smith V. H. (1977) Compact and accurate integral transform wave functions. *Phys. Rev.*, **15**, 1-22.
12. Lain L., Valdermoro C., Torre A., Reguero M. (1987) Spin-adapted reduced Hamiltonian theory. Application the independent pair model to the isoelectronic sequence of lithium. *Phys. Rev. A.*, **35**, 3132-3135.
13. Bogdanovich P. P. (1983) Code for numerical solution of Hartree-Fock equations. *Litovskiy Fizicheskiy Sbornik*, **23**, 18-33.
14. Koga T. (1988) Zero potential energy criterion applied to Hartree-Fock wave functions. *J. Chem. Phys.*, **87**, 4207-4211.
15. Grant I. P. (1994) Relativistic calculation of atomic properties. *Comput. Phys. Commun.*, **84**, 59-77.

HYPERSPHERICAL STURMIAN BASIS FUNCTIONS IN RECIPROCAL SPACE

JOHN S. AVERY
H.C. Ørsted Institute
University of Copenhagen
DK-2100 Denmark

1. Introduction

Reciprocal-space quantum theory, and the theory of hyperspherical harmonics provide us with an elegant method for finding solutions to the Schrödinger equation of an electron moving in the Coulomb field of a many-center charge distribution. This formalism was invented by V.A. Fock [1], who used it to solve the reciprocal-space Schrödinger equation of a hydrogenlike atom. T. Shibuya and C.E. Wulfman [2] were able to extend Fock's formalism so that it could be used in the case of many-center Coulomb potentials; and their methods have been further developed by a number of other authors [3-9].

In the method of Fock, Shibuya, and Wulfman, 3-dimensional reciprocal space is mapped onto the surface of a 4-dimensional unit sphere by the transformation:

$$
\begin{aligned}
u_1 &= \frac{2p_0p_1}{p_0^2+p^2} = \sin\chi \sin\theta_p \cos\phi_p \\
u_2 &= \frac{2p_0p_2}{p_0^2+p^2} = \sin\chi \sin\theta_p \sin\phi_p \\
u_3 &= \frac{2p_0p_3}{p_0^2+p^2} = \sin\chi \cos\theta_p \\
u_4 &= \frac{p^2-p_0^2}{p_0^2+p^2} = \cos\chi
\end{aligned} \tag{1}
$$

C. A. Tsipis et al. (eds.), New Methods in Quantum Theory, 281–296.

where p_0 is related to the energy by

$$E = -\frac{1}{2}p_0^2 \tag{2}$$

Fock was able to show that, with this transformation, the Fourier transforms of the hydrogenlike wave functions can be written in the form:

$$\chi^t_{n,l,m}(\mathbf{p}) = M(p)Y_{n-1,l,m}(\mathbf{u}) \tag{3}$$

where

$$Y_{\lambda,l,m}(\mathbf{u}) = (-i)^l(2l)!!\sqrt{\frac{2(\lambda+1)(\lambda-l)!}{\pi(\lambda+l+1)!}}\ \sin^l\chi C^{l+1}_{\lambda-l}(u_4)Y_{l,m}(\theta_p,\phi_p) \tag{4}$$

is a 4-dimensional hyperspherical harmonic [7,8], (C^α_λ being a Gegenbauer polynomial), and where M(p) is the function

$$M(p) \equiv \frac{4p_0^{5/2}}{(p_0^2+p^2)^2} \tag{5}$$

As the reader may verify, the Fourier-transformed hydrogenlike orbitals shown in Table 1 can be derived from these relations, using the 4-dimensional hyperspherical harmonics given in Table 2, with $p_0 = \zeta/n$. Shibuya and Wulfman extended Fock's method to the case of a many-center Coulomb potential. If the potential has the form:

$$V(\mathbf{x}) = \sum_j Z_j \frac{1}{|\mathbf{x}-\mathbf{X}_j|} \tag{6}$$

then the solutions to the reciprocal-space Schrödinger equation can be expanded in the form:

$$\phi^t(\mathbf{p}) = M(p)\sum_\tau \eta_\tau(\mathbf{u})B_\tau \tag{7}$$

where the basis functions $\eta_\tau(\mathbf{u})$ are defined by

$$\eta_\tau(\mathbf{u}) \equiv \sqrt{\frac{Z_j}{\lambda+1}}\ e^{i\mathbf{p}\cdot\mathbf{X}_j}Y_{\lambda,l,m}(\mathbf{u}) \tag{8}$$

with τ representing the set of indices j, λ, l, m. The coefficients B_τ in the expansion are found by solving the secular equation:

$$\sum_{\tau} \left(S_{\tau,\tau'} - \delta_{\tau,\tau'} p_0\right) B_{\tau'} = 0 \tag{9}$$

where $S_{\tau,\tau'}$ is the matrix of overlap integrals:

$$S_{\tau,\tau'} \equiv \int d\Omega \eta^*_\tau(\mathbf{u}) \eta_{\tau'}(\mathbf{u}) \tag{10}$$

the integral $\int d\Omega$ being taken over the 4-dimensional solid angle [7,8].

TABLE 1.

	$\chi_{n,l,m}(\mathbf{x})$	$\chi^t_{n,l,m}(\mathbf{p})$
$1s$	$\left(\frac{\zeta^3}{\pi}\right)^{1/2} e^{-\zeta r}$	$\frac{4\zeta^{5/2}}{\sqrt{2}\pi(p^2+\zeta^2)^2}$
$2s$	$\left(\frac{\zeta^3}{8\pi}\right)^{1/2} e^{-\zeta r/2}\left(1-\frac{\zeta r}{2}\right)$	$\frac{8(\zeta/2)^{5/2}[(\zeta/2)^2-p^2]}{\sqrt{2}\pi(p^2+(\zeta/2)^2)^3}$
$2p_z$	$\left(\frac{\zeta^5}{32\pi}\right)^{1/2} e^{-\zeta r/2} z$	$\frac{16i(\zeta/2)^{7/2}p_z}{\sqrt{2}\pi(p^2+(\zeta/2)^2)^3}$

TABLE 2. 4-dimentional hyperspherical harmonics

λ	l	m	$\sqrt{2}\pi\ Y_{\lambda,l,m}(\mathbf{u})$
0	0	0	1
1	1	1	$i\sqrt{2}(u_1 + iu_2)$
1	1	0	$-i2u_3$
1	1	-1	$-i\sqrt{2}(u_1 - iu_2)$
1	0	0	$2u_4$
2	2	2	$-\sqrt{3}(u_1 + iu_2)^2$
2	2	1	$2\sqrt{3}u_3(u_1 + iu_2)$
2	2	0	$-\sqrt{2}(2u_3^2 - u_1^2 - u_2^2)$
2	2	-1	$-2\sqrt{3}u_3(u_1 - iu_2)$
2	2	-2	$-\sqrt{3}(u_1 - iu_2)^2$
2	1	1	$2i\sqrt{3}\ u_4(u_1 + iu_2)$
2	1	0	$-i\sqrt{6}\ u_4u_3$
2	1	-1	$-2i\sqrt{3}\ u_4(u_1 - iu_2)$
2	0	0	$4u_4^2 - 1$

2. Evaluation of the hyperangular overlap integrals

Methods for the evaluation of the hyperangular overlap integrals are discussed in references [2], [3], [7] and [8]. The most straightforward method is to express $S_{\tau,\tau'}$ in terms of the functions $W_{i,0}^{l,j}$ studied by Geller [10], and by Harris and Michels [11]. These authors showed that integrals of the form

$$W_{i,0}^{l,j} \equiv \frac{2}{\pi} \int_0^\infty dp \frac{p^{l+2j} j_l(pR)}{(p^2 + p_0^2)^i} \tag{11}$$

can be evaluated by means of the recursion relation:

$$\frac{2l+1}{R} W_{i,0}^{l,j} = W_{i,0}^{l+1,j} + W_{i,0}^{l-1,j+1} \tag{12}$$

starting with the relationship

$$W_{s+1,0}^{l,1} = \frac{R^s}{2^s s! p_0^{s-l-1}} k_{s-l-1}(p_0 R) \tag{13}$$

The functions $k_l(x)$ are the modified spherical Bessel functions given by

$$\begin{aligned} k_{-1} &= \frac{1}{x} e^{-x} \\ k_0 &= \frac{1}{x} e^{-x} \\ k_1 &= \frac{1}{x^2}(1 + x) e^{-x} \\ k_2 &= \frac{1}{x^3}(3 + 3x + x^2) e^{-x} \\ k_3 &= \frac{1}{x^4}(15 + 15x + 6x^2 + x^3) e^{-x} \end{aligned} \tag{14}$$

The modified spherical Bessel functions obey the recursion relations

$$k_{l+1} = \frac{2l+1}{x} k_l + k_{l-1} \tag{15}$$

and

$$k_{-l} = k_{l-1} \tag{16}$$

from which the higher and lower orders can be generated.

Combining equations (8) and (10) we have:

$$S_{\tau,\tau'} = \sqrt{\frac{Z_j Z_{j'}}{(\lambda+1)(\lambda'+1)}} \int d\Omega \; e^{i\mathbf{p}\cdot\mathbf{R}} \; Y_\tau^*(\mathbf{u}) Y_{\tau'}(\mathbf{u}) \tag{17}$$

where $\mathbf{R} \equiv \mathbf{X}_{j'} - \mathbf{X}_j$. The integral in equation (17) can be expressed as a sum of terms of the form:

$$J = \frac{1}{2\pi^2} \int d\Omega \; e^{i\mathbf{p}\cdot\mathbf{R}} \; u_4^\nu h_l(u_1, u_2, u_3) \tag{18}$$

where (from equation (1))

$$h_l(u_1, u_2, u_3) = \left(\frac{2p_0}{p_0^2 + p^2}\right)^l h_l(p_1, p_2, p_3) \tag{19}$$

and where $h_l(p_1, p_2, p_3)$ is an harmonic polynomial of degree l in p_1, p_2, and p_3. In other words, h_l is a homogeneous polynomial for which

$$\left(\frac{\partial^2}{\partial p_1^2} + \frac{\partial^2}{\partial p_2^2} + \frac{\partial^2}{\partial p_3^2}\right) h_l(p_1, p_2, p_3) = 0 \tag{20}$$

In equations (18) and (19), $h_l(u_1, u_2, u_3)$ is the same polynomial, with the p_j's replaced by u_j's. From equation (1), it follows that the 4-dimensional solid angle $d\Omega$ is related to d^3p by

$$d\Omega = \sin^2\chi \; \sin\theta \; d\chi \; d\theta \; d\phi = \left(\frac{2p_0}{p_0^2 + p^2}\right)^3 d^3p \tag{21}$$

Combining these results, and making use of equation (1), we can write:

$$J = \frac{(2p_0)^{l+3}}{2\pi^2} \int d^3p \; e^{i\mathbf{p}\cdot\mathbf{R}} \frac{(p^2 - p_0^2)^\nu}{(p_0^2 + p^2)^{l+\nu+3}} h_l(p_1, p_2, p_3) \tag{22}$$

We now make use of the expansion:

$$e^{i\mathbf{p}\cdot\mathbf{R}} = \sum_{l'=0}^{\infty} i^{l'}(2l+1) j_{l'}(pR) P_{l'}(\hat{p} \cdot \hat{R}) \tag{23}$$

When the angular integration in reciprocal space is performed, only the term where $l = l'$ survives, and we obtain:

$$J = i^l (2p_0)^{l+3} h_l(\hat{R}_1, \hat{R}_2, \hat{R}_3) \frac{2}{\pi} \int_0^\infty dp \, \frac{p^{l+2}(p^2 - p_0^2)^\nu j_l(pR)}{(p_0^2 + p^2)^{l+\nu+3}} \tag{24}$$

J can now be expressed in terms of the integrals of Geller, Harris and Michels. For example, if $\nu = 0$ we have:

$$J = i^l (2p_0)^{l+3} h_l(\hat{R}_1, \hat{R}_2, \hat{R}_3) W^{l,1}_{l+3,0} = \frac{2i^l h_l(t_1, t_2, t_3)}{(l+2)!}(1+t)e^{-t} \tag{25}$$

where

$$\mathbf{t} = (t_1, t_2, t_3) \equiv (p_0 R_1, p_0 R_2, p_0 R_3) \tag{26}$$

In a previous paper [8] we have shown that in general the integral J is given by

$$J = \frac{1}{2\pi^2} \int d\Omega \; e^{i\mathbf{p}\cdot\mathbf{R}} \; u_4^\nu h_l(u_1, u_2, u_3) = \mathcal{F}_{\nu,l}(t) h_l(t_1, t_2, t_3) \tag{27}$$

where

$$\mathcal{F}_{\nu,l}(t) \equiv \frac{i^l \nu! e^{-t}}{2^\nu (2l+1)!} \left[\sum_{s=0}^{[\nu/2]} \frac{2(\nu+2l+1-2s)! F(2s-\nu|2l+2|2t)}{s!(\nu+l+1-s)!(\nu-2s)!} \right.$$
$$\left. -(\nu+1) \sum_{s=0}^{[(\nu+1)/2]} \frac{(\nu+2l+2-2s)! F(2s-\nu-1|2l+2|2t)}{s!(\nu+l+2-s)!(\nu+1-2s)!} \right] \tag{28}$$

In equation (28), $F(a|b|x)$ is a confluent hypergeometric function:

$$F(a|b|x) \equiv 1 + \frac{ax}{b} + \frac{a(a+1)x^2}{b(b+1)2!} + \ldots \tag{29}$$

This gives us a general method for evaluating the overlap integrals $S_{\tau,\tau'}$ after having expressed the product $Y_\tau^* Y_{\tau'}$ as a sum of terms of the form $u_4^\nu h_l(u_1, u_2, u_3)$. If we wish, we can use hyperspherical harmonics of the form shown in Table 3:

$$Y_{\lambda,\mu,|m|,\pm}(\mathbf{u}) = \frac{1}{\sqrt{2}} \left[Y_{\lambda,\mu,m}(\mathbf{u}) \pm Y_{\lambda,\mu,-m}(\mathbf{u}) \right] \tag{30}$$

and in that case, the hyperangular overlap integrals will be real. A few of these integrals are shown in Table 4. The integrals shown in this table were evaluated using equation (27); and the function $\mathcal{F}_{\nu,l}(t)$ is defined by equation (28). Although i appears in the integrals linking the $1s$ and $2p_j$ basis functions, the hyperangular overlap integrals are nevertheless real, since the i combines with another factor i in $\mathcal{F}_{0,1}(t)$. The integral linking the $2p_j$ basis function with itself requires some explanation: From equation (1) we can see that

$$u_1^2 + u_2^2 + u_3^2 = 1 - u_4^2 \tag{31}$$

Thus, for example, if we wish to resolve u_1^2 into polynomials h_l which are harmonic in u_1, u_2 and u_3, we must write

$$u_1^2 = h_3 + (1 - u_4^2)h_0 = \left[u_1^2 - \frac{1}{3}(u_1^2 + u_2^2 + u_3^2)\right] + \frac{1}{3}(1 - u_4^2) \tag{32}$$

and after applying equation (27) we obtain the integral shown in Table 4.

The form of the set of secular equations (9) is interesting, since the eigenvalues are not energies but p_0 values, which are related to the energies by equation (2). For a given state, all of the basis functions correspond to the same value of p_0, and hence to the same energy. They are thus Sturmian basis functions [7,9]. The value of p_0 to which they all correspond gives the correct asymptotic behaviour in direct space to all of the basis functions used to build up a given state. The result is that the convergence of such an expansion is rapid.

As a simple example to illustrate this method, we can consider an electron in the field of two nuclei with charges Z_1 and Z_2. In the lowest approximation, we can represent $\phi^t(\mathbf{p})$ by a linear combination of two basis functions, both with $\lambda = 0$:

$$\phi(\mathbf{u}) \approx \eta_1(\mathbf{u})B_1 + \eta_2(\mathbf{u})B_2 \tag{33}$$

where

$$\eta_1(\mathbf{u}) = \sqrt{Z_1}\, e^{i\mathbf{p}\cdot\mathbf{X}_1} Y_{000} = \frac{1}{\pi}\sqrt{\frac{Z_1}{2}}\, e^{i\mathbf{p}\cdot\mathbf{X}_1}$$

$$\eta_2(\mathbf{u}) = \sqrt{Z_2}\, e^{i\mathbf{p}\cdot\mathbf{X}_2} Y_{000} = \frac{1}{\pi}\sqrt{\frac{Z_2}{2}}\, e^{i\mathbf{p}\cdot\mathbf{X}_2} \tag{34}$$

Using the methods discussed above, we find that:

$$\frac{1}{2\pi^2}\int d\Omega\, e^{i\mathbf{k}\cdot(\mathbf{X}_1-\mathbf{X}_2)} = (1+t)e^{-t} \tag{35}$$

where

$$t \equiv p_0|\mathbf{X}_1 - \mathbf{X}_2| \equiv p_0 R \tag{36}$$

Thus

$$S_{\tau,\tau'} = \begin{pmatrix} Z_1 & \sqrt{Z_1Z_2}\,(1+t)e^{-t} \\ \sqrt{Z_1Z_2}\,(1+t)e^{-t} & Z_2 \end{pmatrix} \tag{37}$$

so that the secular equations, (9), require that p_0 and R satisfy

$$2p_0 = Z_1 + Z_2 \pm \sqrt{(Z_1+Z^2)^2 + 4Z_1Z_2[(1+t)^2e^{-2t} - 1]}$$

$$R = \frac{t}{p_0} \tag{38}$$

Letting the parameter t run from 0 to ∞, we can generate values of p_0 and $E = -p_0^2/2$ and the values of R to which they correspond. In the limit $t = 0$ (and $R_=0$), we obtain the exact ground-state united-atom energy:

$$p_{0+} = Z_1 + Z_2 \qquad E_+ = -\frac{(Z_1+Z_2)^2}{2} \tag{39}$$

while for $t = \infty$, we obtain the exact separated-atom energies:

$$p_{0+} = Z_1 \qquad E_+ = -\frac{Z_1^2}{2}$$

$$p_{0-} = Z_2 \qquad E_- = -\frac{Z_2^2}{2} \tag{40}$$

For intermediate values of t, the energies found from the approximate wave function (33) differ appreciably from the exact energies. However, when a larger number of basis functions are used, the method outlined above is capable of great accuracy. Koga and Matsuhashi [6], for example, were able to obtain 10-figure accuracy by this method in calculations on the H_2^+ ion.

TABLE 3. 4-dimentional hyperspherical harmonics

τ	λ	l	$\|m\|$	$\sqrt{2}\pi\ Y_\tau(\mathbf{u})$
1s	0	0	0	1
$2p_1$	1	1	1	$-2iu_1$
$2p_2$	1	1	1	$-2iu_2$
$2p_3$	1	1	0	$-2iu_3$
$2s$	1	0	0	$2u_4$

TABLE 4. Hyperangular overlap integrals

τ	τ'	$\int d\Omega\ e^{i\mathbf{p}\cdot\mathbf{R}} Y_\tau^*(\mathbf{u}) Y_{\tau'}(\mathbf{u})$
$1s$	$1s$	$\mathcal{F}_{0,0}(t)$
$1s$	$2p_j$	$-2i\mathcal{F}_{0,1}(t)t_j \qquad j = 1,2,3$
$1s$	$2s$	$2\mathcal{F}_{1,0}(t)$
$2p_j$	$1s$	$2i\mathcal{F}_{0,1}(t)t_j \qquad j = 1,2,3$
$2p_j$	$2p_j$	$4\mathcal{F}_{0,2}(t)\left[t_j^2 - \frac{1}{3}t^2\right] + \frac{4}{3}\left[\mathcal{F}_{0,0}(t) - \mathcal{F}_{2,0}(t)\right]$
$2p_j$	$2p_k$	$4\mathcal{F}_{0,2}(t)t_j t_k \qquad j \neq k$
$2p_j$	$2s$	$4i\mathcal{F}_{1,1}(t)t_j$
$2s$	$2s$	$4\mathcal{F}_{2,0}(t)$

3. Many-center non-Coulomb potentials

In a previous paper [12] we have shown that the reciprocal-space Sturmian formalism can be extended, so that solutions can be found to the Schrödinger equation for a charged particle moving in a many-center non-Coulomb potential. If

$$\mathcal{V}(\mathbf{x}) = \sum_j V_j(\mathbf{x} - \mathbf{X}_j) \tag{41}$$

and if $\chi_{n,l,m}$ is a hydrogenlike Sturmian basis function in direct space:

$$\begin{aligned}
\chi_{n,l,m}(\mathbf{x}) &\equiv \frac{1}{(2\pi)^{3/2}} \int d^3p \; e^{i\mathbf{p}\cdot\mathbf{x}} M(p) Y^*_{n-1,l,m}(\mathbf{u}) \\
&= R_{n,l}(r) Y_{l,m}(\theta, \phi) \\
R_{n,l}(r) &= \mathcal{N}(2p_0 r)^l e^{-p_0 r} F(l-1-n|2l+2|2p_0 r) \\
\mathcal{N} &= \frac{2p_0^{3/2}}{(2l+1)!} \left[\frac{(l+n)!}{n(n-1-l)!}\right]^{1/2}
\end{aligned} \tag{42}$$

then solutions to the Schrödinger equation for the particle can be written in the form:

$$\phi^t(\mathbf{p}) = M(p) \sum_\tau e^{i\mathbf{p}\cdot\mathbf{X}_j} Y_{n-1,l,m}(\mathbf{u}) B_\tau \tag{43}$$

where τ stands for the set of indices j, n, l, m. It was shown in reference [12] that the coefficients B_τ can be found by solving the modified set of secular equations:

$$\sum_\tau \left[\mathcal{K}_{\tau'\tau} - p_0 \delta_{\tau'\tau}\right] B_\tau = 0 \tag{44}$$

If we let

$$\begin{aligned}
\mathcal{T}_{\tau'\tau} &\equiv -\frac{\delta_{j',j}}{p_0} \int d^3x \; \chi^*_{n'l'm'}(\mathbf{x}) V_j(\mathbf{x}) \chi_{nlm}(\mathbf{x}) \\
J_{\tau'',\tau'''} &\equiv \int d\Omega \; e^{i\mathbf{p}\cdot(\mathbf{X}_{j'''} - \mathbf{X}_{j''})} Y^*_{\tau''}(\mathbf{u}) Y_{\tau'''}(\mathbf{u})
\end{aligned} \tag{45}$$

then

$$\mathcal{K}_{\tau'\tau} \equiv \sum_{\tau'',\tau'''} C^*_{\tau'\tau''} \sqrt{w_{\tau''}} J_{\tau''\tau'''} \sqrt{w_{\tau'''}} C_{\tau'''\tau} \tag{46}$$

where w_τ and $C_{\tau'\tau}$ are respectively the roots and eigenvectors of the matrix $\mathcal{T}_{\tau'\tau}$:

$$\sum_\tau \left[\mathcal{T}_{\tau'\tau} - w_{\tau''} \delta_{\tau'\tau}\right] C_{\tau\tau''} = 0 \tag{47}$$

In order to illustrate equations (41)-(47), we can consider a homonuclear diatomic molecule with atoms at positions $\mathbf{X}_1$ and $\mathbf{X}_2$ and with a screening function $Ze^{-\xi r}$, so that

$$\mathcal{V}(\mathbf{x}) = -\frac{Z\ e^{-\xi|\mathbf{x}-\mathbf{X}_1|}}{|\mathbf{x}-\mathbf{X}_1|} - \frac{Z\ e^{-\xi|\mathbf{x}-\mathbf{X}_2|}}{|\mathbf{x}-\mathbf{X}_2|} \tag{48}$$

We can calculate the ground state of the molecule in an approximation where we use only two Sturmian basis functions,

$$\begin{aligned} e^{i\mathbf{p}\cdot\mathbf{X}_1} Y_{0,0,0}(\mathbf{u}) &= \frac{1}{\sqrt{2\pi^2}}\ e^{i\mathbf{p}\cdot\mathbf{X}_1} \\ e^{i\mathbf{p}\cdot\mathbf{X}_2} Y_{0,0,0}(\mathbf{u}) &= \frac{1}{\sqrt{2\pi^2}}\ e^{i\mathbf{p}\cdot\mathbf{X}_2} \end{aligned} \tag{49}$$

In this very simple approximation the matrix $J_{\tau'',\tau'''}$ is given by

$$J_{\tau'',\tau'''} = \begin{pmatrix} 1 & (1+t)e^{-t} \\ (1+t)e^{-t} & 1 \end{pmatrix} \qquad t = p_0|\mathbf{X}_1 - \mathbf{X}_2| \tag{50}$$

while

$$C_{\tau',\tau''}\sqrt{w_{\tau''}} = \begin{pmatrix} \sqrt{w} & 0 \\ 0 & \sqrt{w} \end{pmatrix} \tag{51}$$

where

$$w \equiv \frac{Z}{(1+\gamma/2)^2} \qquad \gamma = \xi/p_0 \tag{52}$$

The roots of the secular equation, (44), are then

$$p_0 = w\left[1 \pm (1+t)e^{-t}\right] \tag{53}$$

where w is defined in equation (52). In this approximation the lowest orbital energy of the homonuclear diatomic molecule is then given by

$$E = -\frac{p_0^2}{2} = -\frac{Z^2}{2(1+\gamma/2)^4}\left[1 + (1+t)e^{-t}\right]^2 \tag{54}$$

In the united atom limit ($t = 0$), this corresponds to

$$E_{R=0} = -\frac{(2Z)^2}{2(1+\gamma/2)^4} \tag{55}$$

i.e., the energy of an electron in a screened Coulomb potential with nuclear charge $2Z$. In the separated atom limit, ($t = \infty$), both the roots of equation (53) correspond to the energy of an electron in a screened Coulomb potential with nuclear charge Z.

$$E_{R=\infty} = -\frac{Z^2}{2(1+\gamma/2)^4} \tag{56}$$

We can repeat this calculation for a heteronuclear diatomic molecule, where the effective one-electron potential is given approximately by

$$\mathcal{V}(\mathbf{x}) = -\frac{Z_1\ e^{-\xi_1|\mathbf{x}-\mathbf{R}_1|}}{|\mathbf{x}-\mathbf{R}_1|} - \frac{Z_2\ e^{-\xi_2|\mathbf{x}-\mathbf{R}_2|}}{|\mathbf{x}-\mathbf{R}_2|} \tag{57}$$

Then the equations corresponding to (51) and (52) are

$$C_{\tau',\tau''}\sqrt{w_{\tau''}} = \begin{pmatrix} \sqrt{w_1} & 0 \\ 0 & \sqrt{w_2} \end{pmatrix} \tag{58}$$

and

$$w_j \equiv \frac{Z_j}{(1+\gamma_j/2)^2} \qquad \gamma_j = \xi_j/p_0 \qquad j = 1,2 \tag{59}$$

while the equation corresponding to (53) is

$$p_0 = \frac{w_1 + w_2}{2} \pm \frac{1}{2}\sqrt{(w_1+w_2)^2 + 4w_1w_2\left[(1+t)^2e^{-2t} - 1\right]} \tag{60}$$

where $t \equiv p_0R$, and where R is the internuclear distance. In this illustrative calculation we have used only the two Sturmian basis functions shown in equation (49); but the accuracy could of course be increased by using a larger Sturmian basis set. The formalism discussed here can also be extended to periodic systems, as was shown in reference [13].

Acknowledgement

I am extremely grateful to the Carlsberg Foundation for a travel grant which allowed me to visit Harvard University, where part of the work reported in this paper was performed. I am also very grateful to Professor Dudley Herschbach and his group at Harvard for their advice and encouragement.

References

1. Fock, V.A. (1935) Hydrogen Atoms and Non-Euclidian Geometry, *Bull. Acad. Sci. USSR*, 169; *Z. Phys.* **98**, 145.

2. Shibuya, T. and Wulfman, C.E. (1965) Molecular Orbitals in Momentum Space, *Proc. Roy. Soc. London.* **A286**, 376.

3. Judd, B.R. (1975) *Angular Momentum Theory for Diatomic Molecules*, Academic Press, New York.

4. Monkhorst, H. and Jeziorski, B. (1979) No Linear Dependence or Many-Center Integral Problems in Momentum Space Quantum Chemistry, *J. Chem. Phys.* **71**, 5268-5269.

5. Koga, T. (1985) Direct Solution of the $H(1s)-H^+$ Long-Range Interaction Problem in Momentum Space, *J. Chem. Phys.* **82**, 2022-2025.

6. Koga, T. and Matsuhashi, T. (1988) One-Electron Diatomics in Momentum Space. V. Nonvariational LCAO Approach, *J. Chem. Phys.* **89**, 983-986.

7. Avery, J. (1989), *Hyperspherical Harmonics; Applications in Quantum Theory*, Kluwer, Dordrecht.

8. Avery, J. (1994) Hyperspherical Harmonics, Some Properties and Applications, in *Conceptual Trends in Quantum Chemistry*, E.S. Kryachko and J.L. Calais editors, Kluwer, Dordrecht, pp. 135-169.

9. Avery, J. and Herschbach, D.R. (1992) Hyperspherical Sturmian Basis Functions, *Int. J. Quant. Chem.* **41**, 673-686.

10. Geller, M. (1964) Two-Center Coulomb Integrals, *J. Chem. Phys.* **41**, 4006-4007.

11. Harris, F.E. and Michels, H.H. (1967) Evaluation of Molecular Integrals for Slater-Type Orbitals, *Adv. Chem. Phys.* **13**, 205-266.

12. Avery, J. and Hansen, T.B. (1995) A Momentum-Space Picture of the Chemical Bond, submitted to *Int. J. Quant. Chem.*, (*Roy McWeeny Symposium*).

13. Avery, J., Hansen, T.B., Wang, M. and Antonsen, F. (1995), Sturmian Basis Sets in Momentum Space, *Int. J. Quant. Chem.*, (in press).

INELASTIC SCATTERING WITH COULOMB FORCES: A SEMICLASSICAL S–MATRIX APPROACH

JAN–MICHAEL ROST
Fakultät für Physik, Universität Freiburg
Hermann–Herder–Str. 3, D–79104 Freiburg, Germany
e–mail: rost@phyc1.physik.uni–freiburg.de

Abstract

The development of semiclassical approximations is briefly reviewed. A semiclassical S–matrix theory is described for inelastic scattering of atoms with electrons or photons. As an example ionization of hydrogen by electron impact and double–ionization of helium are presented and the connection with resonance formation below the ionization threshold is discussed.

1. Introduction

Semiclassical approximations have existed almost as long as the quantum theory itself. They have served two rather different purposes: first, to explore the classical limit of quantum mechanics and to gain more insight into the nature of quantum phenomena. Second, to develop a theory that provides reliable approximations for cases in which it is not possible to solve the full problem quantum mechanically. Certainly to the first category belongs the formulation of the semiclassical propagator initiated by van Vleck in 1928 [1] and completed by Maslov [2] and Gutzwiller [3]. Within the second category the WKB–approach has been most successful in different areas of physics. The essentially one–dimensional theory has led to useful results in times where we lacked the computer power to treat complicated, more dimensional problems. For scattering problems a logical application has been the calculation of WKB–phase shifts for elastic scattering, pioneered by Ford and Wheeler [4].

C. A. Tsipis et al. (eds.), New Methods in Quantum Theory, 297–310.

With the improvement of the computers, even the most general semiclassical formulation by van Vleck has become computationally feasible and thus conceptually interesting again. Miller, Marcus and others showed in the seventies that semiclassical approximations for the Green's function itself can lead to remarkable results for reactive scattering in molecular complexes, whose dynamics can not be characterized by short wave lengths [5]. More recent developments include mixed (semi–)classical/quantum representations of the scattering process [6].

For the energy spectrum of bound systems Gutzwiller formulated the "trace formula" [3]. Together with the "cycle expansion", a resummation method, the trace formula was used recently by Wintgen et al. to show that helium can be quantized semiclassically despite the initial failure of the old quantum theory some 60 years ago [7]. Using the van Vleck propagator directly in the time domain Tomsovic and Heller were able to show that the spectra and selected eigenstates of classically chaotic systems can be obtained semiclassically to a good accuracy [8]. These are only a few, although spectacular, examples from a substantial body of work that has advanced semiclassical theories so much in the last few years that it might be justified to speak of a revival of the semiclassical perspective. The revival is fueled by new experiments with excellent energy resolution even for highly excited spectra [9]. Their interpretation needs a theoretical description and understanding of dynamics in the limit of large quantum numbers. This implies a natural demand for further developing semiclassical methods. An impressive example is the progress in understanding the complicated spectrum of hydrogen in the magnetic field achieved by Delos and coworkers in the last few years [10].

Here I will present a semiclassical formulation for inelastic electron–atom scattering derived from the path integral representation of the S–matrix.

2. The classical atom

A classical N–electron atom has some remarkable properties that are important for the semiclassical description of the scattering process. They do not depend on the choice of the coordinate system but are particularly simple to derive in hyperspherical coordinates. The hyperradius $\mathcal{R}^2 = \sum_i r_i^2$, composed of all electron–nucleus distances, measures the overall extension of the atom. All other coordinates transform to a space of $3N - 4$ angles Ω on the hypersphere with radius $\mathcal{R}$. We scale the coordinates, momenta and the hamiltonian H itself with the energy

$$\begin{aligned} \mathcal{R}' &= E\mathcal{R} \\ \mathcal{P}' &= E^{-1/2}\mathcal{P} \\ H' &= E^{-1}H \equiv 1. \end{aligned} \tag{1}$$

The angles and the corresponding generalized momenta are dimensionless and therefore not affected by the scaling. The scaled hamiltonian reads

$$H' \equiv 1 = \frac{\mathcal{P}'^2}{2} + \frac{\Lambda'^2(\Omega)}{2\mathcal{R}'^2} + \frac{C(\Omega)}{\mathcal{R}'} \tag{2}$$

where $\Lambda(\Omega)$ is the grand angular momentum operator, which contains the dependence upon all the momenta in the angles Ω while $C(\Omega)$ can be viewed as an angle dependent generalized charge whose exact form is not important for the present context. For Eq. (2) one can derive the following properties [11]:

(i) As a function of time the hyperradius $\mathcal{R}(t)$ has one extremum that is a minimum. In fact this can be shown for arbitrary masses of the $N+1$ particles.

(ii) In the limit $E \to 0$ the dynamics for any (preserved) total angular momentum L in a two–electron atom is governed by the same effective hamiltonian of $L = 0$. The angular moment has the dimension of action [length]× [momentum]. From Eq. (1) follows that $L' = L\sqrt{E}$. Hence, for any finite L the effective angular momentum is zero in the limit $E \to 0$. Note also that the action scales as $S = S'/\sqrt{E}$ where S' does not depend on the energy E.

(iii) The angle $\theta_{12} = 180°$ between the electrons in a two–electron atom is a fixed point of the classical equations of motion, i.e. if $\theta_{12}(t_0) = 180°$ and $\dot{\theta}_{12}(t_0) = 0$ then $\theta_{12}(t) = \theta_{12}(t_0)$ for all times t.

With *(i)* – *(iii)* we can justify a considerable simplification for electron–atom scattering. (The atom is treated as a core with one active electron). The reason for only using $L = 0$ provided by *(ii)* is very different from elastic scattering with short range potentials where a partial wave analysis shows that only $l = 0$ survives for $E \to 0$. In Coulomb scattering, all angular momenta contribute with a priori unknown weight but the dynamics for each L is determined in the limit $E \to 0$ by the same hamiltonian as for the S–wave. Furthermore, because of *(iii)*, for the total cross section it is sufficient to calculate the process at the fixed point $\theta_{12} = 180°$. Hence, the problem reduces from 12 phase space variables to 4, the two electron–nucleus distances r_i and the conjugate momenta p_i.

3. The scattering amplitude

In the "one–dimensional world" (only one coordinate, the distance r_i describes each electron) the cross section reduces to a probability and is directly proportional to the S–matrix. Clearly, the ability to calculate the absolute cross section is lost in this approximation. However, most experiments gain their absolute measurements only by comparison with a known value in some limit of the cross section. We can either normalize the theoretical curve to the same limit or directly to the relevant experiment at one data point.

A semiclassical S–matrix element is proportional to

$$S_{p_1,p_2;p_1',p_2'}(E) \;\propto\; \sum_{cl.traj.} (\det|\mathcal{A}|)^{-\frac{1}{2}} \exp\left(-\frac{i\Phi(p_1,p_2;p_1',p_2')}{\hbar} - \frac{i\nu\pi}{2}\right),$$

$$\mathcal{A}_{ij} \;=\; \left.\frac{\partial p_i}{\partial r_j'}\right|_{p_j'}, \tag{3}$$

where in the momentum representation p_i and p_i' refer to the final and initial momenta of the electrons. The sum runs over all classical trajectories that describe the inelastic process under consideration. Each trajectory accumulates a phase, which is defined by the classical action $\Phi(p_1,p_2;p_1',p_2') = \int q_1 dp_1 + \int q_2 dp_2$ and a contribution of $\nu\pi/2$ from caustics along the trajectory [3,5]. The true initial state is given as a product of the bound state of the target electron with binding energy ϵ_B and the free projectile state with energy $p'^2/2 = \epsilon' = E + \epsilon_B$. In the inelastic scattering event the projectile energy will change from ϵ' to ϵ.

Conservation of the total energy and the asymptotically fixed energy ϵ' reduce the determinant of the 2×2 matrix $\mathcal{A}$ in Eq. (3) to a one–dimensional derivative of a characteristic final state variable with respect to an initial state variable. The choice is rather arbitrary and consistency among different choices is achieved by the normalization factor which changes accordingly. We find it most convenient to choose the initial position of the projectile r' and its final energy ϵ (while its initial energy ϵ' remains fixed). The scattering amplitude reads now [12]

$$S_{\epsilon,\epsilon'}(E) = \sum_j \sqrt{\mathcal{P}_j(E,\epsilon,\epsilon')} \exp\{\frac{i\Phi_j}{\hbar} - \frac{i\nu_j\pi}{2}\}, \tag{4}$$

with the classical probability for the j–th trajectory of

$$\mathcal{P}_j(E,\epsilon,\epsilon') = \frac{1}{R}\left|\frac{\partial\epsilon}{\partial r'}\right|_{\epsilon'}^{-1} \tag{5}$$

and the normalization R to be discussed in the next section. The sum runs over all classical trajectories j that take the projectile from energy ϵ' to ϵ during the collision.

4. Electron–hydrogen scattering

For pure Coulomb interactions (no perturbation by a core potential) we can invoke the property (i) from section 2 to realize that only a *single* trajectory can fulfill these boundary conditions simultaneously. This is illustrated in Fig. 1 where the function $\epsilon(r')/E$ is plotted for electron–hydrogen scattering. Dependent on the initial condition r' the final energy of the projectile can can be

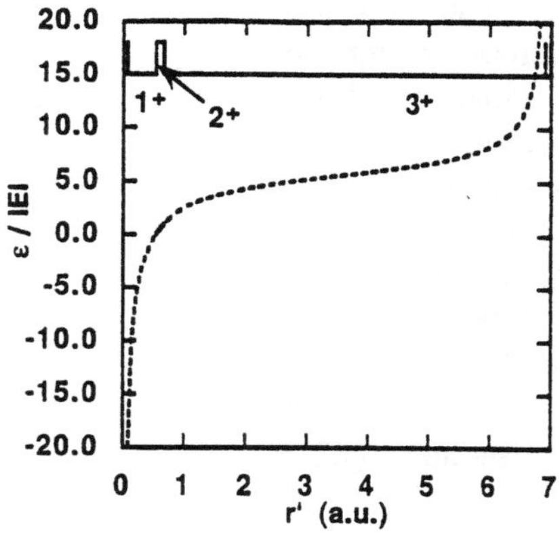

Figure 1: Energy dependence $\epsilon/|E|$ of the projectile on its initial position 1000 a.u.$+r'$, normalized to the total energy $E = 0.1$a.u.. The inverse derivative of this function enters Eq. (4). Intervals $I(i^+)$ correspond to different events, see text.

negative for $r' \in I(1^+)$ (then the projectile has become bound in exchange for the target electron). If $\epsilon > E$ we have an excitation process ($r' \in I(3^+)$, the target electron is still bound after the collision) and for $0 < \epsilon < E$ ($r' \in I(2^+)$) the system has fragmented into the original projectile electron, the target electron and the ionized target. Each point of the function represents a trajectory that connects an initial condition r' with a given final energy ϵ. Since the function is monotonic the connection between final and initial conditions is uniquely established by a single trajectory. This is certainly a remarkable property: it reduces the sum of Eq. (4) to one term. Hence, the *semiclassical* analysis of the threshold dynamics under the simplifications made possible through the properties $(i) - (iii)$ reveals an essentially *classical* behavior [12].

Classically we can calculate the ionization probability as

$$\int_0^E \mathcal{P}(E,\epsilon,\epsilon')d\epsilon = \frac{1}{R}\int_0^E d\epsilon \frac{dr'}{d\epsilon} = \frac{\Delta r'(2^+)}{R}. \tag{6}$$

The normalization is given by all processes that can happen,

$$R = \int_{-\infty}^{+\infty} d\epsilon \frac{dr'}{d\epsilon} = \Delta r'(1) + \Delta r'(2) + \Delta r'(3) \tag{7}$$

The ionization probability of Eq. (6) describes a problem of distinguishable Coulomb particles such as positron–hydrogen scattering [13]. For electron–hydrogen scattering the ionization probability must be modified due to the Pauli–principle which requires the differential scattering amplitude to be symmetric with respect to electron exchange. This could lead to interferences between the two contributions. However, the action is symmetric under electron exchange in the final channel, $\Phi(\epsilon,\epsilon',E) = \Phi(E-\epsilon,\epsilon',E)$ so that the scattering

Figure 2: The total ionization cross section for electron impact on $H(1s)$. The experimental data points are taken from [15]. The calculated cross section (solid) has been normalized to the experimental data at 5.84 eV. The dashed line is the Wannier cross section $\sigma(E) = \sigma_0 E^{1.127}$.

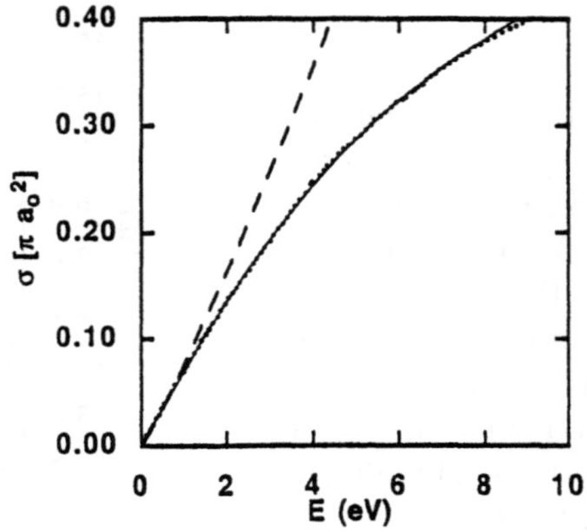

probability

$$P_\epsilon^\pm(E) = \left[\mathcal{P}(E,\epsilon,\epsilon')^{\frac{1}{2}} \pm \mathcal{P}(E,E-\epsilon,\epsilon')^{\frac{1}{2}}\right]^2 \tag{8}$$

is independent of the phase factor in Eq.(4). The "+" sign in Eq. (8) corresponds strictly speaking to the $^1S^e$ partial wave while the "−" sign represents $^3S^e$ symmetry. The total ionization probability $P^\pm(E)$ is now obtained by integrating explicitly over the energy ϵ of one electron,

$$P^\pm(E) = \int_0^{E/2} P_\epsilon^\pm(E)d\epsilon. \tag{9}$$

$P^+(E)$ is in excellent agreement with all data points of the threshold experiment [15] up to excess energies as high as 8 eV (Fig.2) [12]. Written in the form

$$P(E) \propto E^{\alpha(E)} \tag{10}$$

the exponent is $\alpha = 1.127$ in the limit $E \to 0$ as analytically derived by Wannier [14]. However, as can be seen from Fig.2, the Wannier cross section and the (semi)–classical cross section differ for finite E. Note that the semiclassical S–matrix approach is capable of providing a cross section with an absolute energy scale.

5. Photo double–ionization of Helium

This is another fundamental two–electron process where the energy necessary for the electrons to escape the nucleus is provided by a photon. In the present semiclassical description the photo ionization process is divided into (A) the absorption of the photon by one electron and (B) the correlated motion of the electron pair leading to singly or doubly ionized helium. We relate (A) to the total cross section to be calculated analytically in the quasi–classical reflection approximation as known from molecular problems. For (B) the two–electron wavefunction is propagated as already described in sections 3–4. A probe for (B) is the ratio between double and single ionization. It is mainly a consequence of the correlated electron motion following the absorption of the photon by *one* electron.

As before the correlated two–electron dynamics is represented by a semiclassical propagator of total angular momentum $L = 0$ in a restricted two-dimensional configuration space, spanned by the two electron–nucleus distances r_i with fixed interelectronic angle $\theta = \pi$. These approximations may be justified as follows. It is known that helium resonances which differ only by the quantum number of total angular momentum are close in energy (for instance $3s^2, {}^1S^e$ and $3s3p, {}^1P^o$). Hence, the corresponding difference between the S– and P–wave cross section in the double–continuum is also small. It will even be smaller for the *ratio* σ^{++}/σ^{+} which is reasonably approximated by the S–wave cross section.

Total cross sections are dominated by contributions from angles which are fixed points of the classical dynamics. In the case of two electrons the fixed points are $\theta = \pi$ and $\theta = 0$. However, only in the high energy limit, when the electron–electron interaction is a sufficiently weak perturbation, the fixed point $\theta = 0$ will become important. For energies considered here we will calculate σ^{++}/σ^{+} only at the fixed point $\theta = \pi$. Note that we do not restrict the calculation to the threshold region $E \to 0$.

To simulate the helium ground state classically we take an analytic phase space distribution which is close to the various planar periodic orbits presently discussed for the helium atom [16]. We put the two electrons in a completely out of phase motion on a quarter circle of fixed hyperradius $\mathcal{R}_0 \equiv (r_1^2 + r_2^2)^{1/2} = 1.4$ (atomic units will be used unless stated otherwise). This value comes from the average hyperradius (4.15 in energy scaled atomic units, see [16]) divided by the appropriate energy for the ground state (which is for the corresponding classical orbit roughly -2.97 au in WKB quantization [16]). With the interelectronic angle $\theta = \pi$, the resulting phase space distribution is a one parameter manifold dependent on the phase $0 \leq \eta \leq \pi/2$ of the orbit on the circle of radius $\mathcal{R}_0$. However, the phase space distribution to be propagated must be on the final energy shell E and this is achieved by boosting the momentum of one electron to the appropriate new energy shell, $p_1'^2 \to p_1'^2 + 2\omega$ where $\omega = E - E'$. (Unmarked variables refer to the final state and primed variables to the initial state before the propagation). The dipole amplitude in length form including the relevant semiclassical propagator for the dynamics of the electron pair after the absorption reads

$$d(\epsilon, E', E) = \sum_j \sqrt{\mathcal{D}_j(\epsilon, E', E)} \exp[i\Phi_j - i\nu_j\pi/2] \tag{11}$$

with

$$\mathcal{D}_j(\epsilon, E', E) = \frac{2}{\pi} \left|\frac{d\eta'}{d\epsilon}\right|_j (r_1'(\epsilon) + r_2'(\epsilon))^2, \tag{12}$$

the classical probability for the j–th orbit leading to a final energy ϵ of one electron following photon impact. The positions of the electrons before the absorption are $r_1'(\epsilon) = \mathcal{R}_0 \cos\eta'(\epsilon)$ and $r_2' = \mathcal{R}_0 \sin\eta'$, Φ_j and ν_j are defined as in Eq. (3).

The analogon to the normalization R for scattering is now the total ionization probability in the restricted phase space,

$$P_D(E', E) = \int_{-\infty}^{\infty} \sum_j \mathcal{D}_j(\epsilon, E', E) d\epsilon = \mathcal{R}_0^2(1 + \frac{2}{\pi}). \tag{13}$$

For the double ionizing events ($0 \leq \epsilon \leq E$) only *one* trajectory contributes to the sum in Eq. (11) as in the electron–impact ionization. However, once again we must take into account the Pauli principle for the identical electrons, where the action is invariant under electron exchange, $\Phi(\epsilon, E', E) = \Phi(E - \epsilon, E', E)$. Hence, the differential probability for finding one electron with energy ϵ in a double ionized state after photon impact reads

$$P_D(\epsilon, E', E) = \left|\mathcal{D}(\epsilon, E', E)^{\frac{1}{2}} + \mathcal{D}(E - \epsilon, E', E)^{\frac{1}{2}}\right|^2. \tag{14}$$

From Eq. (14) we obtain the probability P_D^{++} for photo double ionization

$$P_D^{++}(E', E) = \int_0^{E/2} P_D(\epsilon, E', E) d\epsilon. \tag{15}$$

The desired ratio of σ^{++}/σ^{+} is now given by $P_D^{++}/(P_D - P_D^{++})$ and is shown in Figure 3. Good agreement is achieved in the intermediate region with the "experimentally recommended data" [18] and with the data by Kossmann et al. [17] near threshold. Note that the present approach reproduces the classical result by Wannier in the limit $E \to 0$, namely, $\sigma^{++} \propto E^{1.056}$. For a realistic judgment of the agreement with the experiment one should keep in mind the approximations of the present approach and the experimental errors of 3 to 5% [18].

To obtain *absolute* cross sections, we calculate first the total photo cross section,

$$\sigma_D(\omega) = \frac{2\pi\alpha}{\omega} \int_{-\infty}^{+\infty} dt < \Psi|\vec{D}^\dagger e^{-iHt} \vec{D}|\Psi > e^{iEt} \tag{16}$$

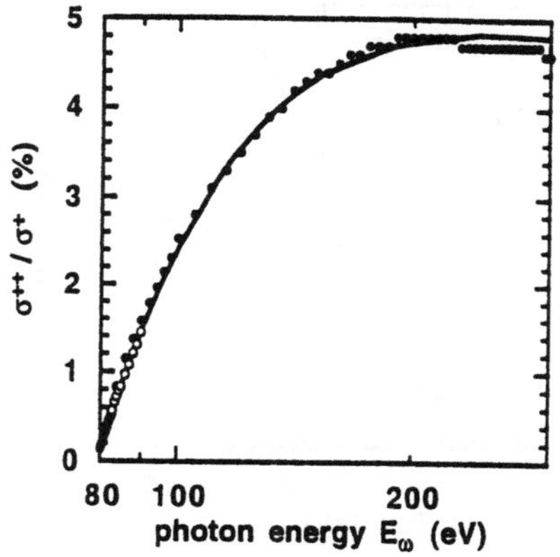

Figure 3: Ratio of double to single photoionization. Solid line: theory according to Eq. (15), experiments: (∘) [17]; (•) "experimentally recommended data" [18]

with the dipole operator $\vec{D}$ and the fine structure constant α. The total final energy is $E = E_i + \omega$, where E_i is the initial energy $E_i = 2.848$ a.u. obtained by the ground state wavefunction $\Psi \propto \exp(-\beta(r_1 + r_2))$ with effective charge $\beta = 2 - 5/16$ [19]. We use the reflection principle which has been developed for photo dissociation in molecules [20], to obtain an analytic expression for σ_D. The propagator $\exp(-iHt)$ in Eq. (16) is approximated classically which reduces the time–integral to a delta–function linking the electron radius to the photon frequency, $r = \omega^{-1/2}$. The result is [21]

$$\sigma_D(\omega) = \frac{1}{\beta^2}\frac{\alpha\pi^2 2^7}{3\sqrt{8}} x^{-\frac{7}{2}} e^{-\sqrt{8/x}} \tag{17}$$

where $x = 2\omega/\beta^2$.

Having completed the calculation of the two separate processes (A) and (B), we can now calculate the absolute photo double ionization cross section from Eqs. (13), (15) and (17)

$$\sigma^{++} = \frac{P_D^{++}}{P_D}\sigma_D. \tag{18}$$

It is shown in Figure 4 where part (a) presents an overview around the maximum on a logarithmic scale to emphasize smaller energies, and part (b) the threshold region.

One might find it surprising that the total cross sections for atomic photo doubleionization far from threshold can be described by an electron motion restricted to the collinear phase space. However, again, this can be understood from a similar behavior below threshold which is reflected by the propensity rules for photoabsorption from the ground state into doubly–excited states of helium [22,23]: There, it is found that predominantly two–electron resonant states are populated whose geometry in a body–fixed frame approaches for high excitation energies a collinear configuration.

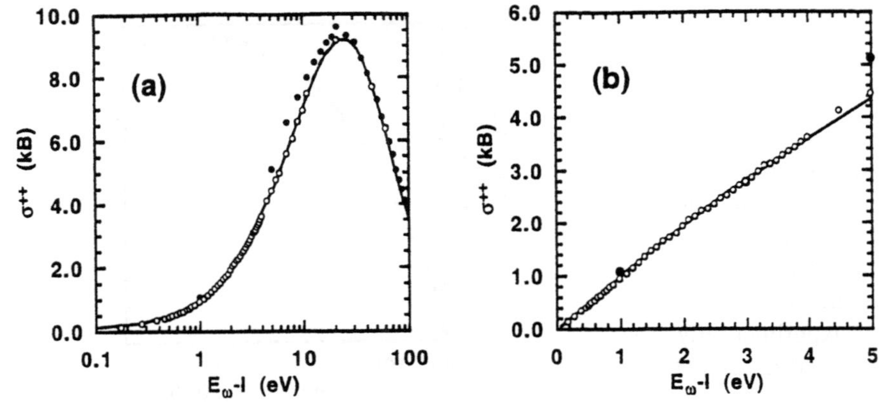

Figure 4: The absolute photo doubleionization cross section as a function of the energy above the double ionization threshold. Coding of the points as in Figure 3. Note that the absolute experimental data from Kossmann et al. [17] have been obtained from $\sigma^{++} = \rho/(\rho+1)\sigma_D$ where $\rho = \sigma^{++}/\sigma^{+}$ and σ_D has been taken from [25]. For a discussion of this procedure see [18].

More importantly from a general perspective, the seemingly oversimplified picture of atomic photo doubleionization as a two step process is confirmed by the quantitative comparison with the experiment. Recently, this two–step process has been independently suggested to understand the distribution of the recoil–momentum of the He^{++}–ion after photo doubleionization [26]. The distribution is directly observable in experiments performed by cold target recoil ion momentum spectroscopy [26].

6. Inelastic scattering below the ionization threshold

If the energy ϵ' of the incident electron beam is lowered the total energy remains negative and no ionization can take place. What happens with the initial conditions which lead to ionization for $E > 0$? The deflection function (Fig. 5) can answer this question. Comparing Fig. 1 with Fig. 5 one sees that *ionizing* orbits get replaced by *chaotic scattering* orbits [27]. Chaos is generated by trajectories which hit the potential boundary since there is not enough energy for double–escape of the electrons. Hence both electrons bounce many times into the nucleus. Thereby, they suffer a time delay compared to electrons following 'direct' trajectories from the intervals 1 and 3. The time delay is the signature of a resonance [28]. What had been ionization above threshold, described by a regular behavior of interval 2^+, becomes resonance formation below threshold, characterized by the irregular scattering in the corresponding interval 2^-. Although it might seem hopeless to deal with a fractal object like the interval 2^- in Fig. 5 we can form the analogy to the total classical ionization cross section. Close to threshold it is justified to neglect all phases since according

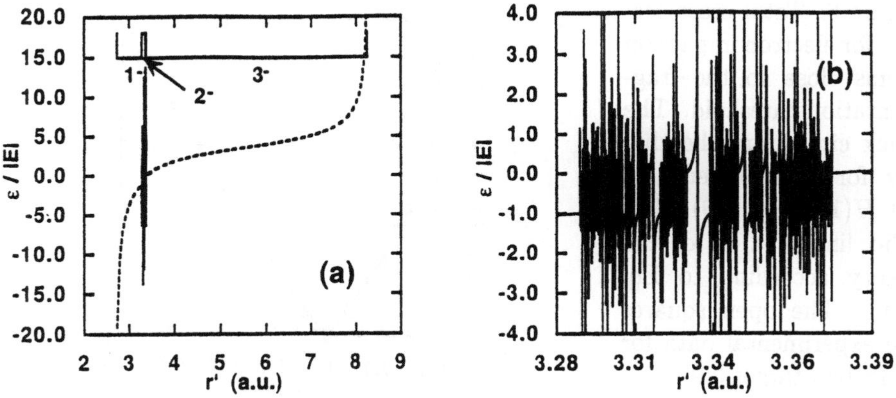

Figure 5: Classical deflection function as in Figure 1, but for $E = -0.1$ a.u. with interval 2^- for chaotic scattering. Part (b) shows a detail of part (a) highlighting the fractal structure.

to the scaling (see (ii) in section 2) they oscillate rapidly as $\exp(-iS'/\sqrt(|E|))$. Hence only the classical probability survives for $E \to 0$. Instead of integrating all contributions from interval 2 over all electron energies ϵ as we did in Eq.(6) for ionization, we integrate now over all contributions from the interval 2^-. The quantity we will obtain is the probability for resonance scattering, that is, the probability that the electron suffered a time delay in the scattering event and, for a short time, an (excited) 3–body complex had been formed. For each ϵ there is not only one but an infinite set of trajectories $\{i\}$ contributing to the sum of Eq. (4). However, to each index i belongs a continuous branch of trajectories with all energies ϵ. This branch yields upon integration over ϵ a small subinterval $\delta_i(r')$. The sum over i recovers then the entire interval $\Delta(2^-) = \sum_i \delta_i$ of chaotic scattering, which replaces $\Delta r'(2^+)$ in Eq. (6):

$$P_{frac}(E) = \sum_i \int_{-\infty}^{\infty} R^{-1} \sum_i \left| \frac{\partial \epsilon}{\partial r'_i} \right|_{\epsilon'}^{-1} d\epsilon = \sum_i \frac{\delta_i r'}{R} = \frac{\Delta r'(2^-)}{R}. \quad (19)$$

Thus, the classical probability for resonance formation below threshold is in exact analogy to the total classical ionization cross section above threshold. Together the probabilities represent 3–body events where both electrons have to participate in the scattering process simultaneously [30]. Figure 6 shows the "3–body" cross section across the ionization threshold. The full circles are calculated according to Eq. (19) while the solid line is a fit with the function

$$\sigma_{3-body}(E) \propto |E|^{1.127}(1 + a|E|^{\frac{1}{2}} + bE). \quad (20)$$

For $E > 0$ Eq. (20) reflects the "classical" result [14,12] that the ionization cross section close to threshold is proportional to the 1.127th power of E. Noting

Figure 6: Total cross section for electron impact on atoms close to the fragmentation threshold. The black circles are the theory for a hydrogen target $H(1s)$ (Eq. (19). The solid line is a fit of the theory according to Eq. (20). The open squares are experimental data for a He($1s^2$) target [29].

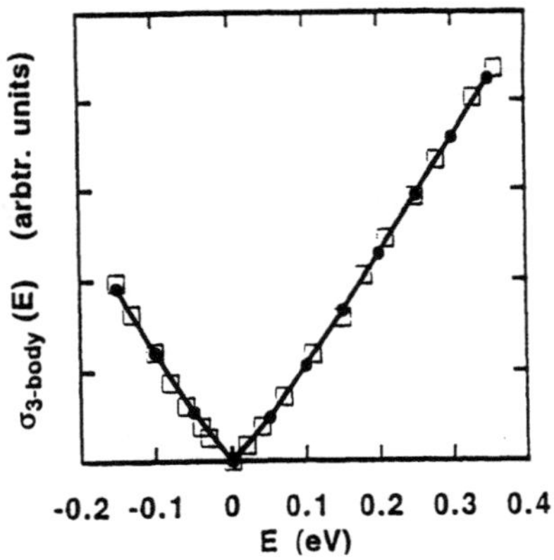

from Fig. 6 and Eq. (20) that this holds also below threshold we can conclude that $|E|^{1.127}$ originates in the 3–body amplitude of the S–matrix and should be independent of the process through which it was activated. The residual dynamics, essentially the density of states, depend on the initial state and the excitation as a whole (by photon or particle impact etc.). However, the residual part varies slowly with energy. Hence, we can take it into account with a Taylor expansion about $E = 0$ in Eq. (20).

In their threshold experiment (electron impact on He) Cvejanović and Read have probed positive as well as negative energies [29]. Within the experimental accuracy the data coincide with our "3–body" cross section although it is for hydrogen and not for helium (Fig. 6). This confirms our concept of a 3–body cross section since the 3–body amplitude in the S–matrix is identical for e–H and e–He scattering where one core electron participates only as a spectator forming a He^+ core which is equivalent to H^+ for the non–core penetrating final states considered here.

7. Summary

In this contribution some special properties of the classical three–body Coulomb system have been described. They lead in the context of a semiclassical S–matrix theory to a surprisingly simple and transparent picture of the ionization of atoms by photon or particle impact. Even more interesting is the picture which emerges from the semiclassical perspective for the two–electron spectrum for increasing energy. For low energies (above the first ionization threshold) resonances are formed. Semiclassically, they are generated by a coherent superposition of many orbits where the phases are substantial for the resonance spectrum. Very close to threshold the phases fluctuate randomly, the resonances disappear and the inelastic cross section may be described essentially classically. In between, phase differences (and this means in the semiclassical context differences in the actions of orbits) play a role to first order which leads to the semiclassical version of Ericson–fluctuations in the inelastic cross section [31]. Work to confirm this interpretation quantitatively and to investigate its consequences is in progress.

Acknowledgment

I am grateful to J. S. Briggs for fruitful discussions. Before his tragic death in the Alpes Dieter Wintgen accompanied this work with his unique support and criticism – Thank you! Financial support from the Sonderforschungsbereich 276 at the University of Freiburg is gratefully acknowledged.

References

1. J. M. van Vleck, Proc. Acad. Nat. Sci. USA **14**, 178 (1928).
2. V. P. Maslov and M. V. Fedoriuk, *Semi-Classical Approximations in Quantum Mechanics* (Boston, Reidel, 1981).
3. M. C. Gutzwiller, *Chaos in Classical and Quantum Mechanics* (Springer, New York, 1990).
4. K. W. Ford and J. A. Wheeler, Ann. Phys. N.Y. **7**. 259 (1959).
5. W. H. Miller, Adv. Chem. Phys. **56**, 38 (1974) and references therein.
6. See the contribution of G. Billing in this volume.
7. D. Wintgen, K. Richter, G. Tanner, CHAOS **2**, 19 (1992) and references therein.
8. S. Tomsovic and E. J. Heller, Physics Today **July**, 38 (1993) and references therein.
9. J. Main, G. Wiebusch, A. Holle, K. H. Welge, Phys.Rev.Lett. **57**, 2789 (1986).
10. J. B. Delos, Few-Body Systems Supp. **7**, 68 (1994).
11. J. M. Rost, J.Phys.B**28**, (in press) (1995).
12. J. M. Rost, Phys.Rev.Lett. **72**, 1998 (1994).
13. J. M. Rost and E. J. Heller, Phys.Rev. **A49**, R4289 (1994).
14. G. H. Wannier, Phys.Rev. **90**, 817 (1953).
15. J. W. McGowan and E. M. Clarke, Phys.Rev. **167**, 43 (1968).
16. P. V. Grujic and N. S. Simonovic, J. Phys. B**28**, 1159 (1995).
17. H. Kossmann, V. Schmidt and T. Andersen, Phys. Rev. Lett. **60**, 1266 (1988); H. Kossmann, Ph.D. thesis, Univ. of Freiburg, (1989).
18. J. M. Bizau and F. J. Wuilleumier, J. Electron Spectrosc. Relat. Phenom., in press (1995).
19. E. Merzbacher *Quantum Mechanics* (Wiley: New York), ch 17 (1970).
20. E. J. Heller, J. Chem. Phys. **68** 2066 (1978).
21. J. M. Rost, M. Pont, R. Shakeshaft, J.Phys.B, to be submitted (1995).
22. H. R. Sadeghpour, Phys. Rev. A **43**, 5821 (1991).
23. J. M. Rost and J. S. Briggs, J. Phys. B **24**, 4293 (1991).

24. J. F. McCann and D. S. F. Crothers, J. Phys. B **19**, L399 (1986).
25. J. A. R. Samson, Z. X. He, L Yin, G. N. Haddad, J. Phys. B **27**, 88 (1994).
26. V. Mergel etal., Euroconference on Coincident studies of electron and photon impact ionisation, book of abstracts, Colmar (1994).
27. J. M. Rost, J. Phys. B**27**, 5923 (1994).
28. B. Eckhardt, CHAOS **3** 613, (1993).
29. S. Cvejanović and F. H. Read, J. Phys. B **7**, 1841 (1974). In this experiment the yield of very low energy electrons is measured. They are interpreted as products of a "three–body" process as discussed in the text.
30. J. M. Rost and D. Wintgen, submitted (1995).
31. T. Ericsson, Phys. Rev. Lett. **5**, 430 (1960).

CUSP CONDITIONS AND ELECTRON CORRELATION

JOHN D. MORGAN III
Department of Physics and Astronomy
University of Delaware
Newark, DE 19716, USA
and
Cornell Theory Center
Cornell University
Ithaca, NY 14853, USA

Abstract. The cusp conditions derived by Kato for the behavior of atomic and molecular wavefunctions in the vicinity of a coalescence of an electron with a nucleus, or of two electrons with one another, play an important role in many areas of atomic and molecular physics. If one wishes to perform even moderately accurate Rayleigh-Ritz variational calculations, it is essential to use basis functions which approximate well the electron-nucleus cusps, and if one wishes to attain 'chemical accuracy' on the order of 10^{-3} to 10^{-4} a.u., it is also important to take account of the electron-electron cusp structure. The attainment of 'spectroscopic accuracy' on the order of 10^{-10} a.u. requires taking account of the higher-order singular structure in the vicinity of three-particle coalescences, which are described by the Fock expansion. After a review of the important developments in our understanding of the singularities in atomic and molecular wavefunctions, I describe how the performance of a basis is enhanced by including more and more of the known cusp behavior into the basis functions. Results are presented both on few-electron atoms using highly sophisticated generalised Frankowski-Pekeris basis functions, and on many-electron atoms and molecules using Kutzelnigg's linear-r_{12} *ansatz*. I also describe how symmetry principles lead to a classification of the types of behavior of the wavefunction which are allowed in the vicinity of an electron-electron coalescence, and how this analysis leads to an understanding of the rates of convergence of partial wave expansions and of the ordering of atomic energy levels in accord with the 'alternating rule', a generalisation to excited states of Hund's First Rule.

C. A. Tsipis et al. (eds.), New Methods in Quantum Theory, 311–338.

Introduction

In the first few decades after the discovery of quantum mechanics, chemists were generally satisfied with relatively simple theoretical methods which gave rough qualitative agreement with experimental observations. Typically the experimental results themselves were not known to high precision, so there was little incentive for theorists to improve the accuracy of their methods, even if such efforts would have been feasible given the limited computational resources of the time. In the 1950's and 1960's the ambitions of theorists and experimentalists alike were raised by two important technological developments: the advent of high-precision laser spectroscopy, and the explosive growth of power of digital computers. During the 1970's and 1980's and 1990's there have continued to be rapid advances in these two areas, so that experimentalists now routinely measure atomic and molecular properties with a precision of a part per million or much better, and they want theorists to provide reliable calculations of comparable accuracy to make possible the full interpretation of their experiments. There is every reason to think that this fruitful symbiosis of high-precision experiment and theory will continue for a very long time indeed.

Theoretical progress has certainly not been a result merely of the increased power of computer hardware. Novel approximation methods and numerical algorithms have been developed, which have produced dramatic increases in the accuracy of results obtainable with a given amount of computer resources. The trends of vectorisation in the 1980's and parallelisation in the 1990's have encouraged theorists to redesign their software to take full advantage of the new developments in computer hardware.

As quantum chemists have striven to attain accuracies of less than a millihartree in their calculations of the energy levels of multi-electron systems, they have often discovered that obtaining higher accuracy requires an enormous increase in the number of basis functions of the type conventionally used: sums of products of one-electron functions centered on a nucleus. The fundamental reason for this ubiquitous slowing-down in the rate of convergence of the energy with respect to the number of basis functions is that the exact many-electron wavefunction, or the first-order correction in Many Body Perturbation Theory (MBPT) or the Coupled Cluster Method (CCM), has singularities, typically discontinuous first derivatives with respect to Cartesian coordinates, which are not present in any finite linear combination of the basis functions, and therefore are approximated relatively slowly as more and more basis functions are added. In light of the increasing efforts devoted to high-precision theoretical calculations of atomic and molecular properties, it is important that the factors which determine the rate of convergence of a numerical method be thoroughly understood, so that the method can be made reasonably efficient by controlling the factors principally responsible for the numerical errors.

The attainment of high accuracy in the widely-used Rayleigh-Ritz variational method is crucially dependent on how rapidly a sequence of approximate wavefunctions, expressed as finite linear combinations of basis functions, converges to the exact wavefunction as the number of basis functions increases. In other words, one wants to know how rapidly finite linear combinations of basis functions converge to the exact solution of the Schroedinger equation in the infinite-dimensional Hilbert space. It is thus intuitively obvious that a key issue is the magnitude of that part of the exact wavefunction which is orthogonal to the fixed finite basis which one is using in the Rayleigh-Ritz variational calculation.

This idea made its appearance in a relatively neglected article, published in 1963 by Charles Schwartz, on the rates of convergence of orthogonal basis set expansions of simple functions of a single variable, where the expansion coefficients are calculated by projecting the exact function onto the basis functions [1]. A crucial idea of quite general importance which emerged from Schwartz's study is that the rate of convergence of such a basis set expansion is determined by how well finite linear combinations of basis functions duplicate the singularity structure of the exact function being approximated. The analytical study of rates of convergence of basis set expansions then received little attention for almost two decades, until around 1980 it was taken up again by Bruno Klahn and me [2], in an outgrowth of our work on trying to understand the relationship between the coefficients of basis functions used in a Rayleigh-Ritz calculation, and those when the exact wavefunction is formally expanded in terms of the same functions. Klahn and I found a more general operator-based formalism for predicting the rate of convergence of a basis set expansion, and we showed how in the case of slowly convergent expansions the expectation values of operators containing large powers of position variables could converge very slowly, or even diverge or converge to incorrect values as the basis approached completeness, even though the expectation values of all powers of the position variables are finite when evaluated with the exact wavefunction. Soon thereafter my colleague at the University of Delaware, Robert N. Hill, found a powerful method for extending the analysis of Klahn and me, and used it to obtain analytically the first two coefficients of the partial wave expansion for the helium ground state energy in terms of integrals over the exact eigenfunction [3]. (One does not, of course, know the exact eigenfunction for helium, but the use of highly accurate trial functions in Hill's formulas yielded numerical results in excellent agreement with the empirical coefficients obtained earlier by Carroll, Silverstone, and Metzger [4].) For a readable overview of the analytical theory (as of 1988) of rates of convergence of Rayleigh-Ritz variational calculations on atoms and molecules, see my contribution to the proceedings of the Versailles NATO Advanced Study Workshop on numerical approaches to the electronic structure of atoms and molecules [5].

Important progress in the practical implementation of these ideas has been made by Werner Kutzelnigg of the Universität Bochum, who in 1985 recognised that the slow convergence of the partial wave expansion for correlation energies arose from the representation of $1 = \langle\Phi|1|\Phi\rangle = \langle\Phi|r_{12}r_{12}^{-1}|\Phi\rangle$ by its partial wave expansion, and proposed a method for calculating 2nd- and higher-order correlation energies using a *linear-r_{12} ansatz* to eliminate the leading source of the error in the truncated partial wave expansions [47]. This method was then used by Kutzelnigg and his coworkers to calculate 2nd- and higher-order correlation energies for a variety of systems, starting with the ground state of the helium atom and ranging up to multi-electronic polyatomic molecules [48, 49, 50, 51, 52, 53, 54, 55]. Recently Noga and Kutzelnigg have extended the *linear-r_{12} ansatz* to the Coupled Cluster Method, thus opening the way to obtaining results of extremely high quality and reliability [56, 57].

At a NATO-ARW held in April 1988 at Versailles, over coffee on the patio with Anne-Marie Martensson-Pendrill, Kutzelnigg and I realised that Hill's analytical formula for the leading term in the partial wave expansion of the energy of the ground state of the helium atom (with the full interaction between the electrons) could be used to calculate the leading term $(-15/256)$ in the partial wave expansion of the second-order term in the $1/Z$-expansion of the ground state energy. Subsequent explorations, carried out primarily at other conferences which we both happened to be attending in 1988 and 1989, revealed that the same held true for other states of helium-like systems of arbitrary symmetry, thus suggesting the possibility of developing a general theoretical framework. Over the course of the next few years we did so, and published our results in 1992 [27].

While developing the framework for calculating the leading coefficients of the partial wave expansions of correlation energies for 2-electron atomic systems, we realised that our techniques could be used to elucidate an intimate, and hitherto poorly understood, connection between the cusp structure of the exact wavefunctions at the electron-electron coalescence and the relative ordering of singlet and triplet energy levels, the subject of Hund's First Rule and its generalization, the 'Alternating Rule'. Our further investigation [58] revealed that although the cusp behavior at the electron-electron coalescence is *not* the cause of the ordering of energy levels of different spin, the two phenomena are indeed closely related, in that each is a consequences of the same factor of $(-1)^{S+L+l_1+l_2}$ in the construction of properly antisymmetrised eigenfunctions of total spin S and total angular momentum L from one-electron functions of angular momentum l_1 and l_2.

1. Analytic Structure of Atomic and Molecular Wavefunctions

Consider the Hamiltonian for a system of N point particles, each of mass m_i and charge q_i, interacting by Coulomb potentials. In units where $\hbar = 1$, the non-relativistic N-body Hamiltonian H is

$$\sum_{i=1}^{N} -\frac{1}{2m_i}\nabla_i^2 + \sum_{j>i=1}^{N} \frac{q_i q_j}{r_{ij}}, \tag{1}$$

where r_{ij} is the distance between particles i and j. Since this differential operator is elliptic, a theorem of Fritz John [6] guarantees that the solutions of the Schrödinger Equation $(H - E)\psi = 0$ are analytic in Cartesian coordinates wherever the potential energy is analytic in these coordinates. However, the Coulomb potential is not analytic where two or more particles coalesce, so there is no reason why the eigenfunctions should be analytic at such coalescences.

1.1. TWO-PARTICLE COALESCENCES: THE KATO CUSP CONDITIONS

In 1957 it was proved by T. Kato [7] that at two-particle coalescences wavefunctions have cusps (discontinuities of derivatives with respect to Cartesion coordinates) which are described by what have come to be known as the *Kato cusp conditions*:

$$\left(\frac{\partial\hat{\psi}}{\partial r_{ij}}\right)_{r_{ij}=0} = \mu_{ij}\, q_i\, q_j\, \psi(r_{ij}=0)\,, \tag{2}$$

where $\mu_{ij} = m_i m_j/(m_i + m_j)$ is the reduced mass of the two-particle subsystem and $\hat{\psi}$ denotes the average of ψ over an infinitesimal sphere centered at $r_{ij} = 0$. Kato's fully rigorous derivation of the cusp conditions is rather long and complex, but in an intuitive way one can easily understand whence they arise by examining the Schrödinger Equation $(H - E)\psi = 0$ for the multiparticle system and transforming the variables $\mathbf{r}_i$ and $\mathbf{r}_j$ into relative and center-of-mass coordinates

$$\begin{aligned} \mathbf{R}_{ij} &= \frac{m_i\mathbf{r}_i + m_j\mathbf{r}_j}{m_i + m_j} \\ \mathbf{r}_{ij} &= \mathbf{r}_i - \mathbf{r}_j\,. \end{aligned} \tag{3}$$

Then as $r_{ij} \to 0$ and all other interparticle distances remain bounded away from 0, the only singularity in the potential energy term in the Schrödinger

Equation is

$$\frac{q_i\, q_j}{r_{ij}}\, \psi\,, \tag{4}$$

and the kinetic energy operator will contain a term

$$-\frac{1}{2\mu_{ij}}\left(\frac{\partial^2}{\partial r_{ij}^2} + \frac{2}{r_{ij}}\frac{\partial}{\partial r_{ij}} + \frac{L_{ij}^2}{r_{ij}^2}\right)\psi\,. \tag{5}$$

We now average expressions (4) and (5) about an infinitesimally small sphere centered at $r_{ij} = 0$, thus eliminating from ψ all except the s-wave component, which is itself annihilated by L_{ij}^2. Assuming that $\partial^2\hat{\psi}/\partial r_{ij}^2$ is bounded, as it is for eigenfunctions of the hydrogen atom, then the coefficients of the $1/r_{ij}$ singularities in eqq. (4) and (5) should cancel, which leads to the Kato cusp condition given by eq. (2).

Since a central theme of this workshop is the dimensional scaling method extensively developed by Dudley Herschbach and his numerous coworkers, in which spatial dimensionality D is treated as a variable parameter while all particles continue to interact by $1/r$ potentials, I here simply mention that the D-dimensional generalisation of the Kato cusp condition is

$$\left(\frac{\partial\hat{\psi}}{\partial r_{ij}}\right)_{r_{ij}=0} = \frac{2}{D-1}\,\mu_{ij}\, q_i\, q_j\, \psi(r_{ij}=0)\ , \tag{6}$$

whose heuristic derivation can be found in [25].

If the wavefunction vanishes at $r_{ij} = 0$, then a higher-order generalisation of the Kato cusp condition can be derived, according to which

$$\psi(\mathbf{r}_{ij}) \simeq Y_l(\Omega)\, r_{ij}^l \left(1 + \frac{\mu_{ij}\, q_i\, q_j}{l+1}\, r_{ij}\right)\ , \tag{7}$$

where $Y_l(\Omega)$ is a spherical harmonic of degree l. A rigorous proof of this leading behavior was provided by M. and T. Hoffmann-Ostenhof and H. Stremnitzer [8].

Let us examine some representative illustrations of electron-nucleus and electron-electron cusps.

(i) At an electron-nucleus coalescence, μ_{ij} is the electron-nucleus reduced mass μ and $q_i q_j = -Z$, so that $\mu_{ij} q_i q_j = -\mu Z$ is negative. A wavefunction which does not vanish at the nucleus decreases as r_{ij} increases away from 0, as shown in Figure 1. The behavior of a hydrogenic s-state eigenfunction provides a typical example of such cusp behavior.

If the wavefunction does vanish at $r_{ij} = 0$, as in a p-state, a d-state, etc., then the wavefunction has a discontinuity not in its first derivative but in

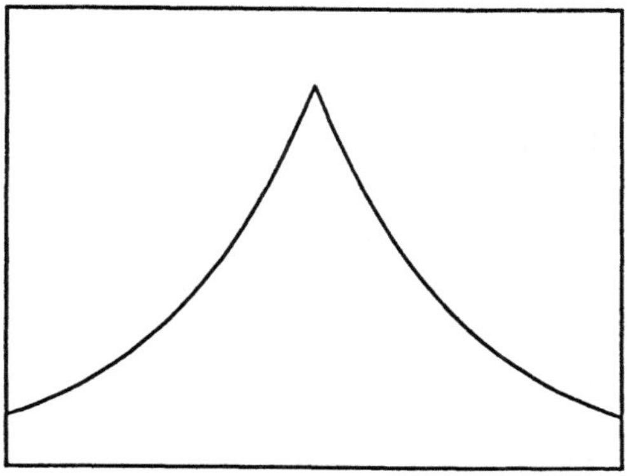

Figure 1. Illustration of an electron-nucleus cusp.

a higher derivative. For example, in a p-state of hydrogen the wavefunction vanishes linearly at $r = 0$, and has there discontinuous 2nd derivatives with respect to Cartesian coordinates, and in a d-state the wavefunction vanishes quadratically at $r = 0$, and has there discontinuous 3rd derivatives with respect to Cartesian coordinates.

(ii) At electron-electron coalescences of electrons in a state with total spin $S = 0$ and natural parity (so that the wavefunction does not vanish) $\mu_{ij} = 1/2$ and $q_i q_j = 1$ is positive, so that the wavefunction ψ displays an inverted cusp behavior given by

$$\psi(\mathbf{r}_{ij}) \simeq 1 + \frac{1}{2} r_{ij} , \tag{8}$$

as illustrated in Figure 2.

In a state with parallel individual spins yielding a total spin $S = 1$, by the Pauli exclusion principle the wavefunction must vanish at $r_{ij} = 0$. In general, it does so linearly, like a p-wave,

$$\psi(\mathbf{r}_{ij}) \simeq Y_1(\Omega)\, r_{ij}^1 \left(1 + \frac{1}{4} r_{ij}\right) , \tag{9}$$

with a discontinuous 2nd derivative with respect to Cartesian coordinates.

In a state with antiparallel individual spins yielding a total spin $S = 0$, but with unnatural parity, the spatially symmetric wavefunction vanishes quadratically at $r_{ij} = 0$,

$$\psi(\mathbf{r}_{ij}) \simeq Y_2(\Omega)\, r_{ij}^2 \left(1 + \frac{1}{6} r_{ij}\right) , \tag{10}$$

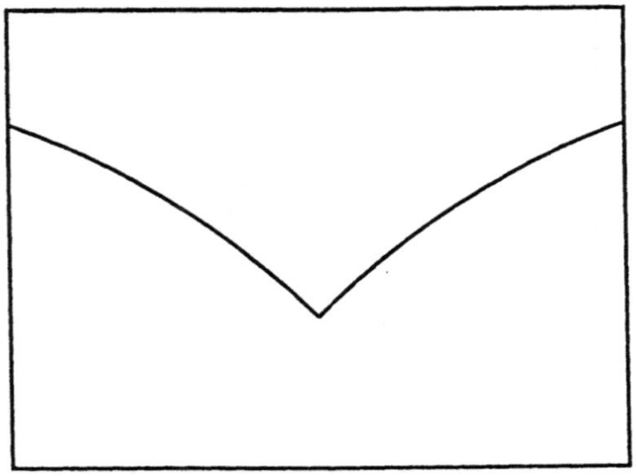

Figure 2. Illustration of an electron-electron cusp for antiparallel spins and natural parity.

with a discontinuous 3rd derivative with respect to Cartesian coordinates. Barring miraculous vanishings in matrix elements of the electron-electron interaction, these are the only possibilities for the electron-electron cusp behavior of an atomic wavefunction describing interacting electrons, since such a wavefunction is an eigenfunction of only total L^2 and one of its components, total S^2 and one of its components, total parity under inversion, and of the symmetry under exchange of two electrons [27, 58].

1.2. THREE-PARTICLE COALESCENCES: THE FOCK EXPANSION

Ternary and higher coalescences of three or more particles are described by an expansion introduced by V.A. Fock [9] in 1954 and subsequently studied by many atomic physicists in the former Soviet Union [10, 11, 12, 13, 14] and the West [15, 16, 17, 18, 19, 20, 21, 22, 23, 24]. The Fock expansion features an expansion

$$\sum_{j=0}^{\infty} \sum_{k=0}^{[j/2]} R^{j/2} (\ln R)^k \, \psi_{j,k}(\Omega) \tag{11}$$

in half-integral powers of the hyperradial variable R and integral powers of $\ln R$, where

$$R = \sum_i r_i^2 \tag{12}$$

is the sum of the squares of the electron-nucleus distances, and the coefficients $\psi_{j,k}(\Omega)$ of the powers of R and $\ln R$ are functions of the hyperspherical angles. The first several $\psi_{j,k}$ are known in closed form. For natural parity singlet states, whose wavefunctions do not vanish at $R = 0$, it is straightforward to derive the results (in $D = 3$ spatial dimensions)

$$\begin{aligned}
\psi_{0,0} &= 1 \\
\psi_{1,0} &= -Z\frac{r_1 + r_2}{R^{1/2}} + \frac{1}{2}\frac{r_{12}}{R^{1/2}} \\
\psi_{2,1} &= -Z\frac{\pi - 2}{3\pi}\frac{\mathbf{r}_1 \cdot \mathbf{r}_2}{R} .
\end{aligned} \tag{13}$$

One can readily see that

$$\psi_{0,0} + R^{1/2}\,\psi_{1,0} = 1 - Z(r_1 + r_2) + \frac{1}{2}r_{12} , \tag{14}$$

as would be expected from the Kato cusp condition. Efforts to solve for $\psi_{2,0}$ were unsuccessful for many years, until in the mid-1980's it was obtained in closed form by Gottschalk, Abbott, and Maslen [21, 22, 23, 24] in terms of Lobachevsky's function and more elementary functions.

In dimensions D different from 3, the above results are only slightly changed: the functional forms remain the same, while the coefficients are multiplied by factors of $(D-1)/2$ and the like.

The convergence properties of Fock's expansion have been studied mathematically by J. Macek [15], by J. Leray [16, 17, 18, 19], and by me [20]. As a result of these studies it is now known that for any value of the energy E (even complex), Fock's expansion does converge for all R to a function which satisfies the Schrödinger Equation, considered as a partial differential equation with no boundary condition at ∞. Extensive comparison of the coefficients of the lower-order terms in Fock's expansion with those obtained from highly accurate variational calculations indicates that for the physically allowed values of E, the physically relevant square-integrable solutions of the Schrödinger Equation are indeed represented by Fock's expansion.

2. Rates of Convergence of Rayleigh-Ritz Variational Calculations on Atoms and Molecules

2.1. THE EFFECT OF SINGULARITIES: A SIMPLE CALCULATION

The effect of the ubiquitous electron-nucleus and electron-electron cusps on the rates of convergence of Rayleigh-Ritz variational calculations on atoms and molecules can be easily understood in terms of a simple heuristic argument, first presented in [25]. To get from configuration space to momentum space we use the Fourier transform, under which a function with a hydrogenic cusp is mapped to a function with p^{-4} fall-off at large momenta:

$$e^{-Zr} \to \frac{1}{(Z^2 + p^2)^2} . \tag{15}$$

Suppose one's finite N-dimensional basis has little weight on a momentum scale beyond $\tilde{p}_N$. The error in the variational energy, ΔE_N, is roughly proportional to (and if the virial theorem holds is exactly equal to) the error in the kinetic energy ΔT_N:

$$\Delta E_N \sim \Delta T_N \sim \int_{\tilde{p}_N}^{\infty} dp\, p^2 \frac{p^2}{2} \left(\frac{1}{(Z^2+p^2)^2} \right)^2 , \tag{16}$$

where the first factor of p^2 comes from the 3-dimensional Jacobian, the second factor of $p^2/2$ comes from the kinetic energy, and the third factor, which behaves as p^{-8} for large p, comes from the density in momentum space. Proceeding to the large $\tilde{p}_N$ limit and ignoring multiplicative constants, we have

$$\Delta E_N \sim \Delta T_N \sim \int_{\tilde{p}_N}^{\infty} dp\, p^2\, p^2\, p^{-8} = \int_{\tilde{p}_N}^{\infty} dp\, p^{-4} \sim (\tilde{p}_N)^{-3} . \tag{17}$$

Since this conference features dimensional scaling, I will remark that in Ref. [25] I showed that the power -3 in eq. (17) is determined by the spatial dimensionality $D = 3$ of the world in which we live.

One readily sees from eq. (17) that the rate at which the error ΔE_N in the variational energy decreases is determined by the rate at which $\tilde{p}_N$ increases with N. Let us now examine two cases: the approximation of the electron-nucleus cusp by a superposition of Gaussians, and the approximation of the electron-electron cusp by a partial wave expansions.

2.2. THE APPROXIMATION OF ELECTRON-NUCLEUS CUSPS BY GAUSSIANS

Because of the ease of doing multi-center integrals, almost all practical quantum chemistry programs for polyatomic molecules use Gaussian basis functions centered on the various nuclei. Since such basis functions, which are analytic everywhere, cannot explicitly represent the cusps at the nuclei in the exact wavefunctions, there naturally arises the issue of the rate of convergence of such expansions. One can study this issue by using a basis of N Gaussian basis functions with 'floating' (i.e., variable) exponents

$$\{\exp(-\zeta_j^{(N)} r^2)\}_{j=1}^{N} \tag{18}$$

in a Rayleigh-Ritz variational calculation of energies of s-states of the hydrogen atom. It was found numerically by Haywood, Nicholson, and Morgan [43] with further theoretical analysis by Kenneth Wilson [44] and independently by Werner Kutzelnigg in the case of 'even-tempered' Gaussian exponents [45], that the errors ΔE_N obey

$$\Delta E_N \sim (\zeta_{\max}^{(N)})^{-3/2} \sim ((\zeta_{\max}^{(N)})^{1/2})^{-3} , \tag{19}$$

where $\zeta_{\max}^{(N)}$ is the largest of the $\zeta_j^{(N)}$. The rate of convergence given by eq. (19) can be explained by observing that under Fourier transformation

$$e^{-\zeta r^2} \to e^{-\frac{1}{4\zeta}p^2} , \tag{20}$$

and thus it is evident that the biggest kinetic energy associated with the Gaussian basis is set by $2\zeta_{\max}^{(N)}$ and that the biggest momentum scale is set by

$$(4\zeta_{\max}^{(N)})^{1/2} = 2\,(\zeta_{\max}^{(N)})^{1/2} \tag{21}$$

and hence that

$$\Delta E_N \sim ((\zeta_{\max}^{(N)})^{1/2})^{-3} . \tag{22}$$

Since $\zeta_{\max}^{(N)}$ grows extremely rapidly with N, one obtains extremely fast convergence of the representation of an s-type function, or more generally of a Slater-Type-Orbital, in terms of a superposition of Gaussians. Hence the representation of electron-nucleus cusps in a Gaussian basis poses no serious obstacle to the attainment of highly accurate variational energies.

2.3. THE APPROXIMATION OF ELECTRON-ELECTRON CUSPS BY A PARTIAL WAVE EXPANSION

Much slower convergence is associated with one of the most commonly used expansions in the theory of the electronic structure of atoms and molecules, the partial wave expansion, in which individual atomic orbitals are expressed as products of radial functions and spherical harmonics. Appropriately antisymmetrized sums of products of the spherical harmonics for the coordinates of each particle can be formed to yield eigenfunctions of total L^2, L_z, S^2, and S_z. To prevent lengthy expressions involving 3-j and 6-j symbols from obscuring the essential physics, let us focus on the partial wave expansion for an S-state of the helium atom:

$$\Psi(\mathbf{r}_1, \mathbf{r}_2) \simeq \Psi_L(\mathbf{r}_1, \mathbf{r}_2) = \sum_{l=0}^{L} f_l(r_1, r_2)\, P_l(\cos\theta) , \tag{23}$$

where the coefficient functions $f_l(r_1, r_2)$ are optimised to minimise the variational energy. The rate of convergence of the variational energy can be estimated by first calculating the angular contribution to the kinetic energy of a Legendre polynomial $P_l(\cos\theta)$, which is easily calculated since it is an eigenfunction with eigenvalue $l(l+1)$ of the squares of the 1-particle angular momentum operators

$$\frac{1}{2}\left(\frac{\mathcal{L}_1^2}{r_1^2} + \frac{\mathcal{L}_2^2}{r_2^2}\right) P_l(\cos\theta) = \frac{1}{2}\, l(l+1)\left(\frac{1}{r_1^2} + \frac{1}{r_2^2}\right) P_l(\cos\theta) . \tag{24}$$

Thus for typical electron-nucleus distances r_1 and r_2 of the order of an a.u. we see that

$$\tilde{p}_L \sim \sqrt{L(L+1)} \sim L+1 \tag{25}$$

and hence

$$\Delta E_L \sim (L+1)^{-3} \,. \tag{26}$$

In 1979 it was found empirically by Carroll, Silverstone, and Metzger [4] that the variational energies obtained from the partial wave expansion converge to the exact result with errors ΔE_L which obey the asymptotic relation

$$\Delta E_L = C_3(L+1)^{-3} + C_4(L+1)^{-4} + \ldots \,, \tag{27}$$

where the constants C_3 and C_4 were numerically found to be approximately equal to 0.0247 and 0.0077, respectively. In 1985 my colleague Robert N. Hill [3] derived explicit formulas for these constants in terms of integrals of the exact two-electron density along the cusp at $r_{12} = 0$:

$$\begin{aligned} C_3 &= 2\pi^2 \int_0^\infty dr\, r^5 \, |\Psi(r,r,0)|^2 \\ C_4 &= \frac{12\pi}{5} \int_0^\infty dr\, r^6 \, |\Psi(r,r,0)|^2 \,. \end{aligned} \tag{28}$$

Numerical evaluation of these integrals using the highly accurate 230-term trial function of Freund, Huxtable, and Morgan [26] to represent the exact ground state $\Psi(r_1, r_2, r_{12})$ yields

$$\begin{aligned} C_3 &= 0.0247419\ldots \\ C_4 &= 0.00774727\ldots \,, \end{aligned} \tag{29}$$

in excellent agreement with the estimates $C_3 \simeq 0.0247$ and $C_4 \simeq 0.0077$ which Carroll, Silverstone, and Metzger [4] had obtained by numerical fitting. (Explicit formulas for the constants C_3 and C_4 in the rates of convergence of the partial wave expansions for all possible singly and doubly excited states of helium were published by Kutzelnigg and Morgan [27] in 1992.) This rate of convergence is rather slow, as can be seen from the table on the facing page, which lists the desired accuracy in the energy, the number $(L+1)$ of partial waves required to obtain it, and $(L+1)^2$, which is proportional to the total number of basis functions:

ΔE_L (a.u.)	$L+1$	$(L+1)^2$
10^{-2}	2	4
10^{-3}	3	9
10^{-4}	6	36
10^{-5}	13	169
10^{-6}	28	784
10^{-7}	60	3,600
10^{-8}	130	16,900

It is clear that an enormous number of basis functions is required to achieve sub-microhartree accuracy using a partial wave expansion. The origin of this slow convergence is the slow representation of the electron-electron cusp at $r_{12} = 0$ by the partial wave expansion, which for any finite L yields an approximate wavefunction analytic in Cartesian coordinates at $r_{12} = 0$.

2.4. BASIS FUNCTIONS WITH THE CORRECT SINGULARITIES

2.4.1. *Specialised Basis Functions for the Helium Atom*

Much faster convergence can be obtained by using a basis which explicitly incorporates linear terms (and higher odd powers) in r_{12}, as was first done in 1929 by Hylleraas in his famous calculation which provided the first clear numerical demonstration that the Schrödinger Equation yielded the correct ground-state energy for a two-electron atom [28]. Hylleraas used a small basis of the form $\phi^H_{n,l,m}(2ks, 2kt, 2ku)$, where

$$\phi^H_{n,l,m}(s,t,u) = s^n\, t^l\, u^m\, e^{-s/2}\,. \tag{30}$$

Here s, t, and u are the Hylleraas variables,

$$\begin{aligned} s &= r_1 + r_2 \\ t &= -r_1 + r_2 \\ u &= r_{12}\,, \end{aligned} \tag{31}$$

which obey the triangular inequalities

$$0 \leq |t| \leq u \leq s \leq \infty\,. \tag{32}$$

The powers n, l, and m are non-negative integers, and k is a flexible scaling parameter. Such a basis is very effective if the number of functions used is not too large, but for large numbers of basis functions numerical linear dependency problems can develop because of the large overlaps between terms with different values of n, l, and m. Another basis which is analytically equivalent but has much better numerical stability properties is the Pekeris [29, 30] basis of products of Laguerre functions

$$L_p(\xi u)\, L_q(\xi v)\, L_r(\xi w)\, e^{-\xi(u+v+w)/2} \tag{33}$$

of perimetric coordinates

$$\begin{aligned} u = &\quad -r_1 + r_2 + r_{12} &&= t + u \\ v = &\quad r_1 - r_2 + r_{12}) &&= -t + u \\ w = &\quad 2(r_1 + r_2 - r_{12}) &&= 2(s-u)\,, \end{aligned} \tag{34}$$

which range independently from 0 to ∞. The indices p, q, and r of the Laguerre polynomials are non-negative integers, and ξ is a flexible scaling parameter which can be varied to optimise the variational energy [31], although for convenience in setting up his matrix equations Pekeris set it equal to its non-optimal value $\sqrt{-E}$ to eliminate the energy term from Schrödinger Equation. In his famous calculation with 1078 such basis functions [30], Pekeris obtained a variational upper bound of

-2.903724375 a.u.,

which is accurate to 2×10^{-9} a.u. Very recently, Cox, Smith, and Sutcliffe [31] have used the same basis but with an optimised non-linear scaling parameter ξ to obtain an upper bound of

-2.903724377017 a.u.,

which is accurate to within 2×10^{-11} a.u. The improvement in accuracy achieved in these calculations with many fewer basis functions than would be needed in a partial wave expansion is attributable to the inclusion in the basis of terms linear in $r_{12} = (u+v)/2$ to describe compactly, instead of with an infinite slowly convergent expansion, the electron-electron cusp.

It also helps to build into the basis terms which describe the singularity in the wavefunction at the 3-particle coalescence, as was first demonstrated in 1966 by Frankowski and Pekeris [32]. To include terms with logarithmic singularities of the type suggested by the Fock expansion, they used scaled basis functions of the form $\phi^{FP}_{n,l,m,i,j}(2ks, 2kt, 2ku)$, where

$$\phi^{FP}_{n,l,m,i,j}(s,t,u) = s^n\, t^l\, u^m\, (s^2+t^2)^{i/2}\, (\ln s)^j\, e^{-s/2}\,. \tag{35}$$

Here s, t, and u are the Hylleraas variables defined in eq. (31) *supra*, and k is a flexible scaling parameter. The index i was 0 or 1, and the other indices n, l, m, and j obeyed the constraints:

$$\begin{aligned} 0 &\leq l \\ 0 &\leq m \\ 0 &\leq n+l+m \\ 0 &\leq j \leq (n+l+m)/2 \,, \end{aligned} \tag{36}$$

with n negative only if $n + l + m \geq 2$, $i = 0$, and $j = 0$. Although the $\ln s$ in Frankowski and Pekeris' basis is not the same as the $\ln R$ in Fock's expansion, the two functions are analytically equivalent since

$$\begin{aligned} \ln R &= \ln\left(\frac{1}{2}(s^2+t^2)\right) \\ &= \ln\left(\frac{s^2}{2}\left(1+\frac{t^2}{s^2}\right)\right) \\ &= 2\,\ln s \;-\; \ln 2 \;+\; \ln(1+t^2/s^2) \,, \end{aligned} \tag{37}$$

which evidently has the same singular structure as $\ln s$ for physical values of t $(0 \leq t^2 \leq s^2)$. With only 246 functions of this form, Frankowski and Pekeris obtained a variational upper bound of

-2.9037243770326 a.u.

for the non-relativistic ground state energy of the helium atom.

Similarly impressive results on excited states of helium and on the H^- ion were obtained by Frankowski using a scaled basis of the form $\phi^F_{n,l,m,j}(2ks, 2kt, 2ku)$, where

$$\phi^F_{n,l,m,j}(s,t,u) = s^n\, t^l\, u^m\, (\ln s)^j \left\{ \begin{array}{c} \cosh(ct/2) \\ \sinh(ct/2) \end{array} \right\} e^{-s/2}, \tag{38}$$

where again s, t, and u are the Hylleraas variables defined in eq. (31) *supra*, and k and c are flexible scaling parameters, which together allow one to model the 'in-out' radial correlation in these loosely bound systems. The $\cosh(ct/2)$ or $\sinh(ct/2)$ is chosen so that t^l times the hyperbolic function has the correct overall exchange symmetry of the state being calculated.

By including some higher-order terms involving negative powers of s and positive powers of $\ln s$ arising from the Fock expansion which Frankowski and Pekeris should not have omitted, adding many more powers of t/s and u/s to reflect electron-electron correlation on short to moderate length scales, and dropping the terms involving $(s^2 + t^2)^{1/2}$ to reduce numerical linear dependence, Freund, Huxtable, and Morgan [26] obtained with 230 functions of the Frankowski-Pekeris type an upper bound of

-2.90372437703407 a.u.

Our calculation was at the time (1984) the most accurate upper bound to the non-relativistic ground state energy of the helium atom, having effected with a smaller number of functions than Frankowski and Pekeris had used about a 35-fold reduction in the error in the variational energy.

In subsequent calculations with a composite basis of 476 functions, 244 of our modified Frankowski-Pekeris functions and 232 Frankowski functions, Baker, Freund, Hill, and Morgan [34] obtained an upper bound of

-2.9037243770341184 a.u.,

in close agreement with Kleindienst and Emrich's value [35]

-2.903724377034119 a.u.

obtained with 448 Frankowski-Pekeris functions with a single length scale. It would be practically impossible to achieve such high accuracy using a partial wave expansion to represent the wave function.

Similarly accurate results for excited states have been obtained by Baker, Hill and Morgan [36] using very compact composite bases of Frankowski-Pekeris functions to represent the three-particle coalescence and Frankowski functions [33], with an open-shell exponential to represent the strong 'in-out' radial correlation in excited states.

I should mention that results of comparable accuracy have been obtained for the ground and excited states of helium by Gordon Drake [37], using very large composite bases of the form

$$r_1^n \, r_2^m \, r_{12}^m \, e^{-\alpha_i r_1 - \beta_i r_2} \; + \; (1 \leftrightarrow 2) \tag{39}$$

with two (or possibly more) sets of exponential parameters (α_i, β_i). Such a basis includes terms which duplicate the true behavior of the wavefunctions at all two-particle cusps, but do not explicitly include the singular terms which appear in Fock's expansion about the three-particle coalescence. To approximate the wavefunction in the vicinity of the three-particle coalescence, one set of exponential parameters (α_i, β_i) increases indefinitely as the number of basis functions is augmented, thus slowly approximating the singularities in the $R \ln R$ terms and the like at $R = 0$. Nonetheless, because this basis does get right the behavior of the wavefunction at all 2-particle coalescences, and the singularity at the 3-particle coalescence is very mild and exists on a set of lower dimension, this basis is capable of producing results, albeit with more basis functions, of an accuracy comparable to those described above obtained with bases which explicitly include the logarithmic terms. I should emphasise that it is of course valuable to have two completely independent theoretical methods which provide results of comparable accuracy, so that one can be sure that both calculations are right, and to provide impetus for both groups to effect further improvements in their methods. The friendly competition between my group and Drake's has been mutually beneficial, and to our experimental colleagues as well, as I proceed to explain now.

As a test of variational methods it is enlightening to compare different methods for doing calculations on the simplified model system of the non-relativistic helium atom with an infinitely massive point-like nucleus, since its Hamiltonian (apart from an overall energy scale factor = 1 a.u.) contains no 'time-dependent' fundamental constants which change with improving measurements. However, for a comparison of theory with experimentally determined transition energies, which are now measured with a precision of a few parts in 10^{10}, it is necessary to include a host of small corrections to account for the finite nuclear mass, relativity and QED effects, and even the non-vanishing nuclear size. There are two components to the errors in such small terms. One component is the error

$$\delta\langle A\rangle \simeq 2\,\langle\psi|A|\delta\psi\rangle\,, \tag{40}$$

which is first-order in $\delta\psi$, in the evaluation of expectation values of various operators A which typically do not commute with the non-relativistic Hamiltonian H, and which therefore can be considerably larger than the error

$$\delta\langle H\rangle = \langle\delta\psi|H - E|\delta\psi\rangle \tag{41}$$

in the non-relativistic energy, which by the Rayleigh-Ritz variational principle is second-order in $\delta\psi$. Since for small $\delta\psi$ the first-order error $\delta\langle A\rangle$ eventually dominates the second order error $\delta\langle H\rangle$ no matter how small is the operator A, to calculate a total relativistic energy with a given accuracy it is necessary to generate variational non-relativistic wavefunctions which are much more accurate than would be needed to calculate a non-relativistic energy of the same accuracy. The other component is the error in the determination of the fundamental constants and nuclear parameters which enter into these correction terms. Assuming that the numerical error in evaluating expectation values can be reduced below the level of the error in the experimentally determined fundamental constants and nuclear parameters, such a comparison of theory with experiment yields an improved measurement of whichever fundamental constant or nuclear parameter is least well determined. For example, the most reliable current estimates of the difference of the squares of the nuclear charge radii of the proton and the deuteron, the ^{3}He nucleus and the α-particle, and the ^{6}Li and ^{7}Li nuclei are derived from a combination of high-precision calculations and table-top experiments on the hydrogen and the helium atoms and the corresponding helium-like lithium ions [39, 40, 38].

Another application of highly accurate trial wavefunctions which include the correct cusp behavior is the use of them in the design of improved exchange and correlations functionals for Kohn-Sham density functional theory, as has recently been done by Umrigar and Gonze [41, 42].

2.4.2. *Kutzelnigg's Linear-r_{12} Method for Arbitrary Atoms and Molecules* For atoms larger than helium, or for molecules with two or more electrons, including the three-particle singularity terms in the basis would lead to intractable integrals. Indeed, it is already a difficult challenge to devise a way of including the electron-electron two-particle cusps in a basis for calculations on multi-electron polyatomic molecules in a computationally efficient way. Fortunately, such a method was introduced in 1985 by Werner Kutzelnigg [47] and then in the late 1980's and early 1990's was further developed by him and his collaborators [48, 49, 50, 51, 52, 53, 54, 55]. In this method, the wavefunction is represented in the form

$$\psi = \left(1 + \frac{1}{k+1} r_{ij}\right) \phi + \chi , \tag{42}$$

where $k = 0$ for an electron pair in a relative singlet state of natural parity, $k = 1$ for a relative triplet state of natural or unnatural parity, and $k = 2$ for a relative singlet state of unnatural parity, and ϕ and χ are represented as sums of products of one-particle functions. This approach increases the rate of convergence of the partial wave expansion from $(L+1)^{-3}$ (for $k = 0$) to $(L+1)^{-7}$, and thereby reduces the error from the millihartree to the microhartree level. The necessary integrals, although of course more complicated than in the conventional partial wave expansion, can still be evaluated efficiently, even for higher levels of Moeller-Plesset theory (Many Body Perturbation Theory), as well as in the Coupled Cluster Method [56]. This method has produced what is currently the most accurate determination of the equilibrium geometry and energy, -1.343835 a.u., of H_3^+ [54, 55], and benchmark results also on the 4-electron systems Be, Li^-, and LiH [57].

2.5. CONCLUSIONS

I should like to conclude this section by taking from my discussion in [5] a summary of the key issues, in order of importance, for performing highly accurate Rayleigh-Ritz variational calculations on atoms and molecules:

0. Have many basis functions concentrated in the correct regions of configuration space and momentum space.

1. Get right the analytic structure at the coalescences of single electrons with the nucleus or the nuclei.

2. Get right the analytic structure at coalescences of pairs of electrons.

3. Get right the analytic structure at coalescences of three or more particles.

∞. Get right the long-distance exponential decay of the wavefunction.

3. Hund's First Rule, the Alternating Rule, and Symmetry Holes

3.1. HUND'S FIRST RULE AND THE ALTERNATING RULE

3.1.1. *Hund's First Rule*

Hund's First Rule [59, 60] states that among many-electron states arising from the same configuration of one-electron orbitals, the ground state has the largest total spin S. This has sometimes been naively 'generalised' to excited states in the form 'among many-electron states arising from the same configuration of one-electron orbitals, the energy of the state decreases as the total spin S of the state increases', but to such a statement there are a plethora of counterexamples. Some of the counterexamples arise from strong configuration interaction, which invalidates the description of a pair of states arising from *the same configuration*, a key proviso in all of Hund's Rules. However, there are many more exceptions which arise even without strong configuration interaction.

3.1.2. *The Alternating Rule*

Consider an atom with a pair of electrons in unfilled shells with angular quantum numbers l_1 and l_2. In 1927 the famous H.N. Russell (of Russell-Saunders coupling) and his collaborator W.F. Meggers [61] formulated an 'Alternating Rule', according to which Hund's First Rule holds for the extremal values of total L in a multiplet and all intermediate values of L in steps of 2:

$$L_{\rm min}, \quad L_{\rm min}+2, \quad L_{\rm min}+4, \ldots, \quad L_{\rm max}-4, \quad L_{\rm max}-2, \quad L_{\rm max},$$

but Hund's First Rule is reversed, with the singlets *below* the triplets, for all the interleaving values of total L:

$$L_{\rm min}+1, \quad L_{\rm min}+3, \ldots\ldots\ldots, \quad L_{\rm max}-3, \quad L_{\rm max}-1\ .$$

The true explanation of the Alternating Rule, like that of Hund's First Rule for ground states of atoms, long remained obscure, though there was no lack of attempted explanations which were either incomplete or simply incorrect.

3.1.3. *Hund's First Rule and Fermi Holes: A Fallacious Explanation*

One venerable explanation of Hund's First Rule runs as follows. In a singlet state, with an antisymmetric spin function and hence a symmetric spatial wavefunction which does not vanish at $r_{12}=0$, the electrons can approach each other more closely than in a triplet state, with a symmetric spin function and hence an antisymmetric spatial wavefunction which must vanish at $r_{12}=0$. Hence one expects

$$\left\langle \frac{1}{r_{12}} \right\rangle_{triplet} < \left\langle \frac{1}{r_{12}} \right\rangle_{singlet} . \tag{43}$$

This enchanting explanation held the field until in the early 1960's extensive accurate calculations on numerous excited states of the helium atom, in particular those by Davidson [62], revealed that in all cases the inequality was reversed:

$$\left\langle \frac{1}{r_{12}} \right\rangle_{triplet} > \left\langle \frac{1}{r_{12}} \right\rangle_{singlet} \quad . \tag{44}$$

This striking contradiction of this superficially plausible argument prompted first the explanation of what had gone wrong (in the triplet states the wavefunction contracts closer to the nucleus, causing an increase in the expectation values of the kinetic energy, the electron-nucleus attraction energy, *and* the electron-electron repulsion energy), and then more attempts to explain why the energy of a triplet should lie below that of a singlet.

3.1.4. *Colpa's Proof of Hund's First Rule for* $(1s)(nl)$ *Configurations*

An important contribution was made in 1972 by J.P. Colpa [63], who provided a rigorous demonstration that at the Hartree-Fock level, the lowest singlet state of the $(1s)(nl)$ configuration, with $l \geq 1$, lies above the lowest triplet state of the same configuration. The basic idea in Colpa's proof is to use the pair of orbitals which minimise the singlet energy in a variational estimate of the triplet energy. This upper bound to the true triplet energy must lie lower than the true singlet energy because of the positivity of the exchange matrix element of the form

$$\langle f(1)g(2)|r_{12}^{-1}|g(1)f(2)\rangle \, , \tag{45}$$

which was proven by Racah [64], by Roothaan [65], and by Slater [66] using the explicit form of the potential r_{12}^{-1}, and more generally holds for any potential which has a positive Fourier transform [67]. Some other purported explanations could be ruled out immediately since they did not take account of the numerous exceptions posed by the Alternating Rule, which was generally neglected, aside from a pair of articles by Warner, Bartell, and Blinder [68] and by Warner and Berry [69] which associated the Alternating Rule with the angular distribution of the pair of electrons.

3.1.5. *The Alternating Rule and the Width of Symmetry Holes: The Correct Explanation*

The study which Kutzelnigg and I undertook of how the nature of the singularity at an electron-electron coalescence determines the rate of convergence of the partial wave expansion led to an unexpected insight into the relationship between the behavior of the two-electron pair function in the vicinity of an electron-electron coalescence and the ordering of atomic energy levels, as reflected in Hund's First Rule for the ground states of atoms and its generalisation by Russell and Meggers to the 'Alternating Rule' for excited states.

To make a long story short, the explanation of the Alternating Rule (and of its special case, Hund's First Rule) by Kutzelnigg and me begins with the observation that those states with total

$$L = L_{\text{min}}, L_{\text{min}} + 2, L_{\text{min}} + 4, \ldots, L_{\text{max}} - 4, L_{\text{max}} - 2, L_{\text{max}}$$

have *natural* parity $= (-1)^L = (-1)^{l_1+l_2}$, whereas those with total

$$L = L_{\text{min}} + 1, L_{\text{min}} + 3, \ldots, L_{\text{max}} - 3, L_{\text{max}} - 1$$

have *unnatural* parity $= -(-1)^L = (-1)^{l_1+l_2}$. We then examine a properly antisymmetrised uncorrelated two-electron wavefunction

$$\begin{aligned} \Phi &= \Phi^S_{L,M,n_1,l_1,n_2,l_2}(\vec{r}_1, \vec{r}_2) \\ &= [2(1 + \delta_{n_1,n_2}\,\delta_{l_1,l_2})]^{-1/2}\Big\{ R_1(r_1)\,R_2(r_2)\,\Omega_{L,M,l_1,l_2}(\omega_1, \omega_2) \\ &\qquad + (-1)^{S+L+l_1+l_2} R_2(r_1)\,R_1(r_2)\,\Omega_{L,M,l_2,l_1}(\omega_1, \omega_2)\Big\}, \end{aligned} \tag{46}$$

where

$$\Omega_{L,M,l_1,l_2}(\omega_1, \omega_2) = \tag{47}$$
$$\sum_{m_1,m_2} (-1)^{l_1-l_2+M}\sqrt{2L+1}\begin{pmatrix} l_1 & l_2 & L \\ m_1 & m_2 & -M \end{pmatrix} Y_{l_1}^{m_1}(\omega_1) Y_{l_2}^{m_2}(\omega_2)\,,$$

as given in eqq. (5.1) and (5.2) of [27]. In determining the behavior of such a wavefunction in the vicinity of $r_{12} = 0$, *i.e.* in the vicinity of $r_1 = r_2$, $\omega_1 = \omega_2$, the crucial issue is whether the two terms in Eq. (36) add together or cancel, *i.e.*, whether the factor $(-1)^{S+L+l_1+l_2}$ is $+1$ or -1, and hence whether $S + L + l_1 + l_2$ is even or odd. If the state has natural parity, so that $L + l_1 + l_2$ is even, then the factor is $+1$ for a singlet state and -1 for a triplet state. If the state has unnatural parity, so that $L + l_1 + l_2$ is odd, the factor is just the opposite: $+1$ for a triplet state and -1 for a singlet state. Thus whether the wavefunction and also its first derivative vanish at the electron-electron coalescence is determined by this factor of $(-1)^{S+L+l_1+l_2}$.

Similarly, the exchange integral for such a state is given by

$$(1 + \delta_{n_1,n_2}\,\delta_{l_1,l_2})^{-1}\,(-1)^{S+L+l_1+l_2} \int dr_1\,dr_2\,r_1^2\,r_2^2 \int d\omega_1\,d\omega_2$$

$$R_1^*(r_1)\,R_2^*(r_2)\,R_2(r_1)\,R_1(r_2)\;r_{12}^{-1}\;\Omega^*_{L,M,l_1,l_2}(\omega_1, \omega_2)\;\Omega_{L,M,l_2,l_1}(\omega_1, \omega_2)\,. \tag{48}$$

Because each Ω is a sum of products of one-electron functions, the presence of cross terms in the exchange matrix element prevents one from merely invoking the earlier result that the exchange matrix element

$$\langle f(\mathbf{r}_1)g(\mathbf{r}_2)|r_{12}^{-1}|g(\mathbf{r}_1)f(\mathbf{r}_2)\rangle$$

is strictly positive. However, a detailed analysis presented in the article by Kutzelnigg and me [58], involving a partial wave expansion for r_{12}^{-1}, the evaluation of the angular integrals in terms of 3-j and 6-j symbols, the discovery by Dr. Volker Termath of an unusual sum rule in Lindner's extensive book on angular momentum theory [70], and the invocation of Racah's little-known proof that for any pair of radial functions R_1 and R_2 the radial exchange integrals

$$G^{(n)} = \int dr_1\, dr_2\, r_1^2\, r_2^2\, R_1^*(1)\, R_2^*(2)\, R_2(1)\, R_1(2)\, \frac{r_<^n}{r_>^{n+1}}\,. \tag{49}$$

after division by $(2n+1)$ form a monotonically non-increasing positive sequence [64], indicates that the integral in eq. (48) is always positive, and hence that the sign of the exchange term is determined solely by the same factor $(-1)^{S+L+l_1+l_2}$ which determines whether the wavefunction and also its first derivative vanish at $r_{12}=0$. Hence there is a very close *association* of the width of the 'symmetry hole at $r_{12}=0$ with the ordering of singlet and triplet energy levels, though neither is the *cause* of the other.

With this explanation of the Alternating Rule as a consequence of the same factor of $(-1)^{S+L+l_1+l_2}$ which determines the width of the 'symmetry hole' at $r_{12}=0$, one can explain why Hund's First Rule works so well for determining the symmetry of the ground state of an atom. The key idea is to examine the symmetry of each pair of electrons in the open shells of a configuration.

To begin, since unnatural parity two-electron states cannot have either $l_1=0$ or $l_2=0$, exceptions to Hund's first rule can occur only in configurations with both $l_1 \geq 1$ and $l_2 \geq 1$. We therefore confine our attention to such configurations.

Consider first the coupling of two equivalent orbitals in a configuration $(nl)^2$, with $l \geq 1$. Coupling them gives rise to a manifold of states with total L ranging from $l-l=0$ to $l+l=2l$. The extremal values, and those in between in steps of 2, are natural parity singlet states, while those in between are unnatural parity triplet states. Because the orbitals are equivalent, for any allowed value of L antisymmetrization constrains the spin function to be *either* a singlet *or* a triplet, but not both. Hence in this case, one never encounters a situation where an unnatural parity singlet lies below an unnatural parity triplet, because there simply are no unnatural parity singlet states of two-electron systems formed from equivalent spatial orbitals.

Next, consider the coupling of two inequivalent orbitals in a configuration $(n_1l_1)(n_2l_2)$, with $l_1 \geq 1$ and $l_2 \geq 1$. Coupling them gives rise to a manifold of states with total L ranging from $|l_1 - l_2|$ to $l_1 + l_2$. Since the orbitals are inequivalent, for each allowed value of L one has *both* singlet and triplet spin states. Hence in such cases there will be unnatural parity singlet states, which can be expected to have the lowest energy in the manifold. Hence there are indeed very many (in fact, infinitely many) exceptions to the naive generalization of Hund's First Rule to excited states of atoms. However, there are very few exceptions involving ground states of atoms, since very few atoms have a ground state configuration in which there are two or more open shells, each with $l \geq 1$.

The lightest element whose ground-state configuration involves two partly filled shells, each with $l \geq 1$, is cerium ($Z = 58$), whose electronic configuration is

$$(1s)^2(2s)^2(2p)^6(3s)^2(3p)^6(4s)^2(3d)^{10}(4p)^6(5s)^2(4d)^{10}(5p)^6(6s)^2(4f)(5d).$$

The $4f$ and the $5d$ electrons can couple to produce two-electron singlet and triplet states with L running from 1 to 5. The quartic symmetry hole in the two unnatural parity singlet states with $L = 2$ and $L = 4$ indicates that one of them should be the ground state, and Hund's second rule suggests that it should be the one with $L = 4$, the $^1\mathrm{G}^o$ state, and indeed it is [71]. Thus this rare violator of Hund's First Rule for a ground state turns out to be perfectly explicable in terms of the Alternating Rule, which as we have seen is associated with a quartic 'symmetry hole' in a two-electron singlet state of unnatural parity.

After cerium, the next case of an atom whose ground state has two incompletely filled shells, each with $l \geq 1$, is gadolinium ($Z = 64$), whose valence configuration is $(4f)^7(5d)$ and whose ground state has maximal spin with the symmetry $^9\mathrm{D}$, resulting from the coupling of the $5d$ orbital to the half-filled $4f$ shell of configuration $^8\mathrm{S}$. Complicated reasons of symmetry prevent the occurrence of a violation of Hund's first rule in this case.

After gadolinium, the next cases of atoms whose ground states have incompletely filled shells each with $l \geq 1$ occur in the actinide series: protactinium ($Z = 91$), uranium ($Z = 92$), neptunium ($Z = 93$), and curium ($Z = 96$). For such heavy elements, in which relativistic effects become important, one should really be doing $j - j$ rather than $L - S$ coupling, so the fundamental premises upon which Hund's rules and the Alternating Rule are based are no longer valid.

One therefore sees that in the first several rows of the periodic table there is only a single violator, cerium, of Hund's First Rule, and this sole violation is readily explicable in terms of the more general Alternating Rule.

3.2. CONCLUSIONS

To conclude this section, I list the key points in this discussion of Hund's First Rule and the Alternating Rule.

(i) Hund's First Rule almost always holds for determining the ground state of an atom.

(ii) There are numerous exceptions to the naive generalisation of Hund's First Rule to excited states of atoms, whereas the Alternating Rule holds quite well for atomic excited states, with only relatively few exceptions caused by strong configuration mixing.

(iii) The Alternating Rule is a consequence of a factor of $(-1)^{S+L+l_1+l_2}$ in the construction of properly antisymmetrised two-electron functions.

(iv) The same factor of $(-1)^{S+L+l_1+l_2}$ determines how rapidly a two-electron function vanishes at the electron-electron coalescence ($r_{12} = 0$).

(v) This common factor of $(-1)^{S+L+l_1+l_2}$ causes a close *association* of the ordering of singlet and triplet energy levels with the width of the 'symmetry holes' in the two-electron functions. Neither phenomenon, however, is caused by the other, and past attempts to 'explain' the ordering of singlet and triplet energy levels as a *consequence* of the behavior of the two-electron function near the electron-electron coalescence are seen to have been misguided.

4. Summary

In summary, the cusp conditions derived by Kato, and their generalisations to higher-order two-particle singularities, as well as the three- and multi-particle singularities described by the Fock expansion, are of great importance in the highly precise calculation of the energies and other properties of atoms and molecules. Approaches which explicitly take into account these singularities can employ many fewer basis functions, and usually generate much more accurate results as well. Moreover, the study of the types of two-electron cusps which occur in excited states of the helium atom has led to a much more profound understanding of the Alternating Rule, a generalization to excited states of Hund's First Rule, for the ordering of singlet and triplet energy levels arising from the same configuration.

5. Acknowledgments

It is a pleasure to thank Constantinos Tsipis, V.S. Popov, Dudley Herschbach, and John Avery for organising such an interesting workshop in such an enjoyable location. I should also like to thank Constantinos Tsipis, Aigle Stiakaki, and Michael Sigalas for their marvellous hospitality while I was finishing this contribution at the U. of Thessalonike. I am further grateful to Dudley Herschbach and to the many members of his research group for the hospitality and support they have extended to me over the past several years. I am also indebted to Cyrus Umrigar and Michael P. Teter, electronic structure experts at the Cornell Theory Center and the Laboratory of Atomic and Solid State Physics of Cornell University, for arranging partial sabbatical support for me in the spring of 1995 on extremely short notice, and to George Cicala and Sally Bould, grievance officers of the U. of Delaware chapter of the American Association of University Professors, without whose selfless endeavors on my behalf this sabbatical would not have been granted.

I should like to thank in approximately chronological order my collaborators in the work described here: Bruno Klahn, Robert N. Hill, Barton D. Huxtable, David E. Freund, James Nicholson, Jonathan D. Baker, Susan E. Haywood, and Werner Kutzelnigg. This work has been supported in part by my NSF grant PHY-9215442.

References

1. Schwartz, C. (1963) Estimating rates of convergence of variational calculations, *Meth. Comp. Phys.* **2**, 241-266.
2. Klahn, B. and Morgan III, J.D. (1984) Rates of convergence of variational calculations and expectation values, *J. Chem. Phys.* **81**, 410-433.
3. Hill, R.N. (1985) Rates of convergence and error estimation formulas for the Rayleigh-Ritz variational method, *J. Chem. Phys.* **83**, 1173-1196.
4. Carroll, D.P., Silverstone, H.J., and Metzger, R.M. (1979) Piecewise polynomial configuration interaction natural orbital study of $1s^2$ helium, *J. Chem. Phys.* **71**, 4142-4163.
5. Morgan III, J.D. (1989) The analytic structure of atomic and molecular wavefunctions and its impact on the rates of convergence of variational calculations, in M. Defranceschi and J. Delhalle (eds.), *Numerical Determination of the Electronic Structure of Atoms, Diatomic and Polyatomic Molecules*, NATO-ASI Series C **271**, Kluwer Academic Publishers, Dordrecht, pp. 49-84.
6. John, F. (1950) The fundamental solution of linear elliptic equations with analytic coefficients, *Commun. Pure Applied Math.* **3**, 273-304.
7. Kato, T. (1957) On the eigenfunctions of many-particle systems in quantum mechanics, *Commun. Pure Applied Math.* **10**, 151-177.
8. Hoffmann-Ostenhof, M., Hoffmann-Ostenhof, T., and Stremnitzer, H. (1992) Electronic wave functions near coalescence points, *Phys. Rev. Lett.* **68**, 3857-3860.
9. Fock, V.A. (1954) Ob uravnenii Shredingera dlya atoma geliya, *Izv. Akad. Nauk SSSR, Ser. Fiz.* **18**, 161-172; English translation: Fock, V.A. (1958) On the Schrödinger equation of the helium atom, *D. Kngl. Norske Videnskab. Selsk. Forh.* **31**,

138-144, 145-152.
10. Ermolaev, A.M. (1958) Razlozheniye volnovikh funktsii mnogoelektronnikh sistem v ryadi Foka, *Vestnik Leningrad. Univ.* **22**, 48-64.
11. Demkov, Yu.N. and Ermolaev, A.M. (1959) Razlozheniye Foka dlya volnovikh funktsii sistemi zaryazhennikh chastits, *Zh. Eksp. Teor. Fiz.* **36**, 896-899; English translation: Fock expansion for the wave functions of a system of charged particles, *Sov. Phys. JETP* **36**, 633-635.
12. Ermolaev, A.M. (1961) K raschetu nerelyativistskikh S-sostoyanii dvukh-elektronnikh atomov i ionov, *Vestnik Leningrad. Univ.* **16**, 19-33.
13. Ermolaev, A.M. and Sochilin, G.B. (1964) Osnovnoye sostoyaniye dvukhelektronnikh atomov i ionov, *Dokl. Akad. Nauk SSSR* **155**, 1050-1053; English translation: The ground state of two-electron atoms and ions, *Sov. Phys. Doklady* **9**, 292-295.
14. Sochilin, G.B.(1969) On the behaviour of the wave function for the 2^3S state of the two-electron atom in the region where both electron-nuclear distances are small, *Intern. J. Quantum Chem.* **3**, 297-301.
15. Macek, J.H. (1967) Application of the Fock expansion to doubly excited states of the helium atom, *Phys. Rev.* **160**, 170-174.
16. Leray, J. (1982) Application à l'équation de Schrödinger atomique d'une extension du théorème de Fuchs, in *Actualités mathématiques: Actes du 6ème Congrès du Groupement des Mathématiciens d'Expression Latine*, Gauthier-Villars, Paris, pp. 179-189.
17. Leray, J. (1983) La fonction de Green de la sphère S^n et l'application effective à l'équation de Schrödinger atomique d'une extension du théorème de Fuchs, in D. Greco (ed.), *Methods of Functional Analysis and Theory of Elliptic Operators*, Liguori Editore, Naples, pp. 165-177.
18. Leray, J. (1983) Application to the Schrödinger Atomic Equation of an Extension of Fuchs' Theorem, in C.P. Bruter, A. Aragnol, and A. Lichnerowicz, *Bifurcation Theory, Mechanics and Physics*, Reidel, Dordrecht, pp. 99-108.
19. Leray, J. (1984) Sur les solutions de l'équation de Schrödinger atomique et le cas particulier de deux electrons, in P.G. Ciarlet and M. Roseau (eds.), *Trends and Applications of Pure Mathematics to Mechanics*, Lecture Notes in Physics **195**, Springer Verlag, Berlin, pp. 235-247.
20. Morgan III, J.D. (1986) Convergence properties of Fock's expansion for S-state eigenfunctions of the helium atom, *Theoret. Chim. Acta* **69**, 181-223.
21. Abbott, P.C. and Maslen, E.N. (1987) Coordinate systems and analytic expansions for three-body atomic wavefunctions: I. Partial summation of the Fock expansion in hyperspherical coordinates, *J. Phys. A* **20**, 2043-2075.
22. Gottschalk, J.E., Abbott, P.C., and Maslen, E.N. (1987) Coordinate systems and analytic expansions for three-body atomic wavefunctions: II. Closed form wavefunction to second order in r, *J. Phys. A* **20**, 2077-2104.
23. Gottschalk, J.E. and Maslen, E.N. (1987) Coordinate systems and analytic expansions for three-body atomic wavefunctions: III. Derivative continuity via solutions to Laplace's equation, *J. Phys. A* **20**, 2781-2803.
24. McIsaac, K. and Maslen, E.N. (1987) Exact wave functions for few-particle systems: The choice of expansion for Coulomb potentials, *Int. J. Quantum Chem.* **31**, 361-368.
25. Morgan III, J.D. (1993) The dimensional dependence of rates of convergence of Rayleigh-Ritz variational calculations on atoms and molecules, in D.R. Herschbach, J. Avery, and O. Goscinski (eds.), *Dimensional Scaling in Chemical Physics*, Kluwer Academic Publishers, Dordrecht, pp. 336-358.
26. Freund, D.E., Huxtable, B.D., and Morgan III, J.D. (1984) Variational calculations on the helium isoelectronic sequence, *Phys. Rev. A* **29**, 980-982.
27. Kutzelnigg, W. and Morgan III, J.D. (1992) Rates of convergence of the partial-wave expansions of atomic correlation energies, *J. Chem. Phys.* **96**, 4484-4508; **97**, 8821 (erratum).
28. Hylleraas, E.A. (1929) Neue Berechnungen der Energie des Heliums im Grundzu-

stande, sowie des tiefsten Terms von Ortho-Helium, *Z. f. Phys.* **54**, 347-366.
29. Pekeris, C.L. (1958) Ground state of two-electron systems, *Phys. Rev.* **112**, 1649-1658.
30. Pekeris, C.L. (1959) 1^1S and 2^1S states of helium, *Phys. Rev.* **115**, 1216-1221.
31. Cox, H., Smith, S.J., and Sutcliffe, B.T. (1994) Some calculations on the ground and lowest triplet states of helium in the fixed-nucleus approximation, *Phys. Rev. A* **49**, 4520-4532.
32. Frankowski, K. and Pekeris, C.L.(1966) Logarithmic terms in the wave functions of the ground state of two-electron atoms, *Phys. Rev.* **146**, 46-49; **150**, 366 (erratum).
33. Frankowski, K. (1967) Logarithmic terms in the wavefunctions of the 2^1S and 2^3S states of two-electron atoms, *Phys. Rev.* **160**, 1-3.
34. Baker, J.D., Freund, D.E., Hill, R.N., and Morgan III, J.D. (1990) Radius of convergence and analytic behavior of the $1/Z$ expansion, *Phys. Rev. A* **41**, 1247-1273.
35. Kleindienst, H. and Emrich, R. (1990) The atomic three-body problem. An accurate lower bo[u]nd calculation using wave functions with logarithmic terms, *Int. J. Quantum Chem.* **37**, 257-269.
36. Baker, J.D., Hill, R.N., and Morgan III, J.D. (1989) High precision calculations of helium atom energy levels, in W. Johnson, P.J. Mohr, and J. Sucher (eds.), *AIP Conference Proceedings* **189**, *Relativistic, Quantum Electrodynamic, and Weak Interaction Effects in Atoms*, American Institute of Physics, New York, pp. 123-145.
37. Drake, G.W.F. (1993) High Precision Calculations for the Rydberg States of Helium, in F.S. Levin and D. Micha (eds.), *Long Range Casimir Forces: Theory and Recent Experiments in Atomic Systems*, Plenum, New York, pp. 107-217, and numerous references therein.
38. Pachucki, K., Weitz, K.M., and Hänsch, T.W. (1994) Theory of the hydrogen-deuterium isotope shift, *Phys. Rev. A* **49**, 2255-2259.
39. Shiner, D., Dixson, R., and Vedentham, V. (1995) Three-nucleon charge radius: A precise laser determination using ^{3}He, *Phys. Rev. Lett.* **74**, 3553-3556.
40. Riis, E., Sinclair, E.G., Poulsen, O. Drake, G.W.F., Rowley, W.R.C., and Levick, A.P. (1994) Lamb shifts and hyperfine structure in $^6Li^+$ and $^7Li^+$, *Phys. Rev. A* **49**, 207-220.
41. Umrigar, C.J. and Gonze, X. (1993) Comparison of Approximate and Exact Density Functionals: A Quantum Monte Carlo Study, in D.A. Browne *et al.* (eds.), *High Performance Computing and its Application to the Physical Sciences, Proceeedings of the Mardi Gras '93 Conference* World Scientific, Singapore, pp. 43-59.
42. Umrigar, C.J. and Gonze, X. (1994) Accurate exchange-correlation potentials and total-energy components for the helium isoelectronic series, *Phys. Rev. A* **50**, 3827-3837.
43. Haywood, S.E., Nicholson, J.D., and Morgan III, J.D. (1986) U. of Delaware preprint, unpublished.
44. Wilson, K.G. (1990) Ohio State University preprint, unpublished.
45. Kutzelnigg, W. (1994) Theory of the expansion of wave functions in a Gaussian basis, *Int. J. Quantum Chem.* **51**, 447-463.
46. Herschbach, D.R. (1986) Dimensional interpolation for two-electron atoms, *J. Chem. Phys.* **84**, 838-851.
47. Kutzelnigg, W. (1985) The r_{12}-dependent terms in the wave function as closed sums of partial wave amplitudes for large l, *Theor. Chim. Acta* **68**, 445-469.
48. Klopper, W. and Kutzelnigg, W. (1987), Moeller-Plesset calculations taking care of the correlation cusp, *Chem. Phys. Lett.* **134**, 17-22.
49. Klopper, W. and Kutzelnigg, W. (1990), MP2-R12 calculations on the relative stability of carbocations, *J. Phys. Chem.* **94**, 5625-5630.
50. Kutzelnigg, W. and Klopper, W. (1991) Wave functions with terms linear in the interelectronic coordinates to take care of the correlation cusp. I. General theory, *J. Chem. Phys.* **94**, 1985-2001.
51. Termath, V., Klopper, W., and Kutzelnigg, W. (1991) Wave functions with terms

linear in the interelectronic coordinates to take care of the correlation cusp. II. Second-order Moeller-Plesset (MP2-R12) calculations on closed-shell atoms, *J. Chem. Phys.* **94**, 2002-2019.

52. Klopper, W. and Kutzelnigg, W. (1991) Wave functions with terms linear in the interelectronic coordinates to take care of the correlation cusp. III. Second-order Moeller-Plesset (MP2-R12) calculations on molecules of first row atoms, *J. Chem. Phys.* **94**, 2020-2030.
53. Klopper, W., Roehse, R., and Kutzelnigg, W. (1991) CID and CEPA calculations with linear r_{12} terms, *Chem. Phys. Lett.* **178**, 455-461.
54. Roehse, R., Klopper, W., and Kutzelnigg, W. (1993) Configuration interaction calculations with terms linear in the interelectronic coordinate for the ground state of the hydrogen triatomic monopositive ion. A benchmark study, *J. Chem. Phys.* **99**, 8830-8839.
55. Roehse, R., Kutzelnigg, W., Jacquet, R., and Klopper, W. (1994) Potential energy surface of the H_3^+ ground state in the neighborhood of the minimum with microhartree accuracy and vibrational frequencies derived from it, *J. Chem. Phys.* **101**, 2244-2249.
56. Noga, J. and Kutzelnigg, W. (1994) Coupled cluster theory that takes care of the correlation cusp by inclusion of linear terms in the interelectronic coordinates, *J. Chem. Phys.* **101**, 7738-7762.
57. Noga, J., Tunega, D., Klopper, W., and Kutzelnigg, W. (1995) The performance of the explicitly correlated coupled cluster method. I. The four-electron systems Be, Li^-, and LiH, *J. Chem. Phys.*, **103**, 309-320.
58. Morgan III, J.D. and Kutzelnigg, W. (1993) Hund's Rules, The Alternating Rule, and Symmetry Holes, *J. Phys. Chem.* **97**, 2425-2434.
59. Hund, F. (1925) Zur Deutung verwicklter Spektren, insbesondere der Elemente Scandium bis Nickel, *Z. f. Phys.* **33**, 345-371.
60. Hund, F. (1927) *Linienspektren und periodisches System der Elemente*, Springer Verlag, Berlin, p. 124.
61. Russell, H.N. and Meggers, W.F. (1927) An analysis of the arc and spark spectra of scandium (Sc I and Sc II), *Sci. Papers Bur. Stand.*, **22**, 329-373, particularly 364.
62. Davidson, E.R. (1965) Single-configuration calculations on excited states of helium. II, *J. Chem. Phys.* **42**, 4199-4200.
63. Colpa, J.P. (1974) Hund's rules reformulated as inequalities and related inequalities for exchange integrals, Slater-Condon parameters, energy differences, and correlation energies, *Mol. Phys.* **28**, 581-585.
64. Racah, G. (1943) Theory of complex spectra. II., *Phys. Rev.* **62**, 438-462, particularly 460-1.
65. Roothaan, C.C.J. (1951) New developments in molecular orbital theory, *Rev. Mod. Phys.* **23**, 69-89, particularly 87.
66. Slater, J.C. (1960) *Quantum Theory of Atomic Structure*, vol. 1, McGraw-Hill, New York, pp. 486-7.
67. Lieb, E.H. and Simon, B. (1977) The Hartree-Fock theory for Coulomb systems, *Commun. Math. Phys.* **53**, 185-194, particularly 191.
68. Warner, J.W., Bartell, L.S., and Blinder, S.M. (1980) Electron correlation and Hund's rule, *Int. J. Quantum Chem.* **18**, 921-925.
69. Warner, J.W. and Berry, R.S. (1985) Hund's rule, *Nature* **313**, 160.
70. Lindner, A. (1984) *Drehimpulse in der Quantenmechanik*, Teubner, Stuttgart, p. 55, first formula in §4.12, with $\hat{j}^2 = 2j + 1$.
71. Martin, W. C., Zalubas, R., and Hagan, L. (1978) *Atomic Energy Levels – The Rare-Earth Elements*, U.S. Govt., Washington, p. 209.

LOCAL-SCALING MANY-ELECTRON DENSITY FUNCTIONAL THEORY

EUGENE S. KRYACHKO
Bogoliubov Institute for Theoretical Physics, Kiev, Ukraine 252143 and Instituto Estructura de la Materia, C. S. I. C., 28006 Madrid, Spain

"A very good thing with the Greeks
is that they do not "refer to the Greeks";
they refer to themselves."
C. F. von Weizsäcker, "Physics and Philosophy"[1]

1. Quidditas of Local-Scaling Density Functional Theory

In the 70 years elapsed since the Thomas [1] and Fermi [2] seminal papers, there has been made the impressive progress in the density functional theory. These 70 years of density functional theory may be roughly divided in three periods: the post-Thomas-Fermi period [3], the one-electron-density one whose start nearly the 40 years ago was initiated by March and Young [4] who first coined the term "density matrix functional theory", and, finally, the Hohenberg-Kohn-Sham period that began at the late seventies after more than decade since the Hohenberg-Kohn theorem [5a] and the Kohn-Sham self-consistent field approach [5b] (see [6-10]) appeared. It was in the middle of the eighties when we thought that in order to make a substantial progress in the mathematical density functional theory, it seemed essential to approach it with a rather new standard of rigor, in some sense, with a relatively fresh mode of consciousness, and with mathematical methods of a more recent vintage than the ones generally employed in this area at that time, on one hand, and, on the other one, that nevertheless reminisce some old ways. The key problem in such an endeavor has been to reformulate the traditional stationary quantum variational principle by means of a

[1] In The Physicist's Conception of Nature. J. Mehra, Ed. D. Reidel, Dordrecht, 1973. p. 737.

C. A. Tsipis et al. (eds.), New Methods in Quantum Theory, 339–358.

meaningful mathematical language as a *bona fide* variational principle that relies on one-electron density as the primary variational variable. Nowadays, one may surely conclude that the approach developed during nearly the last ten years on the basis of local-scaling transformations [9,11-44] lights eventually many of questions in density functional theory which hitherto have been obscured by insufficiency of rigor in this area and is in fact the only one approach that is rigorously derived from quantum variational principle and that shows the concrete computational and well grounded strategies leading, for instance, to the density functional concept of self-consistent field, to the treatment of excited states, and to the concept of fractional occupation numbers.

2. One-Electron Densities and Local Scaling

2.1. MATHEMATICAL PRELIMINARIES

Let us consider an atomic system, for which we pick a spherical coordinate chart with the origin at the nucleus. Any one-electron density can be represented by one-parametric curve when a direction $\mathbf{e_r} \equiv \mathbf{r}/r \equiv \mathbf{e}(\Omega)$ is specified in $\Re$. $\mathbf{e_r}$ is a unit vector defined by a couple of spherical angles $\Omega = (\theta, \phi)$. For a given $\mathbf{e_r}(\Omega)$, let us consider the curves $\tilde{\rho}_1(r)$ and $\tilde{\rho}_2(r)$ resulting from two densities $\rho_1(\mathbf{r})$ and $\rho_2(\mathbf{r})$ such that $\tilde{\rho}_1(r) = \rho_1(r, \mathbf{e_r})$ and $\tilde{\rho}_2(r) = \rho_1(r, \mathbf{e_r})$. These curves are continuously differentiable functions of $r = \mid \mathbf{r} \mid$. That is why they are homotopically equivalent, or in other words, there exists a deformation that maps $\tilde{\rho}_1(r)$ into $\tilde{\rho}_2(r)$. It is written formally as $\tilde{\rho}_2(r) \sim \tilde{\rho}_1(f(r, \mathbf{e_r}))$ where f stands for this deformation and a wavy sign indicates that something to demand the normalization of one-electron densities is missing. The required quantity is precisely the Jacobian of deformation f that acquires the following particular realization: $J(\mathbf{f}(\mathbf{r}); \mathbf{r}) = \mathbf{r}\nabla f^3/3r^3$, with $\mathbf{f}(\mathbf{r}) \equiv f(\mathbf{r})\mathbf{e_r} \equiv f(r, \mathbf{e_r})\mathbf{e_r}$. Now, generalization over all directions in $\Re$ results [11-15,9]

$$\rho_2(\mathbf{r}) = \frac{\mathbf{1}}{\mathbf{3r^3}}\mathbf{r} \cdot \nabla \mathbf{f^3}(\mathbf{r})\rho_1(\mathbf{f}(\mathbf{r})). \tag{1}$$

(1) looks as a first-order nonlinear differential equation for deformation $f(\mathbf{r})$ for given densities ρ_1 and ρ_2. Its solution *does* exist and it is *unique* that enables to define such deformation for any pair of well-behaved densities. We name it a local-scaling transformation, or simply, local scaling [11-15]. In spherical coordinates, along a chosen direction described by $\Omega = (\theta_o, \phi_o)$, this equation can be rewritten merely as

$$\frac{df(r, \theta_o, \phi_o)}{dr} = \frac{r^2 \rho_2(r, \theta_o, \phi_o)}{f^2 \rho_1(f(r, \theta_o, \phi_o), \theta_o, \phi_o)}. \tag{2}$$

In integrated form, this transformation function is

$$f(r,\theta_o,\phi_o) = \left[3\int_o^r \frac{dr\, r^2\rho_2(r,\theta_o,\phi_o)}{\rho_1(f(r,\theta_o,\phi_o)),\theta_o,\phi_o)}\right]^{1/3}. \tag{3}$$

The complete transformation $f(r,\theta,\phi)$ is obtained by solving the above equation over all angular directions. One point deserves to be noticed. If $1/3$ in rhs of (3) is the "building stone" for 3D-space $\Re^3$, $1/n$ does that for $\Re^n$ [17].

2.2. EXAMPLE α: HELIUM APPROXIMATE GROUND-STATE DENSITIES

Let us give one example of local scaling [19]. Pick two densities ρ_E and ρ_K in 3D position space,

$$\begin{aligned} \rho_E(\mathbf{r}) &= (1+S^2)^{-1}\pi^{-1}\{\alpha^3 exp(-2\alpha r) + \beta^3 exp(-2\beta r) \\ &\quad +2S(\alpha\beta)^{3/2} exp[-(\alpha+\beta)r]\}, \\ \rho_K(\mathbf{r}) &= 2\zeta^3\pi^{-1} exp(-2\zeta r) \end{aligned} \tag{4}$$

and their momentum space analogues,

$$\begin{aligned} \gamma_E(\mathbf{p}) &= (1+S^2)^{-1}2^3\pi^{-2}[\alpha^5(p^2+\alpha^2)^{-4} + \beta^2(p^2+\beta^2)^{-4} \\ &\quad +2S(\alpha\beta)^{5/2}(p^2+\alpha^2)^{-2}(p^2+\beta^2)^{-2}], \\ \gamma_K(\mathbf{p}) &= 2^4\zeta^5\pi^{-2}(p^2+\zeta^2)^{-4}. \end{aligned} \tag{5}$$

Parameters α, β, and ζ may be taken at the following values:

$$\alpha = 2.183171, \beta = 1.188531, \zeta = 27/16 = 1.6875, \tag{6}$$

that will be clarified later on. One sees that above densities are spherically symmetric, so that a radial local scaling $f = f(r)$ in the position space and $g = g(p)$ in the momentum space absolutely define the required deformations. These local scalings obey the following integral equations,

$$\int_0^r dx x^2\rho_K(x) = \int_0^{f(r)} dy y^2\rho_E(y), \int_0^p dx x^2\gamma_E(x) = \int_0^{g(p)} dy y^2\gamma_K(y). \tag{7}$$

(7) are equivalent to Eq.(2) and can be analytically handled for the densities listed in (4-5). The resultant nonlinear algebraic equations are then solved numerically and local scalings $f(r)$ and $g(p)$ are plotted in Figure 1 of [19].

3. N-Electron Wave Functions and Local Scaling

3.1. DEFINITION

Concept of local scaling is trivially generalized on Hilbert space $\mathcal{H}_N$ of square integrable N-electron wave functions [11,15,24]. For this purpose, let us pick a "reference" wave function $\Psi_o(\{\mathbf{r}_i, \sigma_i\}_{i=1}^{i=N})$ where σ_i is spin of the ith electron. We define a wave function

$$\Psi_f(\{\mathbf{r}_i, \sigma_i\}) \equiv [\prod_{i=1}^{N} J(f(\mathbf{r_i})]^{1/2} \Psi_o(\{\mathbf{f}(\mathbf{r}_i), \sigma_i\}) \tag{8}$$

as locally scaled "reference" one. Formally, $\Psi_f \equiv \mathcal{F}\Psi_o$. Arbitrariness in choosing Ψ_o ensures the validity of the definition (8) on $\mathcal{H}_\mathcal{N}$. It is obvious that a local scaling that "ties" a given couple of N-electron wave functions matches unambiguously the local scaling transforming the corresponding one-electron densities. However, although any pair of densities are locally scaled, this property no longer holds for an arbitrary pair of N-electron wave functions. To show the latter, we consider the following 2-electron wave functions which, on one side, are rather accurate approximations of the exact and Hartree-Fock ground states of the helium atom, and, on the other side, yield the one-electron densities (4), respectively,

Eckart wave function[45] :

$$\begin{aligned} \Psi_E(\mathbf{r}_1, \mathbf{r}_2) &= (2 + 2S^2)^{-1/2}(\alpha\beta)^{3/2}\pi^{-1}[exp(-\alpha r_1 - \beta r_2) \\ &\quad + exp(-\beta r_1 - \alpha r_2)], \end{aligned}$$

Kellner wave function[46] :

$$\Psi_K(\mathbf{r}_1, \mathbf{r}_2) = \zeta^3\pi^{-1}exp[-\zeta(r_1 + r_2)]. \tag{9}$$

Their momentum-space representatives take on the following appearance,

$$\begin{aligned} \Phi_E(\mathbf{p}_1, \mathbf{p}_2) &= (2 + 2S^2)^{-1/2}2^3(\alpha\beta)^{5/2}\pi^{-2} \\ &\quad \times[(p_1^2 + \alpha^2)^{-2}(p_2^2 + \beta^2)^{-2} + (p_1^2 + \beta^2)^{-2}(p_2^2 + \alpha^2)^{-2}] \\ \Phi_K(\mathbf{p}_1, \mathbf{p}_2) &= 2^3\zeta^5\pi^{-2}(p_1^2 + \zeta^2)^{-2}(p_2^2 + \zeta^2)^{-2}. \end{aligned} \tag{10}$$

It seems being worthwhile to make some notices. The first one. Because local scaling acts on spatial or momentum variables, the singlet spin function has been dropped in Eqs.(9) and (10). Second notice. Kellner wave function is easily arranged into a Slater determinant form together with the singlet spin function. However, the Eckart one does not admit such a representation. Glancing at (8), one may quickly arrive at rather important conclusion that if a "reference" wave function is a Slater determinant, Ψ_f does too for arbitrary local scaling. That is why there does not exist such a local scaling that maps Ψ_K into Ψ_E. This implies that the total N-electron Hilbert space $\mathcal{H}_\mathcal{N}$ is partitioned with respect to local scalings forming a group into disjoint classes named hereafter "orbits". Equivalently, the union of all orbits exhausts Hilbert space,

$$\mathcal{H}_\mathcal{N} = \bigcup_{\varepsilon} \mathcal{O}_\varepsilon. \tag{11}$$

Table I. One-Electron Densities $\rho_E(\mathbf{r})$ and $\gamma_E(\mathbf{p})$ (after Koga et al. [19]).

$\mid \mathbf{p} \mid, \mid \mathbf{r} \mid$	$\gamma_E(\mathbf{p})$ Value	$\gamma_E(\mathbf{p})$ Percent error	$\rho_E(\mathbf{r})$ Value	$\rho_E(\mathbf{r})$ Percent error
0.000	0.25760	0.91	1.75431	0.18
0.050	0.25614	0.91	1.45513	0.44
0.100	0.25177	0.91	1.20203	0.45
0.202	0.23503	0.90	0.82082	0.46
0.301	0.21090	0.88	0.56796	0.47
0.407	0.18078	0.84	0.38615	0.45
0.497	0.15427	0.80	0.27889	0.43
0.607	0.12374	0.72	0.18843	0.39
0.819	0.07642	0.52	0.08994	0.25
1.000	0.04908	0.28	0.04869	0.08
1.221	0.02828	-0.03	0.02354	-0.19
1.492	0.01457	-0.39	0.01002	-0.57
1.649	0.01001	-0.56	0.00620	-0.81
2.014	0.00433	-0.80	0.00212	-1.34
2.226	0.00273	-0.86	0.00117	-1.60
2.460	0.00166	-0.88	0.00061	-1.82
3.004	0.00057	-0.80	0.00015	-2.06
3.669	0.00017	-0.67	0.00003	-1.90
4.055	0.00009	-0.60	0.00001	-1.66

In (11) each orbit $\mathcal{O}_\varepsilon$ is closed respectively to the group of local-scaling transformations, that is, for any pair of Ψ_1 and Ψ_2 from the orbit $\mathcal{O}_\varepsilon$ there exists such local scaling $f_{1\mapsto 2}$ that $\Psi_2 = \mathcal{F}_{1\mapsto 2}\Psi_1$. In other words, if we pick, for instance, Ψ_1 as a "reference" or generating wave function and act on it by all local scalings, we cover the whole orbit $\mathcal{O}_\varepsilon$. One may immediately realize that there are some orbits in $\mathcal{H}_\mathcal{N}$ with $N \geq 2$ which are composed of Slater determinants and some with more rich wave functions. For example, one may deliver two orbits in $\mathcal{H}_2$, one is the *Eckart* orbit $\mathcal{O}_E$ generated by Ψ_E, and the second is the *Kellner* orbit $\mathcal{O}_K$ made by means of Ψ_K. The similar partition into orbits can be carried out within the momentum space picture. Both pictures will be employed below in parallel.

The third notice shows the energy scale[2] of Ψ_K and Ψ_E. With the optimal values (6) of parameters α, β, and ζ, $E_K = -(27/16)^2 = -2.847656$ and $E_E = -2.875661$. As seen, E_K lies above the Hartree-Fock limit energy $E_{HF} = -2.861680$ for the helium atom [47] whereas E_E does below it.

3.2. EXAMPLE β: KELLNER AND ECKART ORBITS

Pick two orbits $\mathcal{O}_E$ and $\mathcal{O}_K$ in $\mathcal{H}_2$. Using the local scaling $f(r)$ [or $g(p)$] that obeys (7), one scales locally Ψ_K (or Φ_K) to that possesses exactly the Eckart one-electron density (or γ_K) and, hence, incorporates some correlation, say, at the level of one-electron densities, but still belongs to the *Kellner* orbit and for this reason remains within the Hartree-Fock approximation. Equivalently, we are in position to construct the following four functions:

$$\begin{array}{lll} \Psi_K & \mapsto & \Psi_K(\rho_E) \Rightarrow \Phi_K(\rho_E) \\ \Phi_K & \mapsto & \Phi_K(\gamma_E) \Leftarrow \Psi_K(\gamma_E). \end{array} \tag{12}$$

Symbol $\mapsto$ indicates local scaling, whereas $\Rightarrow$ ($\Leftarrow$) the Fourier transform FT (FT^{-1}). Using now $\Psi_K(\gamma_E)$ and $\Phi_K(\rho_E)$, we calculate numerically their one-electron densities, $\rho_{KE}(\mathbf{r})$ and $\gamma_{KE}(\mathbf{p})$, respectively, and compare them in Table I with their prototypes, $\rho_E(\mathbf{r})$ and $\gamma_E(\mathbf{p})$. For given $\rho_{KE}(\mathbf{r})$ and $\gamma_{KE}(\mathbf{p})$, one evaluates their moments, $\langle p^n \rangle_{KE}$ and $\langle r^n \rangle_{KE}$ (see Table II).

Table II. One-Electron Moments $\langle p^n \rangle_{KE}$ and $\langle r^n \rangle_{KE}$ (after Koga et al. [19]).

	$\langle p^n \rangle_{KE}$		$\langle r^n \rangle_{KE}$	
n	Value	Percent error	Value	Percent error
-2	2.21414	0.69	5.98956	0.41
-1	1.09856	0.36	1.69040	0.27
0	1.00000	0.00	1.00000	0.00
1	1.39357	-0.32	0.93446	-0.37
2	2.86094	-0.51	1.22833	-0.77
3	8.92808	-0.55	2.12522	-1.13
4	50.4035	-0.66	4.62179	-1.38

Tables I and II demonstrate that the generated $\rho_{KE}(\mathbf{r})$ and $\gamma_{KE}(\mathbf{p})$ look very much like their prototypes. In the position space, the errors in ρ_{KE}

[2]The atomic system of units is employed throughout.

with respect to ρ_E become relatively large for large r, but still remains 2% at most. For the moments, the agreement is pretty well (relative error $< 2\%$), despite the fact that it deviates with increasing n by modulo. In the momentum space, the absolute value of error of γ_{KE} relative to γ_E reaches its maximum equal to 0.91%, and the relative error for moments lies within $\pm 0.7\%$. Overlaps between the wave functions themselves are

$$\begin{aligned} \langle \Phi_K(\gamma_E) \mid \Phi_E \rangle &= \langle \Psi_K(\gamma_E) \mid \Psi_E \rangle = 0.99768, \\ \langle \Phi_K(\rho_E) \mid \Phi_E \rangle &= \langle \Psi_K(\rho_E) \mid \Psi_E \rangle = 0.99767. \end{aligned} \tag{13}$$

(13) shows how the new wave functions constructed within the Hartree-Fock framework become closer to their prototypes that carry a gradual part of correlation. Evaluation of energy of these new functions yields [19]

$$\begin{aligned} E[\gamma_{KE}] &= E[\Psi_K(\gamma_E)] = E[\Phi_K(\gamma_E)] = -2.86000, \\ E[\rho_{KE}] &= E[\Psi_K(\rho_E)] = E[\Phi_K(\rho_E)] = -2.86003, \end{aligned} \tag{14}$$

that is considerably lower the Kellner energy $E_K = -2.847656$ and close to E_{HF}. Energies in (14) can be improved by relaxing parameters α and β,

$$\begin{aligned} E[\gamma_{KE}] &= -2.86031 \,\mathrm{for} \alpha = 2.14767, \beta = 1.21978, \\ E[\rho_{KE}] &= -2.86039 \,\mathrm{for} \alpha = 2.14987, \beta = 1.22448. \end{aligned} \tag{15}$$

3.3. EXAMPLE γ: LOCAL SCALING OF HARTREE-FOCK ATOMS

Consider the single-zeta approximated wave function Ψ_o for the first- and second-row ground-state atoms [47] as the reference one and construct its spherically averaged density

$$\bar{\rho}_o(r) \equiv \int d\Omega \rho_o(\mathbf{r}). \tag{16}$$

ρ_o is the one-electron density associated with Ψ_o. Let $\bar{\rho}(r)$ be the spherically averaged density built via the rule (16) from the near-Hartree-Fock Clementi-Roetti wave function Ψ [47,23]. One then defines a local scaling $s = s(r)$ that maps $\bar{\rho}_o(r)$ into $\bar{\rho}(r)$ as $\bar{\rho}(r) = J(s(r))\bar{\rho}_o(s(r))$. $s(r)$ and $J(s(r))$ for the neon atom are plotted in Figures 1 and 2 of [23].

Applying local scaling $s(r)$ to Ψ_o results in $\Psi_{\bar{\rho}}$ that is interpreted as a possible parent wave function of the near-Hartree-Fock density. Let us take the Fourier transform $\Phi_{\bar{\rho}}$ of $\Psi_{\bar{\rho}}$ and evaluate the associated one-electron momentum density $\gamma_{\bar{\rho}}(\mathbf{p})$ whose spherical averaging gives $\bar{\gamma}_{\bar{\rho}}(p)$. A possible generalization toward improving the single-electron mapping between the position and momentum densities is to use different local scalings for the electrons with different orbital symmetries. The momentum moments $\langle p^n \rangle_{\bar{\rho}}$ are then evaluated and summarized in Table III.

Table III. Momentum moments $\langle p^n \rangle_{\bar{\rho}} (-2 \leq n \leq 2)$ generated from the spherically averaged position density of the near Hartree-Fock quality. $\langle p^o \rangle = 1$ for all atoms. Values in the parentheses are errors in percents relative to the correct moments (after Koga and Yamamoto [23]).

Atom	$\langle p^{-2} \rangle_{\bar{\rho}}$	$\langle p^{-1} \rangle_{\bar{\rho}}$	$\langle p \rangle_{\bar{\rho}}$	$\langle p^2 \rangle_{\bar{\rho}}$
Li	8.8728 (+0.23)	1.7303 (+0.10)	1.6357(+0.03)	4.9564(+0.03)
Be	6.3259 (+0.05)	1.5794 (-0.01)	1.8596(+0.05)	7.2893(+0.04)
B	3.2541 (+0.05)	1.1959 (0.00)	2.1309(+0.05)	9.8153(+0.04)
C	1.9607 (+0.05)	0.95909(-0.00)	2.4114(+0.05)	12.567 (+0.03)
N	1.3009 (+0.05)	0.79964(0.00)	2.6959(+0.04)	15.549 (+0.04)
O	0.93815(+0.05)	0.69411(0.00)	2.9663(+0.04)	18.709 (+0.03)
F	0.70544(+0.05)	0.61148(0.00)	3.2419(+0.04)	22.099 (+0.04)
Ne	0.54820(+0.04)	0.54558(0.00)	3.5209(+0.03)	25.718 (+0.04)
Na	2.9401 (+0.04)	0.79054(-0.18)	3.7051(+0.06)	29.441 (+0.04)
Mg	3.0822 (+0.08)	0.85730(-0.06)	3.8791(+0.04)	33.282 (+0.04)
Al	2.1541 (+0.56)	0.79538(+0.41)	4.0569(+0.03)	37.239 (+0.08)
Si	1.6004 (+0.25)	0.73040(+0.09)	4.2374(+0.07)	41.296 (+0.08)
P	1.2501 (+0.13)	0.67554(-0.02)	4.4172 (+0.11)	45.474 (+0.10)
S	1.0284 (+0.06)	0.63483(-0.07)	4.5906 (+0.15)	49.749 (+0.12)
Cl	0.85977(+0.01)	0.59867(-0.09)	4.7637 (+0.17)	54.131 (+0.14)
Ar	0.72805(-0.02)	0.56206(-0.11)	4.9365 (+0.18)	58.625 (+0.16)

4. Explicit Form of Energy Density Functional

Let us go back to (11) and the orbit structure of Hilbert space $\mathcal{H}_\mathcal{N}$. These orbits are endowed with the characteristic that there are no two wave functions belonging to the same orbit that have the same density. Therefore, for a given orbit, one defines the energy density functional as the restriction of the expectation value of energy taken at those wave functions that belong to this orbit [9]. It becomes evident that, first, there are as many different energy density functionals as the orbits in $\mathcal{H}_\mathcal{N}$, and second, that each density functional depends implicitly of a wave function that generates a given orbit. To find out the energy density functional in explicit form, we embark with the expression for the expectation value of energy, $E[\Phi_g^{[\alpha]}]$ in terms of the 1-matrix $D_g^{1[\alpha]}(x_1, x_1')$, the 2-matrix $D_g^{2[\alpha]}(x_1, x_2; x_1, x_2)$ and the one-electron density $\rho_g(x)$ coming from the wave function $\Phi_g^{[\alpha]}$ that generates αth orbit [3]. The final expression for the αth-orbit energy density functional becomes as follows [9] [4]:

$$\begin{aligned}
E[\Phi_\rho^{[\alpha]}] &\equiv \mathcal{E}_\alpha[\rho(x); \Phi_g^{[\alpha]}] \\
&= \frac{1}{8}\int d^4x \frac{[\nabla_{\mathbf{r}}\rho(x)]^2}{\rho(x)} + \frac{1}{2}\int d^4x \rho(x) \nabla_{\mathbf{r}}\nabla_{\mathbf{r}'}\widetilde{D}_g^{1[\alpha]}(\mathbf{f}(\mathbf{r}), s; \mathbf{f}(\mathbf{r}'), s')|_{x'=x} \\
&+ \int d^4x \rho(x) v(\mathbf{r}) \\
&+ \frac{1}{2}\int d^4x_1 \int d^4x_2 \frac{\rho(x_1)\rho(x_2)\Big(1 + \mathcal{F}_{XC,g}^{[\alpha]}(\mathbf{f}(\mathbf{r}_1), s_1, \mathbf{f}(\mathbf{r}_2), s_2)\Big)}{|\mathbf{r}_1 - \mathbf{r}_2|}.
\end{aligned} \tag{17}$$

Inspecting Eq.(17), one makes the following couple of observations. Number one. The kinetic energy density functional is dropped into two terms. The former one is precisely the familiar von Weizsäcker term that is local and independent of a choice of orbit whereas the latter term is in general the non-local one and does depend on a given orbit. One shows [11] that due to presence of one-third in (3), this term transforms within LDA to the Thomas-Fermi functional. Number two. It is clear that (17) expresses the energy as a functional of the one-particle density $\rho(x)$ within an orbit $\mathcal{O}_\alpha$. Notice that this functional is of a "dynamical type". It is precisely this dynamical character that allows the functional to modify itself so as to maintain the one-to-one correspondence with $E[\Phi_\rho^{[\alpha]}]$. For this reason, Eq.(17) satisfies the condition of functional N-representability.

4.1. EXAMPLE δ: VARIATIONAL PRELIMINARIES

The key problem we are facing now with applying (17) is how to describe the set of one-electron densities properly. For instance, one may define trial densities as in [20,21], viz.,

$$\rho(r) = \sum_{i=1}^{n} c_i r^{a_i} exp(-b_i r^{d_i}). \tag{18}$$

Certainly, this set, say $\mathcal{A}$, does not cover all trial densities for atoms and their ions, however, it reveals pretty good results. We make further a partition of $\mathcal{A}$ into the following subsets:

[3] We use the nottations: $x \equiv \vec{r}, s$ and $\int d^4x \equiv \sum_s \int d^3\vec{r}$.

[4] Because of limit of space for the definition of quantities employed below see Ref.[9].

Table IV. Energy and $\mathcal{A}_9$-density for the helium atom and its isoelectronic series within the Hartree-Fock and Eckart orbits (after Koga [20]).

	H^-		He	
ρ	$E_{HF}[\rho]$	$E_E[\rho]$	$E_{HF}[\rho]$	$E_E[\rho]$
b_1	0.8106	1.0425	2.8024	2.7120
b_2	1.3857	2.0879	3.5822	3.5358
b_3	2.4380	0.4951	5.2275	4.9848
c	8.2572	10.331	1.4190	1.5012
c'	13.704	0.0913	1.5099	2.1147
α		1.0371		2.1926
β		0.2836		1.1900
Energy	-0.4879262	-0.5134555	-2.8616799	-2.8768084
	$(-0.4879297)^a$	$[-0.5133029]^b$	(-2.8616800)	[-2.8756613]

a) Values in parentheses are the Hartree-Fock limit energy [49].
b) Values in square brackets are the original Eckart energy.

$$
\begin{aligned}
n &= 1 \\
&\mathcal{A}_1 : a_1 = 0, d_1 = 1; \mathcal{A}_2 : d_1 = 1; \\
&\mathcal{A}_3 : a_1 = 0; \mathcal{A}_4 : \text{none of above} \\
n &= 2 \\
&\mathcal{A}_5 : a_1 = a_2 = 0, d_1 = d_2 = 1, c = c_2/c_1; \\
&\mathcal{A}_6 : a_1 = a_2, d_1 = d_2 = 1, c = c_2/c_1; \\
&\mathcal{A}_7 : a_1 = a_2 = 0, d_1 = d_2, c = c_2/c_1; \\
&\mathcal{A}_8 : a_1 = 0, d_1 = d_2 = 1, c = c_2/c_1 \\
n &= 3 \\
&\mathcal{A}_9 : a_1 = a_2 = a_3 = 0, d_1 = d_2 = d_3 = 1, c = c_2/c_1, c' = c_3/c_1; \\
&\mathcal{A}_{10} : a_1 = 0, d_1 = d_2 = d_3 = 1, c = c_2/c_1, c' = c_3/c_1; \\
n &= 4 \\
&\mathcal{A}_{11} : a_1 = 0, d_1 = d_2 = d_3 = d_4 = 1, c(c', c'') = c_2(c_3, c_4)/c_1. \quad (19)
\end{aligned}
$$

As seen from (19), all trial densities that belong to $\mathcal{A}$, excepting those from $\mathcal{A}_{3,4,7}$, are precisely linear combinations of generalized Slater-type functions. This allows to handle the integrals in Eq.(7), determining the required local scaling, analytically by means of the formula [48]

$$\int_0^x dt\ t^b exp(-ct) = c^{-(b+1)}\gamma(b+1, cx) \qquad (20)$$

for $x > 0$ and $b > -1$. Here $\gamma(a, x)$ is incomplete gamma function. For a given set of parameters, the energy density functional $E_{HF}[\rho; \text{parameters}]$ has been evaluated numerically using the Romberg integration method, and the parameters have been optimized to minimize the energy via the Powell method of conjugate directions. The results for the optimal energy and optimal density for the He-isoelectronic series are mounted in Table IV. One sees that the five-parameter densities from the subset $\mathcal{A}_9$ compares well with the near-Hartree-Fock limit.

We show further the results for the ground-state Li and Be atoms within the spin-restricted density functional theory [21] and for the Li within its spin-polarized version [25,26]. In the former case, the single-zeta wave function with two exponents ζ_1 and ζ_2 is chosen as that generates a trial orbit. The optimized values for energy are given in Table V. Comparing these data with the known single-zeta and near-Hartree-Fock values clearly demonstrates the improved quality of our density functional variational method. In spin-polarized density functional calculations for the Li atom [26], we have used again the Clementi-Roetti single-zeta wave function to generate a trial orbit. α- and β-densities are optimized separately, within $\mathcal{A}_{11}$ and $\mathcal{A}_9$ subsets, respectively. The result is the following: the optimal density yields the energy -7.431859 that is lower the generator single-zeta energy by 0.013377.

Table V. Energies for the Li and Be atoms within the spin-restricted density functional theory (after Koga [21]).

Generator	Li	Be
(ζ_1, ζ_2)	(2.69063, 0.63961)	(3.68473, 0.95603)
Energy	−7.418482	−14.566740
$\mathcal{A}_8$	(2.99569, 0.73877)	(3.60369, 0.95831)
Energy	−7.423251	−14.563622
$\mathcal{A}_{10}$	(2.73836, 0.66549)	(3.63293, 0.95395)
Energy	−7.431530	−14.568511
$\mathcal{A}_{11}$	(2.56539, 0.62120)	(3.65424, 0.95693)
Energy	−7.431670	−14.569644
HF energy	−7.4327269	−14.573023

Table VI. Initial and optimum expansion coefficients and orbital parameters for Clementi-Roetti-type $1s$ and $2s$ atomic functions for Be (after Ludeña et al. [43]).

j	n_{0j}	α_{0j}	c_{10j}	c_{20j}
		initial values		
1	1	3.000 000 00	0.594 648 81	−0.347 283 13
2	1	6.000 000 00	0.066 072 09	0.671 661 09
3	2	0.500 000 00	0.066 072 09	0.747 007 62
4	2	1.000 000 00	0.132 144 18	0.270 300 52
5	2	1.500 000 00	0.066 072 09	0.671 661 09
6	2	3.000 000 00	0.264 288 36	−0.139 240 47
		optimal values		
1	1	3.497 869 88	0.482 415 35	0.356 699 44
2	1	5.022 981 27	0.130 846 52	0.140 865 80
3	2	0.464 880 31	−0.450 952 74	0.382 571 10
4	2	0.699 961 86	−0.314 141 57	0.264 305 94
5	2	1.836 322 98	0.022 285 31	0.105 865 82
6	2	2.666 255 80	0.134 683 17	0.127 301 36

5. Concept of Self-Consistent Field

The variational principle within the local-scaling density functional theory states that the search for the exact ground-state wave function must be carried out by a combined intra-orbit and inter-orbit minimization [33,9]. Thus, a practical implementation of this variational principle requires that we develop ways of choosing an orbit, of constructing the energy functional within this selected orbit, of attaining the minimum value for the energy within this orbit, and finally, of jumping from one orbit to another so as to repeat in this new orbit the minimization process. Of course, through this repeated intra-orbit and inter-orbit optimization, one should finally arrive at an orbit yielding the exact ground-state energy.

5.1. EXAMPLE ϵ: DETERMINATION OF HARTREE-FOCK WAVE FUNCTION

In the spirit of the local-scaling density functional theory, we deal below with the calculation of the Hartree-Fock wavefunction and energy by means of a constrained variation at fixed density $\rho(\mathbf{r}) = \rho_{HF}(\mathbf{r})$ as well as by intra-orbit and inter-orbit optimizations. Consider an orbit-generating wavefunction, $\Phi_g^{[\mu]}[\{x_i\}]$ constructed from the atomic orbitals $\{\varphi_{g,k}^{[\mu]}(x_i)\}$. For the

Hartree-Fock Be atom, these orbitals are $1s^{T[\mu]} \equiv \phi^{[\mu]}_{\rho,100}(r) = R^{T[\mu]}_{10}(r)/\sqrt{4\pi}$ and $2s^{T[\mu]} \equiv \phi^{[\mu]}_{\rho,200}(r) = R^{T[\mu]}_{20}(r)/\sqrt{4\pi}$. They depend upon the generating orbitals and the initial and final densities. But, as the initial orbitals depend on the expansion coefficients $\{c_{10j}, c_{20j}\}$ as well as and on the primitive function parameters $\{n_{0j}, \alpha_{0j}\}$, it follows that the energy functional can be expressed as

Table VII. Values of the energy density functional $\mathcal{E}_\mu[\rho; \{c_{n0j}, n_{0j}, \alpha_{0j}\}]$ for a single Slater determinant constructed from Clementi-Roetti-type orbitals (after Ludeña et al. [43]).

density	parameters	$E[\rho; \{c_{n0j}, n_{0_j}, \alpha_{0j}\}]$
initial, CRT[a]	initial	−12.911 265 3
Hartree-Fock, CR[b]	initial	−14.411 571 9
Hartree-Fock, CR[b]	optimized	−14.573 020 8
Hartree-Fock, CR[b]	CR[b]	−14.573 020 1[d]
Limiting Hartree-Fock energy[c]		−14.573 023 13

[a]CRT = Clementi-Roetti-type.
[b]CR = Clementi-Roetti, Ref. [47]
[c,d]See Ref. [43].

$$\mathcal{E}[\rho(x); \Phi^{[\alpha]}_g] = \mathcal{E}_\mu[\rho(x); \{c_{10j}, c_{20j}, n_{0j}, \alpha_{0j}\}]. \tag{21}$$

In Table VI, we list the initial and optimal expansion coefficients and orbital parameters for Clementi-Roetti-type [47] $1s$ and $2s$ functions. The optimization procedure was carried out at fixed density $\rho(r) = \rho_{HF}(r)$ of the Clementi-Roetti type, taking as variational parameters the expansion coefficients and the orbital exponents.

Table VIII. Values of Raffenetti-type orbital parameters (after Ludeña et al. [43]).

	$\mathcal{O}^{[1]}_S$	$\mathcal{O}^{[2]}_S$	Raffenetti[a]
$\alpha^{[i]}$	0.545 322	0.551 862	0.582 434
$\beta^{[i]}$	1.281 137	1.318 860	1.318 837

[a] Reference [50].

One may see in Table VII that the choice of initial parameters is not too adequate, energy-wise, as the energy corresponding to these parameters and to the initial density is −12.911 265 3. Keeping these parameters fixed and performing a local scaling from the initial density to the Hartree-Fock Clementi-Roetti density, the energy is substantially improved to −14.411 571 9. The energy that results when one optimizes the parameters, keeping the density fixed and equal to the Hartree-Fock Clementi-

Roetti density, is $-14.573\,020\,8$. This value is slightly lower than that reported by Clementi and Roetti [47], $-14.573\,020\,1$.

Consider now the optimization - by a combined intra- and inter-orbit variation - of the energy density functional given by Eq.(21). For simplicity, we take these orbitals to be locally-scaled Raffenetti-type functions [50]. Hence, the energy density functional becomes explicitly: $\mathcal{E}[\rho(x); \Phi_g^{[\mu]}] = \mathcal{E}_\mu[\rho(x); \alpha^{[\mu]}, \beta^{[\mu]}]$.Values of $\alpha^{[\mu]}$ and $\beta^{[\mu]}$ used in the present calculations are given in Table VIII. It follows from Table IX that the energy corresponding to the orbit-generating wave function is $-13.894\,697\,4$. The intra-orbit minimization is performed from the density $\rho_g(x)$ to the final density $\rho_K(x)$, Eq.(18).

Table IX. Values of the energy density functional $\mathcal{E}_\mu[\rho(x); \{\alpha^{[\mu]}, \beta^{[\mu]}\}]$ (after Ludeña et al. [43]).

density	$\alpha^{[1]}, \beta^{[1]}$	$\alpha^{[2]}, \beta^{[2]}$
ρ_g	$-13.894\,697\,4$	
ρ_K	$-14.522\,012\,3$	$-14.572\,912\,1$
ρ_L		$-14.573\,003\,9$
ρ_{HF}		$-14.573\,020\,8$

This intra-orbit optimization yields an energy value of $-14.522\,012\,3$. Orbit-jumping is accomplished when we carry out an optimization of the Raffenetti parameters α and β at the fixed density ρ_K. The resulting energy is $-14.572\,912\,1$. The new Raffenetti parameters are listed in Table VIII. We have further performed local-scalings from the optimal density $\rho_K(x)$ to the density $\rho_L(x)$ which comes from an approximate Hartree-Fock wave function whose energy is $-14.572\,993$. The energy functional reaches the value of $-14.573\,003\,9$ for $\rho_L(x)$ and $\alpha^{[2]}, \beta^{[2]}$. A closer approximation to the limiting Hartree-Fock value is attained when the local scaling is carried out to the final density $\rho_{HF}(x)$. In this case, the energy is lowered to $-14.573\,020\,8$, a value that compares quite well with the limiting Hartree-Fock value of $-14.573\,023\,13$.

5.2. EXAMPLE ϵ: CI WAVE FUNCTION FOR BERYLLIUM

Let us consider a CI wave function for *Be*. Let us take as our primitive orbitals a single zeta basis set formed by two s functions ($\chi_{01}(r)$, and $\chi_{02}(r)$), and one p function ($\chi_{11}(r)$). It gives rise to the following configurations: $[1s^2 2s^2]$, $[1s^2 2p^2]$, $[2p^2 2s^2]$, and $[1s2s2p^2]$. The orbital parameters defining three different sets of initial orbitals $1s$, $2s$ and $2p$ are given in Table X. The energy density functional corresponding to an orbit-generating wave function such as the one described above is given by $\mathcal{E}[\rho(x); \Phi_{o,g}^{[\mu]}] = \mathcal{E}_\mu[\rho(x); \{C_{I,\bar{\rho}}^{o[\mu]}\}]$ where in rhs we emphasize the fact that the orbit-generating wave function for an orbit $\mathcal{O}_\mu$ could correspond to the optimal wave function obtained in an orbit-jumping minimization from the previous orbit at fixed density $\tilde{\rho}(x)$.The orbitals reported in Table XI are obtained by applying local scalings associated with the final density $\tilde{\rho}(x)$ to the initial orbitals belonging to the above sets A, B and C. The final density can be $\rho_{HF}(x)$, that is, the Hartree-Fock Clementi-Roetti density, or $\rho_{CI}(x)$, the CI density associated with the Esquivel-Bunge 650-term wave function [51]. The density $\rho_K(x)$ appearing also in this Table is the density (18).

Table X. Orbital parameters for the $1s$, $2s$ and $2p$ functions appearing in the CI wave function for Be (after Ludeña et al. [43]).

	$1s$	$2s$			$2p$		
j	c_{10j}	c_{20j}	n_{0_j}	α_{0j}	c_{11j}	n_{1j}	α_{1j}
				Set A			
1	0.997 13	0.243 95	1	3.455 00	1.000 00	2	0.750 00
2	0.012 38	−1.026 46	2	0.970 00			
				Set B			
1	0.997 59	0.204 44	1	3.684 78	1.000 00	2	0.981 40
2	0.012 39	−1.018 24	2	0.956 03			
				Set C			
1	0.997 51	0.210 92	1	3.684 18	1.000 00	2	0.982 50
2	0.012 39	−1.019 49	2	0.974 37			

In Table XII, we list the values of the energy functional $\mathcal{E}_\mu[\rho(x); \{C^{o[\mu]}_{I,\tilde{\rho}}\}]$ evaluated for the coefficients presented in Table XI and for the densities $\rho(x)$ being equal to $\rho_g(x), \rho_{HF}(x), \rho_{CI}(x)$, and $\rho_K(x)$, respectively. Orbit-jumping is a variational process as the wave function is improved at fixed density. The results of these optimizations are given in Table XII. For example, for set A, the initial energy −14.538 796 is improved to the value −14.597 378. Similar lowerings occur for sets B and C. Thus, for set B, we have that the energy goes from −14.599 524 to −14.612 114 and for set C, from −14.599 936 to −14.612 471. Orbit-jumping at fixed optimized density $\rho_K(x)$ leads to the optimal expansion coefficients $\{C^{o[4]}_{I,\rho_K}\}$. For set A, the optimal energy is −14.597 476, for set B, −14.612 121 and for set C, −14.612 495. Considering that the optimal energy for an unscaled function is −14.599 936 (first entry for set C in Table XII), we conclude that local scalings reveal a considerable effect on these CI wave functions. Since the best locally-scaled energy is −14.612 495 (last entry for set C in Table XII), we observe that these transformations improve the energy by −0.012 559.

Table XI. CI expansion coefficients for the approximate ground-state wave functions $\Phi^{[r]}_{0,\rho}(1s)$ of Be for orbits $[i] = 1,\ldots,4$ and densities $\rho = \rho_g$, ρ_{HF}, ρ_{CI} and ρ_K (after Ludeña et al.[43]).

I	$\psi^{CSF}_{I,g}$	$C^{o[1]}_{I,\rho_g}$	$C^{o[2]}_{I,\rho_{HF}}$	$C^{o[3]}_{I,\rho_{CI}}$	$C^{o[4]}_{I,\rho_K}$
		Wave function constructed from set A			
1	$[1s^2 2s^2]$	−0.958 96	−0.957 46	−0.955 58	−0.956 31
2	$[1s^2 2p^2]$	0.283 27	0.288 37	0.294 60	0.292 24
3	$[2s^2 2p^2]$	−0.012 14	−0.010 43	−0.009 33	−0.008 57
4	$[1s 2s 2p^2]$	0.000 21	0.000 04	0.000 03	0.000 05
		Wave function constructed from set B			
1	$[1s^2 2s^2]$	−0.950 49	−0.951 06	−0.951 31	−0.951 14
2	$[1s^2 2p^2]$	0.310 71	0.308 90	0.308 16	0.308 70
3	$[2s^2 2p^2]$	−0.005 20	−0.007 50	−0.006 16	−0.006 31
4	$[1s 2s 2p^2]$	0.000 75	0.000 79	0.000 77	0.000 77
		Wave function constructed from set C			
1	$[1s^2 2s^2]$	−0.949 67	−0.950 09	−0.950 20	−0.950 26
2	$[1s^2 2p^2]$	0.313 21	0.311 86	0.311 55	0.311 36
3	$[2s^2 2p^2]$	−0.005 16	−0.008 47	−0.007 14	−0.007 30
4	$[1s 2s 2p^2]$	0.000 71	0.000 78	0.000 75	0.000 76

5.3. EXAMPLE ζ: DETERMINATION OF KOHN-SHAM ORBITALS AND POTENTIALS FOR BE VIA LOCAL SCALING

Local-scaling density functional theory provides the rigorous tool to determine the Kohn-Sham orbitals and Kohn-Sham potential [5b,52]. It has been shown in [41-44] that by implementing the above density-constrained method, in the context of local scaling, we obtain the same potential reported in the literature. An important point we would like to stress with respect to the present method is that it is based on the constrained minimization of the kinetic energy, and for this reason, it is not necessary to deal at any point of the calculation with the evaluation of 2-electron integrals.

6. Treatment of Excited States

The energy density functional given by Eq.(17) is restricted to the ground state of an N-electron system. The extension of the present formalism to excited states is exposed elsewhere [31,32,9]. We just briefly illustrate it by showing the result of the calculation of the first excited singlet state 2^1S of helium [32]. The resultant energy value obtained there is $E_{1,g} =$

−2.144 140 3. This value, although coming from a very simple reference wave function, is already quite close to the exact energy which is $E_{1,g} = -2.145\,974\,0$ [53]. Thus, again, we show that the local-scaling version of density functional theory leads to upper bounds to the exact energies. Clearly, the exact values can be obtained when the variation is carried out in an orbit which is sufficiently close to the Hohenberg-Kohn orbit, where, by definition, the exact ground- and excited-state wave functions are found.

Table XII. Energy values for the energy density functional $\mathcal{E}_\mu[\rho; \{C^{o[\mu]}_{I,\tilde{\rho}}\}]$ corresponding to the CI wave functions for *Be* (after Ludeña et al.[43]).

density	$\{C^{o[1]}_{I,g}\}$	$\{C^{o[2]}_{I,\rho_{HF}}\}$	$\{C^{o[3]}_{I,\rho_{CI}}\}$	$\{C^{o[4]}_{I,\rho_K}\}$
	From wave function corresponding to set A			
ρ_g	−14.538 796			
ρ_{HF}	−14.589 499	−14.589 526		
ρ_{CI}	−14.595 473	−14.595 548	−14.595 571	
ρ_K	−14.597 378	−14.597 453	−14.597 472	−14.597 476
	From wave function corresponding to set B			
ρ_g	−14.599 524			
ρ_{HF}	−14.609 808	−14.609 833		
ρ_{CI}	−14.609 386	−14.609 385	−14.609 393	
ρ_K	−14.612 114	−14.612 115	−14.612 121	−14.612 121
	From wave function corresponding to set C			
ρ_g	−14.599 936			
ρ_{HF}	−14.609 751	−14.609 803		
ρ_{CI}	−14.611 079	−14.611 091	−14.611 099	
ρ_K	−14.612 471	−14.612 488	−14.612 494	−14.612 495

7. Acknowledgments

The author gratefully thanks Constantinos A. Tsipis and John Avery for the invitation to participate in the NATO Advanced Research Workshop on *New Methods in Quantum Theory*. The NATO Science Committee is acknowledged for the financial support. This work has been partially supported by the DGICYT Projects PB 93 - 0288 - CO2 and SAB 94 - 0271.

References

1. Thomas, L. H. (1927) The calculation of atomic fields, *Proc. Cambridge Phil. Soc.*, **23**, 542.

2. Fermi, E. (1927) Application of statistical gas methods to electronic systems, *Rend. Accad. Naz. Lincei*, **6**, 602.
3. March, N. H. (1957) The Thomas-Fermi approximation in quantum mechanics, *Adv. Phys.*, **6**, 7; Lieb, E. H. (1981) Thomas-Fermi and related theories of atoms and molecules, *Rev. Mod. Phys.*, **53**, 603-641; *Ibid.*, **54**, 311 (E) (1982).
4. March, N. H. and Young, W. H. (1958) Variational methods based on the density matrix, *Proc. Phys. Soc. London*, **72**, 182-192.
5. (a) Hohenberg, P. and Kohn, W. (1964) Inhomogeneous electron gas, *Phys. Rev.*, **136**, B864-871; (b) Kohn, W. and Sham, L. J. (1965) Self-consistent equations including exchange and correlation effects, *Ibid.*, **140**, A1133-1138.
6. Parr, R. G. and Yang, W. (1989) *Density Functional Theory of Atoms and Molecules*, Oxford University Press, Oxford.
7. Dreizler, R. M. and Gross, E. K. U. (1990) *Density Functional Theory*, Springer-Verlag, Berlin.
8. Trickey, S. B. (ed.) (1990) *Density Functional Theory of Many-Fermionic Systems*, *Adv. Quantum Chem.*, Academic Press, New York, Vol. 21.
9. Kryachko, E. S. and Ludeña, E. V. (1990) *Energy Density Functional Theory of Many-Electron Systems*, Kluwer Academic Publishers, Dordrecht.
10. March, N. H. (1992) *Electron Density Theory of Atoms and Molecules*, Academic Press, New York.
11. Petkov, I. Zh., Stoitsov, M. V. and Kryachko, E. S. (1986) Method of local-scaling transformations and density-functional theory in quantum chemistry.I, *Int. J. Quantum Chem.*, **29**, 149-161.
12. Kryachko, E. S., Petkov, I. Zh. and Stoitsov, M. V. (1987) Method of local-scaling transformations and density-functional theory in quantum chemistry.II, *Int. J. Quantum Chem.*, **32**, 467-472.
13. Kryachko, E. S., Petkov, I. Zh. and Stoitsov, M. V. (1987) Method of local-scaling transformations and density-functional theory in quantum chemistry.III, *Int. J. Quantum Chem.*, **32**, 473-489; *Ibid.* **34**, 305-306(E) (1988).
14. Kryachko, E. S. (1984) Some approaches to density functional theory, in J. P. Dahl and J. Avery (eds.), *Local Density Approximations in Quantum Chemistry and Solid State Physics*, Plenum, New York, pp. 207-227.
15. Kryachko, E. S. and Ludeña, E. V. (1987) Many-electron energy density functional theory: point transformations and one-electron densities, *Phys. Rev.*, **A 35**, 957-964.
16. Kryachko, E. S. (1988) The energy density functional theory: present status - reflections, in J. Arponen et al. (eds.), *Condensed Matter Theories*, Plenum, New York, Vol. 3, pp. 209-220.
17. Kryachko, E. S. (1989) Modern developments in the energy density functional theory of many-electron systems, in *Lecture Notes in Chemistry*, Springer, Berlin, Vol. 50, pp. 503-522.
18. Kryachko, E. S. and Koga, T. (1989) Electron densities in momentum and position spaces.I. Mapping relation and density functional aspect, *J. Chem. Phys.*, **91**, 1108-1113.
19. Koga, T, Yamamoto, Y. and Kryachko, E. S. (1989) Electron densities in momentum and position spaces.II. Application of density mapping to the helium ground state, *J. Chem. Phys.*, **91**, 4758-4762.
20. Koga, T. (1990) Point transformations applied to density-functional calculations, *Phys. Rev.*, **A 41**, 1274-1280.
21. Koga, T. (1990) Local scaling transformations applied to density-functional calculations of *Li* and *Be*, *Phys. Rev.*, **A 42**, 3763-3767.
22. Koga, T. (1990) An energy functional of electron-pair density, *J. Chem. Phys.*, **93**, 5856-5861.
23. Koga, T. and Yamamoto, Y. (1990) Mapping between position and momentum densities by the local scaling method, *Phys. Rev.*, **A 42**, 6336-6341.
24. Kryachko, E. S. and Ludeña, E. V. (1991) Formulation of N- and v-representable

density functional theory.I. Ground states, *Phys. Rev.*, **A 43**, 2179-2193.
25. Kryachko, E. S. and Ludeña, E. V. (1991) Formulation of N- and v-representable density-functional theory. II. Spin-dependent systems. *Phys. Rev.*, **A 43**, 2194-2198.
26. Koga, T. and Kryachko, E. S. (1991) Spin-polarized density functional calculations based on local-scaling transformations, *J. Chem. Phys.*, **94**, 2910-2914.
27. Kryachko, E. S. and Ludeña, E. V. (1992) The N-representability problem and the local-scaling version of density functional theory, in A.N. Proto and J.L. Aliaga (eds.), *Condensed Matter Theories*, Plenum, New York, Vol. 7, pp. 229-241.
28. Kryachko, E. S. and Ludeña, E. V. (1992) Local-scaling density functional theory, in S. Fraga (ed.), *Computational Chemistry: Structure, Interactions and Reactivity*, Elsevier, Amsterdam, pp. 136-165.
29. Ludeña, E. V. and Kryachko, E. S. (1992) Local-scaling density functional theory: prospects for applications to the electronic structure of atoms, *Rev. Mex. Astr. Astrofis.*, **23**, 95-106.
30. Kryachko, E. S. and Ludeña, E. V. (1992) Local-scaling version of density functional theory, *New J. Chem.*, **16**, 1089-1098.
31. Kryachko, E. S., Ludeña, E. V. and Koga, T. (1992) Formulation of N- and v-representable density functional theory.III. Excited states, *J. Math. Chem.*, **11**, 325-339.
32. Koga, T. (1991) Local-scaling density-functional theory for excited states, *J. Chem. Phys.*, **95**, 4306-4310.
33. Kryachko, E. S. and Ludeña, E. V. (1992) Formulation of N- and v-representable density functional theory.V. Exchange-only self-consistent field, *Int. J. Quantum Chem.*, **43**, 769-782.
34. Koga, T., Yamamoto, Y. and Ludeña, E. V. (1991) Nonvariational configuration interaction calculations by local scaling method, *J. Chem. Phys.*, **94**, 3805-3807.
35. Koga, T., Yamamoto, Y. and Ludeña, E. V. (1991) Local-scaling density-functional method: intraorbit and interorbit density optimization, *Phys. Rev.*, **A 43**, 5814-5820.
36. Kryachko, E. S. and Koga, T. (1992) Retrospective outlook on computational aspects of energy density functional theory: Clementi's explicit and implicit contributions, *Int. J. Quantum Chem.*, **42**, 591-626.
37. Koga, T. and Matsuyama, H. (1992) Simple and accurate Compton profiles for the helium atom, *Phys. Rev.*, **A 45**, 5266-5269.
38. Kryachko, E. S., Ludeña, E. V. and Koga, T. (1993) The computational local-scaling energy density functional theory, in E. Clementi (ed.), *Methods and Techniques in Computational Chemistry, METECC-94, Vol. B: Medium Size Systems*, STEF, Cagliari, pp. 23-56.
39. Kryachko, E. S. and Ludeña, E. V. (1991) Formulation of N- and v-representable density functional theory. VIII. Relationship between the density-driven and the local-scaling versions, *J. Chem. Phys.*, **95**, 9054-9059.
40. Kryachko, E. S., Ludeña, E. V., López-Boada, R. and Maldonado, J. (1993) The exchange-only self-consistentfield procedure in the local-scaling version of density functional theory: some theoretical and practical considerations, in L. Blum and F. B. Malik (eds.), *Condensed Matter Theories*, Plenum, New York, Vol. 8, pp. 373-384.
41. Ludeña, E. V., López-Boada, R., Maldonado, J., Koga, T. and Kryachko, E. S. (1993) Calculation of the defect kinetic energy in Kohn-Sham theory by means of local-scaling transformations, *Phys. Rev.*, **A 48**, 1937-1943.
42. Ludeña, E. V., López-Boada, R., Maldonado, J., Valderrama, E., Kryachko, E. S., Koga, T. and Hinze, J. (1995) Local scaling version of density functional theory, *Int. J. Quantum Chem. Special Issue "Thirty Years of Density Functional Theory"*.
43. Ludeña, E. V., Kryachko, E. S., Koga, T., López-Boada, R., Hinze, J., Maldonado, J. and Valderrama, E. (1995) Local scaling version of density functional theory: a practical method for rigorous calculations of many-electron systems, in P. Politzer and J. M. Seminario (eds.), *Theoretical and Computational Chemistry: Density Functional Calculations*, Elsevier, Amsterdam.

44. Ludeña, E. V., López-Boada, R. and Maldonado, J. E. (1995) Exact exchange-only density functional theory by means of local scaling transformations, *J. Mol. Str. (Theochem)*, **330**, 33-39.
45. Eckart, C. (1930) The theory and calculation of screening constants, *Phys. Rev.*, **36**, 878-892.
46. Kellner, G. W. (1927) Die ionisierungsspannung des heliums nach der Schrödingerschen theorie, *Z. Phys.*, **44**, 91-109.
47. Clementi, E. and Roetti, C. (1974) Roothaan-Hartree-Fock atomic functions, *At. Data and Nucl. Data Tables*, **14**, 177.
48. Gradshteyn, I. S. and Ryzhik, I. M. (1980) *Table of Integrals, Series, and Products*, Academic Press, New York, p. 317.
49. Clementi, E., Chakravorty, S. J., Corongiu, G. and Sonnad, V. (1990) Independent electron models: Hartree-Fock for many-electron systems in E. Clementi (ed.), *Modern Techniques in Computational Chemistry:* $MOTECC^{TM}-90$, ESCOM, Leiden, 1990, pp. 47-140.
50. Raffenetti, R. C. (1973) Even-tempered atomic orbitals.II.Atomic SCF wavefunctions in terms of even-tempered exponential bases, *J. Chem. Phys.*, **59**, 5936-5949.
51. Esquivel, R. O. and Bunge, A. V. (1987) Accurate electron density and one-electron properties for the beryllium atom, *Int. J. Quantum Chem.*, **32**, 295-312.
52. Talman, J. D. and Shadwick, W.F. (1976) Optimized effective atomic central potential, *Phys. Rev.*, **A 14**, 36-40; Levy, M. and Perdew, J. P. (1985) Hellmann-Feynman, virial, and scaling requisites for the exact universal density functionals. Shape of the correlation potential and diamagnetic susceptibilities for atoms, *Ibid.*, **A 32**, 2010-2021; Aryasetiawan, F. and Stott, M. J. (1988) Effective potentials in density-functional theory, *Ibid.*, **B 38**, 2974-2987; Krieger, J. B., Li, Y. and Iafrate, G. J. (1992) Construction and application of an accurate local spin-polarized Kohn-Sham potential with integer discontinuity: exchange-only theory, *Ibid.*, **A 45**, 101-126; Zhao, Q. and Parr, R. G. (1992) Quantities $T_s[n]$ and $T_c[n]$ in density-functional theory, *Ibid.*, **A 46**, 2337-2343.
53. Pekeris, C. L. (1962) $1^1S, 2^2S$, and 2^3S states of Li^+, *Phys. Rev.*, **126**, 143-145; $1^1S, 2^2S$, and 2^3S states of H^- and of He^+, *Ibid.*, **126**, 1470-1476.

DENSITY FUNCTIONAL THEORY: IMPROVING THE FUNCTIONALS, EXTENDING THE APPLICATIONS

D. R. SALAHUB, E. I. PROYNOV, A. VELA[1] AND E. RUIZ[2]
Département de chimie, Université de Montréal,
C.P. 6128, Succursale Centre-ville,
Montréal, Québec H3C 3J7, Canada

[1] *present address: Departamento de Quimica, Universidad Autonoma Metropolitana-Iztapalapa, A.P. 55-534, México D.F. 09340, México.*

[2] *present address: Departament de Quimica Inorgànica, Diagonal 647, Universitat de Barcelona, 08028 Spain.*

1. Introduction

The foundations of Density Functional Theory (DFT) have been laid, firmed up and shaped throughout several eras, each of which has its own merits and significance. The first *bona fide* density functional goes back to the Thomas-Fermi model [1,2]. This model is still the focus of new research and papers devoted to its development continue to appear [3]. In 1951, Slater [4] proposed an averaging of the conventional Hartree-Fock (HF) potential leading to the famous Xα potential. The corresponding self-consistent field (SCF) equations formed the basis of the Xα method used for electronic structure calculations with remarkable success. After the rigorous foundation of DFT by Hohenberg and Kohn [5] and the Kohn-Sham method [6] it became clear that the Xα method could be viewed alternatively as a particular approximation to the Kohn-Sham DFT approach.

Although not yet completely achieved, the old dream of theoreticians: to calculate many-electron properties without referring explicitly to many-electron wave functions, appears nowadays to be more and more within

C. A. Tsipis et al. (eds.), New Methods in Quantum Theory, 359–394.

reach. Efficient numerical realizations of the KS DFT method yield many molecular and solid state properties: binding and cohesive energies, equilibrium geometries and bulk moduli, spectroscopic characteristics, etc. with a high precision, often comparable to (or even better than) the high-quality *ab initio* correlated methods [7-10,11]. However the current capabilities of the DFT methods seem to reach their limits somewhere in the region of weak inter-molecular interactions (hydrogen bonds, van der Waals interactions, charge transfer complexes and the like) [12,13 and references therein].

The central quantity behind the successes (and the failures) of the DFT approach is electron correlation. Most of the important chemical and solid state phenomena: chemical bonding, reactivity, conductivity, etc., require a treatment of correlation at a high level. In this chapter we review briefly the basic concepts of DFT and devote more attention to the subtle problem of how to treat the electron correlation without many-electron wave functions. The adiabatic connection method (ACM) within the KS DFT is discussed in more detail. This method is of fundamental importance in the field, providing a rigorous route to improving exchange-correlation approximations. Combined with coordinate-scaling and virial relationships within the constrained-search formulation of KS DFT [7,8], the ACM offers a systematic way of progressing towards new, more precise and efficient correlation functionals.

As an illustrative example of how this method can be used in practice, we consider in more detail the derivation of the recently proposed correlation functional [13-16] based on spin-polarized pair correlation functions of the Colle-Salvetti type [17]. Several practical realizations of this functional are then discussed, choosing different exchange counterparts. Selected results of these new exchange-correlation schemes implemented in the Linear-Combination-of-Gaussian-Type-Orbitals (LCGTO)-DFT code deMon-KS [9-11,18], are given, emphasizing the cases that have proven to be difficult for DFT in principle.

2. The Kohn-Sham DFT Approach

The wide popularity and success which DFT enjoys nowadays rests on the basis of several benchmark developments in its background. The cornerstone of contemporary DFT is the 1964 paper of Hohenberg and Kohn (HK) [5], in which two important theorems, of the nature of existence proofs, are formulated and proven, leading to a firm theoretical foundation for DFT. Let us consider an N-electron system described by the Hamiltonian:

$$\hat{H} = \hat{T} + \hat{U}_{ee} + \hat{V}_{ext} \quad , \tag{1}$$

where $\hat{T}$ is the kinetic energy operator, $\hat{U}_{ee}$ is the electron-electron repulsion operator and $\hat{V}_{ext}$ represents some external potential. Then the first HK theorem states that the ground state of this system is fully determined by the corresponding electron density: there is a one-to-one correspondence between the external potential (and therefore the Hamiltonian (1)) and the ground-state electron density. The proof of this theorem is simply by *reductio ad absurdum* based on the minimum-energy principle for the ground state. If one assumes that two external potentials V_{ext} and $V_{ext}^{'}$ (differing by more than a constant) correspond to one and the same ground state density, one arrives at two different Hamiltonians $\hat{H}$ and $\hat{H}^{'}$ having different ground-state wave functions but equal electron density. This situation however leads to a conflict with the variational principle. As a consequence of the first HK theorem, the expectation value of the Hamiltonian (1) and of each operator entering Eq.(1) are representable as unique functionals of the electron density. Intuitively, this consequence of the first HK theorem is not obvious from a physical point of view. It is a straightforward application of definitions to express the total many-electron energy as a functional of the second order reduced density matrix (RDM-2) pertinent to the system [7,8,19]. However, the dynamical equations for the RDM-2 are in a general form of an infinite chain (hierarchy) of coupled equations in which higher order RDM (RDM-3, RDM-4 etc.) are involved [20]. A consequence of the first HK theorem is that for the ground state, the RDM-2 can be represented as a functional of the electron density alone (the diagonal part of the RDM-1).

The many-electron energy functional is conventionally partitioned into several terms:

$$E[n] = T[n] + U_{cl}[n] + \tilde{E}_{xc}[n] + V_{ext}[n], \tag{2}$$

where $U_{cl}[n]$ is the classical part of the electron interaction energy, being a trivial functional of the density:

$$U_{cl}[n] = \frac{1}{2}\int \frac{n(\mathbf{r}_1)n(\mathbf{r}_2)}{r_{12}}\, d\mathbf{r}_1 d\mathbf{r}_2 \quad . \tag{3}$$

The unknowns in Eq.(2) which have to be determined are the kinetic energy, $T[n]$ and the exchange-correlation part $\tilde{E}_{xc}[n]$ of the electron repulsion energy. The second HK theorem proves that the functional (2) reaches its minimum for the exact ground state density, thus providing a variational principle for the functional [5-8]. It is supposed that the variations of $n(\mathbf{r})$ are restricted within the domain of normalized and v-representable densities. This requirement is nontrivial, because not all trial densities necessarily correspond to realizable external potentials. Apart from this restriction, the formal application of the variational calculus to the functional (2) leads to the following Euler-Lagrange equation:

$$\frac{\delta E[n]}{\delta n(\mathbf{r})} = v(\mathbf{r}) + \frac{\delta F[n]}{\delta n(\mathbf{r})} = \mu \quad , \tag{4}$$

where μ is a Lagrange multiplier having the meaning of a chemical potential [7,8] for the electrons. Functional-differentiability of the universal part (the part independent of the nuclear positions)

$$F[n] = T[n] + U_{cl}[n] + \tilde{E}_{xc}[n] \tag{5}$$

is presumed.

The direct application of Eq.(4) in practice is obscure, because the two HK theorems alone do not tell us how to keep the variations $\delta n(\mathbf{r})$ within the domain of v-representable densities. A formulation of DFT that overcomes this difficulty has been given by Levy in his *constrained-search* approach [21,22]. In this approach the direct variation of the density *is avoided* replacing it by a two-step search for the minimum of the functional (2): first search, among all many-electron wave functions $\Psi_{\tilde{n}}$ yielding a given trial density $\tilde{n}$, the one that minimizes the universal part F[n], Eq.(5), of the energy at that density. [The external-potential energy V_{ext} is fixed once the electron density is fixed and can be dropped from the search.] Then repeat this constrained search for various trial densities until the one delivering the absolute minimum of $F[\Psi_{\tilde{n}}]$ is found. Add the respective V_{ext} for that density to achieve the absolute minimum of the whole E[n], Eq.(2). According to the second HK theorem, the minimum occurs at the exact ground state density. The constrained-search procedure formally avoids the v- representability problem, since now the HK theorems and the universal functional F[n] are extended over the domain of all N-representable densities (deduced from N-electron wave functions):

$$F[n] = Min_{\Psi \to n} \langle \Psi \mid \hat{T} + \hat{U}_{ee} \mid \Psi \rangle \quad , \tag{6}$$

and the N-representability conditions of a trial electron density are well known [8,19]. In practice most of the methods aiming at derivations of functionals avoid explicit reference to any particular many-electron wave function Ψ [7,8]. Instead, Green's function and response function techniques are employed [7], as well as *ad hoc* approximations to the first- and the second-order reduced density matrices [8,13-16], to obtain the two basic unknowns- the exchange-correlation and the kinetic energy terms of the functional. The extent to which the so called functional N-representability requirement [19] is violated by the currently used approximate functionals, has not been analyzed in detail up to now. For example, a necessary (but not a sufficient) condition for a functional to be N-representable is to be self-interaction free [19].

An attempt to partly overcome at least one of the two difficulties: finding the explicit kinetic energy functional, was made by Kohn and Sham [6]. They proposed to evaluate the kinetic energy term indirectly, based on a SCF procedure involving a one-electron orbital representation associated with a fictitious (auxiliary) system of "noninteracting" electrons (the KS reference system). The SCF procedure employs a set of N single-particle equations of eigenvalue type:

$$\left[-\frac{1}{2}\nabla^2 + v_{eff}(\mathbf{r})\right] \psi_j = \varepsilon_j \psi_j \quad ,j = 1...M \ ,M \geq N \quad , \tag{7}$$

where an orthonormal manifold of one-electron orbitals ψ is involved, and $v_{eff}(\mathbf{r})$ is a specially chosen local potential (the effective Kohn-Sham (KS) potential):

$$v_{eff}(\mathbf{r}) = v(\mathbf{r}) + \int \frac{n(\mathbf{r}')}{|\mathbf{r} - \mathbf{r}'|} d\mathbf{r}' + v_{xc}(\mathbf{r}) \quad , \tag{8}$$

with the *exchange-correlation potential* defined as:

$$v_{xc}(\mathbf{r}) = \frac{\delta E_{xc}[n]}{\delta n(\mathbf{r})} \quad . \tag{9}$$

The new term E_{xc} appearing in Eq.(9) includes the exchange-correlation energy $\tilde{E}_{xc}$ (see Eq.(2)) derived from the electron-electron repulsion energy expression via the RDM-2, plus an additional kinetic energy term, as specified below. These have been called the "physical" exchange-correlation energy and the "correlation kinetic" term, respectively [9,28,29]. We note, as discussed further below, that these are different quantities, compared to the traditional *ab-initio* exchange and correlation energies, or to the kinetic component of the latter With this choice of the SCF potential, and imposing the condition that the electron density arising from this procedure

$$n(\mathbf{r}) = \sum_j f_j \mid \psi_j(\mathbf{r}) \mid^2 \quad , \tag{10}$$

be properly normalized and *equal to the exact electron density of the actual interacting system*, the SCF solution of Eqs.(7) replaces equivalently the direct solution of the DFT Euler equation (4). Indeed, the latter can be converted to a form corresponding to the same separable Hamiltonian reflected in Eqs.(7), as long as the electron density given by Eq.(10) is equal to the exact density:

$$\mu = v_{eff}(\mathbf{r}) + \frac{\delta T_s[n]}{\delta n(\mathbf{r})} \quad , \tag{11}$$

where v_{eff} is given by Eqs.(8,9) and T_s is the kinetic energy evaluated via the KS system of equations (7) (the so called "KS noninteracting kinetic energy") :

$$T_s = \sum_j f_j \langle \psi_j \mid -\frac{1}{2}\nabla^2 \mid \psi_j \rangle \quad , \tag{12}$$

where f_j in Eqs.(10,12) are orbital occupation numbers. (The realizability of the condition that the SCF electron density, Eq.(10), be equal to the exact one has been confirmed by the method of local scale transformations [19].) Using Eq.(12), the unknown exact kinetic energy functional T[n] can be partitioned into two parts, one of which is already known:

$$T[n] = T_s[n] + T_c[n] \quad . \tag{13}$$

The remainder T_c has the meaning of a correlation kinetic term, since T_s corresponds to a separable Hamiltonian (Eq.(7)). This remainder is the additional term included in the re-defined exchange-correlation energy E_{xc} entering the effective potential Eqs.(8,9):

$$E_{xc}[n] = \tilde{E}_{xc}[n] + T_c[n] \quad . \tag{14}$$

Use of definition (14) of the exchange-correlation energy is necessary to establish the correspondence between the KS system of single-particle equations and the exact DFT Euler equation. However, the so re-defined exchange-correlation energy is already not purely a potential energy term, in contrast to the "physical" one, because it includes a part of the kinetic energy as well, the correlation kinetic term T_c. Let us remember that T_c in Eq.(14) is different from the kinetic energy component of the traditional correlation energy. Hereafter, we will denote the traditional correlation energy by E_{corr}, and its kinetic component by T_{corr}, to distinguish them from the KS quantities, E_c and T_c.

It is worth mentioning that, although the KS equations formally correspond to a separable Hamiltonian, there is an important difference between a real noninteracting system and the KS reference system. The set of single-particle equations (7) are mutually coupled through Eqs.(10,12) and the effective potential v_{eff}, Eq.(8), drives a self-consistent field, similar to that in the HF and Xα SCF methods. However, in contrast to the latter two, the KS SCF scheme is designed to yield the exact ground state energy and the exact density, provided the exact effective potential is known. That is because the solution of Eqs.(7) is equivalent to exactly solving the variational DFT problem. Often the physical meaning of the KS orbitals and orbital energies is questioned, referring to the fact that the relation of these quantities to the exact (or some approximate) many-electron wave function pertinent to the system is obscure [8,9]. Indeed, in the original KS formulation explicit references to many-electron wave functions are avoided. On the other hand, this formulation meets the same problems as the original HK theorems [5]: the v-representability requirement and the restriction to nondegenerate ground states. Both restrictions are relaxed when the KS scheme is considered within Levy's constrained-search approach [21,22]. In addition, the constrained-search analysis clarifies the meaning of the quantities used in the KS scheme. One can formally build a Slater determinant from the KS orbitals (the KS determinant), in analogy to the HF method, although such a determinant is not employed anywhere in the SCF procedure. It has been proven within the constrained-search formulation, that the KS ground-state determinant is the one which yields the exact electron density and minimizes the expectation value of the kinetic-energy operator $\hat{T}$ in Eq.(1) at that density [22,8]. The minimum coincides with the KS SCF kinetic energy T_s. Thus the notion of the KS determinant allows one to distinguish the KS orbitals from the HF orbitals: the HF determinant is the one which minimizes the expectation value of the whole Hamiltonian (1), while delivering the HF electron density. In contrast to the HF method, the actual relationship between the KS determinant and the exact ground state wave function corresponding to Hamiltonian (1) is not so relevant for the KS method. The physical meaning of the KS orbital energies can be associated with their relation to the experimental ionization potentials given by the Janak theorem [23,9]:

$$- I_i = \int_0^1 \varepsilon_i(f_i)\, df_i \qquad , \tag{15}$$

where I_i is the actual energy (*including relaxation effects*) required to remove an electron from the orbital ψ_i, as observed in XPS experiments. It has also been proven that the exact KS HOMO energy equals the negative of the first ionization potential to within a constant shift of the energy scale (for a more detailed discussion on this matter see [9, and references therein]).

The KS orbitals should be treated as a special sort of SCF orbitals *associated with the exact DFT Euler equation*, Eq.(11). Practical applications have demonstrated that the quality of the KS orbital manifold is not less than, and for some purposes even better then the HF orbital manifold, concerning applicability in sum-over-state (SOS) perturbational treatment of magnetic spectroscopic properties [24,11], the orbital spin contamination [11], calculation of ESCA, electron-momentum and electron spectra [25,26] and for the purpose of chemical applicability [9,10]: bond population analysis within MO-LCAO expansions, electron (spin-)density distribution in real and orbital spaces, dipole and higher order moments, (hyper-)polarizabilities [26,27], etc.

Although not used directly in the calculations, the KS determinant (Ψ_{KS}) is a useful notion in the KS DFT formalism: it may be referred to in dividing the exchange-correlation energy, Eq.(14), into separate exchange and correlation parts. The KS exchange-only energy may be defined as [8,28,29]:

$$E_x^{KS}[n] = \langle \Psi_{KS} \mid U_{ee} \mid \Psi_{KS} \rangle - U_{cl}[n] \quad , \tag{16}$$

while the rest of E_{xc} is considered as (Kohn-Sham) correlation energy:

$$E_c[n] = \tilde{E}_{xc} - E_x^{KS} + T_c = \tilde{E}_c[n] + T_c[n] \quad . \tag{17}$$

Since the HF and the KS determinants are different, the KS definition of exchange and correlation energies is different from the HF-based convention. The traditional definition of correlation energy as the difference between the exact (nonrelativistic) many-electron energy and the HF energy, involves several different contributions: electron-electron interaction, kinetic energy, and electron-nuclear interaction :

$$\begin{aligned} E_{corr}[n, n^{HF}] &= E(exact) - E(HF) \\ &= [E_{e-e}(exact) - E_{e-e}(HF)] + [T(exact) - T(HF)] \\ &+ [E_{n-e}(exact) - E_{n-e}(HF)] \quad . \end{aligned} \tag{18}$$

$E_{e-e}(HF)$ contains the usual Coulomb and exchange energies corresponding to the HF determinant and the HF electron density. In addition to these, $E_{e-e}(exact)$ contains those parts of the traditional correlation energy involving the electron-electron repulsion operator:

$$E_{corr}(e-e) = E_{e-e}(exact) - E_{e-e}(HF) \quad , \tag{19}$$

which can be decomposed into two terms. The first term reflects the change in the classical Coulomb interaction due to the change of the electron density from HF to the exact one:

$$E_{corr}^{Coul}(e-e) = \int \frac{[n(\mathbf{r}_1)n(\mathbf{r}_2) - n^{HF}(\mathbf{r}_1)n^{HF}(\mathbf{r}_2)]}{r_{12}} d\mathbf{r}_1 d\mathbf{r}_2 \quad . \tag{20}$$

The second term contains all remaining quantum correlation effects associated with $E_{corr}(e-e)$.

Besides $E_{corr}(e–e)$, the other two components of E_{corr} given by Eq.(18), take into account the changes in kinetic and nuclear-electron energies when the wavefunction changes from the HF determinant to the exact wavefunction and correspondingly, the HF density changes to the exact density. We may call these T_{corr} and $E_{corr}(n-e)$, respectively. Comparing the components of the traditional correlation energy given by Eq.(18) with those that arise in KS-DFT (Eq.(17)), we note:

1. The definition of the KS correlation energy, E_c, involves only one electron density, the exact one.

2. Hence, the KS correlation energy does not contain analogs of the traditional $E_{corr}^{Coul}(e-e)$ and $E_{corr}(n-e)$ components. The contribution from the latter is expected to grow when the number of nuclei increases.

3. The kinetic contribution (T_c) to the KS correlation energy is defined on a different basis than the traditional T_{corr}. One has:

$$T_{corr} = T(exact)[n] - T(HF)[n^{HF}] \quad , \tag{21}$$

$$T_c = T(exact)[n] - T_s[n] \quad , \tag{22}$$

where $T_s[n]$ is the KS noninteracting kinetic energy defined by Eq.(13) and calculated via the KS SCF procedure. Subtracting Eq.(21) from Eq.(22) one obtains:

$$T_c - T_{corr} = T(HF)[n^{HF}] - T_s[n] \quad . \tag{23}$$

Zhao and Parr have evaluated this difference for two-electron systems and for the Be atom [40]. For example, the difference is about 0.005 Hartree (about 0.14 eV, about 3 kcal/mol) for He, and about 0.02 Hartree (about 0.5 eV, about 11 kcal/mol) for Be. We also note that T_{corr} and E_{corr} satisfy the virial theorem:

$$T_{corr} = -E_{corr} \quad , \tag{21a}$$

whereas no such simple relation holds for T_c and E_c [40]. Indeed, Levy and Perdew [29] have shown that:

$$T_c[n] = -E_c[n] + \left.\frac{\partial E_c[n_\gamma]}{\partial \gamma}\right|_{\gamma=1} \tag{22a}$$

where n_γ is the scaled electron density with a coordinate scaling factor γ:

$$n_\gamma(\mathbf{r}) = \gamma^3 n(\gamma x, \gamma y, \gamma z) \quad . \tag{24}$$

The factor γ^3 in Eq.(24) comes from the Jacobian of the scale transformation.

Considering points 1 to 3 and other sources [28,29], it is clear that careful definitions and use of the notion "correlation energy" (or the components thereof) are required. At the level of analysis used above, when it is assumed that the exact functional and hence, the exact electron density, are at hand, there are no contributions to the KS correlation energy arising from correlation induced changes in the density. This contrasts with the traditional definition of correlation, relative to HF, in which such terms are in principle important. Many of the correlation functionals currently in use have been parametrized with respect to the traditional correlation energy, E_{corr}, usually for atoms. While it happens that the differences between E_c and E_{corr} are not very large for some atoms [40], there are indications that such is not the case for molecules in general [16,28,39b]. Clearly, further careful study of these issues is required.

Let us return now to the main focus of this chapter, the KS correlation energy E_c given by Eq.(17). Important exact relations pertinent to E_c and other terms of the DFT energy functional, follow from the combination of the constrained-search formulation of KS DFT with coordinate-scaling many-body arguments [28,7]: if all electron coordinates are uniformly scaled by a parameter γ, then the following relations between scaled and unscaled energy quantities hold true:

$$T_s[n_\gamma] = \gamma^2 T_s[n] \qquad ; \tag{25}$$

$$E_x^{KS}[n_\gamma] = \gamma\, E_x^{KS}[n] \qquad ; \tag{26}$$

$$E_c[n_\gamma] > \gamma\, E_c[n]\,, \qquad \gamma > 1; \tag{27}$$

$$E_c[n_\gamma] < \gamma\, E_c[n]\,, \qquad \gamma < 1, \tag{28}$$

where n_γ is the scaled density, Eq.(24). The KS kinetic energy and the KS exchange energy scale homogeneously, Eqs.(25,26), while the KS correlation energy, Eqs.(17,27,28) ,does not. The HF exchange energy also does not scale homogeneously [28] and relation (26) serves as an important guide to whether one is dealing with the KS type, or with the HF type of decomposition of the energy into exchange and correlation. Concerning finite systems only, Levy has shown that the exact correlation functional $E_c[n]$ must be bounded from below upon an infinite scaling of the electron coordinates when the electron density integrates to a finite total number of electrons N [29]:

$$\lim_{\gamma\to\infty} E_c[n_\gamma] > -\infty\,, \tag{29}$$

where $n_\gamma(\mathbf{r})$ is the scaled density, Eq.(24). This *finite scaling bound* marks a sharp distinction between a finite N-electron system and the infinite (boundaryless) homogeneous electron gas. Most of the currently employed approximations to E_c and especially the local spin-density approximation (LSD) [30] do not satisfy relation (29) and diverge like $-\ln\gamma$ when γ tends to infinity [29,14].

The formal similarity between the KS equations and the HF equations suggests a similarity of their numerical realization. Introducing a Gaussian-type orbital representation [31-34,18] of Eqs.(7) opened the possibility of massive use of the KS DFT in molecular quantum-chemical studies. Highly efficient numerical realizations of the KS DFT scheme are currently available: the program packages DGAUSS [35], DMol [36], NUMOL [37], ADF [38], GAUSSIAN 92 [39], the presently discussed deMon-KS [9-11,18], etc.

3. Correlation Functionals within the Kohn-Sham Scheme

While the dominant part of the kinetic energy in the KS scheme is evaluated self-consistently via the KS orbitals, the exchange-correlation energy has to be derived explicitly as a functional of the density. This is the central problem in the KS DFT, and the success of this theory depends primordially on finding sufficiently accurate approximations for the exchange-correlation functional and potential. In this section we discuss in more detail possible lines of approaching this goal.

As defined by Eq.(17), E_c consists of two terms, the "physical" correlation energy $\tilde{E}_c$ being of pure potential origin, and the kinetic energy remainder T_c. The latter is usually about a half of the absolute value of $\tilde{E}_c$ ($T_c > 0$) [29,40] and it can not just be neglected. The accurate evaluation of T_c is one of the most difficult and demanding parts of the KS methodology. Precise numerical estimations of T_c have been done so far mainly for atoms [40], and for the homogeneous electron gas [52]. More studies are necessary in order to elucidate the magnitude and the importance of this term in a variety of molecules. A method which allows in principle the evaluation of $\tilde{E}_c$ and T_c simultaneously is the coupling-constant integration [30,7,8] borrowed from the diagrammatic many-body theory [41] and adapted to the KS DFT formalism. In this "adiabatic connection" method, a virtual process is considered, in which the actual interelectronic interaction is gradually turned on by introducing a coupling strength parameter in the Hamiltonian:

$$\hat{H}_\lambda = \hat{T} + \lambda \hat{U}_{ee} + \hat{V}_{ext}^{\lambda} \quad . \tag{30}$$

The coupling strength parameter λ varies between 0 and 1 and connects adiabatically the ground state of the KS reference system (the $\lambda = 0$ limit) with the ground state of the interacting N-electron system (the $\lambda = 1$ limit). Continuity of such a virtual process is presumed. Since the electron density at $\lambda = 0$ (the KS reference system) is equal to the exact one, the adiabatic connection between the two systems means that this equality is preserved for any intermediate value of λ, $0 \leq \lambda \leq 1$. According to the HK theorems, this condition can always be provided by introducing a suitable λ -dependent (fictitious) external potential $\hat{V}_{ext}^{\lambda}$ such that for any λ n_λ is equal to the exact density [7,8]. At any intermediate value of λ the KS procedure can be formulated with respect to $\hat{H}_\lambda$. Hence, both the Hamiltonian (30) and the energy functional $E_\lambda[n_\lambda = n]$ depend parametrically on λ and one can apply the Hellmann-Feynman theorem:

$$\frac{dE_\lambda}{d\lambda} = \left\langle \frac{\partial \hat{H}_\lambda}{\partial \lambda} \right\rangle_\lambda = \langle \hat{U}_{ee} \rangle_\lambda + \int n(\mathbf{r}) \frac{\partial}{\partial \lambda} v_\lambda(\mathbf{r})\, d\mathbf{r} \quad , \tag{31}$$

where the constrained-search averaging denoted by $\langle..\rangle_\lambda$ is with respect to the wave function that minimizes the universal part F_λ of H_λ and yields the exact density [7,8]. Taking into account that the classical part of U_{ee} does not depend on λ (because the electron density does not depend on it by construction) and after separating the λ-independent part T_s from T[n], the integration of Eq.(31) with respect to λ gives:

$$E[n] = T_s + U_{cl} + V_{ext} + \int_0^1 d\lambda\, \tilde{E}^\lambda_{xc}[n] \quad , \tag{32}$$

A comparison of Eq.(32) with Eq.(14) leads to the *adiabatic connection formula*:

$$E_{xc}[n] = E_x^{KS}[n] + \int_0^1 d\lambda\, \tilde{E}^\lambda_c[n] = E_x^{KS}[n] + \{\, \tilde{E}_c[n] + T_c[n] \,\} \quad , \tag{33}$$

where the exchange-only energy E_x^{KS}, Eq.(16), is equivalently defined here as the $\lambda = 0$ limit of $\tilde{E}_{xc}$:

$$\lim_{(\lambda=0)} \tilde{E}^\lambda_{xc}[n] = E_x^{KS}[n] \quad ; \tag{34}$$

As pointed out in the previous section, E_x^{KS} is generally different from its conventional HF exchange counterpart, because the corresponding orbitals and Slater determinants are different.

Although the integrand of Eq.(33) contains a potential energy term only, the λ integration yields in effect the correlation-kinetic part T_c as a consequence of the Hellmann-Feynman theorem. Until recently, little has been known about how to incorporate the λ dependence into the "physical" correlation energy, thus leaving the actual magnitude of T_c unknown. However, recently Levy has derived important rules for this procedure, based on coordinate-scaling arguments. Given the "physical" correlation energy $\tilde{E}_c$ (corresponding to the λ=1 limit), the λ-dependent counterpart $\tilde{E}^\lambda_c$ entering Eq.(33) can be obtained via the following coordinate-scaling relationship (λ is the coupling strength parameter):

$$E_c[n] = \int_0^1 \lambda\, \tilde{E}_c[n_\gamma]\, d\lambda \quad , \tag{35}$$

where n_γ is the scaled density, Eq.(24), with a specially chosen scaling parameter $\gamma = 1/\lambda$. The scaling relationship of Levy (35) means that the adiabatic switching on of the electron-electron interaction by varying λ from 0 to 1 is equivalent to a gradual shrinking of the length scale from infinity to its actual value.

Several applications of the adiabatic connection method have been reported in the literature [29a,30,41-47] modeling the physical correlation energy $\tilde{E}_c$ either directly, or by means of a model pair correlation function (CF). The latter approach appears to be particularly promising, because it allows one to use physical arguments and intuition, combined with various known exact properties which this function should obey. The electron CF for an N-electron system is conventionally defined via the diagonal elements of the corresponding RDM-2 as [7,8,19]:

$$\Gamma_{\sigma\sigma'}(\mathbf{r}_1, \mathbf{r}_2) = \frac{1}{2} n_\sigma(\mathbf{r}_1) n_{\sigma'}(\mathbf{r}_2)\, g_{\sigma\sigma'}(\mathbf{r}_1, \mathbf{r}_2) \quad . \tag{36}$$

The normalization condition for RDM-2 leads to the following sum rule for the CF:

$$\int n_\sigma(\mathbf{r}_1)[g_{\sigma\sigma'}(\mathbf{r}_1, \mathbf{r}_2) - 1]\, d\mathbf{r}_2 = -\delta_{\sigma\sigma'} \quad . \tag{37}$$

Once the form of $g(\mathbf{r}_1, \mathbf{r}_2)$ is found, one can obtain its λ-dependent generalization by using again the coordinate-scaling recipes of Levy [29]. Then the following exact relations for g^λ hold true:

$$\lim_{(\lambda \to 0)} g^\lambda_{\sigma\sigma}(\mathbf{r}_1, \mathbf{r}_2) = g^{KS}_{x\sigma}(\mathbf{r}_1, \mathbf{r}_2) \quad ; \tag{38}$$

$$\lim_{(\lambda \to 0)} g^\lambda_{\uparrow\downarrow}(\mathbf{r}_1, \mathbf{r}_2) = 1 \quad , \tag{39}$$

Using electron CFs within the adiabatic connection method is especially convenient when one deals with the spin-polarized case:

$$E^\lambda_c[n_\uparrow, n_\downarrow] = \sum_{\sigma,\sigma'} E^{\sigma\sigma'}_c(\lambda) \quad . \tag{40}$$

In this case each (λ-dependent) spin component $E^{\sigma\sigma'}_c(\lambda)$ is associated with its own spin-dependent CF. From the well-known expression of the physical exchange-correlation energy through the RDM-2 [8,19]:

$$\tilde{E}_{xc} = \sum_{\sigma,\sigma'} \int d\mathbf{r}_1 \int d\mathbf{r}_2 \frac{1}{r_{12}} \Gamma_{\sigma\sigma'}(\mathbf{r}_1, \mathbf{r}_2) - U_{cl}[n] \quad . \tag{41}$$

and using Eq.(36), one can evaluate the corresponding spin-components of E_c by [7,8,14]:

$$E_c^{\sigma\sigma'} = \frac{1}{2}\int d\mathbf{R}\int_0^\infty r dr \langle n_\sigma(\mathbf{R}+\mathbf{r}/2) n_{\sigma'}(\mathbf{R}-\mathbf{r}/2)\rangle \int_0^1 d\lambda [g_{\sigma\sigma'}^\lambda(\mathbf{R},r) - 1] , \tag{42}$$

where for the sake of our further discussion the electron coordinates are changed to the extracular ($\mathbf{R} = (\mathbf{r}_1 + \mathbf{r}_2)/2$) and intracular ($\mathbf{r} = \mathbf{r}_1 - \mathbf{r}_2$) ones. It is taken into account that the ground state energy depends only on the absolute value of the electron-distance vector, and $\langle\ ...\ \rangle$ denotes an average over the angular part of $\mathbf{r}$. The choice of working coordinates is not a trivial procedure in this formalism. For example, the form of the sum rule (37) and the way of solving it are different in different coordinate representations [7,13-16]. In $(\mathbf{R}, \mathbf{r})$ coordinates the sum rule (37) takes the form (for any λ):

$$\int d\mathbf{R}\int_0^\infty r^2 dr \langle n_\sigma(\mathbf{R}+\mathbf{r}/2) n_{\sigma'}(\mathbf{R}-\mathbf{r}/2)\rangle (g_{\sigma\sigma'}^\lambda(\mathbf{R},r) - 1) = -N_\sigma\,\delta_{\sigma\sigma'} \quad . \tag{43}$$

The quantity $\langle ... \rangle$ in Eqs.(42,43) has the meaning of a (spherically averaged) 'Hartree-like' part of the RDM-2, since both the exchange and correlation aspects are embodied in $g_{\sigma\sigma'}(\mathbf{R}, r)$. (Note, that this factorization of the RDM-2 holds true under the assumption that $g_{\sigma\sigma'}$ does not depend on the angular components of the electron-distance vector $\mathbf{r}$.)

To obtain a correlation energy functional of the density one must solve analytically two of the three integrals entering Eq.(42) - those over r and λ. For this purpose, the electron pair CF must be given analytically. This function should in principle be evaluated self-consistently, solving for each system the corresponding quantum hierarchy of coupled equations analogous to those for the RDM [20]. Unfortunately, accurate solutions of these equations are not available in most cases, even numerically, let alone the analytical forms needed to derive a correlation functional. In searching for such analytical representations of $g_{\sigma\sigma'}^\lambda$, a sufficiently large degree of universality is presumed, in the sense that one model function be applicable to a large variety of systems. Such an assumption seems to be reasonable, judging from known analogous situations in statistical physics [7].

One of the most important approximations in KS DFT, the Local Density Approximation (LSD), relies substantially on this universality. Considering the exchange-correlation energy in the form:

$$E_{xc}^{LSD}[n_\uparrow, n_\downarrow] = \int n(\mathbf{r})\varepsilon_{xc}(\mathbf{r})\, d\mathbf{r} \quad , \tag{44}$$

the LSD approximation assumes that the exchange-correlation energy density ε_{xc} of any system coincides locally with that of the homogeneous gas, with the electron-density now being a local function in space. This approximation works surprisingly well for atoms and molecules, and the reasons for this phenomenon have been widely discussed in the literature [7-9,30,48,49]. Let us only mention in this respect the "short-wavelength" hypothesis [48] stating that in any system the interelectronic interaction and correlation at very small length scales behave like those of the homogeneous electron gas. It has been recently proven that this statment is not exact, but seems to be a good approximation in many cases [48]. Concerning the evaluation of the correlation energy within the adiabatic connection method, one should bear in mind that, even for the homogeneous gas, exact analytical results are not yet available, except for limiting values of densities and spin polarizations [50-52]. Beyond the Random Phase Approximation (RPA), quantum Monte Carlo calculations are so far the only available numerical data, covering a wide range of densities. Accurate analytical interpolations of the Monte Carlo results of Ceperley and Alder [53] have been obtained based on Padé-approximant techniques [50-52,54]. Only recently, have such interpolations been properly converted in a way to envisage an accurate analytical form of the homogeneous-gas CF [52]. It was found that the behavior of the electrons in the limit of small and vanishing interparticle distances is of crucial importance. A precise fulfillment of the exact cusp conditions for the CF is essential for reproducing the beyond-RPA numerical data for the homogeneous gas [52]. For inhomogeneous systems the fulfillment of the cusp conditions should be even more essential. In the context of the adiabatic connection method, general λ -dependent cusp conditions have been derived [55], which serve as an important test for different approximations to $g^{\lambda}_{\sigma\sigma'}$. In the spin-polarized case the different spin components of $g^{\lambda}_{\sigma\sigma'}$ obey different cusp conditions:

(i) Cusp conditions for the parallel-spin CF:

$$g^{\lambda}_{\sigma\sigma}(0,\mathbf{R}) = \frac{\partial}{\partial r} g^{\lambda}_{\sigma\sigma}(\mathbf{R},r)\Big|_{r=0} = 0 \quad ; \tag{45}$$

$$\frac{\partial^2}{\partial r^2} g^{\lambda}_{\sigma\sigma}(\mathbf{R},r)\Big|_{r=0} = \frac{2}{3}\frac{1}{\lambda}\frac{\partial^3}{\partial r^3} g^{\lambda}_{\sigma\sigma}(\mathbf{R},r)\Big|_{r=0} \neq 0 \quad ; \tag{46}$$

(ii) Cusp condition for the anti-parallel spin CF:

$$g^{\lambda}_{\uparrow\downarrow}(0,\mathbf{R}) = \frac{1}{\lambda}\frac{\partial}{\partial r} g^{\lambda}_{\uparrow\downarrow}(\mathbf{R},r)\Big|_{r=0} > 0 \quad . \tag{47}$$

Note that the above cusp conditions relate in a rather nontrivial fashion the components of the CF with their first, second and third derivatives over the interelectron distance.

Even if the exact CF of the homogeneous gas were known, it is not necessarily the best choice when the adiabatic connection method is applied to atoms and molecules. The drawbacks of the LSD are well documented [7,8,30,48] and it is often difficult to reach the desired chemical accuracy within this approximation. LSD typically underestimates the exchange by 10-15% in atoms and molecules, while it overestimates the correlation energy by a factor of two [56,57]. A significant improvement of the results within the KS DFT is achieved by the contemporary nonlocal exchange-correlation schemes, correcting the LSD approximation in one or another way for the effects of inhomogeneity, and/or the self-interaction error [57] (the incomplete cancellation of the classical self-repulsion by the LSD exchange term).

In the next section we discuss some results obtained within the so-called Generalized Gradient Approximation (GGA) [48,58-61] and the recently proposed Laplacian corrected schemes [13-16]. Since the GGA functionals have been intensively discussed and reviewed in the literature in the last few years [7,8,48], we concentrate here mainly on the recently developed correlation functional [13-16] employed in the Laplacian corrected schemes. The derivation of this functional is one particular example of a direct implementation of the adiabatic connection formula, Eqs.(33,42). It is generated from a specially chosen family of λ-dependent antiparallel-spin CF of the form: $g^{\lambda}_{\uparrow\downarrow}$ [14]:

$$g^{\lambda}_{\uparrow\downarrow}(\mathbf{R},r) = 1 - \exp(-k_{\uparrow\downarrow}{}^2 r^2)\,[\,F_1(\lambda) + F_2(\lambda)\,] \qquad , \qquad (48)$$

$$F_1(\lambda) = m - \Phi^{\lambda}_{\uparrow\downarrow}(\mathbf{R})\,(m + \lambda r) \qquad , \qquad (49)$$

$$F_2(\lambda) = -\exp(-k^2_{\uparrow\downarrow} r^2)\left[1 - \Phi^{\lambda}_{\uparrow\downarrow}(m + \lambda r) + (\Phi^{\lambda}_{\uparrow\downarrow})^2(1 + \lambda r + \frac{\lambda^2 r^2}{4})\right] \quad (49a)$$

where $k_{\uparrow\downarrow}$ is the "correlation wave vector" (inverse correlation length [14]), and m is a parameter. The function $\Phi^{\lambda}_{\uparrow\downarrow}$ can be determined from the sum rule (43) as shown below. The striking property of this CF is that for any nonzero value of the parameter m it does obey *automatically* the exact cusp condition for the λ -dependent antiparallel-spin CF, Eq.(47),

$$g^{\lambda}_{\uparrow\downarrow}(0,\mathbf{R}) = \frac{1}{\lambda}\,\frac{\partial}{\partial r}\,g^{\lambda}_{\uparrow\downarrow}(\mathbf{R},r)\bigg|_{r=0} = [\Phi^{\lambda}_{\uparrow\downarrow}(\mathbf{R})]^2 > 0 \qquad , \qquad (50)$$

as well as the asymptotic condition (39). However, this pair CF does not obey the cusp conditions for the parallel-spin component and the parallel-spin correlation (beyond the Fermi-hole one) is not taken into account in this model. In many cases one may expect that this type of electron correlation will have a small contribution to the observable properties. The radius of the Fermi hole is usually large enough to keep the electrons with parallel spins apart, so that the shorter ranged Coulomb correlation between them is effectively quenched.

Based on the pair CF, Eqs.(48-49a), first the adiabatic connection formula was resolved to zeroth order in the Taylor- series expansion of the 'Hartree-like' part of the RDM-2 in powers of r about the vector $\mathbf{R}$ (the gradientless limit) [14,15]:

$$\langle n_\uparrow(\mathbf{R}+\mathbf{r}/2)n_\downarrow(\mathbf{R}-\mathbf{r}/2)\rangle \approx 4\pi\, n_\uparrow(\mathbf{R})n_\downarrow(\mathbf{R}) \tag{51}$$

In this limit the sum rule (37) can be simplified as:

$$4\pi\, n_\sigma(\mathbf{R})n_{\sigma'}(\mathbf{R}) \int_0^\infty r^2 dr\, [\, g^\lambda_{\sigma\sigma'}(\mathbf{R},r) - 1\,] = -\delta_{\sigma\sigma'} \quad . \tag{52}$$

Substituting the CF (48-49a) in the sum rule (52) and performing analytically the integration over r one obtains a simple algebraic equation for the unknown factor $\Phi_{\uparrow\downarrow}$ [14]:

$$A\,\Phi^2 + B\,\Phi + C = 0 \quad , \tag{53}$$

$$A = \frac{\sqrt{2}}{4} + \frac{3\sqrt{2}}{64k^2_{\uparrow\downarrow}} + \frac{1}{2k_{\uparrow\downarrow}\sqrt{\pi}} \quad ; \tag{53a}$$

$$B = m(1 - \frac{\sqrt{2}}{4}) + \frac{3}{2\sqrt{\pi}k_{\uparrow\downarrow}} \quad ; C = \frac{\sqrt{2}}{4} - m \quad . \tag{53b}$$

To generalize Eqs.(53,53a-b) to arbitrary λ, one must replace everywhere $k_{\uparrow\downarrow}$ by $k_{\uparrow\downarrow}/\lambda$. The corresponding positive root of Eq.(53) then ensures automatically the fulfilment of the asymptotic limit (39). Unfortunately, using this solution for $\Phi_{\uparrow\downarrow}$ in the CF (48-49a), and then in the adiabatic connection formula, Eq.(42), it is not possible to solve analytically the integral over λ in a closed form, even though the integration over r is available [within the gradientless limit of the 'Hartree-like' part, Eq.(51)]. Nor does the exact form of $\Phi_{\uparrow\downarrow}$ fit to any linear interpolation of the λ-dependent integrand, as suggested for example in the "half-and-half" hybrid methods [62,63]. This aspect of the adiabatic connection method seems to be the bottleneck, and the more realistic CF one uses, the less are the chances to succeed in the analytical derivation of the functional. We have, so far, been able to solve analytically the integral over λ only in the particular case of $m = 2$. In this case, when λ=1 and in the spin-compensated limit

Eqs.(48-49a) recover the Colle-Salvetti CF used in the context of the HF partitioning of the RDM-2 [17]:

$$P_2(\mathbf{r}_1, \mathbf{r}_2) \approx P_2^{HF}(\mathbf{r}_1, \mathbf{r}_2) g^{CS}(\mathbf{R}, \mathbf{r}) \quad , \tag{54}$$

where

$$g^{CS}(\mathbf{R}, \mathbf{r}) = [1 - \varphi(\mathbf{R}, \mathbf{r})]^2 \quad , \tag{55}$$

and

$$\varphi(\mathbf{R}, \mathbf{r}) = exp(-k^2 r^2) \left[1 - \Phi(\mathbf{R}) \left(1 + \frac{r}{2}\right)\right] . \tag{56}$$

Although this CF was first proposed *ad hoc* based on physical reasoning, some dynamical justification of this type of functions has been given based on an analysis of the quantum hierarchy of equations for the correlation functions [20].

Colle and Salvetti used this model CF in the expression for the *conventional correlation energy* (E_{corr}):

$$E_{corr} = \frac{1}{2} \int P_2^{HF}(\mathbf{R}, \mathbf{r}) [\varphi^2(\mathbf{R}, \mathbf{r}) - 2\,\varphi^2(\mathbf{R}, \mathbf{r})] \frac{1}{r} d\mathbf{R} d\mathbf{r} \tag{57}$$

and after some approximations and empirical interpolations they obtained the expression used afterwards as a basis for the Lee-Yang-Parr functional [64]:

$$E_{corr} \approx a \int n(\mathbf{R}) \frac{1 + bW\, exp(-c/k)}{1 + d/k} d\mathbf{R} \quad , \tag{58}$$

where a, b, c, d are fitting constants and W reads:

$$W = 0.3814 n^{-8/3} \left[\nabla_r^2 P_2^{HF}(\mathbf{R} + \mathbf{r}/2 \,;\, \mathbf{R} - \mathbf{r}/2)\right]_{r=0} . \tag{59}$$

In the derivation of Eqs.(58,59) a finite gradient expansion of $P_2^{HF}(\mathbf{R}, \mathbf{r})$ was used [17] in the energy expression, while the gradientless limit of $P_2^{HF}(\mathbf{R}, \mathbf{r})$ about $\mathbf{R}$ was used in the normalization condition for the RDM-2. The latter was then resolved in a form analogous to our Eqs.(53,53a-b) with m=2. This allows us for m=2 to benefit from their interpolation formula [17] for the solution of the normalization condition. Our modification of this formula, to include the required λ dependence, is:

$$\Phi_{\uparrow\downarrow}^{\lambda}(\mathbf{R}) = \frac{\sqrt{\pi}\, k_{\uparrow\downarrow}(\mathbf{R})}{\sqrt{\pi}\, k_{\uparrow\downarrow}(\mathbf{R}) + \lambda\, 1.0085} \quad , \tag{60}$$

where we found the coefficient 1.0085 to provide a slightly better interpolation to the positive root of Eq.(53) (the original value of this coefficient is 1.0 [17]). Unfortunately, we were not able so far to obtain accurate enough interpolations of this kind for other values of the parameter m in Eq.(53), which prevents us from optimizing the model with respect to this parameter. Considering further only the particular case of m=2, the substitution of approximation (53) for $\Phi_{\uparrow\downarrow}$, instead of the exact positive root of Eq.(53), allows already for the integrals required in the adiabatic connection formula to be evaluated analytically [14]. First we perform the integration over r in Eq.(42) at $\lambda = 1$, to obtain the physical correlation energy:

$$\tilde{E_c}^{\uparrow\downarrow}[n_\uparrow, n_\downarrow] = \frac{1}{2}\int_{\Omega_\infty} d\mathbf{R}\, n(\mathbf{R})\, \tilde{\varepsilon}_c^{\uparrow\downarrow}(\mathbf{R}) \qquad , \tag{61}$$

where the correlation energy density within approximation (60) reads:

$$\tilde{\varepsilon}_c^{\uparrow\downarrow}(\mathbf{R}) = -\frac{n_\uparrow(\mathbf{R})n_\downarrow(\mathbf{R})}{n(\mathbf{R})}\,\frac{a_1(\,a_2 + a_3 k_{\uparrow\downarrow})}{k_{\uparrow\downarrow}^2\,(\sqrt{\pi}\,k_{\uparrow\downarrow} + 1.0085)^2} \qquad , \tag{62}$$

and a_i are constants that result from the integration :

$a_1 = 0.392699$; $a_2 = 4.88307$; $a_3 = 12.654$.

The required λ-averaged correlation energy is then obtained by means of the coordinate-scaling representation of the adiabatic connection formula, Eq.(35), which in the considered spin-polarized case reads:

$$E_c^{\uparrow\downarrow}[n_\uparrow, n_\downarrow] = \int_0^1 \lambda\, \tilde{E_c}^{\uparrow\downarrow}[n_{1/\lambda\uparrow}, n_{1/\lambda\downarrow}]\, d\lambda \qquad . \tag{63}$$

The scaled energy in the integrand of Eq.(63) is obtained from Eqs.(61,62) after replacing everywhere the spin-densities n_σ by the corresponding scaled ones $n_{\gamma\sigma}$ [Eq.(24) with $\gamma = 1/\lambda$], the correlation wave vector $k_{\uparrow\downarrow}$ by $k_{\uparrow\downarrow}/\lambda$, and changing the integration variable $d\mathbf{R} \to d(\gamma^3\mathbf{R})$ [14]:

$$\tilde{E_c}^{\uparrow\downarrow}[n_{\gamma\uparrow}, n_{\gamma\downarrow}] = -\frac{\gamma}{2}\int_{\Omega_\infty} \frac{n_\uparrow(\mathbf{R})n_\downarrow(\mathbf{R})\, a_1(a_2 + a_3\,\gamma k_{\uparrow\downarrow})}{k_{\uparrow\downarrow}^2(\sqrt{\pi}\,k_{\uparrow\downarrow}\,\gamma + 1.0085)^2}\, d\mathbf{R} \quad , \tag{64}$$

where $\gamma = 1/\lambda$. After substitution of Eq.(64) in Eq.(63), the analytical integration over λ is readily performed, and the final result for the desired λ-averaged correlation functional is [14,15]:

$$E_c[n_\uparrow, n_\downarrow] = \frac{1}{2}\int d\mathbf{R}\, n(\mathbf{R})\, \varepsilon_c(\mathbf{R}) \qquad , \tag{65}$$

$$\varepsilon_c = \frac{n_\uparrow\, n_\downarrow}{n}\, Q(k_{\uparrow\downarrow}) \qquad , \tag{66}$$

$$Q(k) = -\frac{b_1}{1+b_2\,k} + \frac{b_3}{k}\ln\left(\frac{(b_4+k)}{k}\right) + \frac{b_5}{k} - \frac{b_6}{k^2} \quad , \tag{67}$$

where b_i are numerical constants coming out of the derivation [14]:
b_1=2.763169 ; b_2 =1.757515 ; b_3=1.741397 ; b_4=0.568985 ; b_5=1.572202 ; b_6=1.885389.
One should note that if the last two terms on the r.h.s of Eq.(67) are omitted the remainder bears (in the spin-compensated limit) a close resemblance to the functional of Lie and Clementi [65] derived from physical reasoning. The parallel-spin electron correlation correlation beyond the exchange-only level is not included in Eqs.(65-67).

An important quantity which remains to be determined is the correlation wave vector $k_{\uparrow\downarrow}$. Regarding similar quantities within a different model, Becke [42] has suggested (*ad-hoc*) an expression for the antiparallel-spin correlation length as an average of the parallel-spin ones. Following Becke, we express $k_{\uparrow\downarrow}$ through the corresponding parallel-spin wave vectors [13,16]

$$k_{\uparrow\downarrow} = \alpha_1\,\frac{2k_\uparrow\,k_\downarrow}{(k_\uparrow + k_\downarrow)} \quad , \tag{68}$$

where α_1 is a fitting parameter. Alternative forms of $k_{\uparrow\downarrow}$ can also be used, and finding better approximations for this quantity is one possible line of improving the present model.

Often, models based on CF assume that the correlation wave vectors are functions only of the electron density (gradientless approximations) [17,20]. However, in inhomogeneous systems the correlation lengths should depend also on higher derivatives of the density (so as to "feel" the boundaries of the system) and this is one nontrivial way of incorporating nonlocal corrections in such models [16,42,43,47]. Depending on the chosen form of k_σ in Eqs.(64-68) different correlation energy functionals can be generated. Let us first consider the gradientless limit for $k_{\uparrow\downarrow}$. In this case coordinate -scaling arguments dictate k_σ to be proportional to the local Fermi wave vector (i.e. to $n_\sigma^{1/3}$) [14,15]:

$$k_\sigma = \alpha\,k_{F\sigma} \;= \alpha\,(6\pi^2\,n_\sigma)^{1/3}\,. \tag{69}$$

Substituting Eq.(69) in Eq.(68), one obtains $k_{\uparrow\downarrow}$ in the form:

$$k_{\uparrow\downarrow} = \alpha\,(6\pi^2)^{1/3}\,\frac{2n_\uparrow^{1/3}\,n_\downarrow^{1/3}}{(n_\uparrow^{1/3} + n_\downarrow^{1/3})} \quad . \tag{70}$$

Within the gradientless approximation for $k_{\uparrow\downarrow}$, Eq.(70), the functional (65-67) is of a local type, and the corresponding correlation potential is evaluated via the formula [13]:

$$V_c^\sigma(n_\downarrow, n_\uparrow) = \varepsilon_c(n_\downarrow, n_\uparrow) + n\,\frac{\partial \varepsilon_c(n_\downarrow, n_\uparrow)}{\partial n_\sigma}\ , \tag{71}$$

In spite of its gradientless form, the functional (65-67,70) differs from the LSD approximation in several aspects. First of all it complies to the finite scaling bound (29) inherent to finite systems, while the LSD correlation energy diverges logarithmically upon an infinite coordinate scaling [56-58]. The functional (65-67) also obeys automatically the coordinate-scaling inequalities (27,28). Another salient feature of this functional is that it is self-interaction free as long as the parallel-spin correlation is omitted in the present model.

The functional (65-67) in its gradientless form (70) was combined with a gradientless exchange functional derived from a model Fermi-hole function of Lee-Parr type [66,15]:

$$g^x_{\sigma\sigma}(\mathbf{R}, r) = 1 - \exp(-\,{k_\sigma}^2 r^2)(1 + A\,{k_\sigma}^4 r^4) \quad , A \geq 0 \quad . \tag{72}$$

The one-parameter exchange functional corresponding to approximation (72) has the form [15]:

$$\varepsilon_x(\mathbf{R}) = -C_x(A)\, n^{1/3}\{(1+\xi)^{4/3} + (1-\xi)^{4/3}\} \quad , \tag{73}$$

where the coefficient $C_x(A)$ is given by

$$C_x(A) = \frac{(1 + 2\,A)}{2^{4/3}(1 + \frac{15}{4}\,A)^{2/3}} \quad . \tag{74}$$

The combination of this local exchange with the new correlation functional [named "modified local" (ML1)] was tested on a set of small molecules, and the optimal values of the parameters entering the ML1 scheme were found to be: A=0.72885 ; $\alpha = 0.72$, where the parameter α is introduced by Eq.(70). This scheme gives substantially improved values for the binding energy (E_b) of molecules, compared to the LSD results (with the Vosko, Wilk and Nusair (VWN) version for the correlation energy [54]), within the LCGTO-DFT code deMon-KS [9-11]. With respect to the optimized geometries, the improvement achieved with the ML1 scheme is also noticeable [15]. However, while the parameters of the ML1 scheme were fitted to reproduce the binding energies of test molecules, the output for the total energy and its separate exchange and correlation parts is not as good. In this respect it is interesting to compare the ML1 and LSD schemes. The correlation energy of atoms and molecules is largely overestimated by the LSD approximation. However, the exchange energy is underestimated in this approximation to such an extent that the total LSD energy is usually smaller than the experimentally derived one. In contrast, the correlation energy given by the ML1 scheme is underestimated, while the exchange energy is even slightly overestimated, compared to experiment. Overall, the total energy given by ML1 is much closer to the experimentally derived one, than the LSD energy. Turning again to the binding energies, in many cases the ML1 results are even comparable to the GGA results [15], while the computational effort required for ML1 is comparable to that for VWN. Recently, the ML1 scheme was implemented in the solid state version of deMon ("Bloch-deMon") and the preliminary tests show the same trend: improvement of the cohesive energies and lattice constants [67]. However, for systems known as "difficult" in principle for KS DFT (hydrogen bonds, weak interactions, etc.) one needs a functional with explicit nonlocal corrections.

Within the present formalism nonlocal corrections can be incorporated either in the 'Hartree-like' part of the RDM-2, or in the correlation wave vectors, or in both. Similar to the GGA methods, gradient expansion techniques can be used for this purpose, but this is not the only possibility. We have recently considered an alternative way to obtain nonlocal expressions of k_σ, based on some results of the local thermodynamic (LTD) approach in DFT [16,13]. In the latter, a local temperature-like parameter β (or T_σ) is defined, which is formally related to the SCF-KS kinetic energy density t_s by analogy with classical thermodynamics [7,8]:

$$\frac{1}{\beta_\sigma(\mathbf{r})} \equiv kT_\sigma(\mathbf{r}) = \frac{2}{3}\frac{t_{s\sigma}}{n_\sigma} \quad , \tag{75}$$

where the kinetic energy density is defined via the relation (note the difference with Eq.(44):

$$T_s[n] = \int d\mathbf{R}\, t_s(\mathbf{R}) \quad . \tag{76}$$

The local temperature parameter enters the expression for the spherically-averaged exchange hole used within the LTD in the form [7]:

$$\bar{h}_{\sigma\sigma}(\mathbf{R}, r) \approx n_\sigma(\mathbf{R}) e^{-r^2/\beta_\sigma(\mathbf{R})} \,, \tag{77}$$

which suggests a relationship between β_σ and the parallel-spin wave vectors. A proportional relationship of the form:

$$k_\sigma{}^2 \approx \frac{1}{\beta_\sigma} \approx kT_\sigma(\mathbf{r}), \tag{78}$$

was explored in [16], which leads to a nonlocal expression for the correlation wave vectors entering Eqs.(65-68). This relation has a clear physical meaning: the electron-electron correlation length increases (the wave vector k_σ decreases) when the electrons are "freezing down" and their kinetic energy is decreasing.

To implement the above formulae in the functional (65-68), one has to evaluate with a high precision the kinetic energy density. One should note in this respect that some ambiguity exists in the definition (76) due to the fact that T_s can be equivalently expressed in two different ways [8]:

$$T_s = \sum_i f_i \int \varphi_i(-\frac{1}{2}\nabla^2)\varphi_i d\mathbf{R} = \frac{1}{2}\sum_j f_j \int \nabla\varphi_j \cdot \nabla\varphi_j d\mathbf{R} \quad , \qquad (79)$$

where φ_i are the KS SCF orbitals. In our model we follow the definition of Parr at al. [8] for $t_{s\sigma}$, which in the spin-polarized case reads:

$$t_{s\sigma} = \frac{1}{8}\sum_i \frac{(\nabla n_{i\sigma}.\nabla n_{i\sigma})}{n_{i\sigma}} - \frac{1}{8}\nabla^2 n_\sigma \quad , \qquad (80)$$

where $n_{i\sigma} = \varphi_{i\sigma}^2$ is the ith KS spin-orbital density. After a simple algebra Eq.(80) can be transformed to:

$$t_{s\sigma} = \frac{1}{2}\sum_i (\nabla\varphi_{i\sigma})^2 - \frac{1}{8}\nabla^2 n_\sigma \quad . \qquad (81)$$

The presence of $t_{s\sigma}$ in the nonlocal terms of the correlation functional hinders the derivation of the SCF correlation potential. We have so far implemented this 'Laplacian-corrected', kinetic energy dependent functional in a nonself-consistent fashion, employing the gradientless limit of the potential, Eq.(71). To compensate partly for this deficiency, the correlation parameter α_1 in Eq.(68) is split into two: one used in the nonlocal energy expression (denoted as α_e) and fitted separately from the other one (denoted as α_p) entering the gradientless potential. This gives an additional degree of freedom for adjusting the correlation potential. The parameters were optimized in conjunction with the exchange counterparts used, trying to reproduce as much as possible the experimental binding energies of a set of test molecules, not the experimental (conventional) correlation energies alone. The reason for this is that the experimentally derived correlation energies available in the literature are defined within the HF-based partitioning of the energy. As already mentioned above, this partitioning is different from and does not always match the KS-based partitioning.

The Laplacian-corrected correlation functional was combined with two different GGA exchange counterparts [16,13]. The GGA exchange of Perdew and Wang [58] was derived after a careful real space analysis of the gradient expansion of the exchange hole about its LSD form. Proper cutoffs in real space were introduced in order to reproduce the exact normalization and the sign of the exchange-hole function everywhere in the molecular space. The form of this exchange functional is [58]:

$$E_x^{PW86} = A_x \int d\mathbf{r} n^{4/3}(\mathbf{r})\, F_x^{PW86}(s(\mathbf{r})) \quad , \tag{82}$$

where

$$F_x^{PW86}(s) = \left(1 + 1.296 s^2 + 14 s^4 + 0.2 s^6\right)^{1/15} \quad , \tag{83}$$

and $A_x = -\frac{3}{4}(3/\pi)^{1/3}$. The dimensionless nonlocal variable s is a typical entry in the GGA functionals, and is defined as:

$$s(\mathbf{r}) = \frac{\mid \nabla n(\mathbf{r}) \mid^2}{2n(\mathbf{r})\, k_F(\mathbf{r})} \quad , \tag{84}$$

where k_F is the local Fermi wave vector entering Eq.(69).

The other GGA exchange functional used in the Laplacian-corrected schemes is that of Becke [60]. The nonlocal part of this functional is constructed *ad hoc*, based on dimensional analysis and imposing the requirement that the correct $1/r$ asymptote of the exchange energy density be reproduced. The latter property is very important and reflects the fact that for reference points very far from a finite system the Fermi hole remains "attached" to the system [60]. The form of the exchange functional of Becke [60] is:

$$E_x^B = E_x^{LSD} - \beta \sum_\sigma \int d\mathbf{r} n_\sigma^{4/3} \frac{s_\sigma^2}{(1 + 6\beta s_\sigma \sinh^{-1} s_\sigma)} \quad , \tag{85}$$

where s_σ is a spin-polarized analog of the nonlocal variable s, Eq.(84), and β is a parameter fitted to atomic Hartree-Fock data.

Concerning the GGA correlation functionals, one of the most popular is the functional of Perdew [59]:

$$E_c^P = E_c^{LSD} + \int d^{-1}(\xi) \exp -\Phi\, C_c(n) \frac{\mid \nabla n \mid^2}{n^{4/3}}\, d\mathbf{r} \quad , \tag{86}$$

with

$$\Phi = 1.745\, f[C_c(\infty)/C_c(n)] \frac{\mid \nabla n \mid}{n^{7/6}} \quad , \tag{87}$$

where $d(\xi)$is a spin-interpolating function, $C_c(n)$ is a beyond-RPA gradient-expansion coefficient, f in Eq.(87) is a semi-empirical parameter fitted to the exact correlation energy of the neon atom. Another nonlocal functional widely in use is the Lee-Yang-Par (LYP) functional, based on the Colle-Salvetti expression, Eq.(58,59).

Four different Laplacian-corrected exchange-correlation schemes were constructed and tested in [16,13]:

1. A combination of the Laplacian-corrected correlation functional, Eq.(65-68, 78,81) with the GGA exchange of Becke, Eq.(85), implemented self-consistently (BLAP1);

2. Similar combination with the GGA exchange of Perdew, Eq.(83) (PLAP1);

3. BLAP2- a perturbative realization of the BLAP1 scheme. All nonlocal corrections are included post-SCF and only in the energy expression, using the LSD exchange potential and the gradientless correlation potential (81);

4. PLAP2- a perturbative realization of the PLAP1 scheme, similar to the BLAP2 option.

These schemes were implemented in the LCGTO-DFT program deMon-KS (for details of the code see [9-11,18]). The preliminary tests have shown that in order to synchronize the Laplacian corrected, kinetic energy dependent correlation functional with the GGA exchange counterparts a slight readjustment of the coefficient b_3 in Eq.(67) (governing the contribution of the logarithmic term) is necessary. By varying this coefficient one can monitor the magnitude of the correlation contribution to the binding energies and thus optimize the performance of the functional with respect to a chosen exchange partner. Such a re-scaling *ad-hoc* of the b_3 coefficient is also necessary to absorb to some extent effects of the missing parallel-spin correlation. The DFT modeling of this type correlation (having in most cases only a small effect on the results) is very demanding due to the necessity to carefully synchronize the derivation of the corresponding energy term with the exchange counterpart [47,52,55]. This philosophy of deriving the exchange and parallel-spin correlation together will be followed in future works.

The optimal values of the parameters used in the new schemes BLAP1, PLAP1, BLAP2 and PLAP2 are given in Table 1.

4. Illustrative Examples

Various tests of the Laplacian-corrected schemes, compared to the other nonlocal options of deMon, have been performed in refs.[16,13], but the validation process is still on going. The other nonlocal functionals referred to in these tests are the GGA schemes "Becke exchange-plus-Perdew correlation" (BP) [60,59], "Perdew-Wang exchange-plus-Perdew correlation" [58,59] (PP), and "Perdew-Wang 91" (PW91) [61], the latter being a recent

TABLE 1. Values of the parameters used in the Laplacian-corrected schemes.

OPTION	α_e	α_p	b_3
BLAP1 , PLAP1	1.255	0.394	1.48
BLAP2 , PLAP2	1.27	0.392	1.48

extension and modification of the Perdew-Wang (81) scheme. The GGA functionals were briefly discussed in the previous section, Eqs.(82-87). All the different options were placed under identical numerical conditions (grid, basis sets) using a modified version of the deMon-KS code. Sufficiently large numbers of grid points was used in the calculations. For the sake of subduing side effects of the auxiliary basis set used, one extra iteration was performed at each molecular geometry with a much larger grid, and without analytical fits, as proposed in ref.[68]. The basis set is kept the same for all the options [14,15], and is chosen among the most commonly used ones within the basis set bank of deMon-KS. The same basis set was used during the optimization of the parameters of the Laplacian-corrected schemes.

An overview of the average performance of different functionals in deMon is given in Fig.1 (for binding energies) and Fig.2 (for bond distances). These contain a percentile statistical analysis [16] of the errors in the calculated binding energies and geometries of 40 small and medium sized molecules [13]. Each of the functionals is represented by a box containing 90 % of the data. The solid line inside each box is the median - the value corresponding to 50 % of the data sorted in increasing value of the error. The lower and upper dashed lines represent 25 % and 75 % of the data, respectively. The distance between these two dashed lines is called the interquartile distance (IQD). The smaller the IQD, the narrower the distribution of the data around the median. The agreement with the experiment is on average better, when the median is closer to the zero line. The absolute mean error of each functional is given in the bottom of Figs.1,2 (in kcal/mol). It is seen that the Laplacian-corrected schemes give on average improved results: for the binding energies (Fig.1) the mean errors of BLAP1 and PLAP1 are smaller by about 2-3 kcal/mol, the medians are closer to the zero line, and the IQD is narrower, compared to the GGA options PW91 and BP. The performance of the perturbative versions BLAP2 and PLAP2 is also on average better then the GGA schemes, although slightly worse than BLAP1 and PLAP1 respectively. It is interesting that while the errors for PW91 and BP are located predominantly on the positive side

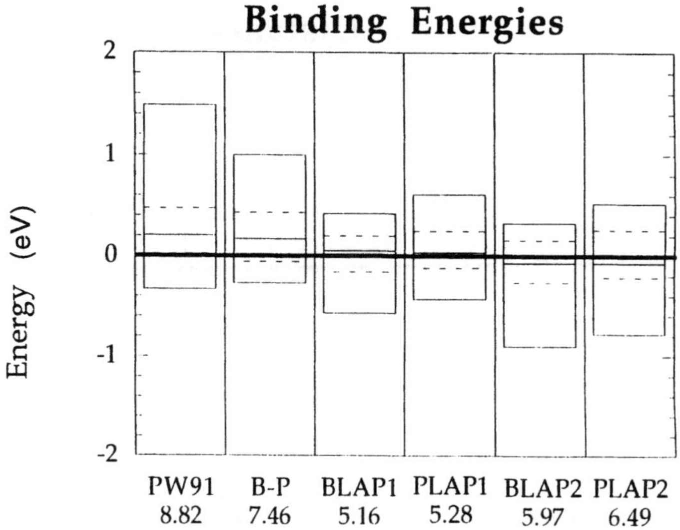

Figure 1. **Percental analysis of the binding energy errors on the set of 40 test molecules considered in ref.[13]. The mean absolute error for each functional is given at the bottom in kcal/mol.**

(overestimated binding energies), those for BLAP2 and PLAP2 show the opposite trend. For BLAP1 and PLAP1 the situation of the boxes is more symmetrical in this respect. Note that in each case the energy difference was calculated between the ground state of the molecule and the sum of the isolated atomic states converged to the appropriate integer orbital occupancies ("nonspherical" atoms). Concerning the bond distances, Fig.2, the mean absolute errors of the Laplacian-corrected schemes BLAP1 and BLAP2, are smaller than those of PW91 and BP, and the medians are closer to the zero line. Comparing the two GGA options involved, while BP gives on average better binding energies, PW91 gives better bond distances. It is interesting that although the BP option has the largest mean error in the bond distances, it has a rather narrow IQD. The GGA options, as well as BLAP1 and PLAP1, tend to overestimate the bond lengths, while BLAP2 (which gives the best results here) is more balanced with respect to the zero line.

A more detailed analysis of the binding energy data has shown [13], that in fact about 10 molecules among the 40 studied are mainly responsible for the relatively large mean errors depicted in Figs.1,2 (between 5.16 and 8.82 kcal/mol). The results for these "difficult" molecules are collected in Table 2. Looking at the absolute mean errors for these 10 molecules, the superiority of the Laplacian-corrected schemes, and especially PLAP1, is clear.

The results for the five aromatic compounds included in Table 2 deserve special attention. The GGA options PW91 and BP fail to reproduce

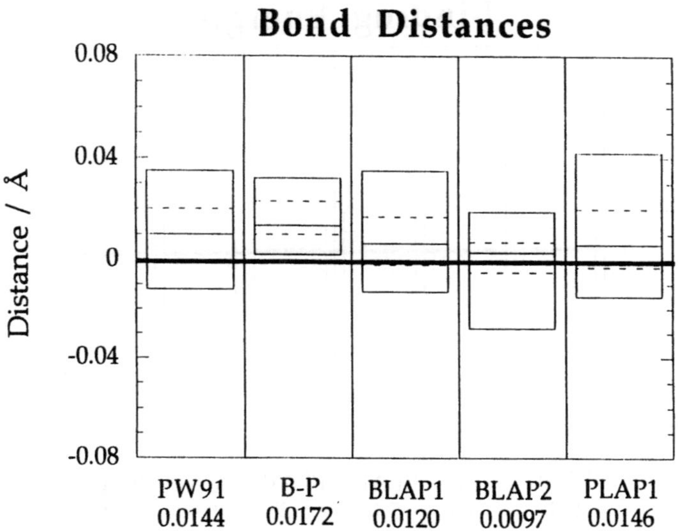

Figure 2. Percental analysis of the bond distance errors on the set of 40 test molecules considered in ref.[13]. The mean absolute error for each functional is given at the bottom.

TABLE 2. Atomization energies (calc-expt., [eV]) of selected "difficult" molecules.

molecule	PW91	B-P	BLAP1	PLAP1	BLAP2	PLAP2	expt[a]
F_2	+0.56	+0.47	+0.42	+0.4	+0.33	+0.28	1.66
O_2	+0.73	+0.57	+0.34	+0.36	+0.20	+0.28	5.21
NO	+0.62	+0.43	+0.39	+0.53	+0.10	+0.31	6.61
N_2O	+1.19	+0.82	+0.62	+0.89	+0.43	+0.78	11.87
H_2O_2	+0.41	+0.43	+0.39	+0.36	+0.20	+0.29	11.79
C_6H_6(benzene)	+1.49	+1.00	-0.39	-0.29	-0.90	-0.78	59.67
C_4H_6(cis-butad.)	+0.65	+0.56	-0.17	-0.06	-0.45	-0.35	44.2
C_4H_5N(pyrolle)	+1.11	+0.72	-0.48	-0.38	-0.84	-0.72	47.09
C_5H_5N(pyridine)	+1.61	+1.05	-0.23	-0.01	-0.70	-0.50	54.13
$C_4H_4N_2$(pyridazine)	+1.02	+0.42	-0.77	-0.43	-1.27	-0.93	48.43
abs.mean.error	0.94	0.65	0.42	0.37	0.54	0.52	
for these 10 mol.	(21.7)	(14.9)	(9.7)	(8.6)	(12.5)	(12.04)	(kcal/m

[a] The experimental values (corrected for ZPE) are taken from ref.[69]

the experimental E_b of these molecules, largely overestimating the binding. The Laplacian-corrected schemes on the contrary tend to underestimate the binding of these compounds, but the absolute errors are much smaller, compared to the GGA ones (except for the BLAP1 and BLAP2 E_b of pyridazine). The performance of PLAP1 on these aromatic systems is relatively better. As a whole, the aromatic compounds seem to be a challenge for the KS DFT methods.

Let us turn now to another example, calculation of relative energies of "difficult" reactions. By difficult we mean here a large discrepancy between the calculated and the experimental relative energies by 10 kcal/mol or more. Table 3 contains 10 such reactions selected among the 29 studied in ref.[13]. Of course, one should keep in mind that for some of these large discrepancies the experimental precision could be in doubt. Overall, however, judging from the absolute mean errors (Table 3), the reaction test confirms the advantage of the Laplacian-corrected schemes when cases that are difficult for DFT are considered. At the same time the relatively large errors for all the options in these cases demand further improvement of the functional and numerical precision. The absolute mean errors with respect to all the 29 reactions studied in [13] are also given in Table 3 (much smaller than those for the difficult reactions alone). In spite of the average improvement, the Laplacian-corrected options gave worse results than the BP scheme for some reactions [13]. The BP option yields on average superior relative energies among the GGA options implemented in deMon, and only this GGA option is included in Table 3.

TABLE 3. Relative energies of selected "difficult" reactions (kcal/mol)

reaction	B-P	BLAP2	BLAP1	PLAP1	expt [a]
$N_2+3H_2 \longrightarrow 2NH_3$	-46.4	-33.8	-35.0	-33.6	-37.0
$H_2O_2+H_2 \longrightarrow 2H_2O$	-76.5	-77.1	-77.8	-75.8	-86.0
$F_2+H_2 \longrightarrow 2HF$	-119.2	-120.7	-120.9	-118.1	-133.0
$O_2+2H_2 \longrightarrow 2H_2O$	-111.5	-112.2	-114.6	-112.2	-125.0
$CO+2CH_4+2H_2O \longrightarrow 3CH_3OH$	+12.9	+29.1	+25.5	24.7	+27.0
$N_2+4NH_3 \longrightarrow 3N_2H_4$	+98.0	+118.2	+113.6	114.9	+107.0
$CHF_3+2CH_4 \longrightarrow 3CH_3F$	+28.6	+24.0	+23.5	+22.0	+32.9
$CF_4+3CH_4 \longrightarrow 4CH_3F$	+42.0	+34.1	+33.3	+30.5	+49.3
$3C_2H_2 \longrightarrow C_6H_6$	-174.5	-150.7	-152.0	-148.5	-146.2 [b]
$3C_2H_4 \longrightarrow C_6H_6+3H_2$	-8.21	-0.6	-0.7	+1.0	-2.3 [b]
mean abs. error for these 10 react.	11.51	8.08	7.36	8.48	
mean abs. error for 29 reactions	6.37	5.49	4.77	5.53	

[a] Data from ref.[70] (corrected for ZPE, at 0°K). [b] Estimated from thermochemical data (corrected for ZPE, at 0°K).

A longstanding challenge for both the DFT and ab initio methods, and for the experimental techniques, is presented by the glycine conformers (a recent detailed review of the problem is given in [71]). It has proven to be especially difficult to reproduce the small energy difference between the two lowest-energy conformers (about 1.0 ± 0.5 [71]), mainly due to the presence of internal hydrogen bonds. A detailed study of the glycine problem employing the Laplacian-corrected exchange-correlation schemes is presented in ref.[72]. The performance of these schemes is very encouraging, especially BLAP1 and PLAP1, yielding conformational energies very close to the experimental ones, within the experimental uncertainty. The GGA options BP, PP and PW91 on the contrary meet substantial difficulties, giving even a wrong order of stability for the first two conformers [72]. BLYP and its hybrid analog B3LYP, give the correct order of the conformer energies, but the energy difference is underestimated.

Further great challenges for KS DFT are in the treatment of weakly interacting systems. The water dimer, for example, has been revisited many times by various DFT and *ab initio* studies [13,73-77] The experimental measurements of the dimerization energy (E_d) still meet difficulties in reaching spectroscopic accuracy and give a too wide range of uncertainty (± 0.7 kcal/mol) [78]. Recently a revised value for the actual E_d was suggested (4.73 ± 0.1 kcal/mol) comparing several different high quality *ab initio* calculations [76,77]. The O-O distance is also problematic. The reproduction of its experimental value (R_{O-O}=2.952 Å, corrected for anharmonicity effects [76,79]) by the current DFT and *ab initio* methods is not satisfactory. Table 4 contains a summary of the best literature results up to date, and the results obtained in ref.[13] with different nonlocal schemes within the modified version of deMon-KS. It is difficult to assess precisely the quality of

TABLE 4. Selected results for the water dimer.

.	PP	BP	BLAP1	PLAP1	MP2 [a]	CEPA-1 [b]	exp [c]
E_d(kcal/mol)	-6.18	-4.74	-3.01	-4.38	-5.44	-4.73	-4.73(±0.1)
R_{O-O}(Å)	2.873	2.892	3.00	2.954	2.911	2.949	2.952

[a] Ref.[75]; [b] Ref.[76]; [c] Revised experimental data from [76,77].

the different options with respect to the dimerization energy, in view of the large experimental uncertainty for this quantity. However, the best estimate (4.73 kcal/mol) would indicate that the energies from BP and PLAP1 are of the right order. The PP result tends to overestimate the binding, while the BLAP1 option gives the opposite trend. Clearly, further work will be required before universal high accuracy (say 0.1 kcal/mol) can be claimed for hydrogen-bonded systems. Concerning the O-O distance, the result of PLAP1 is the best among all the options included in Table 4; BLAP1 has an error (+0.048 Å) similar to that of BP (-0.06 Å).

Recently, another type of weakly interacting systems has proven to be a stringent test for DFT (and *ab initio*): the charge-transfer complexes between ethylene and a halogen molecule [12,80]. The LSD and GGA options overestimate considerably the interaction energy in these systems. The optimized intermolecular distance given by these functionals is too short, probably due to the incorrect long range behavior of their exchange potentials. Table 5 contains results for such systems, the complexes between ethylene and Cl_2, and F_2, as obtained in refs. [12,80]. Among the Laplacian-corrected

TABLE 5. Intermolecular distance d and interaction energy for $C_2H_4\ Cl_2$ and $C_2H_4\ F_2$

.	BP[a]	BLAP1	PLAP1	HH-LYP[a]	MP2[a]	expt[a]
$C_2H_4\ Cl_2$						
d(Cl-plane)[Å]	2.649	2.930	2.870	3.055	3.003	3.128
E_{int}(kcal/mol)	-5.2	-0.1	-1.73	-2.1	-1.6	-1.7 to -2.7
$C_2H_4\ F_2$						
d(F-plane)[Å]	1.922	2.037	2.027	2.945	2.894	
E_{int}(kcal/mol)	-25.7	-4.1	-8.2	-0.5	-0.4	

[a] Data from refs.[12,80]; [b] Results of ref.[13].

options only PLAP1 yields a very good interaction energy for $C_2H_4\ Cl_2$, close to the experimental one. The GGA options overestimate this energy by a factor of two, while BLAP1 substantially underestimates it. Both BLAP1 and PLAP1 give an improved intermolecular distance compared to PP and BP, but still shorter than the experimental one. The hybrid type method "half-and-half" (HH-BLYP) [62,63,39b,80] yields a very good interaction energy for $C_2H_4\ Cl_2$. The intermolecular distance is also good, close to the result of MP2, but still much shorter than the experimental distance. It is interesting that the more refined hybrid scheme B3LYP [63] gives worse results in this case [12,80].

Considering the $C_2H_4\ F_2$ complex, it is difficult to assess the performance of the different methods due to the lack of experimental data. Certainly, the huge relative energies which the GGA options yield for this weakly interacting complex are spurious.

5. Conclusions

Contemporary DFT is becoming a more and more powerful and economic method of electronic structure calculations including electron correlation. In this chapter we have emphasized its applicability to atoms and molecules, but solid state phenomena are also very much a part of DFT territory. Powerful and widely applicable KS DFT algorithms have been developed, requiring less computational effort than CI and other wave function- and Green function-based correlated methods. In CI the exact functional (of the many-electron wave function) is easy to formulate, since the many-electron Hamiltonian is well-known. The difficult part of the CI task is to solve the corresponding variational equations for the many-electron wave function. In KS DFT the "wave-function" part of the problem is simple (if one ignores subtleties associated with the DFT treatment of excited states and the multiplet problem): solve a system of single-particle Hartree-like equations to determine the KS orbitals. These in turn deliver the exact electron density and hence the ground state properties of the system. This is in essence a "single-configuration" problem. The difficulty here is in the functional part, and the progress of this method depends essentially on the development of more and more accurate functionals.

We have shown in this chapter in more detail how a new correlation functional can be derived by applying the adiabatic connection method in a straightforward manner. This of course is not the only possible scenario, but in our opinion it is one of the more advantageous, given the specifics of the KS DFT formalism. The choice of an appropriate pair correlation function at the outset is the decisive step in this scenario. The tremendous difficulties associated with the analytical derivation of this function from first principles, demanding not only a solution of a complicated dynamical hierarchy of equations, but also handling the famous N-representability problem [8,19], force one to look for other alternatives. General, formally exact properties, such as cusp and asymptotic conditions, sum rules, symmetry requirements, coordinate-scaling properties, virial relationships and the like, are indispensable tools for this purpose. Although the "ghost of (functional) N-representability" [19] still remains, as more and more rigorous conditions of this kind are formulated and fulfilled, the constructed models will reflect more and more aspects of reality. The adiabatic connection method within the KS DFT approach has its own technical demands and subtleties. Even if the λ-dependent pair correlation function is reasonably modeled, there is no guarantee that the analytical integration over the coupling-strength parameter will be possible. For the considered family of spin-polarized CF of Colle-Salvetti type, Eqs.(48-49a), we achieved this goal so far only for the particular value of the parameter m=2, and using interpolation (60) instead of the exact solution of the sum rule (53). In the future, it might be possible to use auxiliary functions for a more precise interpolation of λ dependent CFs. For example, this would allow the λ integration to be performed analytically for arbitrary m, using the exact solution of the sum rule.

Combining the Laplacian-corrected correlation functional with efficient GGA-exchange counterparts, we have implemented several new nonlocal exchange-correlation schemes within the LCGTO-DFT program deMon. Extensive tests have shown that these new schemes give on average improved results for various molecular properties, compared to the pure GGA schemes. In this chapter we discussed selected examples which are difficult for DFT. The advantages of the Laplacian- corrected schemes become more clear on these examples. However, further improvement is necessary in order to handle such systems with higher accuracy. Some of the problems could disappear with the development of better exchange functionals than the currently employed GGA. Incorporating Laplacian corrections into the exchange as well is promising, and work along this line is in progress. In its present form the Laplacian-corrected correlation functional also leaves room for improvement: finding a suitable nonlocal representation of the 'Hartree-like' part of the RDM-2, deriving the fully self-consistent correlation potential, optimizing the model with respect to the parameter m, including the parallel-spin correlation, and others.

It might also be worth examining the properties of pair correlation functions in spaces of variable dimensionality (see other chapters in this book). Requiring that an approximate exchange-correlation functional incorporate the correct (exact or interpolated) variation with dimensionality would be in the spirit of the methods exposed in this chapter and we hope that the idea will be pursued.

ACKNOWLEDGMENTS

We thank Professors Tsipis, Popov, Herschbach and Avery for their efforts in organizing this very stimulating workshop and its proceedings. S. Sirois and D. Wei contributed to the applications for glycine and the water dimer, respectively, and we thank them for this. We gratefully acknowledge financial support from NSERC, and computing resources from the Services Informatiques de l'Université de Montréal. We thank Mel Levy for illuminating discussions.

6. References

1. Thomas, L.H. (1927) *Proc. Cambridge Phil. Soc.* **23**, 542.
2. Fermi, E. (1927) *Rend. Accad. Naz. Lincei* **6**, 602.
3. March, N.H. (1992) *Electron Density of Atoms and Molecules*, Academic Press, New York.
4. Slater, J.C. (1951) *Phys. Rev.* **81**, 385.
5. Hohenberg, P., Kohn, W. (1964) *Phys. Rev.* **136**, B864.
6. Kohn, W., and Sham, L.J. (1965) *Phys. Rev.* **140**, A1133.
7. Gross, E.K.U., and Dreizler, R.M. (1990) *Density Functional Theory. An Approach to the Quantum Many-Body Problem*, Springer-Verlag, Berlin.
8. Parr, R.G., and Yang, W. (1989) *Density-Functional Theory of Atoms and Molecules*, Oxford Univ.Press, Oxford.
9. Salahub, D.R., Castro, M.E., and Proynov, E.I. (1994) in G.L. Malli(ed.), *Relativistic and Electron Correlation Effects in Molecules and Solids*, vol.318 of NATO ASI Series B: Physics, Plenum Press, New York, p. 411.
10. Salahub, D.R., Castro, M.E., Fournier, R., Calaminici, P., Godbout, N., Goursot, A., Jamorski, C., Kobayashi, H., Martinez, A., Papai, I., Proynov, E., Russo, N., Sirois, S., Ushio, J., and Vela, A. (1995) in H. Sellers and J.T. Golab(eds.), *Theoretical and Computational Approaches to Interface Phenomena*, Plenum Press, New York, p.187.
11. Malkin, V.G., Malkina, O.L., Eriksson, L.A., and Salahub, D.R. (1995) in J.M. Seminario and P. Politzer(eds.), *Modern Density Functional Theory. A Tool for Chemistry*, Elsevier, Amsterdam, p. 273.
12. Ruiz, E., Salahub, D.R., and Vela, A. (1995) *J. Am. Chem. Soc.* **117**, 1141.
13. Proynov, E.I., Ruiz, E., Vela, A., and Salahub, D.R. (1995) *Int. J. Quant. Chem. Symp.*, submitted.
14. Proynov, E.I., and Salahub, D.R. (1994) *Phys.Rev. B* **49**, 7874.
15. Proynov, E.I., Vela, A., and Salahub, D.R. (1994) Phys. Rev. A **50**, 3766.
16. Proynov, E.I., Vela, A., and Salahub, D.R. (1994) *Chem. Phys. Lett.* **230**, 419; erratum, ibid. (1995) **234**, 462.
17. Colle, R., Salvetti, O. (1975) *Theoret.Chim.Acta* **37**, 329.
18. St-Amant, A., and Salahub, D.R. (1990) *Chem. Phys. Lett.* **169**, 387.
19. Kryachko, E.S., and Ludeña, E.V. (1990) *Energy Density Functional Theory of Many-Electron Systems*, Kluwer Academic Publishers, Dordrecht.
20. Cohen, L., Santhanam, P., Frishberg, C. (1980) *Int. J.Quant.Chem.Symp.* **14**, 143.
21. Levy, M. (1979) *Proc. Nat. Acad. Sci. USA* **76**, 6062.

22. Levy, M. (1982) *Phys. Rev. A* **26**, 1200.
23. Janak, J.F. (1978) *Phys. Rev. B* **18**, 7165.
24. Malkin, V.G., Malkina, O.L., Casida, M.E., and Salahub, D.R. (1994) *J. Am. Chem. Soc.* **116**, 5898.
25. Duffy, P., Chong, D., Casida, M.E., and Salahub, D.R. (1994) *Phys. Rev. A* **50**, 4707.
26. Casida, M.E. (1995) in D.P. Chong(ed.), *Recent Advances in Density-Functional Methods*, World Scientific, Singapore, in press.
27. Guan, J., Casida, M.E., Köester, A., and Salahub, D.R. (1995) *Phys. Rev. B*, in press.
28. a) Levy, M. (1990) *Adv. Quantum Chem.* **21**, 71;
b) Levy, M., and Perdew, J. (1994) *Int. J. Quant. Chem.* **49**, 539.
29. a) Levy, M. (1991) *Phys. Rev. A* **43**, 4637;
b) Levy, M. and Perdew, J. (1985) *Phys. Rev. A* **32**, 2010.
30. Gunnarsson, O., and Lundqvist, B.I. (1976) *Phys. Rev. B* **13**, 4276.
31. Sambe, H., Felton, R.H. (1975) *J. Chem. Phys.* **62**, 1122.
32. Dunlap, B.I., Connolly, J.W.D., and Sabin J.R. (1979) *J. Chem. Phys.* **71**, 3396.
33. Andzelm, J., Radzio, E., and Salahub, D.R. (1985) *J. Comp. Chem.* **6**, 520.
34. Tatewaki, H., and Huzinaga, S. (1979) *J. Chem. Phys.* **71**, 4339.
35. Andzelm, J., and Wimmer, E. (1991) *J. Chem. Phys.* **96**, 1280.
36. Delley, B. (1990) *J. Chem. Phys.* **92**, 508.
37. Becke, A.D., and Dickson, R.M. (1989) *J. Chem. Phys.* **89**, 2993.
38. Baerends, E.J., Ellis, D.E., and Ros, P. (1973) *Chem.Phys.* **2**, 41.
39. a)Johnson, B.G., Gill, P.M., and Pople, J.A. (1992) *J. Chem. Phys.* **98**, 5612;
b) Gill, P.M.W., Johnson, B. G., Pople, J. A., and Frisch, M. J. (1992) *Int.J.Quant.Chem.Symp.* **26**, 319.
40. Zhao, Q., Parr, R.G. (1992) *Phys. Rev. A* **46**, 2337.
41. Von Barth, U., and Hedin, L. (1972) *J. Phys. C* **5**, 1629.
42. Becke, A. (1988) *J. Chem. Phys.* **88**, 1053.
43. Becke, A. (1994) *Int. J.Quant.Chem.Symp.* **28**, 625.
44. Wilson, L.C., and Levy, M. (1990) *Phys. Rev. B.* **41**, 12930.
45. Zhao, Q., Levy, M., and Parr, R.G. (1993) *Phys. Rev A* **47**, 918.
46. Gritsenko, O.V., Rubio, A., Balbás, L.C., and Alonso, J.A. (1993) *Chem.Phys.Lett.* **205**, 348.
47. Gritsenko, O.V., Rubio, A., Balbás, L.C., and Alonso, J.A. (1993) *Phys. Rev. A* **47**, 1811.
48. Burke, K., Perdew, J.P., and Levy, M. (1995) in J.M. Seminario and P. Politzer(eds.), *Modern Density Functional Theory. A Tool for Chemistry*, Elsevier, Amsterdam, p. 29.

49. Langreth, D.C., and Perdew, J.P. (1975) *Solid St. Commun.* **17**, 1425.
50. Wang, Y., and Perdew, J.P. (1991) *Phys. Rev. B* **43**, 8911.
51. Wang, Y., and Perdew, J.P. (1991) *Phys. Rev. B* **44**, 13298.
52. Perdew, J.P., and Wang, Y. (1992) *Phys. Rev. B* **46**, 12947.
53. Ceperley, D.M., and Alder, B.J. (1980) *Phys. Rev. Lett.* **45**, 566.
54. Vosko, S.H., Wilk, L., and Nusair, M. (1980) *Can.J.Phys.* **58**, 1200.
55. Rajagopal, A., Kimball, J., and Banerjee, M. (1978) *Phys. Rev. B.* **18**, 2339.
56. Salahub, D.R. (1987) *Adv. Chem. Phys.* **68**, 447.
57. Perdew, J.P., and Zunger, A. (1981) *Phys. Rev. B* **23**, 5048.
58. Perdew, J.P., and Wang, Y. (1986) *Phys. Rev. B* **33**, 8800.
59. Perdew, J.P. (1986) *Phys. Rev. B* **33**, 8822; Erratum (1986) **34**, 7406E.
60. Becke, A.D. (1988) *Phys. Rev. A* **38**, 3098.
61. Perdew, J.P., Chevary, J.A., Vosko S.H., Jackson, K.A., Pederson, M.R., Sing, D.J. and Fiolhais, C. (1992) *Phys. Rev. B* **46**, 6671.
62. Becke, A.D (1993) *J. Chem. Phys.* **98**, 1372.
63. Becke, A.D (1993) *J. Chem. Phys.* **98**, 5648.
64. Lee, Ch., Yang, W., and Parr, R. (1988) *Phys. Rev. B* **B37**, 785.
65. Lie, G.C., and Clementi, E. (1974) *J. Chem. Phys.* **60**, 1275,1288.
66. Lee, C., and Parr, R.G. (1987) *Phys. Rev. A* **A35**, 2377.
67. Ruiz, E., Kobayashi, H., Proynov, E., and Salahub, D.R. (1995) to be published.
68. Malkin, V.G., Malkina, O.L., Casida, M.E., and Salahub D.R. (1994) *J. Am. Chem. Soc.* **116**, 5898.
69. Clementi, E., and Chakravorty, S.J. (1990) *J. Chem. Phys.* **93**, 2591.
70. Lee, Ch., and Sosa, C. (1994) *J. Chem. Phys.* **100**, 9018.
71. Nguyen, D.T., Scheiner, A.C., Andzelm, J.W., Sirois, S., Salahub, D.R., and Hagler, T. (1995) *J. Am Chem. Soc.*, submitted.
72. Sirois, S., Proynov, E., Salahub, D.R., Nguyen, D., and Hagler, A.T. (1995),to be published.
73. Sim, F., St-Amant, A., Papai, I., and Salahub, D.R. (1992) *J. Am. Chem. Soc.* **114**, 4391.
74. Kim, K., and Jordan, K.D. (1994) *J. Phys. Chem.* **98**, 10089.
75. Frisch, M.J., DelBene, J.E., Binkley, J.C., and Schaefer, H.F. (1986) *J. Chem. Phys.* **84**, 2279.
76. van Duijneveldt-van de Rijdt, J.G., and van Duijneveldt, F.B. (1992) *J. Chem. Phys.* **97**, 5019.
77. Chalashinski, G., and Szczesniak, M.M. (1994) *Chem. Rev.* **94**, 1723.
78. Curtiss, L.A., Frurip, D.L., and Blander, M. (1979) *J. Chem. Phys.* **71**, 2703.
79. Odutola, J.A., and Dyke, T.R. (1980) *J. Chem. Phys.* **72**, 5062.
80. Ruiz, E., Salahub, D.R., and Vela, A. , to be published.

ELECTRON CORRELATION AND THE STRUCTURE OF THE EXCHANGE-CORRELATION POTENTIAL AND THE CORRELATION ENERGY DENSITY IN DENSITY FUNCTIONAL THEORY

EVERT JAN BAERENDS, OLEG V. GRITSENKO
AND ROBERT van LEEUWEN

Afdeling Theoretische Chemie,
Vrije Universiteit, De Boelelaan 1083,
1081 HV, Amsterdam, The Netherlands

Abstract

The relationship is established between key quantities in density functional theory, such as the exchange-correlation energy and potential, and the physics of electron correlation as described by density matrices and exchange and correlation holes. We demonstrate that it is natural to break down the exchange-correlation part of the total Kohn-Sham potential into a kinetic contribution and a screening or hole contribution, and the so-called responses of these potentials. The kinetic and hole potentials immediately afford the exchange-correlation energy density. The response parts do not enter the energy density directly but it is essential to use them in the self-consisten Kohn-Sham potential in order to obtain accurate Kohn-Sham orbitals and an accurate electron density.
The introduced potentials have certain charateristic features (peaks, steps), which carry over to the correlation energy density. The relationship of these features with the physics of electron correlation is demonstrated.

1. Introduction

The current popularity of density functional theory (DFT) in chemistry stems mostly from the combination of accuracy and computational efficiency it affords. This of course has also spurred interest in the way the notoriously difficult problem of electron correlation is treated in this theory. Although the DFT approach is very different from the standard one, which starts from the Hartree-Fock independent particle description and builds increasingly more accurate wavefunctions by inclusion of certain types of substituted determinants, the DFT methods of course also have to treat in some way the physics of electron correlation. As a matter of fact, attempts to explain [1] the success of

C. A. Tsipis et al. (eds.), *New Methods in Quantum Theory*, 395–413.

the local-density approximation (LDA) as well as the derivation of generalized gradient approximations (GGA) [2, 3, 4] rely on known properties of the exchange and correlation holes that describe the correlation between electron motions. We will in the next section introduce some of the basic concepts of DFT trying to tie them in as much as possible with the traditional picture of electron correlation in terms of exchange and correlation holes. It will be possible to identify a number of contributions to the KS potential that each have a clear relation with aspects of electron correlation. The difference between the exchange-correlation energy $E_{XC}[\rho]$) in DFT and the traditional quantum chemical definition of exchange and correlation is stressed. We will subsequently, in section 3, establish the relationship between the behaviour of the exchange and correlation holes and certain characteristics of the effective one-electron Kohn-Sham potential. Interesting features of the KS potential are the relationship with shell structure in atoms (peaks at shell boundaries, different slopes in different shells) and around the bond midpoint (peak at the bond midpoint). In the last section it will be shown that it is possible to generate exchange-correlation energy densities for E_{XC} from accurate (correlated) *ab initio* calculations. Features of the correlation energy density will be related to the physics of electron correlation discussed in the earlier sections.

2. Density functional theory and correlation holes

Correlation between the motion of the electrons can be described transparently with the help of density matrices [5]. The diagonal part of the two-electron density matrix $\Gamma(1,2)$ defined as

$$\Gamma(1,2) = \rho(1)g(1,2)\rho(2) \tag{1}$$

determines the probability to find an electron with coordinates 1 (at position $\vec{r}_1$ with spin s_1) and simultaneously another electron with coordinates 2 (at position $\vec{r}_2$ with spin s_2). The pair-correlation factor $g(1,2)$ in eq.1 embodies the effect of electron exchange and Coulomb correlation. The electron-electron part $\langle W\rangle$ of the total energy

$$E = \langle T\rangle + \langle V\rangle + \langle W\rangle \tag{2}$$

where T is the kinetic energy, V is the potential energy in the external field may accordingly be written as

$$\langle W\rangle = \frac{1}{2}\int \frac{\Gamma(1,2)}{r_{12}} d1d2 = \frac{1}{2}\int \frac{\rho(1)\rho(2)}{r_{12}} d1d2 + W_{XC} \tag{3}$$

where the first term is the Hartree energy of the electrostatic electron repulsion, while W_{XC} is the potential energy of the exchange and correlation

$$W_{XC} = \frac{1}{2}\int \frac{\rho(1)(g(1,2)-1)\rho(2)}{r_{12}} d1d2 \tag{4}$$

$$= \frac{1}{2}\int \rho(1)v^{hole}(1)d1 \tag{5}$$

In eq.5 W_{XC} is expressed as an integral of the potential $v^{hole}(1)$ of the exchange-correlation hole of the density $(g(1,2)-1)\rho(2)$ surrounding the electron at 1.

A central concept in DFT is the exchange-correlation energy E_{XC}, which is different from the traditional quantum chemical definition of the exchange-correlation energy as the sum of the Hartree-Fock exchange energy plus the correlation energy, the latter being traditionally defined as the difference between the exact and Hartree-Fock energies. E_{XC} is defined in the context of the auxiliary system of noninteracting electrons introduced by Kohn and Sham [6], that is related to the exact solution. One may postulate that there exists a local potential $v_s(\vec{r})$ such that the noninteracting electrons moving in this potential, which are thus described by the one-electron equations

$$(-\frac{1}{2}\nabla^2 + v_s(\vec{r}))\phi_i(\vec{r}) = \epsilon_i\phi_i(\vec{r}) \tag{6}$$

yield an electron density

$$\sum_i^N |\phi_i(\vec{r})|^2 = \rho(\vec{r}) \tag{7}$$

that is equal to the exact density of the fully interacting system in a given external potential $v(\vec{r})$. Application of the Hohenberg-Kohn theorem [7] to a system of noninteracting electrons shows that the Kohn-Sham potential v_s if it exists is unique. When a Kohn-Sham calculation has been performed, an alternative way to write the total energy would be

$$E = T_s[\rho] + \int \rho(1)v(1)d1 + \frac{1}{2}\int \frac{\rho(1)\rho(2)}{r_{12}}d1d2 + E_{XC} \tag{8}$$

Since the KS potential is uniquely determined by the density, so are the solutions of the one-electron Kohn-Sham equations. Therefore, the kinetic energy of the electrons described by Kohn-Sham orbitals is a functional of the density:

$$T_s[\rho] = \sum_i^N \int \phi_i^*[\rho](1)(-\frac{1}{2}\nabla^2)\phi_i[\rho](1)d1 \tag{9}$$

This follows of course also immediately from the Hohenberg-Kohn theorem, since this theorem implies that the wavefunction of a nondegenerate system is a functional of the density, and therefore every expectation value is, including that of the kinetic energy. The theorem holds for systems with arbitrary electron interaction, therefore also for the non-interacting Kohn-Sham system. So the first three terms in the r.h.s. of eq.(8), as well as the l.h.s. are defined and this equation therefore defines the socalled exchange-correlation energy E_{XC}. If we compare the equations (8) and (2) for the exact total energy, it is clear that

$$\begin{aligned} E_{XC}[\rho] &= T[\rho] - T_s[\rho] + W_{XC} \\ &= T_{XC} + W_{XC} \end{aligned} \tag{10}$$

where we have used the fact that also the exact kinetic energy $\langle T\rangle$ is a functional of ρ, and have written the difference between the exact kinetic energy and the KS kinetic energy as the exchange-correlation contribution to T (also often simply referred to as the correlation part of the kinetic energy, T_c). It is to be noted that the Kohn-Sham exchange-correlation energy consists of a kinetic part and the pure exchange-correlation part of the exact electron-electron interaction energy. In contrast, the traditional definition $E_X^{HF} + E_{corr}$ contains - sometimes sizable - corrections to the electron-nuclear and Hartree energies due to the difference $\Delta\rho$ between exact and Hartree-Fock densities:

$$
\begin{aligned}
\Delta\rho(1) &= \rho(1) - \rho^{HF}(1) \\
E_X^{HF} + E_{corr} &= E_X^{HF} + E - E^{HF} \\
&= T[\rho] - T^{HF} \\
&\quad + \int \Delta\rho(1)v(1)d1 \\
&\quad + \int \frac{\Delta\rho(1)\rho(2)}{r_{12}} d1d2 + \frac{1}{2}\int \frac{\Delta\rho(1)\Delta\rho(2)}{r_{12}} d1d2 \\
&\quad + W_{XC}
\end{aligned} \tag{11}
$$

The Kohn-Sham definition has the advantage that it only consists of the exchange-correlation corrections to the kinetic energy ($T[\rho] - T_s[\rho]$) and the exchange-correlation part of the exact electron-electron interaction energy (W_{XC}). The terms related to $\Delta\rho$ that are important in the traditional definition, such as the correlation correction to the electron-nuclear energy and the corrections to the Hartree energy may be quite large occasionally, see table 5.1 in ref. [8].

The exchange-correlation energy E_{XC} is usually expressed in DFT via the integral

$$E_{XC} = \int \rho(\vec{r})\epsilon_{xc}(\vec{r})d\vec{r} \tag{12}$$

where ϵ_{xc} is the exchange-correlation energy density per particle. Modelling of ϵ_{xc} with approximate functionals is an essential part of the development of DFT [9, 10]. From 5,10,12 it follows that one can use as ϵ_{xc} just one-half of the exchange-correlation potential corrected with the kinetic term

$$\epsilon_{xc}(\vec{r}) = \frac{1}{2}v^{hole}(\vec{r}) + v_{c,kin}(\vec{r}) \tag{13}$$

$$v_{c,kin}(\vec{r}) = v_{kin}(\vec{r}) - v_{s,kin}(\vec{r}) \tag{14}$$

The kinetic potential v_{kin} can be related to the electron correlation by expressing it in terms of the conditional amplitude [11]

$$\Phi(s_1, 2, ..N|\vec{r}_1) = \frac{\Psi(1, 2, ..N)}{\sqrt{\rho(\vec{r})/N}} \tag{15}$$

The conditional amplitude embodies all information about the correlation between the electrons. Its square describes the probability to find the other electrons with coordinates

2,3,..N when one electron is known to be at $\vec{r}_1$. The kinetic potential is related to the dependence of the conditional amplitude on the changes in the reference position $\vec{r}_1$:

$$v_{kin}([\rho];\vec{r}_1) = \frac{1}{2}\int |\nabla_1\Phi(s_1,2,..N|\vec{r}_1)|^2 ds_1 d2..dN \tag{16}$$

$v_{s,kin}$ is defined analogously to v_{kin} in terms of the one-determinantal wavefunction $\Psi_s(1,2,...,N)$ built from the Kohn-Sham orbitals $\phi_i(\vec{r})$. Note, that in DFT an alternative definition of ϵ_{xc} is often used, according to which the kinetic contribution is taken into account by the so-called coupling-constant integration [12, 13, 14]. We will use definition 13 as a more convenient one to compare with the traditional quantum chemical description of the electron exchange and correlation.

The Kohn-Sham potential may be shown to be equal to

$$v_s(\vec{r}) = v(\vec{r}) + V_{Hartree}(\vec{r}) + v_{xc}(\vec{r}) \tag{17}$$

where

$$v_{xc}(\vec{r}) = \frac{\delta E_{XC}}{\delta\rho(\vec{r})} \tag{18}$$

Taking the functional derivative of E_{XC} of eq.(10) to obtain v_{xc} leads, apart from the sum of v^{hole} and $v_{c,kin}$, to additional terms, which we have called the response potentials [15]

$$v_{xc}(\vec{r}) = v^{hole}(\vec{r}) + v_{c,kin}(\vec{r}) + v_{hole}^{resp}(\vec{r}) + v_{c,kin}^{resp}(\vec{r}) \tag{19}$$

v_{hole}^{resp} in 19 is a measure of the sensitivity of the pair-correlation factor to density variations

$$v_{hole}^{resp}([\rho];3) = \frac{1}{2}\int \frac{\rho(1)\rho(2)}{r_{12}}\frac{\delta g(1,2)}{\delta\rho(3)} d1d2 \tag{20}$$

These density variations may be understood in the following way. If the density changes to $\rho+\delta\rho$, also assumed v-representable, then according to the Hohenberg-Kohn theorem this changed density corresponds uniquely to an external potential $v+\delta v$. For the system with external potential $v+\delta v$ we have the corresponding Kohn-Sham system and the pair-correlation factor $g+\delta g$. So the derivative of g occurring in the response potential may be regarded as the response of g to density changes $\delta\rho$ caused by potential changes δv. The kinetic response potential is

$$v_{kin}^{resp}([\rho];\vec{r}_1) = \int \rho(\vec{r}_2)\frac{\delta v_{kin}([\rho];\vec{r}_2)}{\delta\rho(\vec{r}_1)} d\vec{r}_2 \tag{21}$$

The response potentials do not affect the energy directly, but they do so indirectly since they determine, as part of the Kohn-Sham potential, the SCF density in the Kohn-Sham calculation. The response part of the potential has certainly to be included in an SCF calculation if one wishes to obtain accurate KS orbitals and densities [15].

Like v^{hole} and v_{kin}, the potentials v_{hole}^{resp} and v_{kin}^{resp} can be related directly to the electron correlation in the ground state. Their sum can according to ref. [11] be expressed in terms of the conditional amplitude:

$$v_{hole}^{resp}(\vec{r}_1) + v_{kin}^{resp}(\vec{r}_1) = v^{N-1}(\vec{r}_1) \tag{22}$$

$$= \int \Phi^*(s_1, \vec{x}_2, .., \vec{x}_N|\vec{r}_1) H^{N-1} \Phi(s_1, \vec{x}_2, .., \vec{x}_N|\vec{r}_1) ds_1 d\vec{x}_2 . d\vec{x}_N \tag{23}$$

$$-E_0^{N-1} \tag{24}$$

where H^{N-1} is the Hamiltonian of electrons 2-N (interacting) in the given external potential. The value of v^{N-1} at position $\vec{r}$ is the energy expectation value of the system of $N-1$ electrons described by the conditional amplitude $\Phi(s_1, \vec{x}_2, ..., \vec{x}_N|\vec{r})$ minus the ground state energy of the $N-1$ electron system.

From eqs.13 and 19 the following relation between v_{xc} and ϵ_{xc} can be obtained

$$v_{xc}(\vec{r}) = 2\epsilon_{xc}(\vec{r}) - v_{c,kin}(\vec{r}) + v_{hole}^{resp}(\vec{r}) + v_{c,kin}^{resp}(\vec{r}) \tag{25}$$

In the next section we will study some characteristics of the Kohn-Sham potential and exchange-correlation energy density, which will prove helpful when one wants to model the KS potential as a functional of the density.

3. Characteristics of the Kohn Sham potential

It is useful to make a distinction between the case of closed shell atoms, where electron correlation is predominantly of exchange type, and the remaining (dynamical) correlation is very small compared to the exchange, and the case of dissociating molecules, such as H_2 at large bond distance, where an exchange-only treatment gives large errors and the (nondynamical or near-degeneracy) correlation is large.

Atoms. Turning first to atoms, we may assume that the Kohn-Sham determinantal wavefunction Ψ_s is a reasonable approximation of the exact wavefunction (and closely similar to the Hartree-Fock wavefunction).. It is customary to define as exchange-only DFT or optimized potential model (OPM) the problem of obtaining, for a given external potential, that local potential that generates the orbitals that minimize the energy of a single-determinantal wavefunction. This problem has been addressed by Sharp and Horton [16] and Talman and Shadwick [17]. We will use OPM calculations to obtain a close approximation to Ψ_s and corresponding quantities such as Φ_s and components of the KS potential.
Many of the special features of the potentials are related to a well known property of the exchange hole, which can be written as

$$\rho^{X\,hole}(\vec{r}_1, \vec{r}_2) = (g_s(\vec{r}_1, \vec{r}_2) - 1)\rho(\vec{r}_2) \tag{26}$$

This hole may, in keeping with its definite negative character, be put equal to minus the

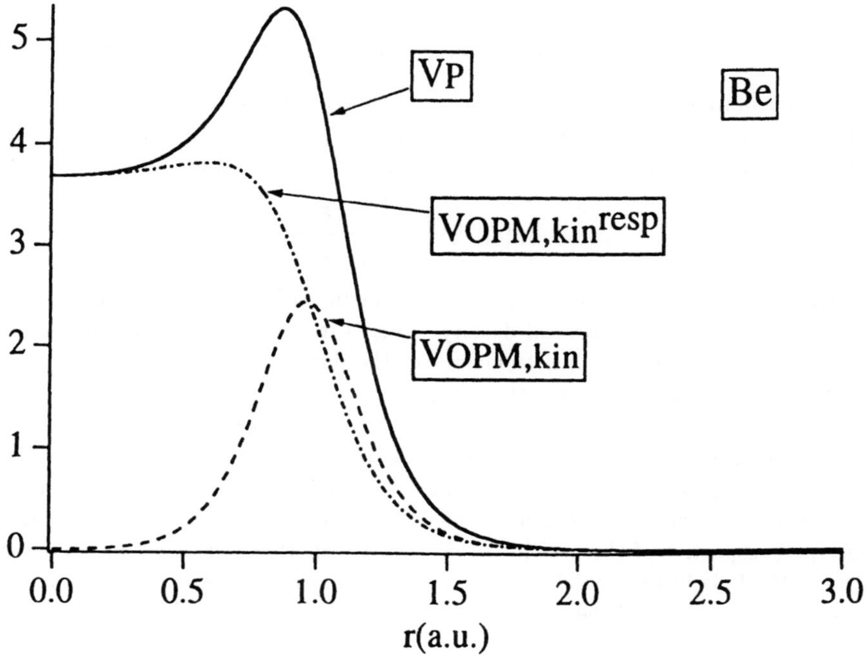

Figure 1. Kinetic potentials (in a.u.) for the *Be* atom

square of a hole amplitude $f(\vec{r}_1, \vec{r}_2)$,

$$\rho^{X hole}(\vec{r}_1, \vec{r}_2) = -\frac{1}{2}|f(\vec{r}_1, \vec{r}_2)|^2 \tag{27}$$

The hole amplitude when the electron is at reference position $\vec{r}_1$, may be written as a linear combination of the occupied orbitals, each orbital occurring with a weight determined by the contribution of that orbital to the total density at the reference point:

$$f(\vec{r}_1, \vec{r}_2) = \sum_{i=1}^{N} \frac{\phi_i(\vec{r}_1)}{\rho^{\frac{1}{2}}(\vec{r}_1)} \phi_i(\vec{r}_2) \tag{28}$$

So if at a position $\vec{r}_1$ the density consists mainly of one orbital density, the hole will have the shape of that orbital density. This is in atoms true to a large degree: within each shell, the density consists mainly of the density of the orbitals of that shell, the orthogonality wiggles of the other orbitals contributing rather little density. In ref. [8] the coefficients in the above expansion of $f(\vec{r}_1, \vec{r}_2)$ are given for the various orbitals $\phi_i(\vec{r}_2)$ in Krypton as a function of $\vec{r}_1$, demonstrating the prevalence of the $\phi_i(\vec{r}_2)$ of a certain shell when $\vec{r}_1$ is in that shell. As a matter of fact, these coefficients are approximately constant when $\vec{r}_1$ moves within the shell, in other words the exchange hole has the same shape irrespective of the position of $\vec{r}_1$ in the shell. However, when $\vec{r}_1$ crosses to another shell, the hole has to switch to the shape that is characteristic for the other shell. This implies that $\nabla_1 \Phi_s$ will be small inside a shell but large at shell boundaries.

We may use the properties of the exchange hole to understand the structure of some of the potentials we have defined above. In figure 1 the potentials $v_{s,kin}$ and $v_{s,kin}^{resp}$ as well as their sum, the Pauli potential v_P (not discussed here), are plotted. Due to the simple one-determinantal nature of Ψ_s, $v_{s,kin}$ can be expressed straightforwardly in terms of the Kohn-Sham orbitals (cf. [18]):

$$v_{s,kin}([\rho]; \vec{r}_1) = \frac{1}{2} \int |\nabla_1 \Phi_s(s_1, \vec{x}_2, ..., \vec{x}_N | \vec{r}_1)|^2 ds_1 d\vec{x}_2 ... d\vec{x}_N = \frac{1}{2} \sum_{i=1}^{N} \left| \nabla_1 \frac{\phi_i(\vec{r}_1)}{\rho^{\frac{1}{2}}(\vec{r}_1)} \right|^2 \tag{29}$$

Note that $v_{s,kin}(\vec{r}_1)$ depends on the gradients of the coefficients in the expansion of the exchange hole amplitude $f(\vec{r}_1, \vec{r}_2)$. The existence of a peak in $v_{s,kin}$ (represented in fig. 1 by $v_{OPM,kin}$) at the boundary between the $1s$ and $2s$ shells at ca. 1 bohr is in agreement with the switching of the exchange hole from $1s$ character to $2s$ character when the reference position $\vec{r}_1$ crosses the boundary between these shells, and the low values of $v_{s,kin}(\vec{r}_1)$ inside the shells agrees with the independence of the hole on the reference position inside the shells. Due to the simple one-determinantal nature of Ψ_s and the correspondingly simple Φ_s it is also possible to derive $v_{s,kin}^{resp}$ from the Kohn-Sham orbitals and energies [15]:

$$\begin{aligned} v_{s,kin}^{resp}([\rho]; \vec{r}_1) &= \int \Phi_s^*(s_1, \vec{x}_2, ..., \vec{x}_N | \vec{r}_1) H_s^{N-1} \Phi_s(s_1, \vec{x}_2, ..., \vec{x}_N | \vec{r}_1) ds_1 d\vec{x}_2 d\vec{x}_3 ... d\vec{x}_N - E_{s,0}^{N-1} \\ &= \sum_{i=1}^{N} (\mu - \epsilon_i) \frac{|\phi_i(\vec{r}_1)|^2}{\rho(\vec{r}_1)} \end{aligned} \tag{30}$$

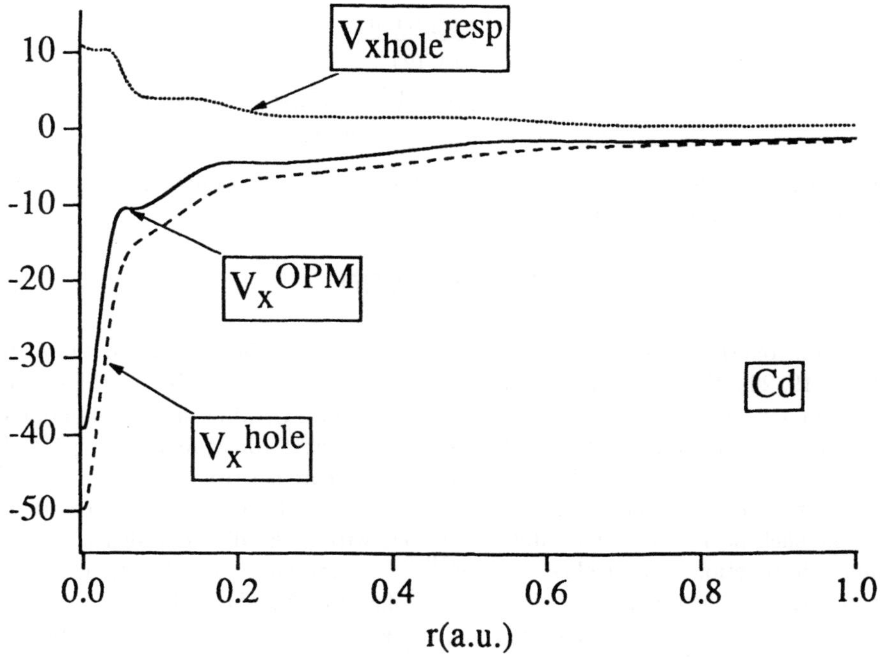

Figure 2. Exchange optimized effective potential and its components (in a.u.) for the *Cd* atom

Here $E^{N-1}_{s,0}$ is the energy of the $N-1$ electron Kohn-Sham system (see [15]) and $\mu = E^N_{s,0} - E^{N-1}_{s,0}$ is equal to the highest occupied one-electron Kohn-Sham energy, ϵ_M, which in turn is equal to the negative of the exact first ionization energy, $\epsilon_M = E^N_0 - E^{N-1}_0$. From our previous discussion it follows that we may expect $v^{resp}_{s,kin}$ to have a step like form. The factors $|\phi_i(\vec{r}_1)|^2/\rho(\vec{r}_1)$ will be constants of the order of magnitude 1 when $\vec{r}_1$ is inside a shell, and the step heights are determined by $(\mu - \epsilon_i)$. This expectation is borne out by the shape of $v^{resp}_{OPM,kin}$ in fig. 1 (see ref. [15] for Kr and Cd).

It is to be expected that in atoms, with predominantly exchange effects and only dynamical (no near-degeneracy) correlation, the potentials v_{kin} and $v_{s,kin}$ and the potentials v^{resp}_{kin} and $v^{resp}_{s,kin}$ will be rather similar, i.e. $v_{c,kin}$ and $v^{resp}_{c,kin}$ negligible. In that case the exchange-correlation potential v_{xc} would be dominated by $v^{hole}_x + v^{resp}_{xhole}$, cf. eq.(19). We will study these latter potentials in atoms in the case of exchange-only DFT, where the OPM method determines the local "exchange-only" potential

$$v^{OPM}_{x\sigma}(\vec{r}) = v^{hole}_{x\sigma}(\vec{r}) + v^{resp}_{xhole,\sigma}(\vec{r}) \tag{31}$$

where $v^{hole}_{x\sigma}$ is the potential of the exchange hole

$$v^{hole}_{x\sigma}(\vec{r}_1) = \int \frac{\rho^{X\,hole}(\vec{r}_2\sigma|\vec{r}_1\sigma)}{r_{12}} d\vec{r}_2 \tag{32}$$

$$\rho^{X hole}(\vec{r}_2\sigma|\vec{r}_1\sigma\prime) = \delta_{\sigma\sigma\prime}\frac{\Gamma_X(\vec{r}_1\sigma,\vec{r}_2\sigma)}{\rho(\vec{r}_1\sigma)} \quad (33)$$

$$= -\frac{\sum_{i,j=1}^{N_\sigma}\phi_{i\sigma}(\vec{r}_1)\phi_{i\sigma}^*(\vec{r}_2)\phi_{j\sigma}^*(\vec{r}_1)\phi_{j\sigma}(\vec{r}_2)}{\rho_\sigma(\vec{r}_1)} \quad (34)$$

Since the exchange energy functional does not have a kinetic contribution, the OPM exchange energy density is just one-half of $v_{x\sigma}^{hole}$

$$\epsilon_{x\sigma}^{OPM}(\vec{r}) = \frac{1}{2}v_{x\sigma}^{hole}(\vec{r}) \quad (35)$$

$v_{x\sigma}^{hole}$ is often referred to as the Slater potential, since it has been analyzed thoroughly by Slater [19, 20]. When calculating the Slater potential from the orbitals, the response part is obtained simply as the difference between the total OPM exchange potential and $v_{x\sigma}^{hole}$. These potentials are plotted in figure 2 for Cd. v_x^{hole} (or equivalently ϵ_x^{OPM}) has a deep well around the nucleus due to exchange of the core electrons. At large distances r v_x^{hole} approaches the Coulombic asymptotics $-\frac{1}{r}$. At intermediate distances it does reflect the shell structure, having different slopes within the different shells, still it is a monotonic function of r. Unlike v_x^{hole}, v_x^{OPM} is a non-monotonic function of r with characteristic intershell maxima. It has been emphasized that these "bumps" are needed to lower the total energy [21]. Interestingly, the maxima originate by the addition of a very simply structured response potential to v_x^{hole}: v_{xhole}^{resp} is flat over the shell regions and exhibits steps in going from one shell to the next. This intriguing step structure of v_{xhole}^{resp} is reminiscent of the step structure found for the response part of v_{kin}. The origin of the step structure in v_{xhole}^{resp} is again related to the special properties of the exchange hole discussed earlier. In order to derive that these properties of the hole lead to the observed step structure in the screening-response potential, we have to consider more closely eq.20 Using for the pair correlation factor in the exchange-only case the well known expression

$$g_{s\sigma}(\vec{r}_2,\vec{r}_3) = 1 - \frac{\sum_{i,j}^{N_\sigma}\phi_i(\vec{r}_2)\phi_i^*(\vec{r}_3)\phi_j(\vec{r}_3)\phi_j^*(\vec{r}_2)}{\rho_\sigma(\vec{r}_2)\rho_\sigma(\vec{r}_3)}, \quad (36)$$

it is clear that taking the derivative with respect to $\rho_\sigma(\vec{r}_1)$ requires knowledge of

$$\frac{\delta\phi_i(\vec{r}_2)}{\delta\rho_\sigma(\vec{r}_1)} = \int\frac{\delta\phi_i(\vec{r}_2)}{\delta v_{s\sigma}(\vec{r}_3)}\frac{\delta v_{s\sigma}(\vec{r}_3)}{\delta\rho_\sigma(\vec{r}_1)}d\vec{r}_3 \quad (37)$$

The derivative of ϕ_i with respect to the Kohn-Sham potential is known in terms of the Kohn-Sham orbitals and one-electron energies:

$$\frac{\delta\phi_i(\vec{r}_2)}{\delta v_{s\sigma}(\vec{r}_3)} = -G_{i\sigma}(\vec{r}_2,\vec{r}_3)\phi_{i\sigma}(\vec{r}_3) \quad (38)$$

where $G_{i\sigma}$ is the following Green's function:

$$G_{i\sigma}(\vec{r}_2,\vec{r}_3) = \sum_{j\neq i}\frac{\phi_j(\vec{r}_2)\phi_j^*(\vec{r}_3)}{\epsilon_{j\sigma}-\epsilon_{i\sigma}} \quad (39)$$

The functional derivative of the Kohn-Sham potential with respect to the density is the inverse density response function:

$$\frac{\delta v_{s\sigma}(\vec{r}_3)}{\delta \rho_\sigma(\vec{r}_1)} = \chi_{s\sigma}^{-1}(\vec{r}_3, \vec{r}_1) \tag{40}$$

It is possible to derive a simple expression for the screening-response potential by application of an "effective energy denominator" type of approximation for the Green's function, as has been done before by Sharp and Horton [16] and Talman and Shadwick [17]. We refer to ref. [22] for details of the derivation. The resulting expression for the screening-response potential becomes

$$v_{xhole,\sigma}^{resp}(\vec{r}_1) = \sum_i^{N_\sigma - 1} w_{i\sigma} \frac{|\phi(\vec{r}_1)|^2}{\rho_\sigma(\vec{r}_1)} \tag{41}$$

where the constant $w_{i\sigma}$ is defined in terms of the expectation value of the local exchange potential $v_{x\sigma}$ over the orbital $\phi_{i\sigma}$ minus a sum over two-electron integrals of exchange type:

$$w_{i\sigma} = \langle \phi_i | v_X | \phi \rangle + \sum_k^{N_\sigma} \langle \phi_{k\sigma} \phi_{i\sigma} | \frac{1}{r_{12}} | \phi_{i\sigma} \phi_{k\sigma} \rangle \tag{42}$$

This is precisely the second term (now to be identified with the response part of the potential) of the approximate expression that had been obtained earlier by Krieger, Li and Iafrate [23] for the local KS potential in the exchange-only case. KLI derived their approximation using the same approximation to the Green's function. It has been demonstrated by Krieger et al. that their approximation is very accurate, while providing at the same time a computationally more tractable way to calculate the potential than the Talman-Shadwick procedure. It is interesting to note that the response part of the exchange potential has the same form as the response part of $v_{s,kin}^{resp}$, eq.(30), leading in both cases to the characteristic step-like behaviour. From the form of the expression 41 we may understand its step character, the height of the steps being governed by the constants $w_{i\sigma}$. We note that the summation in eq. 41 does not include the highest occupied orbital. The step height in the region of the outermost orbital is zero, the potential is therefore short-ranged and does not destroy the $-1/r$ asymptotic behaviour of the hole potential v_x^{hole}.

Figure 3 reveals the effect of Coulomb correlation on atomic potentials by means of comparison of v_x^{OPM} with v_{xc} for the Ne atom. v_{xc} has been constructed from the correlated ρ with the procedure of ref. [24]. As expected, v_{corr}, the difference of these two potentials is quite small. v_{corr} is a non-monotonic function of r in accordance with the theoretical considerations of ref. [25]. It has an absolute minimum at the nucleus due to strong in-out correlation of the reference electron at r = 0 and a local minimum in the intershell region, which opposes the corresponding local maximum of v_{xc}.

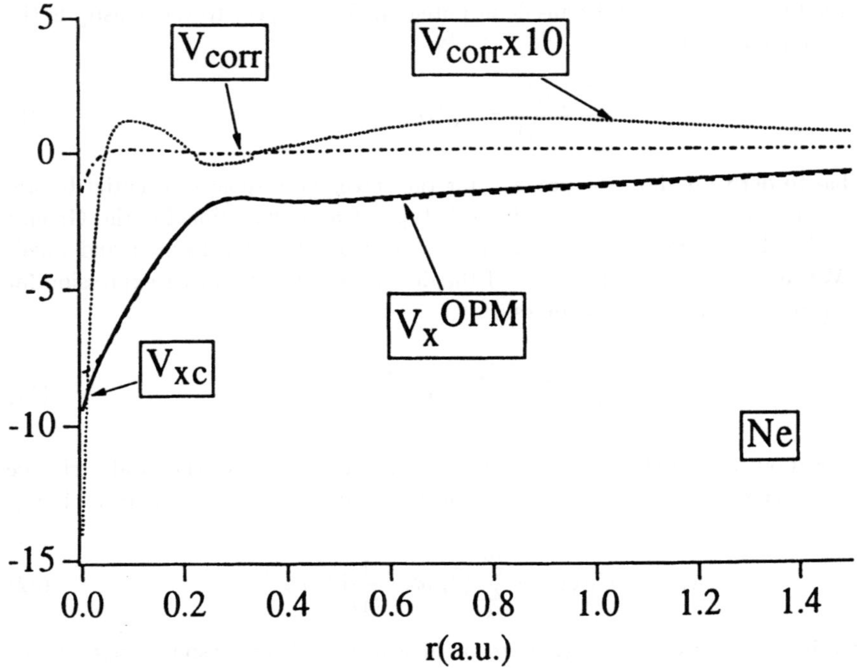

Figure 3. Exchange and exchange-correlation potentials (in a.u.) for the *Ne* atom

Molecules. In molecules we have, in particular when bonds are weak, a situation that is completely different from atoms. Now exchange is no longer dominant, but Coulomb correlation may become equally important. This is exemplified by the case of dissociating H_2. In figure 4 we give plots of the exchange (Fermi) hole, the Coulomb hole, and the total hole [8]. In all cases the reference position $\vec{r}_1$ is at 0.3 bohr to the left of the right H atom. For the Fermi hole the reference position is actually irrelevant: the hole is completely independent of the position of the reference electron and always equal to $-\frac{1}{2}\rho(\vec{r}_2)$. This is a special feature of the two-electron ("single-shell") case. In this case the approximate equality, noted above, of the exchange hole to the shell density for arbitrary reference position in the shell, becomes exact. The conditional amplitude Φ_s desribes a constant hole, i.e.its gradient with respect to the reference position will be zero, so $v_{s,kin}$ is zero. However, when the molecule dissociates and the reference electron is close to the right nucleus, the other electron is known to be around the left nucleus, so the total hole must be $-\rho(\vec{r}_2)$ around the right nucleus and vanishing around the left nucleus (see the figure for $R = 5$ bohr). The Fermi hole however allows the other electron to be with equal probability around either nucleus, which is of course the well known Hartree-Fock failure to describe left-right correlation at large R. The Hartree-Fock density will be much too diffuse, since the other electron partly screens the right nucleus which it should not. The correlation energy contribution $\int \Delta\rho(1)v(1)d1$ (cf. eq. 11) is at large distance actually larger than the correlation contribution to the two-electron energy [8].

The abovementioned features of the Coulomb hole of the H_2 molecule are reflected in the form of the correlation energy density ϵ_c, its potential $\frac{1}{2}v_c^{hole}$ and kinetic v_{kin} components

$$\epsilon_c(\vec{r}) = \frac{1}{2}v_c^{hole}(\vec{r}) + v_{kin}(\vec{r}) \tag{43}$$

$$v_c^{hole}(\vec{r}) = v^{hole}(\vec{r}) + \frac{1}{2}\int \frac{\rho(\vec{r}_2)}{r_{12}}d\vec{r}_2 \tag{44}$$

which are plotted in figure 5 for the equilibrium internuclear distance $R(H-H) = 1.40$ a.u. and in figure 6 for large distance $R(H-H) = 5.0$ a.u. along the bond axis as functions of the distance from the bond midpoint [26]. v_c^{hole} and v_{kin} have been calculated via eqs.16,44 from correlated one- and two-electron density matrices, the latter being obtained from full configuration interaction (CI) calculations of H_2 [11].

One notes the deep wells in $v_c^{hole}(z)$ and $\epsilon_c^{hole}(z)$ around the nucleus and the peaks of $v_{kin}(z)$ and $\epsilon_c^{hole}(z)$ at the bond midpoint. The well of $v_c^{hole}(z)$ reflects the appreciable reduction of the electron-electron repulsion in the atomic regions due to the left-right correlation. The attractive nature of $v_c^{hole}(z)$ will have the effect of making the density more compact around the H nuclei, which is precisely what is required in view of the much too diffuse density that results in an exchange-only treatment [11]. Closer to the bond midpoint, the left-right correlation becomes less important and $v_c^{hole}(z)$ is closer to zero. Although they have the same qualitative features, the functions $v_c^{hole}(z)$ for the equilibrium and large $H-H$ distances differ quantitatively. The minimum of $v_c^{hole}(z)$ for $R(H-H) = 5.0$ a.u. is about twice as deep as that for $R(H-H) = 1.401$, in accordance

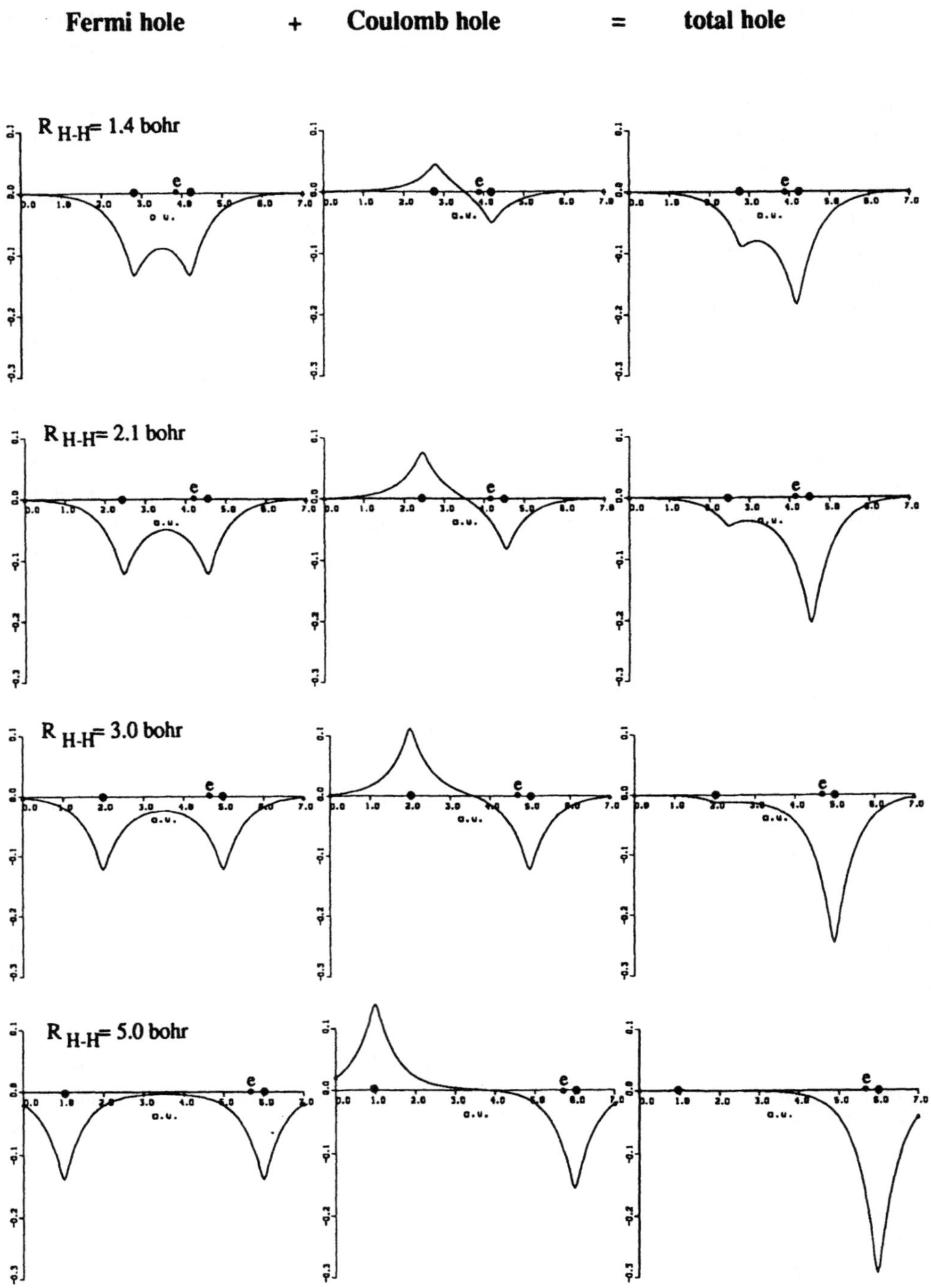

Figure 4. Fermi hole, Coulomb hole and total hole in the H_2 molecule

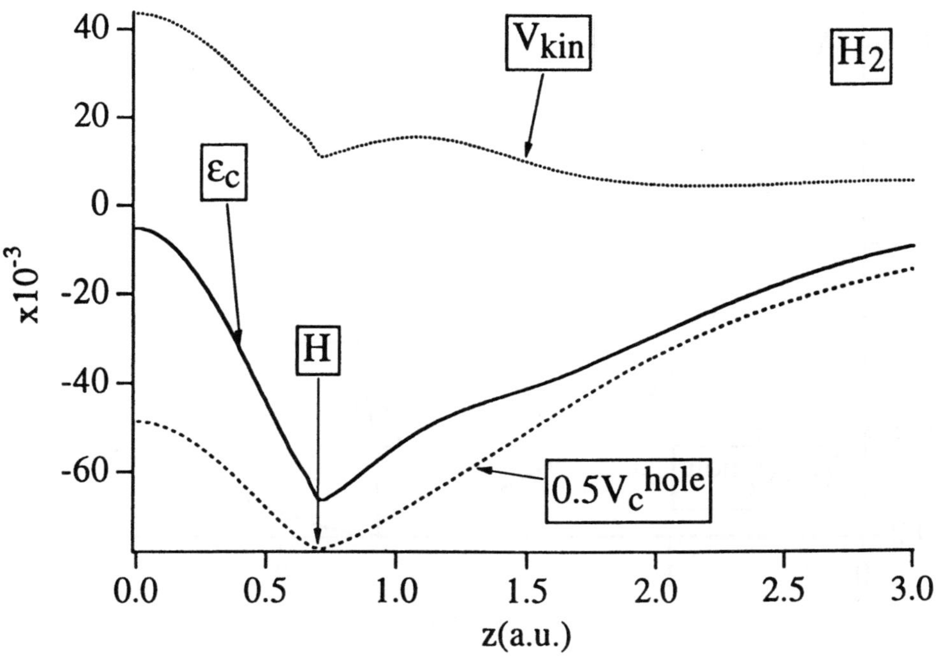

Figure 5. Correlation energy density and its components (in a.u.) for H_2 at equilibrium internuclear distance

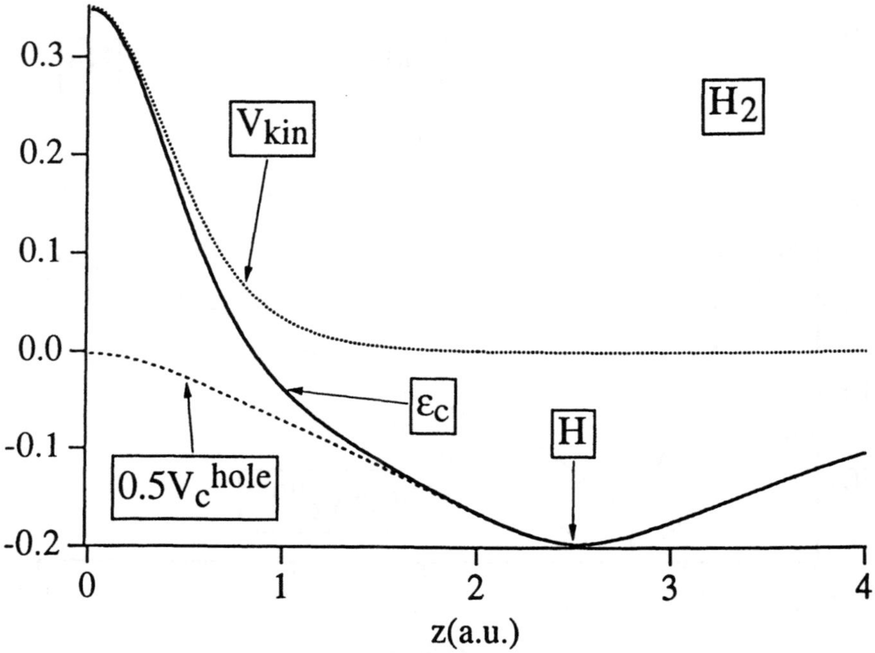

Figure 6. Correlation energy density and its components (in a.u.) for H_2 at large internuclear distance

with the stronger near-degeneracy left-right correlation at large distance. On the other hand, the local maximum of $v_c^{hole}(z)$ at the bond midpoint is much closer to zero for $R(H-H) = 5.0$. Indeed, when the reference electron is at the bond midpoint, which is in this case a region of very low density, we may expect hardly any Coulomb hole, since the symmetrical exchange hole of depth $-\rho(\vec{r})/2$ will in this case already be a good description of the total exchange-correlation hole.

Left-right correlation finds a spectacular way to manifest itself through the bond midpoint peak of $v_{kin}(z)$ [11]. By the definition (29), $v_{kin}(\vec{r})$ is determined by the average rate of change of the conditional amplitude $\Phi(s_1, \vec{x}_2, ..., \vec{x}_N|\vec{r}')$ with changing position $\vec{r}'$ of the reference electron. Φ has the maximal gradient at the bond midpoint $z = 0$, since when the reference electron crosses the bond midpoint, the other electron has to switch quickly from one atom to another due to the left-right correlation. As a consequence, $v_{kin}(z)$ possesses a peak at $z = 0$, which reflects the corresponding "jump" of the conditional amplitude. This peak becomes much higher for large $H-H$ distance, because of increase of left-right correlation in the dissociation limit. We can also mention the non-monotonous behavior of $v_{kin}(z)$ in Figure 5, with a local minimum at the H nucleus. The exact v_{kin} is not necessarily zero at the nucleus in this case, as it is e.g. in He and Be [15], but clearly still exhibits this tendency.

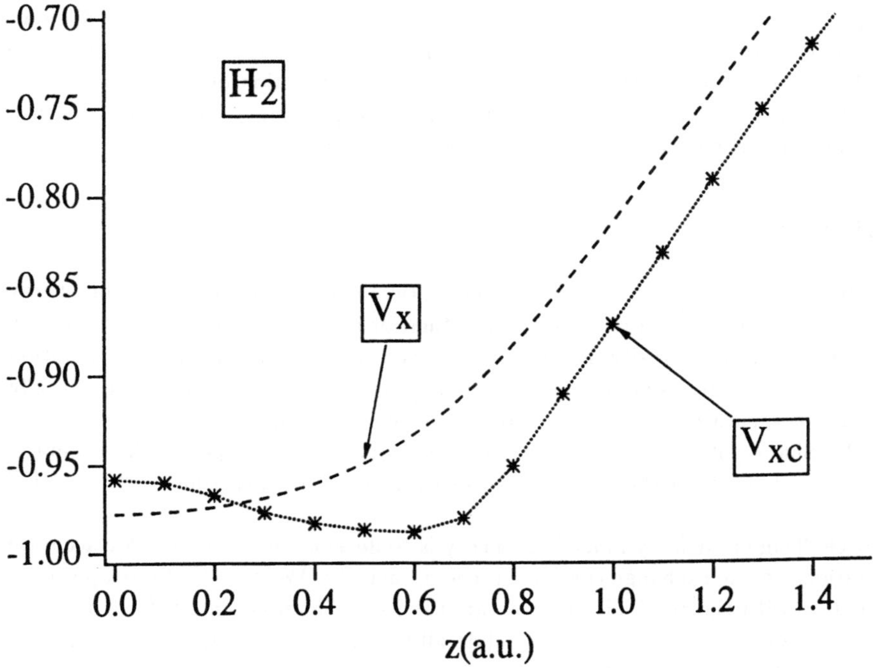

Figure 7. Exchange and exchange-correlation potentials (in a.u.) for the H_2 molecule

The individual features of $v_c^{hole}(z)$ and $v_{kin}(z)$ can be clearly recognized in the plots of the resulting $\epsilon_c(z)$. This is especially true for $R(H-H) = 5.0$ a.u. (See Figure 6). In this case $\epsilon_c(z)$ inherits the bond midpoint peak of $v_{kin}(z)$ and the well around the nucleus of $v_c^{hole}(z)$, with the height of the peak and the depth of the well being very close to those for v_{kin} and $\frac{1}{2}v_c^{hole}(z)$. Because of this, ϵ_c becomes a sign-changing function: it is positive in the bond midpoint region, it changes sign near $z = 1$ a.u. and it is negative at larger z. For the equilibrium geometry, on the other hand, $\epsilon_c(z)$ is everywhere negative (See Figure 5). In this case both $\frac{1}{2}v_c^{hole}(z)$ and $v_{kin}(z)$ have appreciable contributions to $\epsilon_c(z)$ at all z considered. Still, ϵ_c has the same qualitative features, namely, a maximum at the bond midpoint and a well around the H nucleus.

Figure 7 shows the effect of Coulomb correlation on molecular potentials by means of comparison of v_{xc} with the exchange potential v_x for the H_2 molecule at the equilibrium distance. For two-electron systems there is only one occupied Kohn-Sham orbital $\phi_1(\vec{r})$, equal (up to a phase factor) to $\rho^{\frac{1}{2}}(\vec{r})/\sqrt{2}$. Therefore, v_x in this case reduces to the potential of electron self-interaction

$$v_x(\vec{r}) = -\frac{1}{2}\int \frac{\rho(\vec{r}_2)}{r_{12}} d\vec{r}_2 \tag{45}$$

and v_{xc} can be obtained in a straightforward way from the results of the CI calculations [11, 27]. One can see from Figure 7 that v_{xc} and v_x have a qualitatively different behavior in the bond region. v_x is a monotonic function of z with the minimum at the bond midpoint. Contrary to this, v_{xc} is a non-monotonic function of z. The bond midpoint peak of v_{kin} discussed above generates the corresponding peak of the total v_{xc}, while around the nucleus v_{xc} exhibits a well (though not a deep one).

4. Conclusions

We have discussed in this paper the connection between the exchange-correlation energy and potential of density functional theory and their definitions in terms of wavefunction based quantities, such as density matrices, that have traditionally been used by quantum chemists to calculate and analyze electron correlation. This has allowed us to actually calculate, using accurate (correlated) quantum chemical calculations, some components of the Kohn-Sham potential and of the energy density. In this paper we have limited ourselves to an analysis of the special features that can be distinguished in these DFT quantities.

The great challenge in density functional theory is to develop increasingly more accurate approximations to the exchange-correlation energy density. We note that our approach, in combination with the possibility to generate the Kohn-Sham potentials from an exact (very accurate) density, allows one to calculate all the crucial elements needed in a DFT calculation. We therefore are able to judge current approximations for e.g. the exchange and correlation energy densities not just by the total energies they provide but also as functions of $\vec{r}$. More details may be found in ref. [26], where accurate "benchmark" correlation energy densities for selected atoms and molecules are compared to a series of approximate expressions that are currently in use. It appears that there are rather large local deviations in the model energy densities, even if the global numbers, i.e. the energies resulting from integration over all space, are fairly close.

Considering more generally the modelling of exchange-correlation energy densities $\epsilon_{xc}(\vec{r})$, or $\epsilon_c(\vec{r})$ and $\epsilon_x(\vec{r})$ separately, we note that the functions $\epsilon(\vec{r})$ obtained from the correlated $\rho(\vec{r}', \vec{r})$ and $\Gamma(\vec{r}_1, \vec{r}_2)$ can serve as a benchmark for successful models $\epsilon^{mod}(\vec{r})$. The procedure presented in ref. [26] allows to construct $\epsilon(\vec{r})$ for an arbitrary many-electron system, which opens new possibilities for the DFT modelling. When developing a new $\epsilon_c^{mod}(\vec{r})$ or $\epsilon_x^{mod}(\vec{r})$, one can take into account not only E_{xc} estimates or the scaling and asymptotic properties of the E_x or E_c functional, but also the local behavior of the essentially accurate $\epsilon(\vec{r})$ obtained from *ab initio* wavefunctions for a representative set of atomic and molecular systems. For instance, a promising option is to approximate directly the potential $\frac{1}{2}v_c^{hole}(\vec{r})$ and kinetic $v_{kin}(\vec{r})$ components which, as has been shown in the present paper, have a characteristic form. The corresponding work is in progress.

Acknowledgement. EJB wishes to thank the organizers for inviting him to lecture at the NATO workshop on New Methods in Quantum Theory. It has been a most stimulating experience, due to both the science and the people.

References

1. O.Gunnarson, M.Jonson, and B.I.Lundqvist. *Phys. Rev.*, B20:3136, 1979.
2. J.P.Perdew. *Phys. Rev.*, B33:8822, 1986.
3. J.P.Perdew and Wang Yue. *Phys. Rev.*, B33:8800, 1986.
4. A.D.Becke. *Phys. Rev.*, A38:3098, 1988.
5. R.McWeeny. *Methods of Molecular Quantum Mechanics.* Academic Press, 1992.
6. W.Kohn and L.J.Sham. *Phys. Rev.*, 140:1113A, 1965.
7. P.Hohenberg and W.Kohn. *Phys. Rev.*, 136B:864, 1964.
8. M.Buijse and E.J.Baerends. *In: Density Functional Theory of Molecules, Clusters and Solids, pp.1-46, D.E.Ellis, ed.* Kluwer, 1995.
9. R.G.Parr and W.Yang. *Density Functional Theory of Atoms and Molecules.* Oxford University Press, 1989.
10. R.M.Dreizler and E.K.U.Gross. *Density Functional Theory: An Approach to the Quantum Many-Body Problem.* Springer-Verlag, Berlin, 1990.
11. M.A.Buijse, E.J.Baerends, and J.G.Snijders. *Phys. Rev.*, A40:4190, 1989.
12. J.Harris and R.O.Jones. *J. Phys.*, F4:1170, 1974.
13. O.Gunnarson and B.I.Lundqvist. *Phys. Rev.*, B13:4274, 1976.
14. D.C.Langreth and J.P.Perdew. *Solid State Commun.*, 17:1425, 1975.
15. O.V.Gritsenko, R.van Leeuwen, and E.J.Baerends. *J. Chem. Phys.*, 101:8955, 1994.
16. R.T.Sharp and G.K.Horton. *Phys. Rev.*, 30:317, 1953.
17. J.D.Talman and W.F.Shadwick. *Phys. Rev.*, A14:36, 1976.
18. M.Levy and H.Ou-Yang. *Phys. Rev.*, A38:625, 1988.
19. J.C.Slater. *Phys. Rev.*, 81:385, 1951.
20. J.C.Slater. *Quantum Theory of Molecules and Solids. Vol.4. The Self-consistent Field for Molecules and Solids.* McGraw-Hill, 1974.
21. Y.Wang, J.P.Perdew, J.A.Chevary, L.D.Macdonald, and S.H.Vosko. *Phys. Rev.*, A41:78, 1990.
22. R.van Leeuwen, O.V.Gritsenko, and E.J.Baerends. *Z. Physik*, D33:229, 1995.
23. J.B.Krieger, Yan Li, and G.J.Iafrate. *Phys. Rev.*, B45:101, 1992.
24. R.van Leeuwen and E.J.Baerends. *Phys. Rev.*, A49:2421, 1994.
25. M.Levy and J.P.Perdew. *Phys. Rev.*, A32:2010, 1985.
26. P.Süle, O.V.Gritsenko, Á.Nagy, and E.J.Baerends. *J. Chem. Phys.* to be published.
27. O.V.Gritsenko, R.van Leeuwen, and E.J.Baerends. *Phys. Rev.*, A. to be published.

RELATIVISTIC PSEUDOPOTENTIALS AND NONLOCAL EFFECTS

WALTER C. ERMLER[a] AND MARIA M. MARINO[b]

Department of Chemistry and Chemical Engineering
Stevens Institute of Technology, Hoboken, New Jersey 07030, U.S.A.

Abstract

A relativistic effective potential (REP) method is modified to address problems due to nonlocal effects traceable to outer-core electrons. Atomic calculations using small-core REPs explicitly treating outer core electrons form the basis for extracting two-component nodeless valence spinors (NVS) used to generate new large-core REPs. Errors in published large-core REPs are shown to result from the inherent arbitrariness in the choice of match points that define shape-consistent pseudospinors and thus lead to an improper representation of nonlocal effects. These are made salient via calculations on InH with an REP derived from NVSs: the bond length increases relative to that calculated with a published large-core REP and agrees with that obtained using a frozen 4d-subshell small-core REP.

1. Introduction

Procedures based on the use of relativistic effective core potentials (REPs) derived from all-electron *ab initio* numerical Dirac-Fock (DF) atomic wavefunctions afford the ability to reduce the number of electrons that must be treated explicitly to a small enough number that in many instances large basis set full-CI calculations incorporating spin-orbit coupling and other relativistic effects become feasible [1]. The REP method is grounded in the fundamental tenet that atomic core and valence electrons can be formally distinguished and, in fact, treated rigorously from first principles as identifiable sets.

Since REP-based methods rely on the principle of core/valence separability, a choice must be made regarding the number of electrons that will ultimately be treated explicitly. In the procedure developed by Lee, Ermler and Pitzer (LEP) [2], which is the method discussed in this work, there are no restrictions on this choice. For example, in the case of cesium the most practical REP would correspond to a one-electron valence space. But the same atom may also be described by nine-electron and nineteen-electron valence spaces, where the former REP contains the 5s and 5p Cs electrons in the valence space and the latter contains a valence space comprised of 5s, 5p and 4d electrons. A one-electron valence space REP would allow the minimum number of electrons to be treated explicitly and, therefore, would render calculations at, say, the full-CI level of theory possible for clusters of two or more Cs atoms.

C. A. Tsipis et al. (eds.), New Methods in Quantum Theory, 415–423.

In general, there are no restrictions on the number of subshells used to define the valence space. That is, the limit where all subshells are defined as valence is equivalent to an all electron treatment. However, there are restrictions in the number of subshells chosen to define the core space because 'core' and 'valence' are not merely definitions of convenience: valence space orbitals change as a result of chemical processes such as bonding, electronic ionization and excitation, whereas core orbitals do not. Consequently, defining valence orbitals as core orbitals erroneously constrains the former.

Moreover, not all subshells can be placed either in the core or the valence spaces. In many, if not most, cases certain subshells exhibit both core and valence characteristics. These subshells are termed 'outer core' or 'inner valence'. Here the former term is preferred because these subshells are incorporated into pseudopotentials. The general subject of outer core spaces has been discussed in the context of core/valence polarization (and correlation) phenomena in terms of both frozen core [3] and pseudopotential [4] approximations. This study does not treat core/valence polarization effects. Only errors due to nonlocal effects that are revealed in the context of the frozen outer-core Dirac-Fock approximation are addressed.

It has been observed that REPs that include the largest possible number of electrons in the core, large core REPs (LCREPs), keeping in mind the above-mentioned restrictions, generally lead to bond lengths that are too short and dipole moments that are too large when compared to REPs with one or more subshells redefined as valence shells. On the other hand, retaining additional subshells as valence shells, *i.e.* small-core REPs (SCREPs), generally provide sufficient flexibility, because of the greater number of explicitly-treated electrons, so as to lead to good agreement with all-electron calculations [5,6]. Schwerdtfeger *et al.* [5] observed in calculations on InCl that using a SCREP for In led to a bond length in good agreement with all-electron Dirac-Fock results (R_e=2.45 Å in both cases). However, the use of a LCREP led to an appreciably shorter bond length of 2.38 Å.

This study attributes the above-mentioned errors to nonlocal electron repulsion effects traceable to outer-core electrons and presents a modification to the LEP procedure, making possible the derivation of corrected LCREPs that incorporate outer-core subshells. The indium atom is chosen as an example, and results are compared to those due to LCREPs derived using shape-consistent pseudospinors [6].

2. Method and Calculations

The error in the indium LCREP is due to an incorrect accounting for nonlocal effects and it is traceable to the $5d_{3/2}$ and $5d_{5/2}$ pseudospinors used in the Lee, Ermler, Pitzer, Christiansen (LEPC) method [2,7]. This method does not define an outer-core space. Consequently, SCREPs and LCREPs are derived in the same manner. First, an all-electron DF wavefunction is generated [8]. Next, nodes are removed from so-defined valence spinors by means of the shape-consistent pseudospinor procedure [7]. Finally, these nodeless pseudospinors are used to derive the SCREP or LCREP. The quality of the REP is ascertained by comparing atomic results generated using the REP against those due to all-electron DF calculations. This process works well for SCREPs because the error resulting from the shape-consistent approximation is deposited in the core region, *viz.* if the nodes contained in the all-electron DF spinor occur in the core region - which is the case for SCREPs - then the error introduced when these nodes are removed in the shape-consistent procedure remains in the core region. However, if the nodes occur in both the core and outer core (or inner valence) spaces - which is the case for LCREPs - then the error introduced when these nodes are removed in the shape-consistent procedure remains in the core and outer-core regions. This error leads to the incorrect representation of nonlocal effects. In the case of the core region, these errors are nearly always insignificant and, more importantly, can be measured against atomic all-electron DF results to make certain that they are insignificant. In the case of the outer core, however, these errors are appreciable, and they cannot be checked against atomic all-electron DF results to determine their magnitude because of the valence-like properties of this region. *i.e.* to ascertain the magnitude of the errors, the results obtained from LCREP calculations must be checked against both atomic and molecular all-electron DF results because electrons occupying this region exhibit both core (atomic) and valence (molecular) characteristics. It is clear, therefore, that to derive LCREPs devoid of local error the nodes in the corresponding all-electron DF spinors must be removed such that the error resulting from the shape-consistent approximation is deposited solely in the core region. Or, in the case where a spinor's nodes occur in a region corresponding to the outer core, the process by which these are removed must be less radical (*i.e.* change the spinor as little as possible relative to the original) than that of the shape-consistent procedure.

These inaccuracies in the LEPC procedure are remedied by making use of a SCREP to define a new LCREP which explicitly retains, *e.g.* in the case of In, the $4d_{3/2}$ and $4d_{5/2}$ electrons and, consequently, has no 4d subshell nonlocal error. To accomplish this, the SCREP of LaJohn *et al.* [6] is used with a valence basis set of Slater-type orbitals (STO) to derive nodeless atomic spinors (AS), as linear combinations of STOs, *viz.* LCSTO-ASs, which are the 5s, $5p_{1/2}$, $5p_{3/2}$, $5d_{3/2}$, and $5d_{5/2}$ orbitals for In. Nodeless 5s and $5p_{1/2}$ spinors are obtainable from the ground state In wavefunction, whereas the $5p_{3/2}$, $5d_{3/2}$, and $5d_{5/2}$ spinors must be defined in excited state atomic calculations, with the $5p_{3/2}$ spinor coming from a neutral In excited state and the $5d_{3/2}$, and $5d_{5/2}$ spinors from an In^+ state having the $5p_{1/2}$ electron ionized and one of the $5s_{1/2}$ electrons excited first to the $5d_{3/2}$ and then to the $5d_{5/2}$ subshell. (The procedure is identical to that employed in the LECP method for deriving REPs.) The $4f_{5/2}$ and $4f_{7/2}$ are already nodeless, since they are the first spinors of l=3 symmetry. As a result, the calculated numerical Dirac-Fock forms [8] are used to derive the corresponding REPs for $4f_{5/2}$ and $4f_{7/2}$ and no 4f STOs are included in the basis set other than that shown in Table 1, which was added as a polarization function for use in SCF calculations on InH.

The basis set of three 5s, three 5p, four 4d and four 5d STOs (same exponents for the l-1/2 and l+1/2 components of each subshell spinor) shown in Table 1 was used to define the two-component spinors of In. The 5d STOs were omitted from the basis set in the subsequent InH calculations and replaced by the two 5d polarization functions also given in Table 1. The In atom was described as $[core]d_{3/2}{}^4 4d_{3/2}{}^6 5s^2 5p_{1/2}$ for the SCREP and as $[core']5s^2 5p_{1/2}$ for the LCREP. A basis set of four 1s, one 2s and two 2p STOs was used for H (Table 1) in the subsequent InH calculations. All of the STOs were optimized in either atomic or molecular calculations in the presence of the appropriate REP.

The 5s, $5p_{1/2}$ and $5p_{3/2}$ LCSTO-ASs are nodeless as a result of the valence-electron-plus-REP atomic SCF calculations. However, the $5d_{3/2}$ and $5d_{5/2}$ contain one node, since there is a lower-lying spinor of the same symmetry in each case. The node can be removed from each spinor by a myriad of methods [9]. For purposes of the

TABLE 1. Basis Sets of Slater-type Orbitals.

Indium		Hydrogen	
nl	ζ	nl	ζ
5s	7.487	1s	3.500
5s	2.410	1s	1.400
5s	1.391	1s	1.000
		1s	0.800
5p	5.574		
5p	2.018	2s	2.400
5p	1.095		
		2p[a]	1.300
4d	13.23	2p[a]	0.700
4d	4.746		
4d	2.590		
4d	0.967		
5d	1.574		
5d	0.993		
5d	0.736		
5d	0.543		
5d[a]	2.500		
5d[a]	1.200		
4f[a]	3.200		

[a]Polarization functions used in InH SCF calculations.

present study we chose to eliminate the single node in each spinor by constraining the coefficients of the n=5 spinor in such a way that the STOs corresponding to the n=4 spinor (i.e. the 4d STOs) do not contribute. Since the latter is necessarily [2], constrained to be frozen in its ground state form, the coefficients of the n=5 basis functions are zero by definition. Operationally, this is achieved by requiring all coefficients of n=5 STOs in n=4 ASs to be zero at every iteration, thus allowing the n=5 spinor to relax self-consistently while retaining a nodeless form. The l=2 LCSTO-ASs, both prior to and following the constraint on the n=5 ASs, are shown in Table 2.

This procedure removes the node in an outer core AS more accurately than the shape-consistent procedure because the form of the AS is retained as a *bona fide* eigenvector of the valence electron Fock matrix [2] over its entire range (from the atomic nucleus to infinity). In the shape-consistent pseudo-orbital approach the spinor has the form of a power series expansion from some match point to the nucleus. This is an especially important advantage in the present procedure since nodes are removed in the outer core region, which in the shape-consistent pseudo-orbital method, occurs to

TABLE 2. Indium d-Subshell Valence Spinors.[a]

$4d_{3/2}$	$5d_{3/2}$	$5d_{3/2}'$	$4d_{5/2}$	$5d_{5/2}$	$5d_{5/2}'$
-0.0132	0.0021	0.0000	-0.0126	0.0020	0.0000
0.7019	-0.1119	0.0000	0.6831	-0.1106	0.0000
0.3859	-0.0202	0.0000	0.4077	-0.0263	0.0000
0.0105	-0.1674	0.0000	0.0058	-0.1755	0.0000
0.0000	0.2880	0.2338	0.0000	0.2903	0.2352
0.0000	0.6861	0.5983	0.0000	0.6855	0.5967
0.0000	0.2849	0.2469	0.0000	0.2949	0.2486
0.0000	0.0030	0.0020	0.0000	0.0005	0.0003

[a]Based on SCREP frozen 5s and $4d_{3/2,5/2}$ spinors relative to the 4d and 5d STO basis sets of Table 1.

the left (inside) of the match point.

These nodeless two-component ASs (Table 2) are next expanded over the same quadrature as the corresponding numerical Dirac-Fock spinors [8]. Finally, a new LCREP, which we refer to as a nodeless valence spinor REP (NVSREP), is derived in the form of a numerical operator by the identical LEP procedure used to obtain the original SCREP [6]. It is noted that the spinor derived from an atomic calculation as a nodeless spinor, cannot be altered or it is possible that local errors will arise depending on the degree of perturbation relative to the original spinor. Atomic SCF calculations where the $4d_{3/2}$ and $4d_{5/2}$ orbitals of In were frozen in their ground state forms while employing the SCREP were also performed.

3. Results and Discussion

Atomic Hartree-Fock calculations in the context of two-component Dirac spinors and jj-coupling [2] were carried out for the indium atom and cation using the STO basis set shown in Table 1, the SCREP and LCREP of LaJohn *et al.* [6], and the present NVSREP. The spinor energies and positions of the radial maxima are shown in Table 3 along with the corresponding calculated numerical Dirac-Fock (DF) values [8].
It is especially noteworthy that the three REPs all show good agreement with the DF results, both for spinor energies and orbital maxima. The largest discrepancies observed in Table 3 from column to column are of the order of a few millihartrees (6 for the 5s spinor) and 0.02 bohr in the orbital maxima. Thus, it is impossible to determine which form of REP is more accurate based on SCF calculations on the neutral or cationic atom.

The principal difference between the LCREP and NVSREP is traced to the form of the $5d_{3/2}$ and $5d_{5/2}$ pseudo (nodeless valence) spinor. The LCREP pseudospinor was derived using the shape-consistent orbital algorithm of Christiansen *et al.* [7]. In this procedure the outer portion of the spinor is identical to the original

TABLE 3. Indium Atomic Spinor Energies and Orbital Maxima.

	-ε (hartreees)				r_{max} (bohrs)			
	DF	SCREP	LCREP	NVSREP	DF	SCREP	LCREP	NVSREP
In								
4d*	1.027	1.030	-	-	0.93	0.91	-	-
4d	0.991	0.992	-	-	0.94	0.95	-	-
5s	0.393	0.387	0.393	0.387	2.24	2.23	2.23	2.33
5p*	0.202	0.202	0.202	0.204	2.89	2.87	2.87	2.89
5p	0.192	0.192	0.192	0.193	3.00	3.01	3.01	3.01
5d*	0.057	0.057	0.059	0.059	-	-	-	-
5d	0.057	0.057	0.060	0.058	-	-	-	-
In^+								
5d*	0.219	0.215	0.213	0.211	4.70	4.70	4.72	4.73
5d	0.218	0.214	0.212	0.211	4.73	4.73	4.72	4.73
4f*	0.127	0.126	-	0.126	7.80	-	7.80	-
4f	0.127	0.126	-	0.126	7.80	-	7.80	-

large component of the numerical Dirac spinor. The radial portion of the pseudospinor from the so-called match point into the nucleus is fitted such that five derivatives of the spinor are equivalent at the match point. Thus, the shape-consistent pseudospinor has its shape determined by a power series fit from the match point inward. The nodeless valence spinors (forms for l=2 shown in Table 2), on the other hand, are linear combinations of STOs that are determined by an SCF procedure using the SCREP. Particularly, the portion of the $5d_{3/2}$ and $5d_{5/2}$ spinors that lie in the outer core region defined by the filled $4d_{3/2}$ and $4d_{5/2}$ spinors (Table 3) possess shapes that are consistent with actual SCF wavefunctions, as opposed to fits that are required to be smooth from the match point into the nucleus. As a result, the difference between the LCREP and NVSREP is traced solely to the forms of the 5s, $5p_{1/2}$, $5p_{3/2}$, $5d_{3/2}$ and $5d_{5/2}$ nodeless spinors used in the derivation of the REP by the LEP method [2]. This difference is revealed only in molecular calculations and is identified with the errors in the local approximation of the outer core subshells due to the shape-consistent forms chosen for the SCREP. In the present procedure the nonlocal effects due to the $4d_{3/2}$ and $4d_{5/2}$ subshells, which are explicitly present when the SCREP is used, are transferred directly into the NVSREP through the $5d_{3/2}$ and $5d_{5/2}$ SCF nodeless atomic spinors. It is noted that if in the process of forming a pseudospinor the original DF spinor is modified from its exact form from some match point inward towards the nucleus, then its behavior in the outer core region, which possesses valence-like properties, can be changed. The nodeless valence spinors, on the other hand, exhibit properties that closely reproduce the behavior of valence electrons as they penetrate the outer core region.

To demonstrate that these nonlocal effects are properly accounted for in the NVSREP, calculations were performed on a model system - the InH molecule. The three REPs were used in SCF calculations on InH together with the basis sets shown in Table 1. Because the effects are expected to be clearly shown using standard ΛS-coupling for the molecule, spin-orbit coupling was omitted. Averaged REPs were formed as described in [6] by a proper weighted average of the l+1/2 and l-1/2 components of the l>0 components of the In REP. (The $5p_{1/2}$-$5p_{3/2}$ spin-orbit coupling in In is 2129 cm^{-1}, as computed using the SCREP, in comparison with the DF value of 2147 and the experimental value of 2213 [6]. Further, it has been shown that the use of single determinant Hartree-Fock representations in ω-ω molecular coupling leads to erroneous ground state results due to the necessity for a multiconfiguration SCF treatment to ensure proper $p_{1/2}$-$p_{3/2}$ mixing [10]). Results of SCF calculations on InH are given in Table 4. The ground state MO configuration of InH is $1\sigma^2 1\pi^4 1\delta^4 2\sigma^2 3\sigma^2$, with the ten In 4d electrons explicitly represented by the 1σ, 1π and 1δ MOs for the SCREP and SCREP(F) calculations. For a proper direct comparison to SCREP and NVSREP results, the second column [SCREP(F)] of Table 4 presents results due to freezing the three lowest-lying molecular orbitals (MOs) of InH at their atomic (unperturbed) $4d_{3/2}$ and $4d_{5/2}$ forms. The SCREP(F) results, while retaining the 4d electrons explicitly, ensures that only the nonlocal effects are

TABLE 4. InH LCSTO-MO SCF Results.[a]

Property	SCREP	SCREP(F)	LCREP	NVSREP	Exp.[b]
E_{val}[c]	-57.3022	-57.2996	-2.3784	-2.3747	-
$\varepsilon_{1\sigma}$[d]	-1.012	-1.011	-	-	-
$\varepsilon_{2\sigma}$	-0.471	-0.472	-0.475	-0.473	-
$\varepsilon_{3\sigma}$	-0.280	-0.279	-0.281	-0.281	-
$\varepsilon_{1\pi}$	-1.012	-1.011	-	-	-
$\varepsilon_{1\delta}$	-1.017	-1.015	-	-	-
μ[e]	-0.38	-0.40	-0.57	-0.45	-
δ_{In}[f]	-0.05	-0.05	-0.05	-0.04	-
ω_e[g]	1523	1527	1519	1533	1476
$\omega_e x_e$[h]	22.5	22.9	22.0	23.7	25.6
R_e[i]	1.844	1.864	1.811	1.863	1.838

[a]Molecular self-consistent field results in the context of the indicated averaged relativistic effective potential (AREP).

[b]K. P. Huber and G. Herzberg, (1979) *Molecular Structure and Molecular Spectra. IV. Constants of Diatomic Molecules*, Van Nostrand Reinhold, New York.

[c]Total valence electron energy (hartrees) for the experimental bond distance (R_{exp}=1.8383 Å.).

[d]$C_{\infty v}$ molecular orbital energies (hartrees) at R_{exp}.

[e]Dipole moment (D) at R_{exp}.

[f]Mulliken net charge at R_{exp}.

[g]Harmonic frequency (cm^{-1}).

[h]Anharmonicity constant (cm^{-1}).

[i]Calculated equilibrium bond length (Å).

calculated since the outer core cannot undergo a self-consistent field relaxation. It is these results that should be reproduced by the molecular SCF calculations using a large-core pseudopotential derived from the same all-electron numerical DF wavefunctions for In and In^+. Experimental values for spectroscopic constants are shown in Table 4 (for reference only). The calculations presented here were done at the SCF level of theory and are expected to suffer the usual shortcomings due to the neglect of the effects of electron correlation.

The MO energies shown in Table 4 are generally in good agreement for all four calculations. The three lowest-energy MOs correspond to the 4d subshell electrons in the SCREP calculations. It is seen that freezing these MOs using the 4d atomic orbitals leads to, at most, an error of 2 mhartrees, apparently indicating that the direct participation of the 4d subshell in the bonding in InH is nearly negligible at the Hartree-Fock level of approximation. However, while the effect of freezing the inner MOs leads to only a 0.02 D change in the dipole moment and a 4 cm^{-1} increase in the harmonic frequency, it results in an appreciable 0.02 Å increase in the bond length, thus indicating that an analysis based soley on the energy is not sufficient to determine the participation of the outer-core electrons in bonding.

When the ten 4d electrons are included as part of the core Hartree-Fock calculations, using the resulting LCREP should lead to molecular properties that agree with the SCREP(F) results. As seen in Table 4 the LCREP derived using shape-consistent pseudospinors [6] shows an increase of 0.17 D in the magnitude of the dipole moment, and a shortening of the bond length by 0.05 Å in comparison with the SCREP(F) results. This behavior is consistent with that observed for InH by Leininger and Stoll [11] in which a SCREP and a LCREP based on the energy fitting method of Dolg, et al. [12] was used. They calculated values of 1.870 Å, 1510 cm^{-1} and -0.56 D for the bond length, harmonic frequency and dipole moment, respectively, using a 36-core-electron treatment in which the n=4 shell was held frozen SCREP(F). These values compare favorably with our SCREP(F) results, given in Table 4. In addition, results of 1.837 Å, 1510 cm^{-1} and -0.56 D for these properties were obtained by Leininger and Stoll using a 3-valence-electron LCREP, analogous to our LCREP results in Table 4. Although the bond length is somewhat longer than that reported in Table 4, the fact that it exceeds their SCREP(F) by 0.033 Å provides independent proof that LCREPs yield values that are consistently too long. It is also worth noting that this phenomenon is observed both in the present pseudospinor-based REP procedure and that of Dolg et al. [12], in which *ab intio* DF energies for a manifold of states are fitted to an analytic form comprised of Gaussian-type functions.

The InH 4-valence-electron calculation is based on the NVSREP, in which the nodeless valence orbitals derived in atomic SCF calculations were used to generate the pseudopotential. The dipole moment is seen to have decreased to within 0.05 D of the SCREP(F) result, while the bond length agrees with it to 0.001 Å (Table 4). This agreement clearly shows that the nonlocal effects that are misrepresented in the LCREP are faithfully propagated from the SCREP atomic wavefunction - where the outer core electrons are treated explicitly - into the NVSREP.

4. Conclusions

Nonlocal effects are shown in the requisite bond length increase in the NVSREP results relative to an InH molecule described with the shape-consistent LCREP. The NVSREP results are consistent with the analogous SCREP(F) calculation in which the In 4d subshell was held frozen in its atomic SCF form. This is attributed to a more correct accounting for nonlocal effects due outer-core electrons in LCREPs derived from nodeless valence two-component spinors in a manner that involves rigorous "bootstrapping" from the SCREP into the NVSREP.

Acknowledgments

We thank Professors Peter Schwerdtfeger and Hermann Stoll and Dr. Thierry Leininger for engaging in helpful discussions. WCE acknowledges the Alexander von Humboldt Foundation for partial support of this research.

References

[a]Current address: U.S. Department of Energy, Office of Energy Research, Office of Computational and Technology Research, Mathematical, Information and Computational Science Division, ER-31 GTN, Washington, D.C. 20585, U.S.A

[b]Current address: National Coordination Office for High Performance Computing and Communications, 8600 Rockville Pike, B1N30, Bethesda, Maryland, 20894, U.S.A.

1. Marino, M. M., Ermler, W. C., Kern, C. W., and Bondybey, V. E. (1992) *J. Chem. Phys.*, **96,** 3756.
2. Lee, Y. S., Ermler, W. C., and Pitzer, K. S. (1977) . **67**, 5861.
3. Muller, W., Flesch, W. J., and Meyer, W. (1984) *J. Chem. Phys.* **80**, 3297.
4. Stoll, H., Fuentealba, P., Schwerdtfeger, P., Flad, J., v. Szentpaly, L., and Preuss, H., (1984) *J. Chem. Phys.* **81**, 2732; Christiansen, P. A. (1986) *Chem. Phys. Lett.* **127**, 50; Marino, M. M., and Ermler, W. C. (1993) *Chem. Phys. Lett.* **206**, 271.
5. Schwerdtfeger, P., Fischer, T., Dolg, M., Igel-Mann, G., Nicklass, A., Stoll, H., and Haaland, A. (1995) *J. Chem. Phys.* **102**, 2050; and private communication.
6. LaJohn, L. A., Christiansen, P. A., Ross, R. B., Atashroo, T., and Ermler, W. C. (1987) *J. Chem. Phys.* **87,** 2812 .
7. Christiansen, P. A., Lee, Y. S., and Pitzer, K. S. (1979) *J. Chem. Phys.* **71,** 4445.
8. Desclaux, J. P. (1975) *Computer Phys. Commun.* **9,** 31.
9. Szasz, L. (1985) *Pseudopotential Theory of Atoms and Molecules,* Wiley, New York.
10. Christiansen, P. A., and Pitzer, K. S. (1980) *J. Chem. Phys.* **73**, 5160.
11. Leininger, T., and Stoll, H., private communication.
12. Dolg, M., Wedig, U., Stoll, H., and Preuss, H. (1987) *J. Chem. Phys.* **86**, 866.

ONE CENTER EXPANSION IN QUANTUM MECHANICAL MOLECULAR CALCULATIONS

Rimantas J. RAKAUSKAS, Juozas ŠULSKUS
Dept. of Theoretical Physics, Vilnius University,
Saulėtekio 9, 2054 Vilnius, Lithuania

1. Introduction

The Cartesian Gaussian-type functions were first proposed and introduced into quantum chemistry by Boys [1] and linear combinations of them have been extensively used in the last years as basis set orbitals in atomic and molecular calculations. This is due to the fact that all multicenter integrals needed can be evaluated in closed form [2, 3]. But these functions do not form an orthonormal set. For this reason Wulfman [4] proposed the use of the orthogonal complete set of harmonic oscillator wave functions as basis set. Three-dimensional harmonic oscillator functions play the important role in nuclear shell calculations and the mathematical techniques for dealing with HOF have been developed by nuclear physicists.

The mathematical formalism of the Talmi transformation [5] makes it possible to evaluate analytically the molecular multicenter integrals over HOF in closed form [6]. If spherical polar coordinates are used, the molecular integrals over HOF can be represented in terms of, again, HOF, depending only on the distance of the centers and some known quantities as Moshinsky brackets [7], Clebsh-Gordan coefficients and the confluent hypergeometric function [8]. Because the HOF form a complete orthonormal set, HOF can be used as a basis set in LCAO calculations. Moreover, other types of orbitals can be expanded in terms of HOF and molecular integrals over these orbitals can be obtained analytically from integrals over HOF. But with current supercomputers the vast majority of molecular quantum- mechanical calculations are carried out with CGF basis sets by using packages POLYATOM/2, GAUSSIAN92, HONDO7 [9], GAMESS.

The usage of CGF in determination of adiabatic potential energy surfaces (PES) for molecular dynamics purposes have problems, mainly due to the immensely great number of CGF to be used, because CGF, and also HOF, do not resemble physically meaningful one-electron wave functions for the one-nucleus one-electron Coulomb problem, like exponentially decaying hydrogenic or Slater-type functions.

The usage of STF leads to a correct representation of the behaviour of wave functions at very short or very long distances from the nucleus, but difficulties arise during the evaluation of multicenter integrals with STF for polyatomic molecules. The contraction of

C. A. Tsipis et al. (eds.), New Methods in Quantum Theory, 425–435.

mixed basis sets, using CGF, HOF and STF, allows a reduction of the number of basis functions used for solving problems dealing with electron-molecule scattering, molecular dynamics, large amplitude molecular vibrations, excited electronic states, photoionization of molecules.

2. Transformation formulas

Analytical transformation of Slater type AO localised in center with radius vector $\vec{R}_b$ to the center $\vec{R}_a$ (Cartesian coordinate systems of centers a and b are parallel) have the form [10]:

$$r_b^{n-1} Y_{l,m}\left(\Omega_{\vec{r}_b}\right)\exp(-\xi r_b) = \sum_{l',m',l'',m''} \sqrt{\frac{(2l+1)(2l''+1)}{4\pi(2l'+1)}} \times$$

$$\times C(l,m;l'',m'';l',m') R^{(\mathrm{STF})}_{nll'l''}(\xi, r_a, R_{ab}) Y_{l'',m''}\left(\Omega_{\vec{R}_{ab}}\right) Y_{l',m'}\left(\Omega_{\vec{r}_a}\right), \tag{1}$$

were $C(l,m;l'',m'';l',m') = \begin{bmatrix} l & l'' & l' \\ 0 & 0 & 0 \end{bmatrix} \begin{bmatrix} l & l'' & l' \\ m & m'' & m' \end{bmatrix}$ - multiplication of Clebsh-Gordan coefficients. $R^{(\mathrm{STF})}{}_{nll'l''}(\xi, r_a, R_{ab})$ - radial part of transformation. It depends on quantum numbers n, l, l', l'', parameter ξ and AO localisation vector $\vec{R}_{ab}$:

$$R^{(\mathrm{STF})}{}_{nll'l''}(\xi, r_a, R_{ab}) = (-1)^l \pi^{\frac{3}{2}} \xi^{-(n+1)} (\xi r_a)^{-(l'+1)} (\xi R_{ab})^{-(l''+1)} \times$$

$$\times 2^{-(n+l+l'+l'')} \frac{\Gamma\left(\frac{n-l}{2}+1\right)\Gamma\left(\frac{n-l-1}{2}+1\right)\Gamma(n+l+2)}{\Gamma(n+1)} \times$$

$$\times \left\{ \exp\left[-\xi\left|r_a - R_{ab}\right|\right] \sum_{\alpha=0}^{n+l''} \sum_{\beta=0}^{n+l'} A_{\alpha\beta}(nll'l'')(\mp\xi r_a)^{\beta}(\pm\xi R_{ab})^{\alpha} - \right.$$

$$\left. - \exp\left[-\xi\left|r_a + R_{ab}\right|\right] \sum_{\alpha=0}^{n+l''} \sum_{\beta=0}^{n+l'} A_{\alpha\beta}(nll'l'')(\xi r_a)^{\beta}(\xi R_{ab})^{\alpha} \right\}, \tag{2}$$

were upper signs in $\{\mp\}$ belongs to $R^{(STF)}{}_{nll'l''}(\xi, r_a, R_{ab})$ with $r_a \le R_{ab}$ and lower - for the case $r_a > R_{ab}$. $A_{\alpha\beta}(\ldots)$- elements of the matrix $\|A\|$ with dimension $(n+l''+1)\times \times(n+l'+1)$:

$$A_{\alpha\beta}(nll'l'') =$$

$$= \sum_{k=0}^{\left[\frac{n-1}{2}\right]} \sum_{q=0}^{l'} \sum_{v=0}^{l''} \sum_{k=0}^{s} \frac{(-1)^{(s+k)} 2^{(q+v-k)} \Gamma(2l'-q+1)\Gamma(2l''-v+1)}{\Gamma\left(\frac{n-l}{2}-k+1\right)\Gamma\left(\frac{n-l-1}{2}-k+1\right)} \times$$

$$\times \frac{\Gamma(n+k+1)\Gamma(l-l'-l''+2k+q+v+1)}{\Gamma\left(l+k+\frac{3}{2}\right)\Gamma(k+1)\Gamma(q+1)\Gamma(v+1)\Gamma(k+1)} \times$$

$$\times \frac{1}{\Gamma(l'-q+1)\Gamma(l''-v+1)\Gamma(q-k+1)\Gamma(z+1)\Gamma(n-s-z+1)} \times$$

$$\times \frac{1}{\Gamma(l-l'-l''+2k+q+v-s+k+1)}, \tag{3}$$

were $s = q + v + n - \alpha - \beta$, $z = \alpha - v$ with conditions $0 \le s \le n$, $0 \le z \le n - s$.

The transformation to another center of HOF

$$\varphi^{\text{HOF}}(\vec{r}_A) = N^{\text{HOF}}(\alpha; n,l,m) \times$$

$$\left|\vec{r} - \vec{R}_A\right|^l L_n^{l+\frac{1}{2}}\left(2\alpha r^2\right) Y_{l,m}\left(\Omega_{\vec{r}_A}\right) \exp\left(-\alpha\left|\vec{r} - \vec{R}_A\right|^2\right),$$

$$N^{\text{HOF}}(\alpha; n,l,m) = (2\alpha)^{\frac{l}{2}+\frac{5}{4}} \left[\frac{n!}{\alpha\Gamma\left(n+l+\frac{3}{2}\right)}\right]^{\frac{1}{2}}. \tag{4}$$

is of analogous form with STF, Eq. (1).

Radial part for the case of HOF $R^{(\text{HOF})}_{nll'l''}(\xi, r_a, R_{ab})$ takes the form:

$$R_{nll'l''}^{(\mathrm{HOF})}(\xi,r_a,R_{ab})=(-1)^{(n+l+l'')}i^{(l+l'+l'')}(2\pi)^{\frac{3}{2}}(2\alpha)^{-\left(l+\frac{3}{2}\right)}\alpha^{\frac{1}{2}(l+l'+l''+3)}\times$$

$$\times r_a^{\left(l'+\frac{1}{2}\right)}R_{ab}^{\left(l''+\frac{1}{2}\right)}\sum_{m=0}^{n}\frac{(-1)^m}{2^m m!}\binom{n+l+\frac{1}{2}}{n-m}\times$$

$$\times\sum_{s=0}^{\infty}(-1)^s\frac{\Gamma\left(s+m+\frac{1}{2}(l+l'+l''+3)\right)}{s!\Gamma\left(s+l'+\frac{3}{2}\right)}\times$$

$$\times{}_2F_1\left(-s,-\left(s+l'+\frac{1}{2}\right),l''+\frac{3}{2},\frac{R_{ab}^2}{r_a^2}\right)\alpha^s r_a^{2s}. \tag{5}$$

The analytical expansion of SGF in the case of linear molecules takes a form:

$$\left|\vec{r}-\vec{R}_A\right|^{2n_A+l_A}Y_{l_A,m_A}\left(\Omega_{\vec{r}_A}\right)\exp\left(-\alpha\left|\vec{r}-\vec{R}_A\right|^2\right)=(2l_A+1)!!\,(2l_A+1)^{1/2}\left(\frac{z_A}{R_A}\right)^{L+l_A}\times$$

$$\times R_A^{2n_A}\exp\left(-\alpha\left(r^2+R_A^2\right)\right)R_A^{l_A}\sum_{L=|m_A|}^{\infty}Y_{L,m_A}(\Omega_{\vec{r}})\frac{1}{(2L+1)^{1/2}}V_{L,m_A}^{(l_A,m_A)}(r), \tag{6}$$

$$V_{L,m_A}^{(l_A,m_A)}(r)=\left\{\sum_{l=0}^{n_A}\left[(2l+1)\sum_{t=l}^{2n_A-l}T_{l,t}^{(2n_A)}\left(\frac{r}{R_A}\right)^t\right]\times\right.$$

$$\times\sum_{l_2=0}^{l_A}\frac{(2l_2+1)}{(2l_A-2l_2+1)!\,(2l_2+1)!!}\left(\frac{r}{R_A}\right)^{l_A-l_2}C(l_A,m_A;l_2,0;l_A-l_2,m_A)\times$$

$$\left.\times\sum_{l_e}(2l_e+1)i_{l_e}(2\alpha R_A r)\left[\sum_{L'}C(l,0;l_A-l_2,m_A;L',m_A)C(L',m_A;l_e,0;L,m_A)\right]\right\}, \tag{7}$$

were $T_{l,t}^{(2n_A)}=(-1)^l\dfrac{(2n_A+1)}{(t-l)!!(t+l+1)!!(2n_A-t-l)!!(2n_A-t+l+1)!!}$, $i_{l_e}(2\alpha R_A r)$ - modified Bessel function,

The program for determination of the radial part of the expansion, Eq. (7), for any l_A, m_A and L was written using symbolic computation system REDUCE [11] (appendix). In table 1 we present the analytical expressions of $V_{L,m_A}^{(l_A,m_A)}(r)$ for some SGF's. The radial part of the expansion $V_{L,m_A}^{(l_A,m_A)}(r)$ is a sum of modified Bessel functions multiplied by the polynomials of $\left(\frac{r}{R_A}\right)$, were r is the distance between the origin of coordinate system and electron and R_A is the distance from the origin of coordinate system and center A. The methods of evaluation of Bessel functions are well developed.

TABLE 1. The radial parts $V_{L,m_A}^{(l_A,m_A)}(r)$ of the expansion Eq. (7)

Bessel function, Eq. (7)	Polynomial at Bessel function	Bessel function, Eq. (7)	Polynomial at Bessel function
$n_A=0,\ l_A=1,\ m_A=0$		$n_A=0,\ l_A=2,\ m_A=0$	
i_{L-1}	$\frac{1}{3}L\left(\frac{r}{R_A}\right)$	i_{L-2}	$\frac{L(L-1)}{10(2L-1)}\left(\frac{r}{R_A}\right)^2$
i_L	$\frac{1}{3}(2L+1)$	i_{L-1}	$\frac{2}{15}L\left(\frac{r}{R_A}\right)$
i_{L+1}	$\frac{1}{3}(L+1)\left(\frac{r}{R_A}\right)$	i_L	$\frac{L(L+1)(2L+1)}{(2L+3)(2L-1)}\left(\frac{r}{R_A}\right)^2 + \frac{1}{15}(2L+1)$
$n_A=1,\ l_A=0,\ m_A=0$		i_{L+1}	$\frac{2}{15}(L+1)\left(\frac{r}{R_A}\right)$
i_{L-1}	$-2L\left(\frac{r}{R_A}\right)$	i_{L+2}	$\frac{(L+1)(L+2)}{10(2L+3)}\left(\frac{r}{R_A}\right)^2$
i_L	$(2L+1)\left(\frac{r}{R_A}\right) \quad +(2L+1)$		
i_{L+1}	$-2(L+1)\left(\frac{r}{R_A}\right)$		

3. Conclusions

The use of both the analytical and the numerical approach for the transformation of Slater-type, spherical Gaussian and harmonic oscillator functions allows the effective determination of molecular multicenter integrals, one center expansions of molecular wave functions and molecule-electron interaction potential with necessary accuracy, enables to use vector processors effectively.

The transformation formulas (Eq. 5) can be applied for Cartesian Gaussian functions transformation after preliminary transformation of CGF to HOF basis. The algorithms developed for multicenter integrals with STF [10] can be applied for HOF and the evaluation of multicenter integrals of mixed basis sets (STF and CGF) may be performed.

4. References

1. Boys, S. P. (1950) *Proc. Roy. Soc.* **A200,** 542.
2. Shavitt, L. (1963) *Methods in Computational Physics* **2,** 1.
3. Saunders, V. R. (1983) *Methods in Computational Molecular Physics*
4. Wulfman, (1960) *J. Chem. Phys.* **33,** 1567.
5. Talmi, I. , De-Shavit, A. (1963) *Nuclear Shell Theory.* Academic,NY
6. Vanagas, V. (1967) *Teor. i eksper. khem.* **3,** 24 (in Russian).
7. Moshinsky, M. (1959) *Nucl. Phys.* **13,** 104.
8. Maratis, D. K. (1979) *J. Chem. Phys.* **71,** 917.
9. Dupuis, M. , Watts, J. D., Villar, H. O., Hurst, G. J. B. (1989) *Comp. Phys. Commun.* **52,** 415.
10. Aleksandrov, B. S., Bolotin, A. B., Pošiūnaitė, N. P., Rakauskas, R. J., Shugurov, V. K. (1981) *Multicenter integrals* Vilnius, Vilnius University (in Russian).
11. Rayna, G. (1987) *REDUCE software for algebraic computation* New York, Springer-Verlag

APPENDIX. The REDUCE program for generation of the radial parts of expansion (7)

```
ON DIV$
OFF EXP$
      PROCEDURE CL1230(J1,J2,J)$
%
% J1  J2  J  - CLEBSH-GORDON COEFFICIENT. J1,J2,J - ARE NUMBERS.
% 0  0  0
%
% A.P.JUCYS, A.A.BANDZAITIS...
%
BEGIN SCALAR G,SG,C$
G:=(J1+J2+J)/2$
SG:=(-1)**(G-J)$
C:=((2*J+1)*F1(J1+J2-J)*F1(J1-J2+J)*F1(-J1+J2+J)/F1(J1+J2+J+1))
**(1/2)*F1(G)/(F1(G-J1)*F1(G-J2)*F1(G-J))*SG$
RETURN C
END$
```

```
        PROCEDURE CLNUM(J1,M1,J2,M2,J,M)$
%
% J1  J2  J - CLEBSH-GORDON COEFFICIENT. J1,... ,M- ARE NUMBERS.
 % M1  M2  M
 %
% A.P.JUCYS, A.A.BANDZAITIS...  FORMULA (13.1A)
%
 BEGIN SCALAR C$
 C:=0$
 IF (M1+M2 NEQ M) OR (J1+J2 < J) THEN RETURN C$
 ZMIN:=MAX(0,J2-J-M1,J1+M2-J)$
 ZMAX:=MIN(J1+J2-J,J1-M1,J2+M2)$
 C:=(F1(J1+J2-J)*F1(J1-J2+J)*F1(-J1+J2+J)*
    F1(J1+M1)*F1(J1-M1)*F1(J2+M2)*F1(J2-M2)*FC1(J+M)*
    F1(J-M)*(2*J+1)
   /F1(J1+J2+J+1))**(1/2)*
 << FOR Z:=ZMIN:ZMAX DO
 << SUM( (-1)**Z/
      ( F1(Z)*F1(J1+J2-J-Z)*F1(J1-M1-Z)*F1(J2+M2-Z)*
        F1(J-J2+M1+Z)*F1(J-J1-M2+Z) )) >> >>$
RETURN C
END$
        PROCEDURE CLPAR(J1,M1,J2,M2)$
%
% J1  L+J2  L  - CLEBSH-GORDON COEFFICIENT WITH PARAMETER L.
% M1    M2  M
%
% A.P.JUCYS, A.A.BANDZAITIS...  FORMULA (13.1A)
%
 BEGIN SCALAR C,C1,C2$
 M:=M1+M2$
 IF (M1+M2 NEQ M) THEN RETURN C$
 ZMIN:=MAX(0,J2-M1)$
 ZMAX:=MIN(J1-M1,J1+J2)$
     L1IA:=MIN(J2-J1,J1+J2+1)$
     L2IA:=MAX(J2-J1,J1+J2+1)$ L21:=L2IA-L1IA$
%
 C:=0$
   FOR Z:=ZMIN:ZMAX DO
  <<
I1IA:=MIN(J2-M2,J2+M2-Z)$
I2IA:=MAX(J2-M2,J2+M2-Z)$ I21:=I2IA-I1IA$
    J1IA:=MIN(M,-J1-M2+Z)$
    J2IA:=MAX(M,-J1-M2+Z)$ J21:=J2IA-J1IA$
K1IA:=MIN(-M,-J1-M2+Z)$
K2IA:=MAX(-M,-J1-M2+Z)$ K21:=K2IA-K1IA$
C:=C+ ( (-1)**Z*((F1(J1+J2)*F1(J1-J2)*F1(J1+M1)*F1(J1-M1))
      **(1/2)/
    (F1(Z)*F1(J1+J2-Z)*F1(J1-M1-Z)*F1(M1-J2+Z))) ) *
%
```

```
<<      (2*L+1) * << IF Z EQ 0 THEN 1
     ELSE
     << FOR I:=1 STEP 1 UNTIL Z  PRODUCT(L+J2+M2-Z+I) >> >> *
%
     <<  IF I21 EQ 0 THEN 1
     ELSE
     <<  IF I2IA EQ (J2-M2) THEN
     << FOR G:=1:I21 PRODUCT(L+J2+M2-Z+G) >>
     ELSE
     <<  FOR G:=1:I21 PRODUCT(1/(L+J2-M2+G)) >> >> >> *
%
    <<  IF J21 EQ 0 THEN 1
     ELSE
     <<  IF J2IA EQ M THEN
     << FOR G:=1 STEP 1 UNTIL J21  PRODUCT(L-J1-M2+Z+G) >>
     ELSE
       IF J2IA EQ (-J1-M2+Z) THEN
     <<  FOR G:=1:J21  PRODUCT(1/(L+M+G))  >> >> >> *
%
     <<  IF K21 EQ 0 THEN 1
     ELSE
     <<  IF K2IA = (-M) THEN
     << FOR G:=1:K21  PRODUCT(L-J1-M2+Z+G) >>
     ELSE
     <<  FOR G:=1:K21  PRODUCT(1/(L-M+G)) >> >> >> *
%
     <<  IF L2 EQ 0 THEN 1
     ELSE
     <<  IF L2IA = J2-J1 THEN
     << FOR G:=1:L21  PRODUCT(2*L+J1+J2+1+G) >>
     ELSE
     << FOR G:=1:L21  PRODUCT(1/(2*L+J2-J1+G)) >> >> >>
 >>**(1/2) >>$
%
 RETURN C
END$
      PROCEDURE F1(N)$
% N! - FACTORIAL OF N.
 BEGIN SCALAR F$
 F:=1$
 IF N <= 1 THEN RETURN F
 ELSE  FOR I:=2 STEP 1 UNTIL N DO F:=F*I$
 RETURN F
 END$
      PROCEDURE F2(N)$
% N!! - FACTORIAL OF N.
 BEGIN SCALAR F$
 F:=1$
 IF N <= 1 THEN RETURN F
 ELSE  FOR I:=N STEP -2 UNTIL 1 DO F:=F*I$
```

```
 RETURN F
 END$
;EMD; % end of CLEBSH-GORDON coeff. file
  %
  %
%
%*****************************************************************
% Here input of "NA,LA and MA" is necessary                  *
%                                              *
% For example:                                       *
% NA:=0; LA:=1; MA:=0;                                    *
%*****************************************************************
 ARRAY COR(LA),TTT(NA);
 OPERATOR JLE,JJ,JJ1;
   FOR L2:= 0 STEP 1 UNTIL LA DO
 << COR(L2):= RR**(LA-L2) *
   ( (2*L2+1)/ (F2(2*LA-2*L2+1)*F2(2*L2+1))) *
   CL0(LA,L2)**2 >>;
%
   FOR LL:= 0 STEP 1 UNTIL NA DO
 << CLL:=0;
   FOR TT:=LL STEP 2 UNTIL 2*NA-LL DO
 << CLL:=CLL + (-1)**LL*F1(2*NA+1)*RR**TT/
   (F2(TT-LL)*F2(TT+LL+1)*F2(2*NA-TT-LL)*F2(2*NA-TT+LL+1)) >>;
   TTT(LL):=CLL*(2*LL+1) >>;
%
 A:=0;
 FOR LL:=0 STEP 1 UNTIL NA DO % loop l
 << ALL:=TTT(LL);
   AAL2:=0;
 FOR L2:= 0 STEP 1 UNTIL LA DO  % loop  l2
 << AL2:=COR(L2);
   LLE1:=LL+LA-L2;
   LLE0:=LL-LA+L2;
    B:=0;
    FOR LLE:= -(LL+LA-L2) STEP 2 UNTIL (LL+LA-L2) DO  % loop le
 << D:=0;
   LP0:=LL+LA-L2; LP1:=MAX(ABS(LLE),LL-LA+L2);
   FOR LP:=LP0 STEP -2 UNTIL LP1 DO               % loop l'
 << D:=D+
 CL1230(LL,LA-L2,LP) * CLNUM(LL,0,LA-L2,MA,LP,MA)
 * CLPAR(LP,0,LLE,0) * CLPAR(LP,MA,LLE,0) >>; %end of loop l'
   B:= B + (2*(L+LLE)+1) * JLE(L+LLE) *  D >>; %end of loop le
   AAL2:=AAL2 + AL2*B >>;       % end of loop l2

   A:= A + AAL2*ALL
   >>;                     % end of loop l
```

```
%***************************************************************
%                                                 *
% output of radial part of expansion              *
%                                                 *
%***************************************************************
 FACTOR RR;
  OFF EXP;
   ON MCD;
    ON GCD;
    FOR LE0:=-(NA+LA):(NA+LA) DO
 << FOR LLE:=-(NA+LA):(NA+LA) DO
 << JJ(L+LLE):=0 >>;
    JJ(L+LE0):=1;
    B:=SUB(JLE=JJ,A);
    JJ1(L+LE0):=B >>;
%OUT NL12;
    WRITE "NA= ", NA, " LA= ", LA,"  MA= ", MA;
    FOR LLE:=-(NA+LA):(NA+LA) DO
 << WRITE "JJ1( ", L+LLE, " ):= ",  JJ1(L+LLE) >>;
%SHUT NL12;
 ;EMD;
% end of radial part expantion
```

APPENDIX 2. Transformation matrix from cartesian Gaussian functions to real spherical harmonics

(ijk)	(nlmq)
	i+j+k=1
(100)	(0111)
(010)	(011-1)
(001)	(0100)
	i+j+k=2
(200)	$\frac{\sqrt{5}}{3}(1000) - \frac{1}{3}(0200) + \frac{1}{\sqrt{3}}(0221)$
(020)	$\frac{\sqrt{5}}{3}(1000) - \frac{1}{3}(0200) - \frac{1}{\sqrt{3}}(0221)$
(002)	$\frac{\sqrt{5}}{3}(1000) + \frac{2}{3}(0200)$
(110)	(021-1)
(101)	(0211)
(011)	(021-1)
	i+j+k=3
(300)	$\frac{\sqrt{21}}{5}(1111) + \frac{1}{\sqrt{10}}(0331) - \frac{\sqrt{3}}{5\sqrt{2}}(0311)$
(030)	$\frac{\sqrt{21}}{5}(111-1) - \frac{1}{\sqrt{10}}(033-1) - \frac{\sqrt{3}}{5\sqrt{2}}(031-1)$
(003)	$\frac{\sqrt{21}}{5}(1100) + \frac{1}{5}(0300)$
(210)	$\frac{\sqrt{7}}{\sqrt{15}}(111-1) + \frac{1}{\sqrt{2}}(033-1) - \frac{1}{\sqrt{30}}(031-1)$
(201)	$\frac{\sqrt{7}}{\sqrt{15}}(1100) + \frac{1}{\sqrt{3}}(0321) - \frac{1}{\sqrt{5}}(0300)$
(120)	$\frac{\sqrt{7}}{\sqrt{15}}(1111) - \frac{1}{\sqrt{2}}(0331) - \frac{1}{\sqrt{30}}(0311)$
(021)	$\frac{\sqrt{7}}{\sqrt{15}}(1100) - \frac{1}{\sqrt{3}}(0321) - \frac{1}{\sqrt{5}}(0300)$
(102)	$\frac{\sqrt{7}}{\sqrt{15}}(1111) + \frac{2\sqrt{2}}{\sqrt{15}}(0311)$
(021)	$\frac{\sqrt{7}}{\sqrt{15}}(111-1) + \frac{2\sqrt{2}}{\sqrt{15}}(031-1)$
(111)	(032-1)

DISTRIBUTED GAUSSIAN BASIS SETS: SOME RECENT RESULTS AND PROSPECTS

S. WILSON
Rutherford Appleton Laboratory
Chilton, Oxfordshire OX11 0QX, England

Abstract. Recent years have witnessed a growing interest in the systematic implementation of finite basis set expansions using Gaussian basis sets in both non-relativistic and relativistic molecular electronic structure studies. A promising development is the distributed basis set in which the exponents and the distribution of the Gaussian functions are generated according to empirical prescriptions. A particularly simple distributed basis set is the Gaussian Cell model in which the functions are arranged on a regular lattice, with one function centred on each lattice point and all functions having the same exponent. Not surprisingly, this simple model does not achieve high accuracy. However, recent work on one-electron diatomic and triatomic systems has demonstrated that, with simple refinements, it can be very significantly enhanced and accuracies approaching 1 μHartree can be achieved for the total energy. Alternatively, a Laplace transform can be used to obtain an integral representation of the elliptical function in terms of Gaussian functions. This Laplace transform provides a prescription for the generation of a systematic sequence of distributed basis sets of Gaussian functions from a sequence of atomic even-tempered sets. The prospects for distributed basis sets in Hartree-Fock studies and beyond, for both non-relativistic and relativistic molecular electronic structure calculations, are surveyed briefly.

1. Introduction

The controlled reduction of basis set truncation errors is an essential prerequisite of a quantitative quantum chemistry and recent years have witnessed a growing interest in the systematic implementation of the algebraic ap-

C. A. Tsipis et al. (eds.), New Methods in Quantum Theory, 437–461.

proximation[1, 2], *i.e.* finite basis set expansions, in both non-relativistic and relativistic studies.

Basis set truncation effects are widely recognized as the main source of error in contemporary molecular electronic structure calculations. Over thirty years ago in 1963, Schwartz[3] emphasized that

> *... an orderly plan of attack is... essential for any mathematical analysis of the convergence rate. The old habit of picking the "best" (chosen by art) choice of a fixed number of terms is to be discarded if one wants to see how the problem converges.*

He continued

> *One must choose a set of basis functions and then gradually add more ... in some systematic manner.*

Ruedenberg and his co-workers[4] pointed out that even-tempered basis sets[1] afforded a systematic method for generating atomic orbital basis sets[5–9]. Feller and Ruedenberg[10] and Schmidt and Ruedenberg[11] showed that sequences of such basis sets can be constructed that systematically approach a complete set and thus approach the Hartree-Fock limit. Wilson[1] pointed out that, in spite of this progress,

> *The vast majority of contemporary molecular calculations adopt a pragmatic approach ... The art of selecting a basis set is based on previous experience in treating similar systems using basis sets of comparable quality.*

Huzinaga[12] has suggested that calculations in which the basis set truncation error is not controlled should be termed

> *quasi-empirical.*

The same point has been made by Davidson and Feller[13] who state that

> *the smaller the basis set the more ab initio calculations assume an empirical flavor.*

Clementi[14] has complained that

> *In the seventies a "McDonald's Fast Food" mentality had spread among the users of quantum chemistry programs, and the idea of optimizing a basis set ... was abandoned because of the heavy labour involved*

and as a result the

[1] Even-tempered Gaussian basis sets were first suggested by R. McWeeny in his *D. Phil.* thesis (Oxford, 1948).

chemical literature ... was flooded with computations which essentially compared experiments with basis set superposition error data.

Recently, Taylor[15] has again emphasized the importance of the error associated with basis set truncation

The ultimate accuracy of any calculation is determined by the one-particle basis set. This is one of the most obvious, yet one of the most ignored, observations about quantum chemical calculations. For example, it is in general just not possible to get the right answer for the right reasons using a DZP [double-zeta plus polarization] *basis. This is not an argument against using such sets, but their limitations must be kept constantly in mind.*

For about fifty years the algebraic approximation in the form of the Linear Combination of Atomic Orbitals (L.C.A.O.) method was regarded as the poor relation of the finite difference methods which have been employed in atomic structure calculations since the pioneering work of Hartree[16] in the 1930s. In the 1970s, it was recognized that the algebraic approximation facilitated the integration over the continuum that arises in the description of electron correlation by means of many-body theories[17]. In the 1980s, it was shown that in studies of the relativistic atomic and molecular electronic structure problem, the algebraic approximation affords a representation of not only the positive energy branch but also the negative energy branch of the relativistic spectrum[18]. This facilitates the study of not only a relativistic many-body perturbation theory but also the covering theory, quantum electrodynamics[19]. Systematic implementation of the algebraic approximation has allowed, for example, the Dirac-Hartree-Fock energies of atoms to be determined to an accuracy which matches that achieved in finite difference calculations; typically $10^{-8} - 10^{-9}$ Hartree[20]. For molecules the situation is far less satisfactory[21] and, even in non-relativistic molecular structure calculations, the basis set truncation error is often seen as the largest source of error[1].

Over the past decade, fully numerical Hartree-Fock calculations for diatomic molecules have been reported using first finite difference methods[22–25] and then finite element techniques[26, 27]. Most of these calculations have been formulated within the framework of non-relativistic quantum mechanics. They provide invaluable benchmarks against which finite basis set calculations can be compared and facilitate the appraisal of methods suggested for the systematic development of molecular basis sets[28–35].

Atom-centred basis sets have been most widely used in molecular calculations but, in calculations designed to match the accuracy achieved in numerical Hartree-Fock studies of diatomic molecules, it has recently been shown that they can be usefully supplemented by off-centre sets[30–33]. Indeed, by including bond centred functions in a systematically constructed basis set for the ground state of the nitrogen molecule it has been possible to obtain an energy that is within a few μHartree of the numerical result. The success of these calculations suggests investigation of the construction of basis sets including off-atom basis functions in more general terms.

A promising development is the distributed basis set[36, 37] in which the exponents and the distribution of the Gaussian functions are generated according to empirical prescriptions. It should be noted that Gaussian functions are particularly well suited for such a procedure since, unlike exponential basis functions, they do not introduce a cusp. This approach is distinct from the use Gaussian lobe functions[38] and the Floating Spherical Gaussian Orbital (F.S.G.O.) model[39–41] in that there is no attempt to mimic higher harmonics directly and non-linear optimization is avoided. This allows the use of large basis sets resulting in high precision. In this article, we describe two distributed Gaussian basis sets which have been discussed recently.

A particularly simple distributed basis set is the Gaussian Cell model, introduced by Murrell and his coworkers in 1974[42, 43], in which the functions are arranged on a regular lattice, with one function centred on each lattice point and all functions having the same exponent. Not surprisingly, this simple model does not achieve high accuracy. However, recent work on one-electron diatomic and triatomic systems have demonstrated how, with simple refinements[44–46] it can be very significantly enhanced and accuracies approaching 1 μHartree can be achieved.

An alternative method for generating a distributed basis set is based on a Laplace transform. The 'natural' basis functions for diatomic molecules are elliptical functions. A Laplace transform can be used to obtain an integral representation of an elliptical function in terms of Gaussian functions. This Laplace transform provides a prescription for the generation of a systematic sequence of distributed basis sets of Gaussian functions from a sequence of atomic even-tempered sets[47].

In section 2, a brief overview of methods for constructing basis sets for atomic and molecular electronic structure calculations is provided giving particular attention to systematic approaches. The theoretical aspects of the distributed basis set approach are discussed in section 3 concentrating on the Gaussian Cell model and its extensions and on distributions determined by a Laplace transform of elliptical functions. Some recently obtained results are described in section 4 whilst in the final section, section 5, the prospects for the distributed Gaussian basis set method are assessed.

2. Basis Set Construction

2.1. ATOMIC BASIS SETS

Atomic basis sets have traditionally been designed by invoking the variation principle so as to obtain the most accurate energy values with the smallest number of basis functions. More recently, there has been a growing interest in the development of sequences of basis sets which formally approach completeness as the number of functions is increased and which, with the power of contemporary computing machines, can be used to obtain atomic Hartree-Fock energies to a very high precision. This approach is illustrated in Figure 1 where a systematic sequence of even-tempered basis sets of Gaussian-type functions is used to calculate the Hartree-Fock energy for the ground state of the F^- ion[48]. In this Figure, the notation $A[B]$ indicates that system A is studied using a sequence of basis sets originally designed for B. $[B']$ indicates that a diffuse function is added to basis sets $[B]$. $[B'']$ indicates that two diffuse function of each symmetry type are added to the basis set $[B]$. It can be seen that provided one diffuse function of each symmetry type is added to the basis set designed for the neutral species, the anion can be accurately described within the Hartree-Fock model within the algebraic approximation.

In relativistic atomic structure calculations systematically constructed sequences of basis sets can yield Dirac-Hartree-Fock energies of an accuracy comparable with that obtained in finite difference studies. This is illustrated in Table 1 for the ground state of the argon atom[19]. Relativistic effects are at their most important for the heavy elements. For such systems it is important to take account of the finite size of nuclei. It is only in recent years that it has become widely recognized that once the point nucleus model is abandoned in favour of the more physical finite nucleus model the Gaussian function becomes a more appropriate choice of basis function[21].

2.2. MOLECULAR BASIS SETS

Molecular basis sets are almost invariably constructed from atomic basis sets. In 1960, Nesbet[49] pointed out that molecular basis sets containing only basis functions types necessary to reach the atomic Hartree-Fock limit, the so-called isotropic basis set, cannot possibly account for polarization in molecular binding. Two approaches to the problem of designing molecular basis sets can be identified:

a) the addition of atom-centred polarization functions to the atomic basis sets

b) the addition of off-centre functions of the same symmetry as the atomic Hartree-Fock basis set.

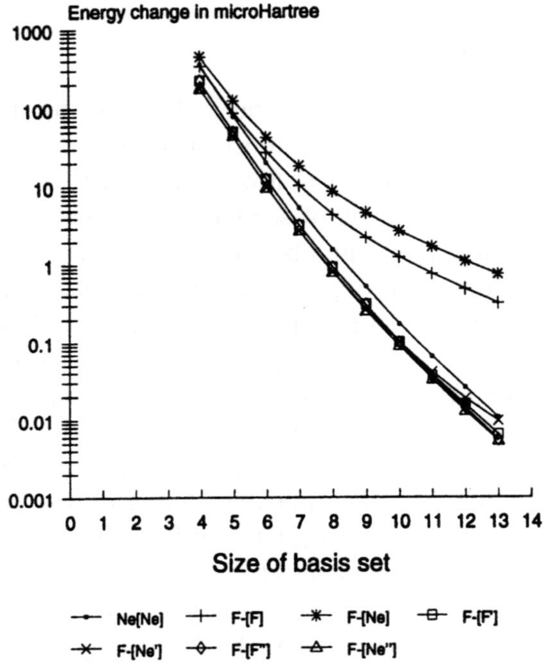

Figure 1. Convergence of the matrix Hartree-Fock energy for the F^- ion with increasing size of basis set. See text for details.

In the early sixties, Reeves, Harrison and Fletcher[50–52] established important elements of the current 'mainstream' usage of Gaussian basis sets. They concluded that[53]

> *no significant advantage was gained by letting the basis functions float away from the nuclear centres, that orbital exponents should be distributed approximately in geometric progression, and that it is not advantageous to optimize separate sets of Gaussians for the description of individual shells.*

These conclusions support approach *a)* and form the foundation of the majority of current mainstream practice (see the review by Shavitt[53] for a masterly overview). However, these conclusions were drawn from calculations using basis sets consisting of a small number of functions ($\sim 10 - 10^2$). Today, the so-called 'direct' algorithms[54], which avoid the storage of the

TABLE 1. Matrix Dirac-Hartree-Fock calculations for the *Ar* ground state using a systematic sequence of exponential-type basis sets[†].

N	α	β	$-E$
9	0.500	1.550	528.64394819
10	0.484	1.516	528.68336415
11	0.470	1.486	528.68431173
12	0.458	1.462	528.68428973
13	0.447	1.440	528.68443068
14	0.437	1.421	528.68443246
15	0.429	1.404	528.68444856
16	0.421	1.389	528.68444990
17	0.413	1.376	528.68445050
$E_{\text{finite difference}}$			528.68445077

[†]All energies are given in Hartree units.
Taken (with correction) from the work of Quiney, Grant and Wilson[19].

two-electron integrals over the basis functions, have enabled calculations using in excess of 10^3 to be carried out on a workstation and calculations using basis sets of a size approaching 10^4 to be seriously contemplated. It is, therefore, timely to reconsider approaches *a)* and *b)* in the light of these developments. We note that these approaches are not mutually exclusive in that hybrid prescriptions for basis set design can be adopted.

The idea of using of bond functions in molecular structure calculations is an old one[55]. Indeed, a number of authors have advocated the use of basis functions located at the bond centre in both self-consistent field and in calculations taking account of correlation effects[56–65]. Such functions were introduced to provide the same effects as the addition of a set of polarization functions but at a considerably lower cost. Davidson and Feller[13] point out that

> *For first-row diatomics the introduction of a single (s,p) set at the centre of the bond provides* 90% *of the energy lowering obtained with a single set of d functions at the nuclear centres.*

In 1980, Čársky and Urban noted that general rules for selecting exponents and positions of the bond functions are still lacking[66].

In previous work on the ground state of the nitrogen molecule[30, 31] a basis set designated $[30s15p15d15f : 27s12p10d\ bc]$ (the notation $[A : B\ bc]$ is used where A denotes the atom-centred basis sets and B the bond-centred basis sets) was found to lead to a total energy that is within 2.3 μHartree of the finite difference result. This basis set contains a total of 510 atom centred basis functions, 255 functions on each atomic centre, and 113 functions in the bond centred subset, making the total number of basis function 623. Of these 250 atom-centred functions contribute to the description of the ground state of the nitrogen molecule in the Hartree-Fock ansätz as do 93 of the bond centred functions.

Given the success of calculations using off-atom basis functions, we were led to investigate approach *b)* in more detail. Very recently, Moncrieff and Wilson[67] have demonstrated that the Hartree-Fock ground state of the N_2 molecule can be described to an accuracy of ~ 50 μHartree using a basis set of s and p functions only. The functions were distributed along the line passing through the nuclei. They also considered electron correlation effects using these basis sets. In other recent work, distributed basis sets for molecular electronic structure calculations have been considered in rather more general terms[44–46] and this work will be discussed in section 3.

2.3. SYSTEMATIC GENERATION OF ATOMIC BASIS SETS

In the algebraic approximation, the atomic self-consistent field orbitals

$$\phi_i(n,l,m) = f_{n\ell}(r)\, r^\ell\, Y_\ell^m(\theta,\varphi) \tag{1}$$

are approximated as a linear combination of N primitive basis functions, $\chi_{k\ell m}^N$,

$$\phi_i^N(n,\ell,m) = \sum_{k=1}^{N} c_{k,n\ell}^N\, \chi_{k\ell m}^N \tag{2}$$

These basis functions can be written

$$\chi_{k\ell m}^N = R_{k\ell}(r)\, r^\ell\, Y_\ell^m(\theta,\varphi) \tag{3}$$

where the $Y_\ell^m(\theta,\varphi)$ are normalized spherical harmonics. If the basis functions are chosen to be Gaussian-type functions then the radial factor has the form

$$R_{k\ell}(r) \;=\; N_\ell(\zeta_{k\ell}^N)\exp(-\zeta_{k\ell}^N\, r^2) \tag{4}$$

with normalization constant

$$N_\ell(\zeta_{k\ell}^N) = \{(\zeta_{k\ell}^N)^{2\ell+3}\, 2^{4\ell+7}/\pi(2\ell+1)!!^2\}^{\frac{1}{4}} \tag{5}$$

Klahn[68] has presented a generalized Müntz-Szász theorem[69–74] which essentially states that a set of Gaussian-type functions is asymptotically complete in $L\,(R^+)$

$$\lim_{N\to\infty} \Gamma_\ell(N) = \infty, \quad \forall\ell \tag{6}$$

where

$$\Gamma_\ell(N) = \sum_{k=1}^{N} \frac{\zeta_{k\ell}^N}{1 + \left(\zeta_{k\ell}^N\right)^2}, \quad \forall \ell \tag{7}$$

Now it has been observed that exponents which have been carefully optimized for atoms often form a good approximation to a geometric progression[4, 50, 51]

$$\zeta_{k\ell}^N = \alpha_\ell^N \left(\beta_\ell^N\right)^k, \; \beta_\ell^N > 1, \; k = 1, 2, ..., N \tag{8}$$

Conversely, if it is assumed that the exponents do form a geometric progression and the parameters α_ℓ^N and β_ℓ^N are optimized for atoms then there is found to be little lost in accuracy[4, 5]. Basis sets developed in this way are termed even-tempered basis sets and open up the possibility of constructing the large and flexible basis sets that are inevitably required for calculations of high precision.

As the number of basis functions, N, is increased we require that our basis set approach a complete set. The generalized Müntz-Szász theorem[68] can be used to show that this is not the case if α_ℓ and β_ℓ are held fixed as N is increased. One possible choice which does lead to a complete set as $N \to \infty$ is

$$\alpha_\ell^N \to 0 \tag{9}$$

$$\beta_\ell^N \to 1 \tag{10}$$

$$\left(\beta_\ell^N\right)^N \to \infty \tag{11}$$

These limits ensure that the lowest exponent for each ℓ value tends to 0

$$\zeta_{\min,\ell} = \alpha_\ell \beta_\ell \to 0 \tag{12}$$

the largest exponent for each ℓ tends to ∞

$$\zeta_{\max,\ell} = \alpha_\ell \left(\beta_\ell\right)^N \to \infty \tag{13}$$

and the spacing between the exponents is successively reduced

$$\beta_\ell \to 1 \tag{14}$$

The limits can be guaranteed by generating successive basis sets according to the following empirical recursions[75]

$$\alpha_\ell^N = \left[\frac{\beta_\ell^N - 1}{\beta_\ell^{N-1} - 1}\right]^{a_\ell} \alpha_\ell^{N-1}, \; a_\ell > 0 \tag{15}$$

and

$$\ln \beta_\ell^N = \left[\frac{N}{N-1}\right]^{b_\ell} \ln \beta_\ell^{N-1}, \; -1 < b_\ell < 0 \tag{16}$$

which are essentially the expressions first given by Schmidt and Ruedenberg[11].

The prescription given above for generating atomic basis sets in a systematic fashion is, of course, not unique. Amongst the alternative, but closely related approaches, are the universal basis sets[76] and universal systematic sequences of basis sets[75], geometric basis sets[77], and well-tempered basis sets[78]. The approach of Haywood and Morgan[79] should also be mentioned.

2.4. SYSTEMATIC GENERATION OF MOLECULAR BASIS SETS

Most work on the design of molecular basis sets has concentrated on the development of polarization sets which can be added to atomic basis sets to described the molecular environment. Our recent work comparing finite difference with finite basis set calculations has demonstrated that a sub-μHartree level of accuracy for the total Hartree-Fock energy can be achieved when a systematic design of a molecular basis set is undertaken[28–35]. It should be emphasised that there is no unique solution to the problem of designing molecular basis sets. Formally, a one-centre expansion can be made but this will be poorly convergent for heavy atoms located off the expansion centre[2].

3. Distributed Gaussian Basis Sets

Off-atom functions effectively introduce linear combinations of higher harmonics centred on the atom. Partial wave expansions of an off-centre $1s$-type Gaussian function have been discussed, for example, by Christoffersen *et al*[80] and by Kaufmann and Baumeister[81].

A modified form of the Rayleigh expansion[82] may be written

$$\exp\left(\mathbf{r}_1 \cdot \mathbf{r}_2\right) = \sum_{\ell=0}^{\infty} (2\ell+1) \; i_\ell\left(r_1, r_2\right) \; P_\ell(\cos\gamma) \tag{17}$$

in which $i_\ell\left(x\right)$ is a modified Bessel function and $P_\ell\left(\cos\gamma\right)$ is a Legendre polynomial of order ℓ. γ is the angle between the directions $\mathbf{r}_1$ and $\mathbf{r}_2$. Using the expansion (17) in an s-type Gaussian function, $\exp\left(-\zeta\left|\mathbf{r}-\mathbf{r}_\lambda\right|^2\right)$, with $\mathbf{r}_1$ assigned to $\mathbf{r}$ and $\mathbf{r}_2$ assigned to $\mathbf{r}_\lambda$, gives

$$\exp\left(-\zeta\left|\mathbf{r}-\mathbf{r}_\lambda\right|^2\right) = \tag{18}$$

$$\exp\left[-\zeta(r^2+r_\lambda^2)\right] \sum_{\ell=0}^{\infty} (2\ell+1)\; i_\ell(2\zeta\, r\, r_\lambda)\; P_\ell(\cos\gamma)$$

The addition theorem for surface spherical harmonics allows equation (18) to be written

$$\exp\left(-\zeta\left|\mathbf{r}-\mathbf{r}_\lambda\right|^2\right) = \tag{19}$$

$$4\pi\, \exp\left[-\zeta(r^2+r_\lambda^2)\right] \sum_{\ell=0}^{\infty} \sum_{m=-\ell}^{+\ell} i_\ell(2\zeta\, r\, r_\lambda)\, Y_{\ell m}(\Omega_r)\; Y_{\ell m}^*(\Omega_{r_\lambda})$$

which is the expansion of an off-centre Gaussian function.

Gaussian functions are appropriate functions for electronic structure calculations not only because of the widely recognized fact that they lead to molecular integrals which can be evaluated efficiently and accurately but also because they do not introduce a cusp into the approximation for the wave function at a physically inappropriate point. It should also be noted that Gaussian functions are suitable for the description of wave functions in the vicinity of nuclei once the point nucleus model is abandoned in favour of a more realistic finite nucleus model.

3.1. THE GAUSSIAN CELL MODEL

The original Gaussian cell model[42, 43] is a distributed Gaussian basis set in which the basis functions have a single common exponent and are arranged on a regular lattice. The normalized basis functions may be written

$$\chi_m(\mathbf{r}) = \left(\frac{2\zeta}{\pi}\right)^{\frac{3}{4}} \exp\left(-\zeta\left|\mathbf{r}-\mathbf{r}_\lambda\right|^2\right) \tag{20}$$

where ζ is the orbital exponent. $\mathbf{r}_{\{\lambda\}}$ is a point on a regular lattice defined by

$$\mathbf{r}_{\{\lambda\}} = \{\pm m_x\lambda_x, \pm m_y\lambda_y, \pm m_z\lambda_z\},$$

$$m_x = 1,2,\ldots,n_x;\, m_y = 1,2,\ldots,n_y;\, m_z = 1,2,\ldots,n_z \tag{21}$$

n_x, n_y, n_z are taken to be odd and the lattice point $\{0,0,0\}$ is taken to coincide with the mid-point of the bond. It will be assumed that

$$\lambda_x = \lambda_y = \lambda_z = \lambda \tag{22}$$

Furthermore, the nuclei will be taken to lie on the z-axis and positioned on one of the lattice points, so that

$$\lambda = \frac{\frac{1}{2}R}{i},\; \left\{1 \le i \le \frac{1}{2}(n_z-1)\right\} \tag{23}$$

where i is an integer. Murrell *et al*[42, 43] emphasized the importance of ensuring the coincidence of a lattice points with each of the nuclear positions.

3.2. LAPLACE TRANSFORM OF AN ELLIPTICAL FUNCTION

The 'natural' basis functions for calculations on diatomic molecules are elliptical functions. They have been employed in diatomic molecular electronic structure studies by a number of authors[83–92] and may be defined as follows

$$\frac{1}{\sqrt{2\pi R^3}} \exp(-\alpha\lambda) \exp(-\beta\mu) \lambda^s \mu^t [(\lambda^2 - 1)(1 - \mu^2)]^{\frac{|\nu|}{2}} \exp(\imath\nu\varphi) \tag{24}$$

where α and β are screening parameters and s and t taken integer values $0, 1, 2, ...$ (λ, μ, φ) are the usual prolate spheroidal coordinates $\lambda = \frac{1}{2R}(r_a + r_b)$, $1 \leq \lambda \leq \infty$; $\mu = \frac{1}{2R}(r_a - r_b)$, $-1 \leq \mu \leq 1$; φ, $0 \leq \varphi \leq 2\pi$. $2R$ is the nuclear separation and r_a (r_b) is the distance from nucleus $a(b)$. $\nu = 0, \pm 1, \pm 2, ...$ determines the angular symmetry of the function: $\sigma, \pi, \delta, ...$ Elliptical basis functions have been employed in electronic structure calculations for diatomic molecules and have been shown to afford high accuracy[28, 88, 90–92].

The approximation of elliptical functions by a finite expansion of Gaussian-type functions has been investigated recently[47]. Attention was restricted to the use of $1s$ Gaussian-type functions to approximation the simplest elliptical basis functions with $s = t = 0$ and $\nu = 0$, that is

$$\chi(\lambda, \mu; \alpha, \beta, R) = \frac{1}{\sqrt{2\pi R^3}} \exp(-\alpha\lambda) \exp(-\beta\mu) \tag{25}$$

The $1s$ Gaussian-type functions centred on the point p are now written as

$$(2\eta/\pi)^{\frac{3}{4}} \exp(-\eta r_p^2) \tag{26}$$

where η is a screening parameter. Expressing χ in terms of the coordinates $r_a\,(= R(\lambda + \mu))$ and $r_b\,(= R(\lambda - \mu))$ gives a product of two exponential functions

$$\chi(r_a, r_b; \zeta_a, \zeta_b, R) = \frac{1}{\sqrt{2\pi R^3}} \exp(-\zeta_a r_a) \exp(-\zeta_b r_b) \tag{27}$$

where the screening constants for the exponential functions, ζ_a and ζ_b, are related to those for the elliptical function, α and β, by

$$\zeta_a = \frac{1}{2R}(\alpha + \beta) \tag{28}$$

and

$$\zeta_b = \frac{1}{2R}(\alpha - \beta) \tag{29}$$

An exponential function can be related to an integral over Gaussian functions by means of the Laplace transform[82]

$$\exp(-\zeta r) = \frac{\zeta}{2\sqrt{\pi}} \int_0^\infty d\eta\, \eta^{-\frac{3}{2}} \exp(-\frac{\zeta^2}{4\eta}) \exp(-\eta r^2) \tag{30}$$

The discrete representation of this integral may be written

$$\exp(-\zeta r) \approx \sum_{p=1}^{N} c_p \exp(-\eta_p r^2) \tag{31}$$

and this relation has been used by Silver[93] and, more recently, by Kutzelnigg[94] to develop approximations for exponential functions in terms of Gaussian functions. In the work of Silver[93], the exponents η_p were chosen to be the points of a Gaussian quadrature scheme and the coefficients, c_p, are related to the corresponding weights. On the other hand, Kutzelnigg[94] took the exponents, η_p, to be an even-tempered set.

Use of the Laplace transform for the elliptical function provides an exact representation of χ by means of a continuous distribution of Gaussian functions over the variables η_a and η_b:

$$\begin{aligned} \chi(r_a, r_b; \zeta_a, \zeta_b, R) &= \frac{1}{\sqrt{2\pi R^3}} \frac{\zeta_a \zeta_b}{4\pi} \int_0^\infty d\eta_a \int_0^\infty d\eta_b\, \eta_a^{-\frac{3}{2}} \eta_b^{-\frac{3}{2}} \\ & \quad \exp\left(-\frac{1}{4}\left[\frac{\zeta_a^2}{\eta_a} + \frac{\zeta_b^2}{\eta_b}\right]\right) \exp(-\{\eta_a r_a^2 + \eta_b r_b^2\}) \end{aligned} \tag{32}$$

Equation (32) can be converted into a finite expansion in terms of Gaussian-type functions by performing the double integration with any one of a number of numerical quadrature schemes. In the expansion

$$\chi(r_a, r_b; \zeta_a, \zeta_b, R) \approx \sum_{p=1}^{N} \sum_{q=1}^{N} c_{ap}\, c_{bp} \exp(-\eta_p r_a^2) \exp(-\eta_q r_b^2) \tag{33}$$

the parameters η_p and c_p are specified by the chosen quadrature. Using the Gaussian product theorem[95], (33) may be written

$$\chi(r_a, r_b; \zeta_a, \zeta_b, R) \approx \sum_{p=1}^{N} \sum_{q=1}^{N} c_{ap} c_{bp} \gamma_{pq} \exp(-\eta_{pq} r_{pq}^2) \tag{34}$$

where the coefficients γ_{pq} are given by

$$\gamma_{pq} = \exp\left(-\frac{\eta_p \eta_q}{\eta_p + \eta_q} R^2\right) \tag{35}$$

and the screening parameters are

$$\eta_{pq} = \eta_p + \eta_q \tag{36}$$

The coordinate r_{pq} is defined with respect to the point (x_{pq}, y_{pq}, z_{pq}) where

$$x_{pq} = \frac{\eta_p x_a + \eta_q x_b}{\eta_p + \eta_q} \tag{37}$$

with similar relations for y_{pq} and z_{pq}.

4. Some Recent Results

4.1. THE GAUSSIAN CELL MODEL REVISITED

The original Gaussian Cell model basis set, with one function centred on each lattice point and all functions having the same exponent, may be written

$$S_{Gaussian\ Cell\ Model} = S_1^{(\lambda_1)} \oplus S_1^{(\lambda_2)} \oplus \ldots \oplus S_1^{(\lambda_N)} \tag{38}$$

or

$$S_{Gaussian\ Cell\ Model} = \sum_{p \in \{L\}} \oplus S_1^{(\lambda_p)} \tag{39}$$

where

$$S_1^{(\lambda_p)} = \left\{(2\zeta/\pi)^{\frac{3}{4}} \exp(-\zeta r_p^2)\right\} \tag{40}$$

The convergence behaviour of this method is illustrated in Table 2 where some results recently obtained by Ralston and Wilson[44] are reproduced. Not surprisingly, this simple model does not achieve high accuracy; indeed, the lowest energy reported in Table 1 is in error by ~ 7705 μHartree. However, more recent work on one-electron diatomic[44, 45] and triatomic[46] systems has demonstrated that, with simple refinements, it can be very significantly enhanced and accuracies approaching ~ 1 μHartree can be achieved for the total energy.

4.2. MOLECULAR LATTICE BASIS SETS

It is clear that the major failing of the Gaussian Cell model is its lack of flexibility particularly in the regions close to the nuclei. However, methods

TABLE 2. Calculations for the H_2^+ ground state using a regular cubic lattice distributed basis set†.

Lattice	N	i	E	ζ	λ
$5 \times 5 \times 5$	125	1	−0.582 046	1.407 4	1
$5 \times 5 \times 5$	125	2	−0.577 133	0.618 4	$\frac{1}{2}$
$7 \times 7 \times 7$	343	1	−0.591 606	1.545 2	1
$7 \times 7 \times 7$	343	2	−0.587 718	0.675 2	$\frac{1}{2}$
$7 \times 7 \times 7$	343	3	−0.586 792	0.616 3	$\frac{1}{3}$
$9 \times 9 \times 9$	729	1	−0.592 429	1.556 8	1
$9 \times 9 \times 9$	729	2	−0.593 043	0.907 2	$\frac{1}{2}$
$9 \times 9 \times 9$	729	3	−0.592 283	1.556 8	$\frac{1}{3}$
$9 \times 9 \times 9$	729	4	−0.590 744	0.747 4	$\frac{1}{4}$
$11 \times 11 \times 11$	1331	2	−0.594 929	1.174 9	$\frac{1}{2}$
Exact			−0.602 634		

†All energies are given in Hartree units.
The internuclear distance is set at its equilibrium value of 2.0 bohr.

for constructing basis sets for accurate atomic Hartree-Fock calculations are well established. Ralston and Wilson[44], therefore, introduced the molecular lattice basis sets which may be written

$$S_{Molecular\ Lattice\ Basis\ Set} = S_1^{(\lambda_1)} \oplus S_1^{(\lambda_2)} \oplus \ldots \oplus S_1^{(\lambda_{q-1})} \oplus S_{N_{atom}}^{(\lambda_q)} \oplus S_1^{(\lambda_{q+1})} \ldots \oplus S_1^{(\lambda_N)} \tag{41}$$

or

$$S_{Molecular\ Lattice\ Basis\ Set} = \left(\sum_{p \in \{L'\}} \oplus S_1^{(\lambda_p)} \right) \oplus \left(\sum_{q \in \{A\}} \oplus S_{N_{atom}}^{(\lambda_q)} \right) \tag{42}$$

where atomic basis sets are centred on lattice points which coincide with nuclei and a single function is centred on each of the lattice points which do not coincide with nuclei with all of these functions having the same exponent. If $\{A\}$ denotes the atomic nuclei and $\{L\}$ the full set of lattice points then

$$\{L'\} = \{L\} - \{A\} \tag{43}$$

The atomic basis set may be chosen to be even-tempered sets.

$$S_{N_{atom}}^{(\lambda_p)} = \left\{ (2\zeta_k/\pi)^{\frac{3}{4}} \exp(-\zeta_k r_p^2); \zeta_k = \alpha_{(N_{atom})} \beta_{(N_{atom})}^k; k = 1, 2, \ldots, N_{atom} \right\} \tag{44}$$

TABLE 3. Calculations for the H_2^+ ground state using a molecular regular cubic lattice distributed basis set†.

Lattice	N	i	E	ζ	λ
$5 \times 5 \times 5$	143	1	−0.602 462	0.691 9	1
$5 \times 5 \times 5$	143	2	−0.602 462	0.518 2	$\frac{1}{2}$
$7 \times 7 \times 7$	361	1	−0.602 508	0.738 1	1
$7 \times 7 \times 7$	361	2	−0.602 536	0.688 1	$\frac{1}{2}$
$7 \times 7 \times 7$	361	3	−0.602 532	0.622 8	$\frac{1}{3}$
$9 \times 9 \times 9$	747	1	−0.602 515	0.718 4	1
$9 \times 9 \times 9$	747	2	−0.602 593	0.831 7	$\frac{1}{2}$
$9 \times 9 \times 9$	747	3	−0.602 586	0.732 5	$\frac{1}{3}$
$9 \times 9 \times 9$	747	4	−0.602 570	0.673 8	$\frac{1}{4}$
$11 \times 11 \times 11$	1349	1	−0.602 517	0.706 6	1
$11 \times 11 \times 11$	1349	2	−0.602 607	0.880 1	$\frac{1}{2}$
$11 \times 11 \times 11$	1349	3	−0.602 597	0.748 8	$\frac{1}{3}$
Exact			−0.602 634		

†All energies are given in Hartree units.
The internuclear distance is set at its equilibrium value of 2.0 bohr.

The convergence behaviour of these molecular lattice basis sets is illustrated in Table 3 where results, taken from the work of Ralston and Wilson[44], show that this approach can yield an accuracy of ~ 27 μHartree for the H_2^+ ground state at its equilibrium geometry. In these calculations, N_{atom} was set to 10 and the atomic even-tempered parameters were taken from the tables of Schmidt and Ruedenberg[11].

4.3. SYSTEMATICALLY EXTENDED MOLECULAR LATTICE BASIS SETS

The molecular lattice basis set, although a considerable improvement on the original Gaussian Cell model, does not afford an accuracy approaching the 1 μHartree level even for the simplest of molecules, the hydrogen molecular ion. The molecular lattice even-tempered basis set was, therefore, introduced[45]. This is defined as

$$S_{Molecular\ Lattice\ Even-tempered\ Basis\ Set} = \qquad (45)$$
$$S_{N_{lattice}} \oplus S_{N_{lattice}} \oplus \ldots \oplus S_{N_{lattice}} \oplus S_{N_{atom}} \oplus S_{N_{lattice}} \cdots \oplus S_{N_{lattice}}$$

or

$$S_{Molecular\ Lattice\ Even-tempered\ Basis\ Set} = \qquad (46)$$

TABLE 4. Calculations for the H_2^+ ground state using a $5 \times 5 \times 5$ molecular regular cubic lattice of systematically constructed distributed basis sets with $i = 2^{\dagger}$.

$N_{lattice}$	N	M	E_0	δ
1	143	143	−0.602052	582
2	266	255	−0.602592	42
3	389	368	−0.602621	13
4	512	476	−0.602629	5
5	635	591	−0.602630	4
6	758	705	−0.602629	5
7	881	820	−0.602630	4
8	1004	933	−0.602630	4
9	1127	1047	−0.602630	4
10	1250	1165	-0.602630	4
Exact			−0.602634	

†All energies are given in Hartree units and energy differences in μHartree. The internuclear distance is set at its equilibrium value of 2.0 bohr.

$$\left(\sum_{p \in \{L'\}} \oplus S_{N_{lattice}}^{(\lambda_p)} \right) \oplus \left(\sum_{q \in \{A\}} \oplus S_{N_{atom}}^{(\lambda_q)} \right)$$

where atomic basis sets are located on each of the nuclei and "lattice" basis sets are centred on each lattice point. The "lattice" basis sets are identical and may be taken to be even-tempered.

$$S_{N_{lattice}}^{(\lambda_p)} = \left\{ (2\zeta_k/\pi)^{\frac{3}{4}} \exp(-\zeta_k r_p^2); \zeta_k = \alpha_{(N_{lattice})} \beta_{(N_{lattice})}^k; k = 1, \ldots, N_{lattice} \right\} \tag{47}$$

The convergence behaviour of these systematically extended molecular lattice basis sets are displayed in Table 4 for a $5 \times 5 \times 5$ lattice where it can be seen that the lowest energy is in error by just 4 μHartree. The convergence pattern for three difference lattices are shown in Figure 2.

Calculations using a molecular lattice distributed basis set have also been carried out[46] the linear H_3^{2+}, a simple polyatomic system for which the results of finite element studies have recently been published[96].

4.4. LAPLACE-TRANSFORM BASED DISTRIBUTED BASIS SETS

The prescription for distribution of the basis set obtained by considering the analytical relation between elliptical functions and Gaussian functions

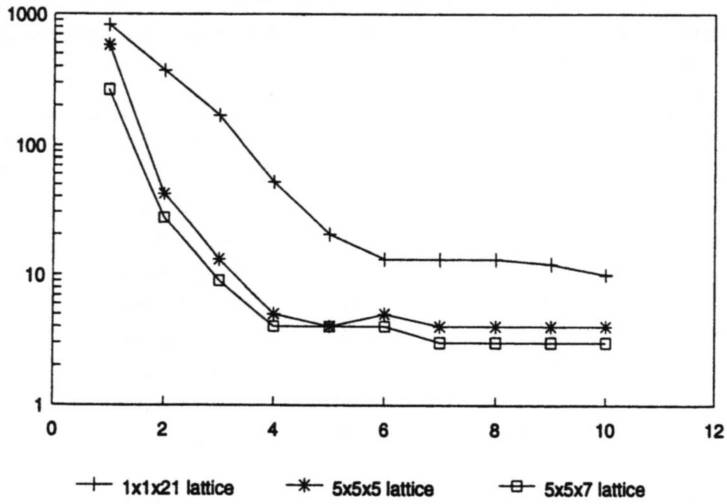

Figure 2. Convergence of the ground state energy of the hydrogen molecular ion for a distributed Gaussian basis set of s-type functions using a regular linear lattice and a regular cubic lattice together with systematic sequence of even-tempered functions on each lattice point.

using a Laplace transform provides an alternative technique for the construction molecular distributed basis sets. It provides a method for generating systematic sequences of molecular basis sets from systematic sequences of atomic basis sets. Here, we shall consider a systematic sequence of even-tempered atomic basis sets.

The exponents for an atom-centred even-tempered basis set of Gaussian-type functions are defined by the geometric series

$$\eta_p = \alpha\beta^p, \ p = 1, 2, ..., N \tag{48}$$

For this choice of generating formula, the exponents in the expansion (34) are given by

$$\eta_{pq} = \alpha(\beta^p + \beta^q), \ p, q = 1, 2, ..., N \tag{49}$$

and, for a diatomic system with nuclei placed on the z-axis, the coordinate r_{pq} is defined with respect to the point $(0, 0, z_{pq})$ where

$$z_{pq} = \frac{\beta^{p-q} z_a + z_b}{\beta^{p-q} + 1}, \ p \geq q$$

$$z = \frac{z_a + \beta^{q-p} z_b}{1 + \beta^{q-p}}, \; q \geq p$$
$$p, q = 1, 2, ..., N \tag{50}$$

In this way, a systematic sequence of atomic even-tempered basis sets can be used to generate a sequence of molecular basis sets. Convergence of the calculated energy of the hydrogen atom and the hydrogen molecular ion observed with a systematic sequence of even-tempered basis sets of $1s$ Gaussian functions is displayed in Figure 3. For the H_2^+ ion the energy obtained with the basis set distributed along the line passing through the nuclei according to the Laplace transform of an elliptical functions is compared with that obtained with atom-centred $1s$ Gaussian functions. The atom-centred basis set exhibits smooth behaviour with increasing number of functions but is not capable of describing the H_2^+ ground state[49]. The Laplace transform based distribution, on the other hand, displays a smooth convergence pattern with increasing basis set size and monotonically decreasing energy values.

5. Prospects

Basis sets capable of supporting high precision calculations (*i.e.* total energies approaching the 1 μHartree level of accuracy) for molecules within the Hartree-Fock model are now a reality. Such basis sets can be efficiently constructed if the usual approach in which basis functions are taken to be atom centred is abandoned and functions centred on, for example, the bond centre are introduced. The distributed basis set concept provides a general approach to the construction of basis sets which include off-atom functions. Just as the introduction of even-tempered basis sets showed that it was not necessary to exhaustively optimize exponents if schemes are devised for their systematic extension (indeed, for atoms such an approach can yield an accuracy comparable with that achieved in finite difference calculations), so the introduction of distributed basis sets provides systematic schemes for the placing of the functions in space. The two distributions described in this article, the Gaussian Cell model and its extensions and the Laplace transform of an elliptical function, have been shown to be capable of an accuracy approaching 1 μHartree.

Basis sets capable of supporting high precision electron correlation calculations are under development[97]. The accurate approximation of the solutions of the Hartree-Fock problem for a molecule is a necessary precursor for any treatment of correlation effects. Not only is the correlation energy defined with respect to the total Hartree-Fock energy but the solutions of the Hartree-Fock equations define the quasi-particles in terms of which a description of the many-body system is developed[98–104].

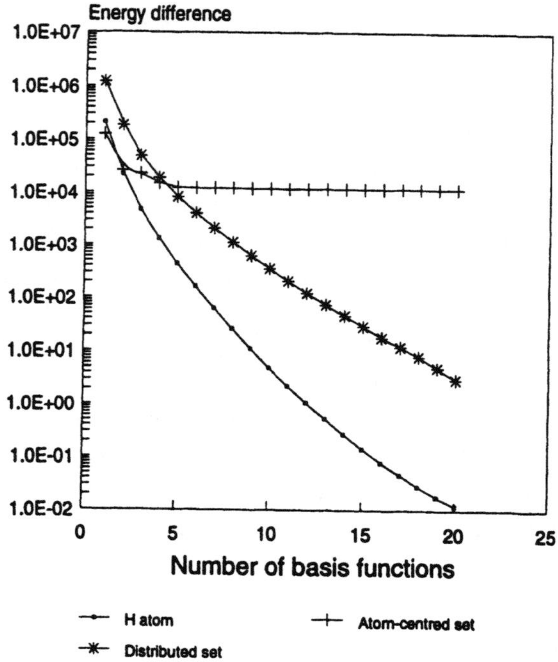

Figure 3. Convergence of the calculated energy of the hydrogen atom and the hydrogen molecular ion observed with a systematic sequence of even-tempered basis sets of 1s Gaussian functions. For the molecular systems the basis sets are either atom-centred or distributed along the line passing through the nuclei according to the Laplace transform of an elliptical functions.

Basis sets capable of supporting high precision in non-relativistic studies can be easily modified to serve as a basis for relativistic molecular structure calculations. Essentially, additional contracted basis functions are required in the relativistic case to describe the contraction of the single particle state functions in the core region whilst additional diffuse functions are needed to account for the expansion of the valence single particle state functions. In relativistic electronic structure calculations within the algebraic approximation matched basis sets must be employed for the large and small components of the wave function. The large and small components of the relativistic wave function must be approximated in terms of basis functions which satisfy the condition[19]

$$\chi^S_{i\kappa} \propto \left(\frac{d}{dr} + \frac{\kappa}{r}\right) \chi^L_{i\kappa} \tag{51}$$

There is a one-to-one correspondence between the large and small component basis functions. If the large component basis set contains s functions then the small component basis sets must contain p functions with the same exponents; if the large component basis set contains p functions then the small component basis set must contain functions which are a linear combination of s and d functions; in general, if the large component basis functions have the form

$$\chi^L_{im\kappa} = N_{i\kappa} r^{\ell} \exp(-\zeta_i r^2) \Omega^{\mathrm{sgn}(\kappa)}_{jm}(\theta, \varphi) \tag{52}$$

then the small component basis functions have the form

$$\chi^S_{im\kappa} = \imath N_{i\kappa} \left((\ell + \kappa + 1) - 2\zeta r^2\right) r^{\ell-1} \exp(-\zeta_i r^2) \Omega^{-\mathrm{sgn}(\kappa)}_{jm}(\theta, \varphi) \tag{53}$$

where κ is the usual angular quantum number, $2j = 2|\kappa| - 1$, $-j \leq m \leq j$. The restriction of the large components basis set to functions of low symmetry when using distributed basis sets imposes corresponding restrictions on the small component basis set[105].

By restricting basis sets to functions of low symmetry distributed basis sets lead to efficient integral evaluation algorithms; an essential ingredient of the 'direct' self-consistent field procedures[54] which are required for both non-relativistic and relativistic calculations using large basis sets. Furthermore, when basis sets are systematically constructed and extended there is considerable scope for integral approximation.

Finally, it should be mentioned that distributed Gaussian basis sets provide a natural vehicle for the determination of point charge models of molecular systems which are required for hybrid quantum mechanical/classical calculations. Many of the techniques suggested by Hall and his coworkers[106–108] for obtaining point charge models from floating spherical Gaussian orbital electronic structure calculations can be applied to the results of distributed Gaussian basis set studies[109]. Hybrid methods offer considerable scope for the study of molecules in different environments, *e.g.* on surfaces, in solution and in cages.

Acknowledgement

I would like to express my gratitude to Professor C.A. Tsipis, Aristotle University of Thessaloniki, Director of the NATO Advanced Research Workshop on *New Methods in Quantum Theory*, for asking me to deliver the plenary lecture from which this article is derived. I would also like to acknowledge numerous conversations about finite basis set methods, the algebraic approximation and its accurate implementation with Dr. B.H. Wells, Dr. J.W. Thompson and Dr. D. Moncrieff. Discussions with Dr. B.J. Ralston about the Gaussian Cell model were most useful as was correspondence with Dr. A.S. Shalabi about negative ions. Most of the calculations described in this article were carried out on the Turbolaser field test machine at the Rutherford Appleton Laboratory.

References

1. S. Wilson, in *Methods in Computational Molecular Physics*, ed. G.H.F. Diercksen and S. Wilson, Reidel, Dordrecht (1983)
2. S. Wilson, Adv. Chem. Phys. **67**, 439 (1987)
3. C. Schwartz, Meth. Comput. Phys. **2**, 241 (1962)
4. K. Ruedenberg, R.C. Raffenetti and R. Bardo, in *Energy structure and reactivity, Proc. 2nd. Boulder Research Confer. in Theoret. Chem.*, Wiley, New York (1973)
5. R.C. Raffenetti, J. Chem. Phys. **58**, 4452 (1973)
6. R.C. Raffenetti, J. Chem. Phys. **59**, 5936 (1973)
7. R.C. Raffenetti and K. Ruedenberg, J. Chem. Phys. **59**, 5978 (1973)
8. R. Bardo and K. Ruedenberg, J. Chem. Phys. **59**, 5956, 5966 (1973)
9. R. Bardo and K. Ruedenberg, J. Chem. Phys. **60**, 918 (1974)
10. D.F. Feller and K. Ruedenberg, Theoret. chim. Acta. **52**, 231 (1979)
11. M.W. Schmidt and K. Ruedenberg, J. Chem. Phys. **71**, 3951 (1979)
12. S. Huzinaga, Comput. Phys. Repts. **2**, 279 (1985), 3951 (1979)
13. E.R. Davidson and D.F. Feller, Chem. Rev. **86**, 681 (1986)
14. E. Clementi, S.J. Chakravorty, G. Corongiu and V. Sonnad, in *Modern Techniques in Computational Chemistry*, ed. E. Clementi, p.83, ESCOM, Leiden (1990)
15. P.R. Taylor, in *European Summer School in Quantum Chemistry*, Lecture Notes in Chemistry **58**, p. 399, edited by B.O. Roos, Springer-Verlag, Berlin (1992)
16. D.R. Hartree, *The Calculation of Atomic Structures*, Wiley, New York (1957)
17. S. Wilson and D.M. Silver, Phys. Rev. A**14**, 1969 (1976)
18. H.M. Quiney, I.P. Grant and S. Wilson, Physica Scripta **36**, 460 (1987)
19. H.M. Quiney, I.P. Grant and S. Wilson, in *Many-Body Methods in Quantum Chemistry,* Lecture Notes in Chemistry **52**, 307, edited by U. Kaldor, Springer-Verlag, Berlin (1989)
20. S. Wilson, in *The Effects of Relativity in Atoms, Molecules and the Solid State*, edited by S. Wilson, I.P. Grant and B.L. Gyorffy, Plenum Press, New York (1991)
21. L. Laaksonen, I.P. Grant and S. Wilson, J. Phys. B: At. Mol. Phys. **21**, 1969 (1988)
22. L. Laaksonen, P. Pyykkö and D. Sundholm, Comput. Phys. Rept. **4**, 313 (1986)
23. P. Pyykkö, in *Numerical Determination of the Electronic Structure of Atoms, Diatomic and Polyatomic Molecules*, ed. M. Defranceschi and J. Delhalle, Kluwer, Dordrecht (1989)
24. J. Kobus, Chem. Phys. Lett. **202**, 7 (1993)
25. J. Kobus, Comput. Phys. Commun. **78**, 247 (1994)
26. D. Heinemann, B. Fricke and D. Kolb, Phys. Rev. **A38**, 4994 (1988)
27. D. Heinemann, A. Rosen and B. Fricke, Physica Scripta **42**, 692 (1990)
28. B.H. Wells and S. Wilson, J. Phys. B: At. Mol. Phys. **18**, L731 (1985)
29. B.H. Wells and S. Wilson, J. Phys. B: At. Mol. Opt. Phys. **22**, 1285 (1989)
30. D. Moncrieff and S. Wilson, J. Phys. B: At. Mol. Opt. Phys. **26**, 1605 (1993)
31. D. Moncrieff and S. Wilson, Chem. Phys. Lett. **209**, 423 (1993)
32. J. Kobus, D. Moncrieff and S. Wilson, J. Phys. B: At. Mol. Opt. Phys. **27**, 2867 (1994)
33. J. Kobus, D. Moncrieff and S. Wilson, J. Phys. B: At. Mol. Opt. Phys. **27**, 5139

(1994)

34. S. Wilson, D. Moncrieff and J. Kobus, *TlF* ($^1\Sigma^+$) *: Some preliminary electronic structure calculations*, Rutherford Appleton Laboratory Report RAL-94-082 (1994)
35. J. Kobus, D. Moncrieff and S. Wilson, J. Phys. B: At. Mol. Opt. Phys. *(to be submitted)*
36. S. Wilson and D. Moncrieff, Molec. Phys. **80**, 461 (1993)
37. D. Moncrieff and S. Wilson, Molec. Phys. **82**, 523 (1994)
38. H. Preuss, Z. Naturforsch. **11a**, 823 (1956); H. Preuss, Z. Naturforsch. **19a**, 1335 (1964); *ibid.* **20a**, 17, 21 (1965); J.L. Whitten, J. Chem. Phys. **39**, 349 (1963); F. Driessler and R. Ahlrichs, Chem. Phys. Lett. **23**, 571 (1973); V.R. Saunders, in *Methods in Computational Molecular Physics*, edited by G.H.F. Diercksen and S. Wilson, Reidel, Dordrecht
39. A.A. Frost, J. Chem. Phys. **47**, 3707 (1967)
40. A.A. Frost, J. Chem. Phys. **47**, 3714 (1967)
41. B. Ford, G.G. Hall and J.C. Packer, Intern. J. Quantum Chem. **4**, 533 (1970)
42. L.M. Haines, J.N. Murrell, B.J. Ralston, and D.J. Woodnutt, J. Chem. Soc. Faraday Transactions II **70**, 1794 (1974)
43. B.J. Ralston, D. Phil. Dissertation, University of Sussex, 1974
44. B.J. Ralston and S. Wilson, J. Molec. Struct. THEOCHEM *(accepted for publication)*
45. S. Wilson, J. Molec. Struct. THEOCHEM *(accepted for publication)*
46. S. Wilson, Intern. J. Quant. Chem. *(submitted for publication)*
47. S. Wilson, J. Phys. B: At. Mol. Opt. Phys. *(submitted for publication)*
48. A.S. Shalabi and S. Wilson, J. Molec. Struct. (THEOCHEM) *(accepted for publication)*
49. R.K. Nesbet, Rev. Mod. Phys. **32**, 272 (1960)
50. C.M. Reeves, J. Chem. Phys. **39**, 1 (1963)
51. C.M. Reeves and M.C. Harrison, J. Chem. Phys. **39**, 11 (1963)
52. R. Fletcher and C.M. Reeves, Comput. J. **6**, 287 (1963)
53. I. Shavitt, Israel J. Chem. **33**, 357 (1993)
54. J. Almlof, K. Faegri and K. Korsell, J. Comput. Chem. **3**, 385 (1982)
55. H. Preuss, Z. Naturforsch **19**a, 1335 (1964); ibid **20**a, 17, 21 (1965)
56. S. Rothenberg and H.F. Schaefer III, J. Chem. Phys. **54**, 2765 (1971)
57. W. Butscher, S. Shih, R.J. Buenker and S.D. Peyerimhoff, Chem. Phys. Lett. **52**, 457 (1977)
58. D. Neisius and G. Verhaegen, Chem. Phys. Lett. **66**, 358 (1979)
59. L. Lawlor, K. Vasudevan and F. Grein, Chem. Phys. Lett. **75**, 79 (1980)
60. P. Mach and O. Kysel, J. Comput. Chem. **6**, 312 (1985)
61. J.M.L. Martin, J.P. Francois and R. Gijbels, J. Comput. Chem. **10**, 152 (1989)
62. J.M.L. Martin, J.P. Francois and R. Gijbels, J. Comput. Chem. **10**, 875 (1989)
63. J.M.L. Martin, J.P. Francois and R. Gijbels, Theoret. chim. Acta **76**, 195 (1989)
64. F.M. Tao and Y.K. Pan, J. Chem. Phys. **97**, 4989 (1992)
65. F.M. Tao and Y.K. Pan, Molec. Phys. **81**, 507 (1994)
66. P. Čársky and M. Urban, *Ab initio Calculations. Methods and Applications in Chemistry*, Lecture Notes in Chemistry **16** (Springer-Verlag, Berlin)
67. D. Moncrieff and S. Wilson, Molec. Phys. *(in press)*
68. B. Klahn, J. Chem. Phys. **83**, 5748 (1985)
69. H. Müntz, Festschrift H.A. Schwartz (1914), p.303
70. O. Szász, Math. Ann. **77** (1926)
71. R. Courant and D. Hilbert, *Methoden der Mathamatischen Physik*, Vol. 1, p.86 (Springer, 1931)
72. R.E.A. Paley and N. Wiener, American Mathematical Society Colloquium Publications, Vol. XIX, pp. 32-36 (New York, 1934)
73. A. Schönhage, *Approximationstheorie*, pp. 49-52, (de Gruyter, Berlin, 1971)

74. J.R. Higgins, *Completeness and Basis Properties of Sets of Special Functions*, pp. 95-97, (Cambridge University Press, Cambridge, 1977)
75. S. Wilson, Theoret. chim. Acta **57**, 53 (1980); S. Wilson, Theoret. chim. Acta **58**, 31 (1980)
76. D.M. Silver, S. Wilson and W.C. Nieuwpoort, Intern. J. Quantum Chem. **14**, 635 (1978); D.M. Silver and W.C. Nieuwpoort, Chem. Phys. Lett. **57**, 421 (1978); D.M. Silver and S. Wilson, J. Chem. Phys. **69**, 3787 (1978); S. Wilson and D.M. Silver, Chem. Phys. Lett. **63**, 367 (1979); S. Wilson and D.M. Silver, J. Chem. Phys. **72**, 2159 (1980); S. Wilson, in *Specialist Periodical Report: Theoretical Chemistry* **4**, 1 (Royal Society of Chemistry, London, 1981); S. Wilson and D.M. Silver, J. Chem. Phys. **77**, 3674 (1982); S. Wilson, Molec. Phys. **43**, 1489 (1983); D. Moncrieff and S. Wilson, J. Phys. B: At. Mol. Opt. Phys. **27**, 1 (1994)
77. E. Clementi and G. Corongiu, Chem. Phys. Lett. **90**, 359 (1982); *Geometric Basis Sets for Molecular Computations, IBM Tech. Rept. POK-11 (1982);* E. Clementi, S.J. Chakravorty, G. Corongiu and V. Sonnad, in *Modern Techniques in Computational Chemistry*, ed. E. Clementi, ESCOM, Leiden (1990)
78. S. Huzinaga, Comput. Phys. Repts. **2** (1985) 281
79. The work of S. Haywood and J.D. Morgan III is described in: J.D. Morgan III in *Numerical Determination of the Electronic Structure of Atoms, Diatomic and Polyatomic Molecules*, edited by M. Defranceschi and J. Delhalle, Proceedings of NATO Advanced Research Workshop, April 1988, Kluwer, Dordrecht (1989)
80. R.E. Christoffersen, D. Spangler, G.G. Hall and G.M. Maggiora, J. Amer. Chem. Soc. **95** (1973) 8526
81. K. Kaufmann and W. Baumeister, J. Phys. B: At. Mol. Opt. Phys. **22** (1989) 1
82. M. Abramowitz and I. Stegun, *Handbook of Mathematical Functions*, Dover, New York (1970)
83. F.E. Harris, J. Chem. Phys. **32** (1960) 3
84. D.D. Ebbing, J. Chem. Phys. **36** (1962) 1361
85. F.E. Harris and H.S. Taylor, J. Chem. Phys. **38** (1963) 2591
86. H.S. Taylor, J. Chem. Phys. **39** (1963) 3375
87. H.S. Taylor and F.E. Harris, Molec. Phys. **6** (1963) 183
88. S. Wilson, J. Phys. B: At. Mol. Phys. **16** (1983) L429
89. D.M. Bishop and L.M. Cheung, J. Chem. Phys. **78** (1982) 1396
90. B.H. Wells & S. Wilson, J. Phys. B: At. Mol. Phys. **19** (1985) 17
91. J.W. Thompson and S. Wilson, J. Phys. B: At. Mol.Opt. Phys. **23** (1990) 2205
92. J.W. Thompson and S. Wilson, J. Phys. B: At. Mol.Opt. Phys. **25** (1992) 621
93. D.M. Silver, Molec. Phys. **22** (1971) 1069
94. W. Kutzelnigg, Intern. J. Quantum Chem. **51** (1994) 447
95. I. Shavitt, in *Methods in Computational Physics* **2** (1963) 1
96. S. Hackel, D. Heinemann, D. Kolb and B. Fricke, Chem. Phys. Lett. **206** (1993) 91
97. D. Moncrieff and S. Wilson, *work in progress*
98. N.H. March, W.H. Young and S. Sampanthar, *The many-body problem in quantum mechanics*, Cambridge University Press (1967)
99. I. Lindgren and J. Morrison, *Atomic many-body theory*, (Springer-Verlag, Berlin, 1982)
100. S. Wilson, *Electron correlation in molecules*, (Clarendon Press, Oxford, 1984)
101. F.E. Harris, H.J. Monkhorst and D.L. Freeman, *Algebraic and diagrammatic methods in many-fermion theory*, (Oxford, University Press, New York, 1992)
102. S. Wilson, Comput. Phys. Repts. **2**, 389 (1985)
103. T.E. Sorensen, W.B. England, D.M. Silver and E.O. Steinborn, Theoret. chim. Acta. **84**, 1 (1992)
104. T.E. Sorensen, W.B. England and D.M. Silver, J. Phys. B: At. Mol. Opt. Phys. **22**, L539
105. S. Wilson, *work in progress*

106. G.G. Hall, Chem. Phys. Lett. **20** (1973) 501
107. A.D. Tait and G.G. Hall, Theoret. chim. Acta **31** (1973) 311
108. G.G. Hall, Adv. At. Mol. Phys. **20** (1985) 41
109. S. Wilson, *work in progress*

EXACT *AB INITIO* QUANTUM CHEMISTRY

JAMES B. ANDERSON
Department of Chemistry, The Pennsylvania State University, University Park, Pennsylvania 16802, U.S.A.

1. Introduction

For systems of a few electrons the Schrödinger equation may be solved without systematic error. The method used is exact in that it requires no physical or mathematical assumptions beyond those of the Schrödinger equation. It is not variational and it does not give an upper bound to the energy. It gives *the* energy directly along with an error bar corresponding to a statistical or sampling error.

The term 'exact' is used in the sense described by Zabolitzky [1]. The value of π is not known exactly but it may be computed to any desired precision. A calculation method is exact if there exist finite amounts of computer time such that a result may be determined with any arbitrary precision.

To judge the accuracy of the calculations it is not necessary to compare the results of calculations with the results of experimental measurements. Neither 'calibration' nor 'benchmarking' is required. In fact, comparing results with those of experiments allows one to judge the accuracy of the experiments.

We summarize here recent quantum Monte Carlo calculations making use of an exact cancellation scheme [2, 3] which overcomes the problem of treating the nodes. The node structure evolves naturally with the determination of the wavefunction. We report results for several systems: the molecular ion H_3^+, the molecule H_2 (including nuclear motion), the intermediate H-H-H (the potential energy surface for the reaction $H + H_2 \rightarrow H_2 + H$), the stable dimer He-He, the unstable pair H-He, the cluster He_3, the cluster He_4, and the molecule LiH.

For the record we note that fixed-node quantum Monte Carlo calculations [4–6], which are not exact and are not described here, give the

C. A. Tsipis et al. (eds.), New Methods in Quantum Theory, 463–468.

lowest-energy solutions for systems with larger numbers of electrons: CH_4 [7], HF [8,9], F-H-H [8,9], N_2 [10], solid N_2 [10], and many others. For example, for the F-H-H system the fixed-node Monte Carlo energies are about 10 kcal/mol above exact, but 40 kcal/mole below the lowest analytic variational energies.

2. The Exact Cancellation Method

The exact cancellation scheme we have used [2,3] is based on an earlier scheme by Arnow et al. [11] and incorporates some of the best features of fixed-node methods, released-node methods, and other cancellation methods. The method takes full advantage of the symmetric and antisymmetric properties of wavefunctions and allows pairwise cancellations, multiple cancellations, and self-cancellations of wavefunction samples.

The method is most easily described for the case of the first excited state of a particle in a one-dimensional box of length L having walls of infinite potential. The exact solution is the wavefunction $\psi = A \sin(2\pi x/L)$ with (typically) the wavefunction positive to the left and negative to the right with the node at the center.

If a collection of positive and negative wavefunction samples is caused to evolve in imaginary time using the Green's function approach to solving the Schrödinger equation, the positive particles and the negative particles 'diffuse' to fill the entire box, multiplying within the box and disappearing at the edges. The positive and negative particle distributions separately approach the nodeless ground-state wavefunction $\psi = A \sin(\pi x/L)$.

However, when positive and negative samples in close proximity are allowed to cancel each other, the system evolves to overlapping distributions with positive samples penetrating the negative distribution and negative samples penetrating the positive distribution. If cancellation is carried out on the basis of overlapping Green's functions for the next moves of the samples, if the numbers of positive and negative samples are maintained equal and, if cancellation is reasonably efficient, the difference (i.e., positive minus negative) in the two distributions evolves to the wavefunction of the first excited state of the particle-in-a-box.

For higher dimensions and for molecular systems the procedure is essentially the same. The ground-state wavefunction for electrons (fermions) is antisymmetric to permutation of electrons of the same spin in the same way that the wavefunction of the first excited state of the particle-in-a-box is antisymmetric to inversion through the center of the box. In the higher dimensions the antisymmetry restricts but does not specify the nodal hypersurfaces.

In order to determine energies the wavefunctions are sampled repeatedly. In determining energies with high accuracy it is useful to take advantage of any prior knowledge of the wavefunction in the form of a trial wavefunction for importance sampling. The energy is obtained as the weighted average of local energies for a large number of samples. If the number is very large and the variance in the local energies is very low, the uncertainty in the computed energy will be low.

3. The Molecular Ion H_3^+

The first quantum Monte Carlo calculation for a molecular system was that for H_3^+ in 1975 [4]. That calculation gave an energy of -1.344 ± 0.013 hartrees for the equilateral triangle equilibrium configuration of the ground state. The wavefunction for the ground state of H_3^+ has no node, but use of the exact cancellation method allows very large steps and a corresponding reduction in the serial correlation of samples and the local energies. Using exact cancellation the most recent calculation [12] gave an energy of $-1.343\ 835 \pm 0.000\ 001$ hartrees. The uncertainty indicated is one standard deviation (σ) and is approximately equal to the probable error. This was the first quantum calculation of any type to achieve an accuracy of 1.0 microhartree in the total energy of a polyatomic system.

4. The Molecule H_2 (Four-Body)

The ground state of H_2 is also nodeless. The energy of the complete four-body system, without the Born-Oppenheimer or any other adiabatic approximation, has been determined in two quantum Monte Carlo calculations. The first [13] in 1991 gave an energy of $-1.164\ 024 \pm 0.000\ 009$ hartrees corresponding to a dissociation energy of $36\ 117.9 \pm 2.0\ \text{cm}^{-1}$ (including relativistic and radiative corrections). The second [14] in 1995 gave an energy of $-1.164\ 0239 \pm 0.000\ 0009$ hartrees corresponding to a dissociation energy of $36\ 117.84 \pm 0.20\ \text{cm}^{-1}$ (also including relativistic and radiative corrections). Again, the uncertainty indicated is one standard deviation. The most recent experimental value for the dissociation energy is $36\ 118.11 \pm 0.08\ \text{cm}^{-1}$ [15].

5. Potential Energy Surface for $H + H_2 \rightarrow H_2 + H$

Exact quantum Monte Carlo calculations [16, 17] have produced a potential energy surface for the reaction $H + H_2 \rightarrow H_2 + H$ accurate to within ± 0.01 kcal/mol at the saddle point and within ± 0.10 kcal/mol or better elsewhere on the surface.

The uncertainty in the Monte Carlo result for the H-H-H saddle point configuration is $\pm 0.000\ 014$ hartrees or $\pm 3\ \text{cm}^{-1}$. Clementi et al. [18] recently speculated that an analytic variational calculation for H-H-H accurate to within 10 cm^{-1} "would call for somewhere around 3500 years" on a computer similar to the one we used. Ours took three months.

As Clementi et al. [16] pointed out, there are many opportunities for significant reductions in the computational effort for analytic variational calculations and it may be possible to "reduce the task to 10 to 20 years" or even less.

6. The He-He Interaction

For the dimer He-He exact quantum Monte Carlo calculations [19] yield energies accurate within ±0.000 0003 hartrees or ±0.10 K at the equilibrium separation of 5.6 bohr. For total energy the quantum Monte Carlo calculations are more accurate than the lowest-energy variational calculations [20] by a factor of 12,000.

Nevertheless, the calculated potential energy curves - the energies relative to separated atoms - for the Monte Carlo and the lowest-energy variational calculations are in excellent agreement with each other and with an experimental-theoretical compromise curve [21] based mostly on experimental data.

7. The H-He Interaction

The van der Waals attraction for the H-He pair is even weaker than that for He-He. Exact quantum Monte Carlo calculations [22] for H-He similar to those for He-He yielded a potential energy curve with a well depth of 6.8 ± 0.5 K at the equilibrium distance of 6.8 bohrs.

8. Helium Clusters He_3 and He_4

Exact quantum Monte Carlo calculations for the helium trimer [23] and the tetramer [24] have also produced very high accuracies in total energies and interaction energies. In general, the results show the interactions to be very close to pairwise additive for internuclear distances in the region of the pair minima and greater.

9. Potential Energy Curve for LiH

The potential energy curve for LiH has been determined in quantum Monte Carlo calculations [25, 26] based not on cancellation but on a simplified 'released-node' or 'transient' calculation. The movement of positive and negative wavefunction samples is carried out using the Green's function approach described above, but no attempt is made at cancellation. Instead, the calculation is begun with samples from the output of a fixed-node calculation. These proceed toward the boson ground-state distribution, but after several dozen iterations, the difference (i.e., positive minus negative) in the two distributions approaches the fermion ground-state distribution. With additional iterations the difference shrinks and is eventually lost in statistical noise.

The calculated energy [25] for an internuclear separation of 3.015 bohr is $-8.070\ 21 \pm 0.000\ 05$ hartrees, a value lower than that of the lowest-energy variational calculation, more accurate than that of prior Monte Carlo calculations, and in excellent agreement with the non-relativistic energy of $-8.070\ 21$ hartrees from experimental measurements. The calculated curve [26] supports the experimentally predicted 24 vibrational levels for LiH, one more than the 23 given by the lowest-energy variational curve.

10. Acknowledgement

The research was supported by the National Science Foundation (Grant No. CHE-8714613) and by the Office of Naval Research (Grant No. N00014-92-J-1340).

References

1. Zabolitzky, J. G. (1985) Monte Carlo Solution to the Many-Body Schrödinger Equation, *Prog. Part. Nucl. Phys.* **16**, 103-138.
2. Anderson, J.B., Traynor, C.A., and Boghosian, B.M. (1991) Quantum Chemistry by Random Walk: Exact Treatment of Many Electron Systems, *J. Chem. Phys.* **95**, 7418-7425.
3. Anderson, J.B. (1995) Exact Quantum Chemistry by Monte Carlo Methods, in S. R. Langhoff (ed.), *Quantum Mechanical Electronic Structure Calculations with Chemical Accuracy*, Kluwer Academic Publishers, Dordrecht, pp. 1-45.
4. Anderson, J.B. (1975) A Random Walk Simulation of the Schrödinger Equation: H_3^+, *J. Chem. Phys.* **63**, 1499-1503.
5. Anderson, J.B. (1976) Quantum Chemistry by Random Walk: H ^2P, H_3^+ D_{3h} $^1A_1$, H_2 $^1\Sigma_g^+$, H_2 $^3\Sigma_u^+$, Be ^1S, *J. Chem. Phys.* **65**, 4121-4127.
6. Anderson, J.B. (1995) Fixed-Node Quantum Monte Carlo, *Int. Rev. Phys. Chem.* **14**, 85-112.
7. Garmer, D.R. and Anderson, J.B. (1987) Quantum Chemistry by Random Walk: Methane, *J. Chem. Phys.* **86**, 4025-4029.
8. Garmer, D.R. and Anderson, J.B. (1987) Quantum Chemistry by Random Walk: Application to the Potential Energy Surface for $F + H_2 \rightarrow HF + H$, *J. Chem. Phys.* **86**, 7237-7239.
9. Garmer, D.R. and Anderson, J.B. (1988) Potential Energies for the Reaction $F + H_2 \rightarrow HF + H$ by the Random Walk Method, *J. Chem. Phys.* **89**, 3050-3056.
10. Mitas, L. and Martin, R.M. (1994) Quantum Monte Carlo of Nitrogen: Atom, Dimer, Atomic and Molecular Solids, *Phys. Rev. Lett.* **72**, 2438-2441.
11. Arnow, D.M., Kalos, M.H., Lee, M.A., and Schmidt, K.E. (1982) Green's Function Quantum Monte Carlo for Few Fermion Problems, *J. Chem. Phys.* **77**, 5562-5572.

12. Anderson, J.B. (1992) Quantum Chemistry by Random Walk: Higher Accuracy for H_3^+, *J. Chem. Phys.* **96**, 3702-3706.
13. Traynor, C.A., Anderson, J.B., and Boghosian, B.M. (1991) A Quantum Monte Carlo Calculation of the Ground State Energy of the Hydrogen Molecule, *J. Chem. Phys.* **94**, 3657-3664.
14. Chen, B. and Anderson, J.B. (1995) Improved Quantum Monte Carlo Calculation of the Ground State Energy of the Hydrogen Molecule, *J. Chem. Phys.* **102**, 2802-2805.
15. Balakrishnan, A., Smith, V., and Stoicheff, B.P. (1992) Dissociation Energy of the Hydrogen Molecule, *Phys. Rev. Lett.* **68**, 2149-2152.
16. Diedrich, D.L. and Anderson, J.B. (1992) An Accurate Monte Carlo Calculation of the Barrier Height for the Reaction $H + H_2 \rightarrow H_2 + H$, *Science* **258**, 786-788.
17. Diedrich, D.L. and Anderson, J.B. (1994) Exact Quantum Monte Carlo Calculations of the Potential Energy Surface for the Reaction $H + H_2 \rightarrow H_2 + H$, *J. Chem. Phys.* **100**, 8089-8095.
18. Clementi, E., Corongiu, G., Bahattacharya, D., Feuston, B., Frye, D., Preiskorn, A., Rizzo, A., and Yue, W. (1991) Selected Topics in *ab Initio* Computational Chemistry in Both Very Small and Very Large Chemical Systems, *Chem. Rev.* **91**, 679-700.
19. Anderson, J.B., Traynor, C.A., and Boghosian, B.M. (1993) An Exact Quantum Monte Carlo Calculation of the He-He Interaction Potential, *J. Chem. Phys.* **99**, 345-351.
20. Liu, B. and McLean, A.D. (1989) The Interacting Correlated Fragments Model for Weak Interactions, Basis Set Superposition Error, and the Helium Dimer Potential, *J. Chem. Phys.* **91**, 2348-2359.
21. Aziz, R.A. and Slaman, M.J. (1991) An Examination of *ab Initio* Results for the Helium Potential Energy Curve, *J. Chem. Phys.* **94**, 8047-8053.
22. Bhattacharya, A. and Anderson, J.B. (1994) An Exact Quantum Monte Carlo Calculation of the H-He Interaction Potential, *Phys. Rev.* **A 49**, 2441-2445.
23. Bhattacharya, A. and Anderson, J.B. (1994) The Interaction Potential of a Symmetric Helium Trimer, *J. Chem. Phys.* **100**, 8999-9001.
24. Chen, B. and Anderson, J.B. (1995) The Interaction Potential of the Helium Tetramer, to be published.
25. Chen, B. and Anderson, J.B. (1995) A Simplified Released-Node Quantum Monte Carlo Calculation of the Ground State of LiH, *J. Chem. Phys.* **102**, 4491-4494.
26. Chen, B. and Anderson, J.B. (1995) A Quantum Monte Carlo Calculation of the Potential Energy Curve for LiH, to be published.

COMMENTS ON COMPUTATIONAL CHEMISTRY, IN GENERAL AND QUANTUM CHEMISTRY, IN PARTICULAR

ENRICO CLEMENTI AND GIORGINA CORONGIOU

METECC Laboratoire, University L. Pasteur,
3, rue de l'Université, 67084 Strasbourg, France

Abstract: The field of computational chemistry is composed by many subfields, which must be connected one to the other to ensure reliability and relevance to laboratory chemistry not only for small molecular systems but also for large ones. The evolution in industrial productivity has created a well defined path which must be followed by computational chemistry to be relevant. Thus, the need for Global Simulations today and Cognisant Simulations in the near future, when seventh generation computers will be used. Concerning quantum chemistry, a number of problems still await a satisfactory answer. One is the many body problem, where in our opinion much can be done using effective potentials not explicitly dependent on the electronic density. We report on the Coulomb-Hole Hartree-Fock effective potential, and we introduce the J- and the K- functionals, two new effective potentials which provide very reasonable correlation energies. In this pragmatic way we show that the old density functional approach, recently rediscovered by chemists, furnishes only one among the many possible solutions to the many body problem. We then consider the problem of the choice of representations, in particular Schrödinger versus Dirac; we present arguments in favour to a switch to the Dirac representation for any molecular system with or without heavy atoms.

1. Definition of computational chemistry

Computational chemistry, as a specific aspect of chemistry, in particular, and of science, in general, is a relatively new field of research and development. Computational chemistry includes several "sub fields", like quantum chemistry, statistical molecular mechanics, molecular modelling, molecular graphics and visualisation, information retrieval from chemical data banks, artificial intelligence and expert chemical systems. These sub fields are well defined, when relatively old and established, less so for the most recent ones. Thus *quantum chemistry* , the earliest sub field, includes the so called ab initio and semiempirical quantum mechanical techniques, aiming mainly to wave function and energy (including derivatives) computations; *statistical molecular mechanics* includes a) classical and quantum Monte Carlo, aiming at the determination of time independent statistical distributions either of electrons in atomic or very small molecular systems or of atoms and molecules in very complex systems described either as quantum or classical bodies, respectively; b) classical molecular dynamics, aiming at the

C. A. Tsipis et al. (eds.), New Methods in Quantum Theory, 469–499.

derivation of classical trajectories (with force fields obtained either i) from analytical potentials - either semi empirical or fitted from quantum mechanical ab-initio data-potentials- or ii) computed directly from approximated quantum mechanical solutions) and c) statistical quantum dynamics. *Micro-dynamics and cellular automata* are two related techniques, which, however, should not be included in an operational definition of statistical molecular dynamics of molecular systems. The sub field of *molecular modelling* is rather hill defined in its current literature use, being used at times to include also molecular dynamics; below we shall restrict the use of the term *molecular modelling* to molecular mechanics, molecular docking, and structure/activity relationships. The sub field *of molecular graphics and visualisation* is very much in evolution and includes the traditional 2D and 3D graphics, both static and time dependent, rendering and virtual vision; possibly more than a sub field molecular graphics and visualisation should be classified as a very specialised technique, which is leading to the development of specific hardware, languages and application software. *Information retrieval from chemical data banks* includes the formation of the data banks for structures, formulas, syntheses, physical-chemical and pharmaceutical characterisations, and the automated preparation of documents and statistics. Finally, *artificial intelligence and expert chemical systems*: this is in itself a very large and recent sub field of computational chemistry which includes also neural nets, factor analyses and molecular similarity. cluster analysis, pattern recognition, computer oriented chemical synthesis, and aspects of laboratory automation.

We have most likely omitted to explicitly mention one or more aspects of particular relevance to some computational chemist! One point, however, seems to emerge very clearly: computational chemistry is a vast array of activities generated to respond to practical applied needs; computational chemistry is also an attempt to pragmatically measure oneself with the enormous complexity of chemistry, using the modern tools of information technology. Perhaps we should add an observation: most likely today, 1995, still there exist no higher learning institution where computational chemistry, fully and comprehensively, as above defined, is taught to chemistry students! Apparently the technological advancements have out-stripped University's ability to cope with this fast evolution.

2. Evolution in industrial productivity

In the following sections of this chapter we shall limit ourselves mostly to quantum chemistry and molecular dynamics, with a few comments about micro dynamics. But, before this technical part, we wish to start with a few broad historical considerations, since computational chemistry is one among the emerging disciplines which clearly illustrates the limitations of traditional science. This done we shall narrow it down to science, to chemistry and finally to computational chemistry. We wish to note that part of what follows has been published by us in several papers, but in a rather scattered way; here we wish to collect this material into a more unified presentation.

Let us briefly recall the main steps of the historical evolution in the productivity of goods [1-3], which clearly includes also the industrial productivity of chemicals, one of the main factors needed to understand the evolution of chemistry as a science.

Whereas in the very distant past, science, as today defined, was obviously one among the less relevant forms of predictive knowledge, now in our present time the socio-economical structure would most likely collapse without science and derived technologies. Indeed, in the distant past productivity, a measure related to survival ability, was mainly limited to hunting, at first, and agriculture, later. Nearer to our time, in the second half of the eighteenth century, the beginning of modern science and the population increase brought about the *first industrial revolution*, namely automated factories and production processes, which did make use of man as an integrated component of machines. This situation evolved from the crude beginnings of the late part of the eighteenth century, to the immediate post-war, in the late 1940's. At this time the first electronic computers did appear, but they were too unfriendly to the users and therefore they were relegated to a few very specialised tasks, often related to fundamental science or high technology. Instead, mechanical robots, since easily adaptable to the many needs of the industry, were becoming the backbone of the automated production lines; this marks the beginning of the *second industrial revolution*. Today, the computer has become much more versatile and user-friendly and it is being integrated into the robot, yielding the intelligent-robot of the first generation, and bringing about the beginning of the *third industrial revolution*. The above use of the word "revolution" is quite realistic, as we can surmise from even a superficial browsing at the history in the 1750-2000 A. D. time interval, in particular at the interdependence between the evolution of productivity and markets and the social structures in different countries, or the interdependence among production processes, in a geographical, economical, environmental, national and international way. Perhaps we can already foresee the next and *fourth industrial revolution*, which could result by the merging of intelligent-robot production lines with networks and communication satellites, using modern communication technology: high level of productivity could be achieved by very detailed supervision of intelligent-robot production lines supervised, if needed, even from great geographical distances, and independently from national boundaries. The draw back of this daring evolution is the increasingly wide gap between a minority which understands, uses and exploits these new technologies and the vast majority which is left out from these innovative socio-economical processes.

3. Evolution of the computer use in scientif applications

In this section we provide with a short reminder on the *evolution of the electronic computers* [1-3]. We recall that computer architecture and technology has much changed in the last fifty years and, correspondingly, large changes have taken place in the way computers are used. Presently, we are near the end of the sixth generation of electronic computers.

The *first generation* (1945-1955) uses vacuum tubes and static memories (ENIAC, EDSAC, and IBM-650, as examples). The *second generation* (1955-1965) has transistor technology, magnetised core memory, tapes and punched cards as I/O and the first high level language, FORTRAN (IBM-7094 is an example of computer for this generation). The *third generation* (1965-1975) can be typified by medium to large scale integrated circuits, channels, magnetic disks, availability of system software, virtual memory, and time sharing. The *fourth generation* (1975-1985 or

1987) had efficient vector processors, a beginning of parallel hardware (multiprocessors), very large system integration, VLSI, memories up to gigabits, top performance up to about one gigaflop, a full spectrum of products from super computers to workstations, to personal computers, PC, and new operating systems UNIX oriented; FORTRAN remains strong for scientific applications, having survived replacement attempts by PASCAL, COBOL, PL, PROLOG, even if C languages are being more and more utilised. Networks are becoming of common use and satellites communications are a reality. The nearly monopolistic position of IBM is eroded and CRAY becomes the leader in scientific computers, with competent and aggressive newcomers from Japan's computer industry. The *fifth computer generation*, officially born in 1984, with a MITI announcement in Japan, proposes the merging of "knowledge processing" with super computers and is followed by an explosion of new efforts, like in expert systems, neural machines, artificial intelligence, voice and pattern recognition, etc. etc. But widespread commercial hardware realisations do not appear, and the "Fifth Generation" remains mainly a landmark for new concepts, rather than a true computer generation, in the way we have been accustomed. The *sixth computer generation* (1986-1997) is characterised by the diffusion of workstations well priced and with excellent performance relative to the super computers of the fourth generation, and by the widespread acceptance of parallel programming and related hardware. The latter is supplied either in the form of cluster of workstations with relatively slow communications, or in specially well integrated processors, like those offered by CRAY, the T3D machine, or attempted by the Thinking Machine Corporation, the CM5, in particular. The Japanese trust initially appears somewhat weaker, and Europe is once more attempting to enter into the computer hardware market, despite the many failures of the past. However truly massive parallelism, with many thousand of nodes, is in practice not achieved. UNIX dominates. The peak performance is from one gigaflop to about one hundred gigaflops, with a larger and larger gap between peak performance and sustained performance; message-passing oriented system software are generally available. Networks are available easily and everywhere; the PC's are becoming more and more powerful and assuming characteristics previously available only on workstations. Graphics and animation are used interactively, and virtual reality and multi-media are common aspects of the computer world. The *seventh generation* (1997-2005) will finally bring into concrete software-hardware realisations the promises of the fifth generation; in addition, new materials might perhaps become available for component manufacture, including organic materials, adding flexibility to the computer industry which will aim at offering teraflop products. The progress in communication theory and technology, as well as the deeper understanding in the mechanisms of brain activity in storage and information retrieval, and in voice and pattern recognition, will make the new products very friendly even to the non specialist. The computer and its software will gradually become *the most important tool* for communication, education, industrial and public management, industrial product innovations, medicine, from diagnoses, to surgery, and to hospital administration, economical planning for city, states, international communities, environmental research and planning, and in many more aspects of life down to both mass and individual entertainment.

The one hundred year time interval from 1945 to 2045 is changing the structure of human society more deeply (and likely irreversibly) than previously accomplished in the last 3000 of human history. But it will not be a success story without problems and errors. A main limitation for the next decade will likely result from insufficient preparation and awareness at the top level of management in the educational, political, financial and administrative structures.

Having briefly defined the evolution of computers, the main tool for computational chemistry and for the rationalisation of industrial productivity planning, we pass now to some aspects of the *evolution of computational chemistry*. It is not surprising that computational chemistry, which started in a few laboratories, notably (but not exclusively) at the LMSS laboratory at the University of Chicago with R. S. Mulliken and C. C. Roothaan, at the SSMT group at M.I.T. with J. C. Slater, and at the University Chemical Laboratory of Cambridge with S. F. Boys, follows a pattern common to other fields of science, for example, fluid dynamics of air masses, or of water in seas and oceans. In general, we can distinguish three distinct stages, which we shall designate as computerisation, global simulation and cognisant simulation, respectively.

In the first stage - *computerisation*- the machine execution of a computer program replaces the operations by hand of the scientist in solving a single, well defined equation. The "equation" is generally a model used to represent some aspect of the dynamics of matter. Examples of computer programs used in the early development of quantum chemistry are POLYMOLE [4], IBMOL [5], or the computer codes developed for example by S. F. Boys [6], or by Roothaan's group [7].

In the second stage - *global simulation*-, a number of computer programs are linked together, each of them representing one computerisation step, the ensemble representing the solutions of a problem requesting more than one model. An example would be: 1) use a quantum chemistry code to solve for the energy at many configurations of two interacting molecules, A and B; 2) use of suitable routines to obtain an interaction potential of a predetermined form by fitting the previously computed interaction energies; 3) use these interaction potentials to compute molecular dynamics trajectories of a liquid composed of many molecules of type A and B; 4) use these trajectories to obtain the X-rays and neutron scattering structure factors, infrared and the Raman spectra of the liquid. Note that in step 1) matter is represented as an ensemble of nuclei and electrons, whereas in step 3) matter is represented as an ensemble of classical particles with given mass and velocity, at a given temperature, and in a given volume.

It is apparent that global simulation represents an operational way to re normalise complex problems. Today there are relatively few codes which are organised for global simulations, namely in such a way that the input of a code, representing a given model of matter, is obtained as the output of another code, which often represents matter at a different level of resolution. The METECC-94 [8] series of volumes represent attempts in this direction and it follows previous similar efforts, namely MOTECC-89 [9], MOTECC-90 [10] and MOTECC-91 [11]; METECC-95 [12] follows this path. We should recall that the performance of the computers of the sixth generation makes, for the first time, the global simulation approach feasible.

With time computational chemists will have to address greatly more difficult and complex tasks, where, for example, data banks are researched as a step of the global process and interactive graphics might become essential to visualise complex environments, or where the use of expert systems to find optimal paths-solutions becomes essential. Therefore, gradually, aspects of artificial intelligence will be used more and more in executing a global simulation, and this will be particularly efficient in conjunction with the computers of the seventh generation. In this way we enter the third stage- *cognisant simulation.* It follows that one of the challenges of today computational chemistry is to evolve from the computerisation level, where it is now predominantly confined, toward the global simulation and then to the cognisant simulation levels. In this contest, it is of some interest to comment on the somewhat old-fashion debate, confronting "computing molecules" versus "understanding molecules". Manifestly, we all wish to reach understanding and knowledge, and therefore the debate is focused on the most efficient methodology to reach a common goal. Generally, the advocates of "understanding" pretend to operate from one step higher than merely obtaining from the computer a "computed value" of some molecular property. However, we should realise that the global and the cognisant simulations are becoming more and more the equivalent of the so-called "understanding", and are becoming more and more valid methodologies even for that level of chemical complexity, where the traditional "understanding" has somewhat failed. Possibly, at the root of the t debate there is from one side the postulate that scientific equations do indeed represent reality, whereas from the other side there is the postulate that science offers only models, namely it can merely "simulates" reality.

In the early sixties it was customary to talk of computations, and from this the term "computational chemistry" did originate. Perhaps, from a semantic viewpoint, it would have been preferable to talk of "Simulations and Models in Chemistry". In this context, I would like to restate my old conviction "we can simulate everything".

In a global simulation, today, and in a cognisant simulation, tomorrow, we move from "small" to "large" systems, since we move from high to low resolution. One more observation prompted from history: periods of intensive logical and intellectual innovation bring about attempts to organise and to systematise; thus Aristotle's work, or Thomas of Aquino rewriting of Aristotle or, nearer to us, the French Encyclopáedia efforts: the global simulations and cognisant simulations are a modern way to answer an ancient human need.

We conclude with the statement that *a major challenge for computational chemistry is to speed up the evolution from Computerisation to Cognisant Simulations.*

4. A definition of "large molecule"

Whereas for a given molecular system the number and type of atoms are well defined quantities, the concept of "large molecules" in the contest of computational chemistry varies both with time and space, namely with the evolution (time) of computer hardware and software and with the financial disposability and other specifications for each computational chemist (space and time). This fact brings about, year after year, discussions and even debates on the best definition for what constitutes "a large molecule" for a computational chemist. However, with a bit of common sense, it is clear that there must be an upper limit to the definition (the lower limit is of little general interest), and that the term "molecule" should really be understood as "ensemble of computations on a molecular system, performed to solve a well defined chemical or physical-chemical problem". Thus an ensemble of many small molecules must be included into the definition of "large molecule". We should in addition stress that a *large number* of computations on a small system can be computationally as demanding as *one* computation on a large system. Thus the problem of "computation of large molecules" reverts itself quickly into the problem of high performance computing for numerically intensive tasks.

However, to the "true chemist", the problem is rather simple, namely it can be redefined within the following question "can computational chemists deal with the same type of systems which are daily of interest in the laboratories or - to the least - with systems very similar in complexity". Here, the meaning of the term "complexity" is vastly more broad than simply the number of atoms and electrons of the system under consideration. It involves, for example, variation of temperature, time, pressure, environment (from solvent, to pH, to radiation), it means, alternatively, spectroscopy, thermodynamics, syntheses, including catalysis, enzyme studies, problems on inorganic or genetic materials, structure-activity relations, super molecules, solid state reactions, namely ... "chemistry".

With these observations, and in order to be pragmatic (and, necessarily, somewhat simplistic) we shall report the definition of large molecules one of us (E.C.) has recently presented at the Braunlage symposium [13], warning, however, that the quantitative definition of "large molecule" depends on the means available at a defined time to a specific computational chemist. It appears, therefore, that we must specify some boundary condition to our definitions; we shall consider the four years time span from today to the end of 1998, and assume that by that deadline "super computers" of parallel type with at least 64 to 128 nodes, 128 MB random access memory per node, will be easily available and intensively used at least in the most industrialised Countries.

Within these conditions, the expression "large molecule" is assumed to refer to systems requiring computations with about 4 to 5 thousand functions in the basis set, for post Hartree-Fock ab-initio, non-relativistic methods, 1 to 2 thousand functions for ab-initio Fock-Dirac type four components methods, again 1 to 2 thousand functions for problems requiring geometry optimisation and mapping of energy surfaces. In terms of the number of electrons these computations are for systems with up to a few hundred electrons, explicitly considered.

This forecast is therefore not too exciting, since still far from the needs of every day chemistry. We recognise that ab-initio techniques (either with CI or MR-CI or MC-SCF methods or perturbation techniques) can be routinely performed with a variety of computer codes, but the high dependence on the basis set offsets most of the gains. Very long ago we have pointed out that the Hartree-Fock dependence on the basis set is often less than the theoretical limit, namely the forth power of the number of the basis functions [14]; more recently [15] for large molecules we have shown to be even less than the third power, but one should stress that the more complete the basis set is, the slower the decrease from the forth power will be. In this respect a long linear (or about linear) molecule with a minimal or a small basis set can easily show a pleasant low-power dependency, which is, however most deceitful, if used as an illustration of a more general rule. (In recent years this observation has been rediscovered over and over). The rather negative comment on the practical usefulness of ab-initio methods, should not be taken as an advise to dump it all as useless; on the contrary, the DFT, the CH-HF (see below) and semi-empirical techniques need much post Hartree-Fock ab-initio testing, for example, on excited states, or for molecular geometry far away from equilibrium, or for unusual expectation values (different from the energy). The ab-initio techniques provide the obvious and necessary benchmarks for many simplified approaches, and will remain the basic methods to be used to test fundamental concepts both in atomic and molecular systems.

As long ago realised by the solid state physics community but by very few computational chemists, the only techniques which seem to be practical for large molecules are those using *effective hamiltonian* either operating on electronic density or wave functions. Concerning density functionals (called "statistical methods" in the early quantum mechanics literature) there are essentially two main avenues, i) either the merge of the exchange and correlation corrections in some empirically tested functional of the electronic density (this is the classical path called Density Functional Theory, DFT) or ii) the addition of a correlation term to the exchange operator term. This alternative avenue was not considered by solid state physics, because of the high computational cost in evaluating exchange matrix elements; on the contrary, in quantum chemistry the alternative is not expensive with gaussian basis sets expansion for the orbitals. In those techniques where the effective Hamiltonian operates on the wave function, the correlation correction term is a functional of the electron electron distance (this is the case for the Coulomb-Hole-Hartree-Fock, CH-HF, and the J- K- functionals [16-19] discussed below in this work).

One could notice that a good number of solid state techniques has been developed for solving problems related to bulk matter, thus neglecting surface states, and initially with limited reliability for amorphous materials. It happens that molecules, including large ones, are essentially not far from amorphous surfaces. It is therefore to be expected that simple DFT approaches, for example with local potentials, will not fit well chemical problems; naive solid state techniques will need a bit of tailoring to be more reliable, when used for atomic and molecular computations. Therefore one would hope that computational chemists will develop appropriate effective potentials optimally suited to their needs, and not necessarily linked to density functionals, at least those used today.

The effective hamiltonian techniques extend the number of electrons we can compute, relative to ab-initio, post Hartree-Fock methods. However computations with hundreds of atoms remain unfeasible; presently, this task can be tackled only with semi-empirical approaches. Concerning the latter, we have witnessed dramatic improvements in the last decade, due to the increased performance of computational means, and to the availability of most efficient techniques; this has been well illustrated at the Braunlage symposium [13]. We have estimated [13] that "large molecules" for semi-empirical methods, will deal with systems in the 1000 to 4000 atom range, a great jump over ab-initio post-Hartree-Fock or even DFT and CH-HF.

Above we have considered present days quantum chemistry, limited, by and large, to absolute zero temperature and developed in a timeless frame, both rather serious limitations to much of real chemistry. Molecular dynamics fills in this gap, but the foundation is classical rather than quantum mechanical and atoms rather than electrons are "the particles" of the system. Schematically, three types of techniques are used to represent the interactions between atoms; the forces acting on each atom are either quantum mechanically computed at each time step, or are computed from analytical force fields obtained either from fitting quantum chemistry energies and derivatives or by fitting laboratory data. In all cases the molecular dynamics computation yields classical trajectories based on solving

Newton equations for the particles of the systems (either atoms, or fragments of molecules or molecules). It should be stressed that the trajectories are highly dependent on the quality of the force field, and are described by giving the position co-ordinates at very short time steps (10^{-15} sec.). Thus the number of time-steps becomes the second basic parameter to define the complexity of molecular dynamics problems (the first one being the number of atoms). If one is considering problems with very long lifetime, of the order of several nanoseconds or microseconds, then the force field can average short time fluctuations, and one can use Langevin rather than Newton mechanics. We have proposed [13] that within the previously assumed time period (next four years) and with the computational facilities previously defined, Newton type molecular dynamics, using quantum mechanical force field, will be able to solve molecular systems with up to a few hundred atoms and up to 100 picoseconds or up to a few million atoms and up to one nanosecond when using analytical potentials at each time step. The time scales expand by about a factor of ten if one selects to use Langevin rather than Newton mechanics. The "up to a few hundred atoms" molecular computations with quantum mechanical forces (e.g. Car-Parrinello method) are generally based on pseudo potentials in the inner shells and density functional methods for the valence electrons. There is the possibility to combine these computations with those which make use of analytical potentials; part of the macromolecule, the one of main chemical interest, as the active site of a protein, is treated with quantum mechanics, the rest of the molecule with analytical potentials. This suggestion is not new and an experiment was presented for an enzymatic reaction involving papaine long ago [20]. Today the computers performance is greatly improved and, in addition, there is the experience both with Car-Parrinello [21] method and with very similar approaches [22] which, interestingly, are designed to more closely follow Born-Oppenheimer surfaces. We expect to see a strong growth for these methods, which will be used especially in reaction chemistry and biochemistry.

Computational chemistry generally deals with systems at equilibrium conditions, a rather artificial situation in real systems; these conditions can be removed, but then the systems must be characterised by boundary conditions. Micro-dynamics, using either Newton (or Langevin) equations [23-26] is still rather unfamiliar to most computational chemists, however, it represents the logical link between discrete and continuous systems (the latter being the domain of fluid dynamics). We shall not expand on micro-dynamical computations, but we must to the least mention that this field represents the upper limit of a large molecular system. In time, when computational chemistry will more and more overlap with chemical engineering ("a must" in near future chemical industry) micro-dynamics will become a very standard aspect of molecular modelling and of computational chemistry.

In the following of this work we shall consider mainly recent or new solutions to the correlation energy problem, where, however, the computational complexity is kept near to what one experiences in Hartree-Fock computations. We shall also recall some recent forecast relative to the use of four components Dirac-Fock functions.

These are two topics which are central to the development of computational chemistry for large molecular systems. We shall then add a detailed example of global simulation where starting with atomic wave functions we create a good basis set for molecular computations, then with molecular ab initio quantum mechanics we obtain force fields for molecular interaction, which are used to describe matter in the solid and in the liquid states. The example is concluded by recalling MD simulations for non-equilibrium systems, specifically the flow past an obstacle; these examples show that MD simulations are capable to solve hydrodynamics problems that, generally, are studied with Navier-Stokes equations. In the last section, we shall add a short comment on the expectations from parallel computer architecture.

5. Two stumbling blocks to quantum chemistry: the many body problem and the choice of the hamiltonian

An old stumbling block for quantum chemistry is its inability to find a solution for the correlation energy problem; another is the need to deal with an hamiltonian which recognises both spin and relativistic effects, namely the need to realistically face the limitations of the Schrödinger representation, today still the most popular approach in quantum chemistry.

Let us briefly comment first on the many body problem. In the Schrödinger equation the energy is obtained by operating with an appropriate hamiltonian on a wave function. There are two basically different ways to attempt to obtain the exact energy, namely, an energy which is at least as accurate as the "best potential experiment." (Note that in this way we have placed experimental measurements at the foundation of our scientific knowledge; alternatively, we could have ask for logical self consistency with a set of axioms, either assumed to be "true" or only logically interesting).

The *traditional way of quantum chemistry* to attempt a solution of the many body problem is to use the quantum equivalent of the classical hamiltonian (or some evolution of the same) and to progressively improve on the energy (variationally), with more and more parameters in the wave functions. Thus, one progresses from Hartree-Fock to larger and larger configuration interactions, to multi-SCF-configurations, to explicit inclusion of inter electronic distance terms, added to configuration interaction (the Hylleraas configuration interaction) or added to multi-SCF-configurations (an alternative not yet explored). Parallel to these variational approaches we have witnessed perturbational approaches notably from MP2, to MPN, to Coupled Clusters. The uniform characterisation of these approaches is that the parameter space grows much faster than our computational ability; rather simple complexity arguments could prove this point, without the need of any numerical experimentation. On the other hand the experimental limitations and the need to asses the reliability limits of the Schrödinger representation did advise to place much energy on this otherwise loosing path.

The *alternative way* explored to attempt to obtain a "reliable value" for the energy of a system is the use of an effective hamiltonian. Note that also in the traditional quantum chemistry approach the wave function is often obtained via the use of effective potentials (indeed in the Hartree-Fock method, the Coulomb and the Exchange operators are examples of effective potentials); however the final energy is computed by operating with the "proper hamiltonian" on the final wave function. In the second alternative, the final energy is assumed to be the one computed directly from the effective hamiltonian, operating on an approximated operand, like a wave function of "pre-determined complexity".

Let us assume that the exact wave function Ψ can be decomposed as the combination of two parts, designated as $\Psi(1)$ and $\Psi(2)$ respectively, and let us assume that $\Psi(1)$ satisfies two conditions:
1) it corresponds to a wave function of "pre-defined complexity" chosen for an effective potential approach
2) it represents a reasonable approximation to Ψ. Then, we write the Schrödinger equation

$$E = \langle \Psi \mid H \mid \Psi \rangle \tag{1}$$

and postulate that

$$E = E(1) + E(2) \tag{2}$$

where

$$E(1) = \langle \Psi(1) \mid H \mid \Psi(1) \rangle \tag{3a}$$

and

$$E(2) = \langle \Psi(2) \mid H \mid \Psi(2) \rangle + \Psi(1), \Psi(2) \text{ cross terms} \tag{3b}$$

with the very important condition that $E(2) << E(1)$, namely $E(2)$ corresponds to a relatively small perturbation on the system. If we could find an effective hamiltonian $H_{(eff)}$ such that we can approximate $E(2)$ as

$$E(2) = \langle \Psi(1) \mid H_{(eff)} \mid \Psi(1) \rangle \tag{4}$$

then

$$E = \langle \Psi(1) \mid H + H_{(eff)} \mid \Psi(1) \rangle \tag{5}$$

However, a solution to Eq. (4) represents a most difficult task, when approached quantum mechanically, since this would require knowledge and fulfilment on all the necessary and sufficient conditions to be satisfied by the equation. An alternative, which is pragmatically used in most effective potential proposals, including density functional theory, is to find approximated mathematical expressions, which, numerically, aim to satisfy Eq. (5). Note that the above derivation holds both for Ψ and for ρ , the exact total electronic density, for which we can assume that there exists an approximated density $\rho(1)$ corresponding to a wave function of pre-determined complexity. The Hartree-Fock wave function is particularly well adapted to be selected as the wave function of pre-defined complexity, because it is computationally very simple and at the same time a most reasonable zero order approximation to the exact Ψ, especially if we expand to very few determinants as many as needed to account for the non dynamical correlation energy. Note that the closer $\Psi(1)$ is to Ψ, the more reasonable becomes the approximation implied in Eq(4) and Eq(5).

The strength of this approach over the traditional one, is that one works with *a pre-determinate complexity*, $\Psi(1)$ or $\rho(1)$, imposed as a constraint from the very start. The pre-determination is generally introduced on physical arguments, hopefully reasonable. The necessary shortcoming of this alternative is that it is generally most difficult to asses a priori the validity or even the practical value of the effective hamiltonian; indeed, if this could be done rigorously, one would have solved exactly the many body problem. Thus one is compelled to make use of semi empirical parameters, either as part of a functional form which is formally and theoretically derived from first principles, or of a functional form which corresponds to nothing more than an intuitive, but hopefully well grounded, guess. Another unpleasant aspect is that the computed total energy can be lower than the exact one, even when we use a variational technique; this unpleasant feature is a direct demonstration that we are dealing more with a mathematical than a physical "short-cut" to the exact energy. In other words, in practice, we can not solve have Eq. (5), but only approximate it:

$$E \sim \langle \Psi(1) \mid H + H_{(eff)} \mid \Psi(1) \rangle \tag{6}$$

In conclusion, Eq. (4) can be indirectly solved as a numerical rather than quantum-mechanical approximation, and only when $\Psi(1)$ is very near to Ψ, namely when $\Psi(1)$ is nearly the correct solution for the H operator. It appears that we shall have to accept that exact solutions are more a part of the mathematical universe, than of the physical one.

We could note that, in a way, the traditional quantum chemistry approach has failed to provide an operational solution for large molecules (and therefore a technique of interest to much of the chemical industry) likely because it had insufficient work contacts with computational physics and in particular with computational solid state physics. Hopefully now, after that the majority of quantum chemists have learned to appreciate and/or use the density functional computer packages, since long known to solid state physics community, quantum chemistry has the opportunity to experiment, improve and use density functionals and/or other different effective hamiltonian methods. At the same time quantum chemists should continue in the traditional path (perhaps with a somewhat different prospective, since now part of the pressure to be "relevant" has been released) solving with correct quantum mechanics more and more complex problems, taking advantage of the evolution in computer technology, and preparing benchmarks necessary in testing empirical avenues.

Fundamental problems left unanswered for a long time (possibly because some of us were pressed by the desire to prepare optimal quantum chemical codes) should be faced, however. Among these we must remind ourselves that the hamiltonian used in the Schrödinger representation is not exact, but, like everything in science, it represents only a model (therefore it makes not much sense to insist in the use of it to the bitter end) which ignores basic facts, like that electrons have spins and can reach velocities non negligible relative to the velocity of light; on the other hand the Dirac alternative has drawbacks too since it is de facto used for many electron systems but was derived for only one electron (in an electromagnetic field).

It is of interest to recall that the use of the effective hamiltonian in quantum chemistry is very old, even if the terminology "effective hamiltonian" was often not used; indeed one can go back to Hückel or to all quantum chemical semi empirical formulations; note that we can always rewrite the semi empirical hamiltonian, as in Eq. (6). It is therefore, somewhat unfortunate that in the past quantum chemistry literature, the term "semi empirical" has eclipsed the appreciation that we are dealing with effective hamiltonian approximations, and that most ab-initio methods can be reformulated as effective hamiltonian, whenever the exact wave function is not obtained. The pre-determined complexity level for the wave function was, however, very clearly enunciated in all "semi empirical" proposals.

6. Back to one of the early effective hamiltonian in quantum chemistry

Today, it is becoming more and more apparent to most quantum chemists that the traditional quantum chemistry approaches are at a dead end for even moderately large systems. Anticipating this trend, in the early sixties [27] Clementi did propose to deal with the correlation problem in large molecules by introducing a Coulomb Hole, an effective potential, for pairs of electrons with opposite spins, using an *indirect* functional of the electronic density and constraining the wave function to be a single Slater determinant. The proposal was tested for atoms and ions. Notice that any effective potential formulation in terms of integrals on basis set is, by construction of the Fock matrices, an "indirect" function of the electronic density. This early work was soon followed by other papers, at times with a modification of the density functional introduced by Wigner, revisited by Gombas, at times with the Coulomb Hole. After much work to improve and expand the limits of ab initio quantum chemistry [3] Clementi in 1972, concluded [14] on the need for quantum chemistry to use some effective hamiltonian as the *only reasonable way to compute non-trivially small chemical systems.*

In the following we shall recall on some recent progress for the Coulomb-Hole-Hartree-Fock technique and summarise our work in the area of effective potentials. Note that up to the early sixties these approximations were known as "statistical methods" in recognition of the pioneering Thomas-Fermi work; the term "density functional" was also used, for example in the review volume by Gombas on pseudo potentials for atoms.

By the end of the 1930 decade, the basic formalism on which today quantum chemistry is based, was clearly enunciated. The milestones are the "single-particle" approximation by Hartree [28] and Fock [29], the many body approach by Hylleraas [30] and the density approach (at the time called "statistical approach") by Thomas [31] and by Fermi [32]. Dirac's electron theory was also enunciated [33] but hardly applied to electronic systems of chemical interest. The road open by Hartree-Fock and Hylleraas was later widened by the multi-configuration approach [34], a variational improvement on the many-determinant CI. In this contest the Heitler-London [35] work remains somewhat isolated; there is the intriguing possibility that, whereas the papers from Refs. 28-34 provide the base of today quantum chemistry, perhaps a substantially modified Heitler-London approach might be included into the foundations of tomorrow theoretical chemistry. Alternatively stated, theoretical chemistry can not develop in a balanced fashion without greater attention to a formalism, where the different types of bonds, bond-energies, and particularly chemical formulas are fully recognised as main theoretical and quantum chemical objectives.

As it is known, the "single-particle model", the HF-model, is formally expressed by a wave function of pre-determined complexity, the single Slater determinant wave function. In a way the correspondence between the single-particle model and assumption of *one* Slater determinant is rather unphysical and, in retrospect, it represents somewhat of a misfortune in the development of quantum chemistry! Indeed, the choice of *one* determinant brings about an uneven accuracy in the wave functions and in the corresponding energies both for atomic and molecular systems. To make matter worse, the energy difference between "one-Slater determinant" and "exact non relativistic" wave functions, has been taken as the accepted definition for the many-body effect, namely it defines the correlation energy. Whereas it is well known that often the one-determinant wave function provides a reasonable approximation to many atomic and molecular electronic systems, degeneracy effects, curve crossing, either for transition states or at molecular dissociation, can bring about errors as large as the neglect of many-body corrections. In this context we recall that there are dynamical and non-dynamical correlation effects and that *the true electron-electron many-body interaction relates to the dynamical correlation part only*, whereas the non dynamical part should be considered as part of Hartree-Fock-type, "single-particle" model.

The route initiated by Hartree and Fock, in order to obtain accurate wave functions, is by now rather clearly understood and can be condensed into three rules:

a. use one or as many determinants as needed to secure correct spin and angular momentum [36], in practice solve for restricted HF,

b. use as many additional determinants to ensure *removal of non-dynamical correlation* [37-39],

c. add to this the very, very large CI expansion, or better the multi-configuration expansion, or whenever possible Hylleraas CI expansions [40] to solve for the *dynamical correlation* (or any "simpler" large expansion approach).

Unfortunately, the third task is computationally nearly impossible for more than a few dozen electrons at the CI level, or for more than 3 or 4 electrons at the Hylleraas-CI level. In general, the above methods are valid for relatively few electron systems or for those many-electron systems, where we need an accurate representation for only a *small* subset of electrons. We wish to stress that the above methods do, however, represent very important achievements, since the few-electron high accuracy test cases available in literature, provide most important benchmarks. The above three rules have a perturbation equivalent, which lacks the safety of lower bounds, but might be computationally more efficient; note that perturbation techniques are close to effective potential techniques.

A very different approach to many-electron systems was taken by Thomas [31] and Fermi [32], possibly because they were interested not in the detailed description of each electron (and related excitation process) in the n-electron system, but rather in the overall distribution of the electronic density associated to the n-electrons. Thus, one is lead to solve the Laplace rather than the Schrödinger equation, attempting to compute the many-body interaction without specific reference to one electronic contribution after another.

An attempt to merge the two approaches was proposed by E. Wigner [41], who solves for a Hartree-Fock function and then uses the HF density to approximate at once the many-body correction. In this way, the "correlation energy term" was shaped into what today is called a *density functional*, and its solution did require fitting one of the few soluble many-electron systems, namely the homogeneous gas of electrons. The work of Wigner can be considered as one of the first density functional application in the physics literature (unless one prefers to select Thomas and Fermi work, for this "first"); note that the Wigner's correlation energy term does not include the exchange correction, being the latter correctly accounted for by the Hartree-Fock approximation.

Along Wigner's lines, in 1963, Clementi [27] proposed to disregard Wigner's perturbation approach but to use an effective Hamiltonian, as in Eq. (6), where the correlation energy is obtained by adding to the Hartree-Fock-type electronic interaction a new term, which represents an average Coulomb-hole to account for many-body corrections for electron pairs with opposite spins. This was done in analogy to the "pre-correlation" energy gained via the exchange term for electron pairs with parallel spin. Formally, this correction is accomplished by replacing the $1/r_{ij}$ operator with an effective operator, a *hard* Coulomb hole, expressed via terms which were parametrized using available values of correlation energies from atoms and molecules. This approach in which the electron pairs with opposite spin were assumed to be surrounded by a small region with an infinitely repulsive electron-electron potential was called Coulomb-Hole-Hartree-Fock, CH-HF, for short. In Clementi's approach [27] the form of the wave function was initially chosen to be a single Slater determinant, but later it was expanded to as many determinants [37-39] as needed to account for degeneracy and other non-dynamical effects. A more refined approach followed [17] and it makes use of a new Coulomb-hole effective effective potential of the form

$$-[\exp(-\omega_{ij}\, r^{2ij})] / r_{ij} \qquad (7)$$

In this new form [17], the *soft*-Coulomb-hole replaced the previous hard-Coulomb-hole [27] . We recall that the technique of the hard-Coulomb-hole was applied to determine the correlation energy for the atoms He to Ar; in addition the Coulomb-Hole-Hartree-Fock wave function was found to be particularly well suited to describe the polarizability of atomic systems [27]. We recall that the possibility to transfer atomic correlation energies to molecular systems has been often considered since the early days of computational quantum chemistry; details can be found in Refs. 42-44.

It is interesting to determine which numerical value ω must assume in order to yield total energy values which account for the correct correlation; we have answered this by fitting for the available correlation energy for the ground state atoms He, Ne, Ar, Zn, and Kr. We have obtained the following values for ω: 27.3, 640.0, 2490.4, 5650.0 , 95000.0 respectively. These values can be fitted nicely with the function $\omega = 7Z^2$ · It is interesting to note that this function scales as the total energy for atoms, and this gives us a clue for the ω parametrisation.

The underlying approximation in the Coulomb-Hole-Hartree-Fock is in the assumption that there exist a zero-order solution, which has, percentually, the same reliability at any point on the Born-Oppenheimer surfaces; therefore, it assumes the existence of a well defined "reference function".

In specific problems, like the case of incorrect Hartree-Fock dissociation limits, or of near degeneracy, the problem has a rather clear-cut solution. However, often the introduction of several determinants not only eliminates degeneracy effects, but brings about also a small contribution to the dynamic electron-correlation correction. In such cases the Coulomb-Hole-Hartree-Fock approximation can yield energies more stable than the correct ones. In a very pragmatic and somewhat naive way, we could take advantage of this inherent weakness, and relate the *single* determinant CH-HF energy to an *upper* bound and the energy from the *short linear* combination of CH-HF determinants, to a *lower* bound.

In full detail, we recall that in the Hartree-Fock model the total energy expression is obtained by operating on a Slater-determinantal function with the standard Hamiltonian

$$H = \Sigma_i (-1/2\, \Delta_i^2 - \Sigma_A Z_A / r_{iA} + \Sigma_j 1/ r_{ij}) \qquad (8)$$

where i and j refer to electrons and Z to the A-th nucleus of the molecule. The Slater determinant is the well known antisymmetrized product of one electron-functions (atomic or molecular orbitals), which are solution of the eigenvalue equations, obtained variationally in a self-consistent procedure. In solving for the one particle equations, the electron-electron repulsion hamiltonian term is replaced by two effective potential terms, the well known Coulomb and exchange operator, J and K, defined as

$$J_{ab} = \langle \phi_a (1) \phi_b (2) \mid 1/r_{12} \mid \phi_a (1) \phi_b (2)\rangle \tag{9}$$

$$K_{ab} = \langle \phi_a (1) \phi_b (2) \mid 1/r_{12} \mid \phi_a (2) \phi_b (1)\rangle \tag{10}$$

where the ϕ's are the one-electron-functions, the spin orbitals; we recall that the K term exists only for spin orbitals with the parallel spins.

With this in mind, the Hartree-Fock energy for an open-shell system can be written, following the classical paper by Roothaan [36], as

$$E = 2\Sigma_k H_k + \Sigma_{kl}(2J_{kl} - K_{kl}) + \\ + f [\Sigma_m H_m + f \Sigma_{mn} (2aJ_{mn} - bK_{mn}) + 2 \Sigma_{km} (2J_{km} - K_{km})] \tag{11}$$

Here, k and l refer to orbitals of closed shells and m and n are open shell atomic or molecular orbitals. The constants a and b are the vector coupling coefficients of Roothaan, the number f is, in general, the fractional occupation of the shell.

In the Coulomb-Hole-Hartree-Fock approximation we replace the operator $1/r_{ij}$ with an effecting operator $f(1/r_{ij})$, designated as Ω; alternatively stated we retain the standard form of both the Coulomb and Exchange effective operators and we add a new effective operator Ω; the new operator, when acting on the single determinant, places each electron inside a Coulomb Hole, which in essence does not allow to electron i to penetrate into the space coordinates of electron j within a spherical volume -centred around electron j- and with a parametrised sphere radius; this is tantamount to introduce a symmetric Coulomb hole around each electron, when two electrons interact via the Coulomb term. In this way the total energy is given as:

$$E = 2\Sigma_k H_k + \Sigma_{kl} [2 (J_{kl} - \Omega_{kl}) - K_{kl}] + \\ + f\{\Sigma_m H_m + f \Sigma_{mn} [2a (J_{mn} - \Omega_{mn}) - bK_{mn}] + \\ + 2\Sigma_{km} [2(J_{km} - \Omega_{km}) - K_{km}]\} \tag{12}$$

Initially [27], this condition was imposed on the J integrals, and the Coulomb hole was a square potential "hard" Coulomb hole. Subsequently a "soft" hole was used and applied to both Coulomb and the Exchange interactions (where for simplicity of notation we use again the symbol Ω for the new effective potential even if the form of the Coulomb hole in the following expression is the one of Yukawa-type, reported above for ω):

$$E = 2\Sigma_k H_k + \Sigma_{kl} [2 (J_{kl} - \Omega_{kl}) - (K_{kl} - \Omega_{kl})] + \\ + f\{\Sigma_m H_m + f \Sigma_{mn} [2a (J_{mn} - \Omega_{mn}) - b(K_{mn} - \Omega_{mn})] + \\ + 2\Sigma_{km} [2(J_{km} - \Omega_{km}) - (K_{km} - \Omega_{kn})]\} \tag{13}$$

A more refined soft-Coulomb-Hole-Hartree-Fock method would assume that ω is an explicit function of the electronic interaction energy of the electrons i and j, and that there should be *two different values* for ω, one for the Coulomb interactions and a second for the Exchange interactions. Indeed, the main dynamic correlation error in the Hartree-Fock approximation is the lack of the equivalent of a Fermi hole around electron pairs with unparallel spins, but there is still some error due to the inability of the Exchange repulsion to fully correlate pairs with parallel spin. Computations on excited states are expected to bring out quantitatively this detail.

Since the CH-HF can only account for the dynamical part of the correlation energy, the wave function is represented by a variationally obtained linear combination of few Slater determinants, as many as needed to account for degeneracy, near degeneracy and incorrect dissociation. We recall three typical effects which contribute to non dynamic correlation: the atomic near degeneracy of the type $1s^22s^22p^n$ and $1s^22s^02p^{n+2}$, the degeneracy of molecular curve crossing in transition states, and the need for more than one determinant for a correct description of atomic dissociation products. We stress that, in all such situations, the linear combination of determinants is very short (from two to about a dozen) and well defined.

In Table 1, we report the computed correlation energies according to Ref. 28 (hard hole) for the atoms from Z=1 to Z= 18, and, in Table 2, the computed correlation energies from the soft Coulomb Hole for atoms from Z=1 to Z=54, are used to obtain first ionisation potential; in the Table we present the Hartree-Fock, the experimental and the CH-HF first ionisation potentials [17].

In Table 3, we report the dissociation energies for a few molecules at the equilibrium geometry, comparing experiments with Hartree-Fock and the Coulomb-Hole-Hartree-Fock approximation [19]. The results, even if not "excellent", are "encouraging".

We note that the parametrisation of ω for molecular computations has been the subject of a few revisions, non one considered satisfactory, since too many empirical terms are used.

Considering the dissociation products for cases like HeH^+ the single determinant is equally good both for the equilibrium geometry and for the dissociation limits. Not always this is the case. We have already stressed that for some cases it will be necessary to start with a more-than-one determinant approach, since at dissociation, there is a bond breaking not correctly described by a single determinant wave function (of either HF or CH-HF type) in the restricted (Hartree-Fock) formalism as discussed in Refs. 38 and 39. The F_2 molecule provides a classical example of the problem for the dissociation products: we recall that because of its large bond distance, the bonding $^1\Sigma_g$ and the antibonding configuration $^1\Sigma_u$ are intermixed even at the equilibrium geometry. Another classical case is provided by H_2, where again there is intermixing between the $1\sigma^2_g$

TABLE 1. Total correlation energy (in a.u.) of ground state neutral atoms with two different parametrisations [17,18] for ω

Atom	HF Total Energy	Correlation Energy from Ref. 17	Correlation Energy from Ref. 18	Reference Correlation Energy**
He	- 2.8617	0.044	0.0436	0.0420
Li	- 7.4327	0.046	0.0460	0.0453
Be*	- 14.573	0.092	0.0966	0.0943
B*	- 24.529	0.103	0.126	0.1248
C*	- 37.689	0.115	0.156	0.1564
N	- 54.401	0.130	0.176	0.1883
O	- 74.809	0.196	0.247	0.2580
F	- 99.409	0.268	0.321	0.3248
Ne	- 128.547	0.347	0.397	0.3912
Na	- 161.859	0.375	0.405	0.3965
Mg*	- 199.615	0.442	0.449	0.4394
Al*	- 241.877	0.467	0.471	0.4706
Si*	- 288.854	0.492	0.505	0.5057
P	- 340.719	0.517	0.527	0.5409
S	- 397.505	0.585	0.600	0.6062
Cl	- 459.482	0.658	0.667	0.6683
Ar	- 526.817	0.736	0.740	0.7261
Zn	- 1777.814	1.740	1.533	(1.6975)
Kr	- 2751.996	2.262	1.902	(1.8909)
Sr	- 3131.481	2.436	2.000	(1.9772)
Cd	- 5463.841	3.453	2.819	(2.7253)
Xe	- 7228.444	4.010	3.122	(3.0877)

*) Two-configurations CH-HF.
**) From Ref. 54; values in parentheses from accurate numerical MP2 data (G. Corongiu, unpublished).

and the $1\sigma^2_u$ configurations, extremely evident at the dissociation limits, where the $1\,\sigma^2_u$ configuration prevails.

We recall, that the work in the two papers (Refs. 38, 39) represents a detailed discussion both on the use of few determinants to cope with the dissociation products and on the use of density functionals to cope with the correlation problem. The same has been done in the CH-HF approximation, providing not only correct dissociation products but also a notable improvement in the values of the molecular dissociation energy.

TABLE 2. **First ionisation potential (in eV) from Hartree-Fock, HF, Coulomb-hole-Hartree-Fock, CH-HF and experimental data**

Z	HF	CH-HF	Exp.	Z	HF	CH-HF	Exp.
3	5.34	5.39	5.39	29	6.41	7.03	7.73
4	8.04	9.28	9.32	30	7.63	9.29	9.39
5	7.93	8.17	8.30	31	5.47	5.64	6.00
6	10.79	11.07	11.26	32	7.43	7.66	7.90
7	13.96	14.29	14.53	33	9.53	9.81	9.81
8	11.89	13.61	13.62	34	8.41	9.75	9.75
9	15.72	17.44	17.42	35	10.78	12.14	11.81
10	19.85	21.58	21.56	36	13.27	14.67	14.00
11	4.95	5.18	5.14	37	3.73	4.18	4.18
12	6.61	7.95	7.65	38	4.67	6.03	5.70
13	5.50	5.69	5.99	39	5.74	6.07	6.38
14	7.65	7.91	8.15	40	5.05	6.65	6.84
15	10.02	10.32	10.49	41	5.66	6.13	6.88
16	9.02	10.51	10.36	42	5.87	6.34	7.10
17	11.80	13.32	12.97	43	5.29	7.16	7.28
18	14.78	16.35	15.76	44	5.92	6.55	7.37
19	4.01	4.35	4.34	45	5.92	6.63	7.46
20	5.12	6.49	6.11	46	6.67	8.23	8.34
21	5.35	6.81	6.54	47	5.91	6.75	7.58
22	5.51	7.06	6.82	48	6.90	8.64	8.99
23	5.81	7.35	6.74	49	5.05	5.22	5.79
24	5.90	7.24	6.77	50	6.83	7.04	7.34
25	5.90	7.67	7.44	51	8.66	8.91	8.64
26	6.28	8.03	7.87	52	7.62	8.82	9.01
27	7.78	7.92	7.86	53	9.62	10.84	10.45
28	7.61	7.82	7.64	54	11.68	12.93	12.13

Shortly after the CH-HF proposal, in 1965, there is a somewhat similar proposal by Kohn and Sham [47], the well known K-S density functional approximation. In the K-S work, the effective Hamiltonian operates on a single determinant and it contains a term which deals not only with the correlation energy, as proposed by Wigner and later Clementi, but also with a simplified expression for the exchange energy following Slater proposals [48]; the two effective potential terms, the exchange and the correlation, are merged into one, the so-called exchange-correlation term. The exchange-correlation term is chosen to be an explicit functional of the electronic density, and requires empirical parameters obtained by fitting the energies of many body corrections for an homogeneous gas of electrons. To appreciate the connections between exchange energy and

TABLE 3. Dissociation energy (in a.u.) from laboratory data, Hartree-Fock, HF, and Coulomb-hole-Hartree-Fock, CH-HF, with only one determinant

Molecule	D_e (exp)*	D_e (HF)*	D_e (CH-HF)
Li_2	0.0419	0.0062	0.0443
F_2	0.0610	-0.0492	-0.0075
NaH	0.0779	0.0343	0.0709
CL_2	0.0922	0.0396	0.0557
LiH	0.0926	0.0546	0.0966
CHF	0.0959	0.0226	0.0628
B_2	0.1102	0.0320	0.0908
AlH	0.1106	0.0878	0.1242
MgO	0.1303	-0.0476	0.0421
BH	0.1316	0.1025	0.1401
NaCl	0.1569	0.1176	0.1699
HCl	0.1698	0.1303	0.1819
BeO	0.1738	0.0674	0.1854
H_2	0.1746	0.1335	0.1687
LiCl	0.1793	0.1408	0.2157
NaF	0.1830	0.1089	0.1734
O_2	0.1915	0.0427	0.1152
LiF	0.2209	0.1487	0.2233
HF	0.2249	0.1610	0.2174
C_2	0.2337	0.0267	0.1446
PN	0.2367	0.0661	0.2286
BF	0.2900	0.2279	0.3058
CS	0.2914	0.1661	0.2908
H_2S	0.2946	0.2141	0.3188
SiO	0.3065	0.1882	0.3154
N_2	0.3642	0.1872	0.3434
H_2O	0.3736	0.2598	0.3685
PH_3	0.3958	0.2752	0.4169
CO	0.4131	0.2895	0.4127
H_2O_2	0.4333	0.2296	0.3875
NH_3	0.4796	0.3235	0.4810
HCN	0.4994	0.3241	0.5127
SiH_4	0.5211	0.4134	0.5647
C_2H_2	0.6501	0.4777	0.6871
CH_4	0.6761	0.5287	0.6985
CH_3OH	0.8258	0.6014	0.8383
C_2H_4	0.9059	0.6876	0.9505
C_2H_6	1.1473	0.8879	1.1982

*) From Ref. 11, Chapter 8, Table IX.

correlation energy, the reader is referred, for example, to a paper by J.C. Slater [48]. We recall that the Exchange energy can be written in terms of the electronic density, a development which has lead to the Xα - approximation; it is of interest to note that appropriate parametrisation of the multiplicative factor in the Slater (or Gombas) exchange energy expression, can provide either an approximation to obtain Hartree-Fock energies or, with a different parametrisation, to experimental energies, essentially the Kohn-Sham energy.

Comparing the Coulomb-Hole Hartree-Fock with the Kohn-Sham approximation, we note that the two have notable differences, the main one -already noted- is that in the CH-HF method the exchange contribution is computed correctly and therefore excited states for a given electronic configuration are intrinsically available from this method. Another difference is that the restriction to a single determinant is removed in the CH-HF method, when the zero order wave function calls for more than one determinant wave function [37-39]. It is of interest to recall that in today literature there is often a tendency whereby the Kohn-Sham approximation is presented as being directly "derived" from the Hohenberg-Kohn theorem; we would like to note that it is well known that the Hohenberg-Kohn theorem does not provide any necessary and sufficient condition for the construction of the Kohn-Sham density functionals. Thus the Kohn-Sham equations, as today implemented, do not belong to the class of ab initio teoretical methods, but are simply a specific example of a well chosen but *empirical effective potential.*

7. Two novel effective potentials: The J-functional and the K-functional

To improve on the Hartree-Fock energy, we can either increase the electronic kinetic energy, since correlation is equivalent to avoidance of electron-electron collision, a process which is obtained by speeding up the electron and at the same time by introducing a change in its orbit, whenever we are on a collision path. Alternatively and equivalently (virial theorem) we could reduce the Coulomb repulsion [49] present in the Hartree-Fock model, following the experience in passing from the Hartree to the Hartree-Fock models. Likely, because we are accustomed to work in a time independent framework, and because of the popularity of the coordinate rather than in the momentum space, tradition has favoured the coulomb interaction.

The simplest way to decrease the Coulomb energy is to assume that the electrons repel each other not in vacuo, $1/r_{ij}$, but in a dielectric medium [50], namely $1/(\varepsilon r_{ij})$. It is very easy to experiment with this hypotheses [49], and to determine the value of ε which will yield an energy equal to the sum of the Hartree-Fock and the correlation energies. The value for $1/\varepsilon$ obtained for the atoms from Z=2 to Z= 54 is a regular smooth function starting with a value of about 0.94 which at Z=10 has increased to about 0.985 and then gradually increases toward a value a bit smaller than 1.0. It is not difficult to use these data to determine a parameter for the many center integrals in a molecule. Note that this approach is a "revisited" proposal, which did originally appear in the early semiempirical literature of the molecular π electron spectroscopy. However, it should be stressed that an algorithm which yield essentially correct values of total correlation energy is unsatisfactory, unless it also provides a correct distribution of the correlation effects for each electron pair, and between electron pairs. Bethe-Goldstone [51] basic ideas can not be ignored. In Table 4 we report the pair-wise contribution to the correlation energy computed with the ε values fitted for the Ne and the Ar ground state atoms as above discussed; we can see that the correlation energy is far from approximately equal for all *direct pairs* (i.e. the two electrons in the 1s, 2s, 3s pairs or the three direct pairs in the 2p and 3p shells), thus the distribution of the pair correlation energies is unacceptably incorrect..

To improve we would have to transform ε into a many parameter variable, ε_{ij}, aiming to increase the value for inner shells and decrease for the valence shells. This complex parametrization takes away much of the appeal from this very simple approximation.

TABLE 4. Distribution of pair correlation energies (in a.u.) in neon and argon ground states with the factor ε^{-1} The pair correlation for the 2p electrons are lumped in one value; same for the 3p electrons.

	Ne			Ar				
	1s	2s	2p	1s	2s	3s	2p	3p
1s	043			.039				
2s	.041	.007		.043	.008			
3s				.013	.012	.002		
2p	.121	.077	.101	.142	.089	.036	.126	
3p				.035	.032	.023	.096	.028

We have therefore experimented by reducing the numerical value of the coulomb integrals; after some trail we have temporarily settled on the following form, which replaces the standard Hartree-Fock integral J(pq,rs) where p, q, r and s are basis of the set with which we expand the atomic or molecular orbital:

$$J'(pq,rs) = J(pq,rs)\ [1 - f\ (Z, n, J)] \tag{14}$$

where

$$f(Z,n,J) = A\ [\ 1+ (Z-n)B]\ [\ J(pq,rs)^{c}\ (1+ZD)]^{(-1)} \tag{15}$$

where A, B, C, and D are numerical constants valid for any system and equal to 0.02959, 0.28306, 0.43988, and 0.64629, respectively; Z and n are the atomic number and the number of electrons in the atom. The terms in n and Z are introduced to yield the correct dependence of the atomic correlation energy along isoelectronic series and for increasing values of n for the neutral atoms.
The total energy for the J-functional effective is given by the expression:

$$E = 2\Sigma_k H_k + \Sigma_{kl}\ (\ 2J'_{kl} - K_{kl}) + f\ [\ \Sigma_m H_m + \\ + f\ \Sigma_{mn}\ (2aJ'_{mn} - bK_{mn}\ ) + 2\ \Sigma_{km}\ (2J'_{km} - K_{km})\ ] \tag{16}$$

where J' is the previously given modified coulomb integral.

Computationally, the task to transform an Hartree-Fock code into a J-functional code is trivial. In Table 5 we report the correlation energies computed for the ground state atoms from the J-functional. The computed correlation energies [49] are compared with those by Davidson and Chakravorty [54], which are similar to those proposed long ago by Clementi and Veillard [52], more recently revised by Rizzo *et al.* [53].

TABLE 5. Atomic correlation energies (in au) from the J-functional.

Atom	J-Funct.	Ref. 54	Atom	J-Funct.	Ref. 54
He	-0.030	-0.042	F^+	-0.246	-0.261
Li	-0.044	-0.045	Ne^{+2}	-0.251	-0.265
N	-0.182	-0.188	Ne^+	-0.312	-0.326
O	-0.238	-0.258	Na^+	-0.318	-0.330
F	-0.303	-0.325	Ar^+	-0.695	-0.690
Ne	-0.376	-0.391	Xe^+	-2.220	-1.864
Na	-0.410	-0.396	Kr^+	-3.609	-3.071
P	-0.580	-0.541	Ni^{+10}	-0.774	-0.871
S	-0.633	-0.606	Ni^{+11}	-0.719	-0.820
Cl	-0.689	-0.668	Ni^{+12}	-0.667	-0.757
Ar	-0.740	-0.726	Ni^{+13}	-0.617	-0.681
Zn	-1.858	-1.697	Ni^{+17}	-0.455	-0.432
Kr	-2.284	-1.891	Ni^{+18}	-0.428	-0.409
Cd	-3.308	-2.725	Ni^{+19}	-0.348	-0.351
Xe	-3.672	-3.088	Ni^{+20}	-0.275	-0.291
Li^+	-0.032	-0.043	Ni^{+21}	-0.211	-0.225
O^+	-0.188	-0.194	Ni^{+25}	-0.052	-0.053
F^{+2}	-0.193	-0.199	Ni^{+26}	-0.038	-0.046

Note that the Coulomb hole Hartree-Fock brings about a correction which is large for very small inter electronic distances and particularly well suited for corrections in the inter electronic cusp regions; the J-functional brings about a correction which survives at very large inter electronic distances, being it of coulomb type. Therefore, one would suspect that a more appropriate effective hamiltonian would be a combination of the Coulomb-hole Hartree-Fock and of the J-functional (with a somewhat different parametrization relative to the one given above).

Next, we report on yet another functional [49], which is appropriate for intermediate inter electronic distances, larger than those of interest at the inter electronic cusp, but shorter than coulomb type. In this respect we recall that it is somewhat intuitive that one should tune up the effective potential to a given inter electronic range of distances. Alternatively stated, it seems rather unlikely that a correction representing van der Waals interactions will maintain the same

functional form and long range dependence as a correction devised for very short inter electronic distances, namely in the cusp region.
The main physical idea underling the K-functional, is that we wish to create around electron pairs with opposite spin the same type of repulsive interaction available for electron pairs with parallel spin; in other words we wish an "exchange type correction" for electron pairs with opposite spin.
Formally, we can define a spin rotation matrix R which transforms α spin into β spin and β spin into α spin.

$$R_\sigma = \begin{pmatrix} 0 & -1 \\ 1 & 0 \end{pmatrix} \tag{17}$$

In the unrestricted Hartree-Fock representation, see for example R. McWeeny [55], a one electron function is of the form

$$\Psi_i(x) = \Psi_i^\alpha(r)\,\alpha + \Psi_i^\beta(r)\,\beta \tag{18}$$

Operating with the rotation matrix R_σ on the one electron function, we have

$$\Psi_i(x') = R_\sigma \Psi_i(x) = \Psi_i^\alpha(r)\,\beta - \Psi_i^\beta(r)\,\alpha \tag{19}$$

If we now define the exchange-type operators K' and K" as

$$K'(i)\,\Psi_i(x_1) = \int g(1,2)\,\rho(x_1,x_2)\,\Psi_i(x_1')\,dx_2 \tag{20}$$

and

$$K''(i) = R_\sigma^{+}\,K'(i) \tag{21}$$

then

$$\Sigma_i < \Psi_i(x) \mid K'' \mid \Psi_i(x) > = \int g(1,2)\,\rho(x_1,x_2)\,\rho(x'_1,x'_2)\,dx_1\,dx_2 \tag{22}$$

Note that for closed shell $\rho(x_1,x_2) = \rho(x'_1, x'_2)$ and therefore the effect of K" is the same as the effect of the usual exchange operator K. Instead, when all the electrons have the same spin, the integral is zero. To correlate electron pairs both with same and opposite spins, we use an effective operator K"' defined as

$$K''' = 1/2\,(K + K'') \tag{23}$$

The programming implementation of the K" integrals is a very simple task, if one selects to use the unrestricted Hartree-Fock method; however, for the restricted Hartree-Fock, one must define new vector coupling coefficients.

The next problem with the use of Coulomb hole operator K"', is to find a parametrization which will yield the desired correlation energies.

Recalling that in a collision problem the scattering cross section is function of the electron velocity, v, we can solve our problem by selecting an effective operator of the form f(v)K"' and parametrize f(v) as

$$f(v) = a(\chi_1 + \chi_2 + \chi_3 + \chi_4)^b / I^c \tag{24}$$

where $\chi_1, ..., \chi_4$ are the four orbital exponents of the gaussian basis set, I is the value of the two-electron Coulomb integral (with the same basis functions) and a, b and c are three parameters to be optimised. From numerical experimentation the optimised values of a, b and c are 0.065, 0.25, and 0.45, respectively.

In Table 6, we report the computed correlation energies of all the (closed or open shell) atoms of the second and third row of the periodic.

TABLE 6. Atomic energies (in au) with K-functional

Atom	Unrestricted energy[a]	K-functional correlation[a]	Restricted energy	Reference correlation[b]	Total energy error[c]
Li	-7.4315	-0.0526	-7.4327	-0.0453	-.0073
Be	-14.5729	-0.0808	-14.5730	-0.0944	.0135
B	-24.5293	-0.1116	-24.5291	-0.1248	.0130
C	-37.6899	-0.1459	-37.6886	-0.1564	.0091
N	-54.4043	-0.1833	-54.4009	-0.1883	.0016
O	-74.8140	-0.2478	-74.8094	-0.2580	.0057
F	-99.4113	-0.3168	-99.4093	-0.3248	.0059
Na	-161.8577	-0.4061	-161.8589	-0.3965	-.0096
Mg	-199.6144	-0.4385	-199.6146	-0.4394	.0010
Al	-241.8767	-0.4678	-241.8767	-0.4706	.0027
Si	-288.8546	-0.5016	-288.8544	-0.5057	.0039
P	-340.7192	-0.5392	-340.7188	-0.5409	.0013
S	-397.5065	-0.5986	-397.5049	-0.6062	.0051
Cl	-459.4828	-0.6605	-459.4821	-0.6683	.0070
He	-2.8616	-0.041	-2.8616	-0.0420	.001
Ne	-128.5471	-0.406	-128.5471	-0.3912	.015
Ar	-526.8175	-0.725	-526.8175	-0.7261	.001
Zn	-1777.8481	-1.628	-1777.8481	(-1.6975)	(-0.069)
Kr	-2752.0550	-2.031	-2752.0550	(-1.8909)	(0.140)
Cd	-5465.1331	-2.635	-5465.1331	(-2.7253)	(0.090)
Xe	-7232.1384	-3.106	-7232.1384	(-3.0877)	(0.018)

In parenthesis, numerical MP2 from G. Corongiu (unpublished).
a) From Ref. 49. b) From Ref. 54. c) Difference between the fifth and the third column. When the unrestricted energy (column 2) is lower than the numerical restricted energy (column 4) the difference between the fourth and the second column is added.

Notice that the reference correlation energy is defined as the difference between the exact energy and the restricted Hartree-Fock energy. Therefore, in comparing the values of the correlation energy obtained with the K-functional with the reference ones (sixth column) we have subtracted the difference between restricted and unrestricted energy.

The method appears to be able to evaluate, with reasonable accuracy, the dynamical correlation in closed and open-shell atoms and molecules. The three parameters have been optimised using the reference correlation energy of small closed-shell atoms (He, Ne, Ne^{8+} and Ar). The reasonable results obtained for heavy atoms, open-shell atoms and closed and open-shell molecules, together with the reasonable values obtained for the pair correlation, provide an indication on the reliability of the proposal.

The K-functional technique above presented has been also applied to molecular cases. In Table 7, we show the computed dissociation energy for the hydrides of the second row atoms and other diatomic molecules.

TABLE 7. Molecular dissociation energies (in au): Computed (Hartree-Fock, HF, and K-functional, K-functl, percent error and experimentall)

Molecule	HF	HF % Error	K-functl	K-functl % Error	Exp.
LiH	.055	41	.093	1	.093
BeH	.080	-	.106	-	.08-.09
BH	.102	22	.142	8	.132
CH	.091	32	.137	2	.134
NH	.077	38	.131	5	.125
OH	.111	35	.166	-2	.170
FH	.161	28	.218	-3	.225
HCl	.125	26	.180	6	.170
CO	.287	31	.402	-3	.413
PN	.070	70	.219	-8	.237
CS	.164	44	.286	-2	.291

The method appears to be able to evaluate, with reasonable accuracy, the dynamical correlation in closed and open-shell atoms and molecules. The three parameters have been optimised using the reference correlation energy of small closed-shell atoms (He, Ne, Ne^{8+} and Ar). The reasonable results obtained for heavy atoms, open-shell atoms and closed and open-shell molecules, together with the reasonable values obtained for the pair correlation, provide an indication on the reliability of the proposal.

8. Comment on Relativistic Effects and on Dirac-Fock formalism

Relativistic effects [56-58] are essential in the study of molecules containing heavy atoms. The more widely used methods for evaluating relativistic effects on molecules are perturbation techniques and the relativistic effective core potential (RECP) method [59]. Only in the last decade there have been attempts to evaluate relativistic effects in a more rigorous way [60-73]. Codes for Dirac-Fock calculations on closed and open-shell molecules are described in this volume. The method we would like to comment in this short section is the one based on the Dirac-Coulomb hamiltonian (called also the Dirac-Fock hamiltonian), in which the Coulomb field due to the nuclei and to the other electrons is added to the exact Dirac hamiltonian of a free electron. Even if the use of this hamiltonian involves many approximations (neglect of nuclear motion and spin, Breit correction, higher order quantum electrodynamics terms, more than two-body interactions) still it has be shown that the main relativistic effects are well described.

Here we wish to call attention to one aspect, namely the advisability to use the Dirac-Fock hamiltonian in all quantum chemical problems, whether the molecule contains light or heavy atoms.

We have already mentioned that the Schrödinger representation makes use of an hamiltonian which neither recognises the existence of the electron spin nor of relativistic effects; it seems therefore that there are not compelling reasons to retain the Schrödinger representation, and to replace it with Dirac-Coulomb. Historically, however, there where practical obstacles which prevented even to consider this course of action, two in particular: the numerical instability associated with the solution of the Dirac equations (the variational collapse) and the added complexity one must face in passing from the two component spin-orbital to the four component spinor. The realisation that the large and small components are not independent but coupled (for example by the kinetic balance constraints, see Ref. 70-73), the increased computational performance, demonstrated particularly by the advent of commercial parallel computers, the availability of public domain general Dirac-Fock codes for molecular systems, are changing the old paradigms, and we should consider actively the realistic possibility to drop the Schrödinger representation in favour of the Dirac representation for all molecular systems, whether with high or low Z atoms. Concerning the increased numerical complexity we recall that in principle the computational time and disk space, for a traditional non-direct SCF code, increases by a factor of 4^4 relative to non-relativistic Hartree-Fock computation with the same basis set in the large component. This factor is a combined factor resulting from the doubling of the basis set due to the kinetic balance, and a second factor of two resulting from spin orbit. To offset this increase, as we have stressed in few papers [74-76], many integrals, especially in the small components, can be neglected even for high Z numbers. Thus the factor 4^4 is only an upper limit and for low Z atoms the number of integrals (with value smaller than a pre-assigned threshold) is not much larger than in the Hartree-Fock approximation. Recently, we have revisited the energy surface for the NH_3 + HCl reaction [77-79], using several methods and codes. The Coupled Cluster and the MR-CASSCF give the correct value of the binding energy of the complex; the MP2 limit and the K-functional are about one Kcal/mole too high, the density functionals both overestimate the binding, with the LDA about twice the correct value, and the GGA one and half times. The Hartree-Fock, when close to the limit, being without correlation correction, as expected, underestimates. We should note that the complex binding energy, in the gas phase, is small, about 6 Kcal/mole.

The computer time distribution, comparing different methods (and codes) is very interesting and is given in Table 8 A first comment is related to relativistic vs. non relativistic computations. In the MOREL2 code, comparing the CPU and the disk space we see that the 4^4 factor is reduced to 1.2^4 for the disk space and 1.4^4 comparing total CPU time. With modern parallel computers these factors are becoming less and less of an obstacle. We also recall that, with strict equivalence, there is a correlation error in the Hartree-Fock solution, and an equivalent error in the Dirac-Fock solutions; however all we have learned in the past fifty years about Schrödinger correlation correction, it can be transferred to Dirac correlation correction. This is for example shown by the inclusion of the K-functional in a Dirac-Fock code.

Two more advantage are implicit in our proposal for a general adoption of the Dirac representation. First we can define the correlation correction in a much sound way, since we do not have any more to refer to a non existing exact-non-relativistic solution. Secondly we do not have neither to work at two major sets of molecular codes, nor to consider anymore a representation which ignores both spin and relativistic effects and pretends to be used as "the" hamiltonian for chemistry.

TABLE 8. Elapsed times to perform a single-point calculation on a IBM RS/6000-560 workstation. In the basis set column, the two numbers correspond to the total number of primitive and contracted function, respectively

Method	Program	Elapsed time	Basis set
DFT-LDA	MOLECOLE-DFT	2 min.	150/78
DFT-GGA	MOLECOLE-DFT	3 min.	150/78
DFT-LDA	MOLECOLE-DFT	4 min.	155/155
DFT-GGA	MOLECOLE-DFT	5 min.	155/155
SCF	MOLECOLE	5 min.	150/78
SCF	MOLECOLE	100 min.	155/155
SCF-KF*	MOLECOLE	102 min.	155/155
SCF	MOLECOLE	4 hours	242/180
HF limit	MOLECOLE	2 days†	380/380
MP2	MOLECOLE	9 hours	242/180
MP2 limit	MOLECOLE	2 days†	380/380
CASSCF	MOLCAS	2 hours	328/124
MR-CASSCF	MOLCAS	6 hours	328/124
Coupled Cluster	ACES II	~ 8 hours**	328/166
Dirac-Fock	MOREL-1	7 hours	155/155***
Dirac-Fock	MOREL-2‡	20 min.	155/155***
Dirac-Fock-KF*	MOREL-1	7 hours	155/155***
Dirac-Fock-KF*	MOREL-2‡	22 min.	155/155***

* With K-functional for correlation energy corrections.

** Data obtained using IBM RS/6000-580 workstation, assumed to be 1.33 times faster than IBM RS/6000-560.

*** Basis functions for the large components; two-electron integrals smaller than pre-assigned threshold are not written but nevertheless computed in the present MOREL code version.

† The timings for these computations (over one day on an IBM mainframe with large number of disks and of main memory) is fully dominated by the elapsed time needed to retrieve about 15 Gigabytes containing the integral files; a direct-SCF approach is clearly needed for this size computation, preferably coupled with use of parallel hardware.

‡ The estimated time for a MOREL version (MOREL-2) with the ability to avoid the computation of two-electron integrals smaller than a pre-assigned threshold is $5 \times 1.4^4 = 19.16$ min.

The data in Table 7 are also of interest because different methods and codes are used. The density functional stands out for its speed, but, most unfortunately, at least for the case studied, it turns out also to be the less reliable. The K-functional is also fast and appears to be more reliable than the density functional. We have not data for the J-functional for the same molecular system, since the molecular J-functional code is not ready, but there is no reason for thinking that the method works for atoms and not for molecules. The dramatic increase in computer time within the same approximation (HF or MP2) due to increase of the basis set size is well understood, but it provides a demonstration why very seldom limit computations are actually performed. The post-Hartree-Fock computations have all indicated that the methods are reliable but too expensive, prohibitively so for even marginally larger systems than NH_3 + HCl.

9. Conclusion

It appears that it is advisable to concentrate on effective potentials, and with this paper I hope to have demonstrated that there are many unexplored possibilities, within the vast field of effective potential approximations. Distances of interest at the electronic cusp, equilibrium internuclear distances, short range exchange forces, longer range coulomb interactions, dipole-dipole interactions (and wan der Waals forces) seems to call for an effective potential formalism far more sophisticated than present days density functionals. Twenty years were needed for the chemistry community at large to discover the literature on density functionals; I hope and wish that much less will be needed to produce effective potentials both computationally efficient and chemically reliable. I am very confident, since from one hand from the very start quantum chemistry did enter the semi-empirical avenue (with effective hamiltonian approximations at times rather extreme, but essentially adequate to the goal) and from the other hand quantum chemists have developed a most impressive ability in dealing with very complex computational problems; these two lines of effort, kept less apart, are all we need.

References

1. E. Clementi, "Introduction to MOTECC-89", MOTECC-89, E. Clementi, Ed., ESCOM Press, Leiden, 1989.
2. E. Clementi, "Supercomputers and Supercomputing in Chemical Research and Development", Theoretical Biochemistry & Molecular Biophysics, D.L.Beveridge & R. Lavery, Eds., Adenine Press, Albany, N.Y., 1990.
3. E. Clementi, Int. J. Quantum Chem.**42**, 547 (1992).
4. See for example J. P. Wright, "Solid State and Molecular Theory Group", M.I.T., Technical Report (1963).
5. E. Clementi and R. D. Davies, J. Comp. Phys. **1**, 223 (1966).
6. S. F. Boys and G. B. Cook, Rev. Mod. Phys. **32**, 285 (1960).
7. B. Ransil, Rev. Mod. Phys. **32**, 239 (1960).
8. E. Clementi, Ed., METECC-94, Volumes A, B, and C, STEF Press, Cagliari, 1993.
9. E. Clementi, Ed., MOTECC-89, ESCOM Press, Leiden, 1989.
10. E. Clementi, Ed., MOTECC-90, ESCOM Press, Leiden, 1990.
11. E. Clementi, Ed., MOTECC-91, ESCOM Press, Leiden, 1991.
12. E. Clementi and G. Corongiu, Eds., METECC-95, STEF Press, Cagliari, 1995
13. International Workshop on "Electronic Structures Methods for Truly Large Systems: Moving the Frontiers in Quantum Chemistry", August 1994 in Braunlage, Germany. The proceedings of the meeting will be collected in a special volume of the International Journal of Quantum Chemistry (1995).
14. E. Clementi, Proc. Natl. Acad. Sci. USA **69**, 2942, (1972).
15. E. Clementi and G. Corongiu, see Table III in Chapter 6 in MOTECC-89, E. Clementi, Ed., ESCOM Press, Leiden, 1989.
16. For early review papers on CH-HF applications, see E. Clementi. BM J. Res. Dev. **9**, 1, (1965); and E. Clementi, Chem. Rev. **68**, 34, (1968).

17. S. Chakravorty and E. Clementi, Phys. Rev. A **39**, 2290, (1989).
18. E. Clementi and D. W. M. Hoffmann, Chapter 1, Volume B, METECC-94, E. Clementi, Ed., STEF Press, Cagliari, 1993.
19. E. Clementi and D. W. M. Hoffmann, Int. J. Quantum Chem. **52**, 849, (1994); see also Ref. 49.
20. G. Bolis, M. Ragazzi, D. Salvaderi, D. Ferro and E. Clementi, Int. J. Quantum Chem. **14**, 815, (1978).
21. M. Parrinello, "First Principle Molecular Dynamics", Chapter 14, MOTECC-90, E. Clementi, Ed., ESCOM Press, Leiden, 1990.
22. D. A. Estrin, G. Corongiu, and E. Clementi, Chapter 12, Vol B, METECC-94, E. Clementi, Ed., STEF Press, Cagliari, 1993.
23. M. Wojcik, G. Corongiu, J. Detrich, M. N. Monsour, and E. Clementi, *Supercomputer Simulations in Chemistry*, Lecture Notes in Chemistry,Vol. **44**, M. Dupuis, Ed., Springer, Berlin, 1986.
24. D.C.Rapoport and E. Clementi, Phys. Rev. Letters **37**, 695, (1986).
25. L. Hannon, G. C. Lie and E. Clementi, J. Stat. Phys. **51**, 965, (1988).
26. G. H. Givens and E. Clementi, J. Chem. Phys. **90**, 7376, (1989).
27. Note: Results on the atoms from He to Ar were first reported by E.Clementi a) at the Symposium Molecular Structure andSpectra, Columbus, Ohio, June 1963; b) at the Harvard-MIT Colloquium in Physical Chemistry, October 1963; c) Theoretical Chemistry, Gordon Conference 1964; d) E. Clementi, IBM Journal Res. and Dev **9**, 1 (1965); e) E.Clementi, Chem. Rev. **68**, 34 (1968); f) E. Clementi, Int. J. Quantum Chem. **3**, 179 (1969).
28. D.R. Hartree, Proc. Cambridge Phil. Soc. **24**, 89 (1928).
29. V. Fock, Z. Physik. **61**, 126 (1930).
30. E.A. Hylleraas, Z. Physik **54**, 347 (1929).
31. L.H. Thomas, Proc. Camb. Phil. Soc. **23**, 542 (1927).
32. E. Fermi, Rend. Accad. Lincei **6**, 602 (1927).
33. P.A.M. Dirac,*The Principles of Quantum Mechanics*, Oxford Press, 1930.
34. D.R. Hartree, W.Hartree, and B.Swirles, Phil. Trans.Roy. Soc. A**238**, 229 (1939).
35. W. Heitler and F. London, Z. Physik. **44**, 455 (1927).
36. C.C.J. Roothaan, Rev. Mod. Phys. **32**, 179 (1960); *ibid*., **44**, 3050 (1965).
37. E. Clementi and A. Veillard, J. Chem. Phys. **44**, 3050 (1965).
38. G.C. Lie and E. Clementi, J. Chem. Phys..**60**, 1275 (1974).
39. G.C. Lie and E. Clementi, J. Chem. Phys. **60**, 1288 (1974).
40. J. S. Sims and S. Hagstrom, Phys. Rev. A .**4**, 908 (1971); J. Chem. Phys. **55**, 4699 (1971).
41. E. Wigner, Phys. Rev. **46**, 1002 (1934).
42. E. Clementi, J. Chem. Phys. **39**, 1 (1963).
43. C. Hollister and O. Sinanoglu, J. Am. Chem. Soc. **88**, 13, (1966).
44. E. Clementi and S. Chakravorty, J. Chem. Phys. **93**, 2591 (1990)
45. E. Clementi and S. J. Chakravorty, *A Comparative Study of Density Functional Models to Estimate Molecular Atomization Energies*, IBM Research Report, KGN-215, 1990.
46. Preliminary results have been reported in Ref. 19.
47. W. Kohn and L. J. Sham, Phys. Rev. **136**, A1133 (1965); W. Kohn and L. J. Sham, Phys. Rev. A . **40**, 1133 (1965).
48. J. C. Slater, Phys. Rev. **81**, 3835 (1965), and references therein given.
49. L. Pisani, L. De Windt, and E. Clementi, Int. J. Quantum Chem. (in press).
50. Note: The use of a dielectric constant in a semiempirical π electron integrals approximation has been pioneered by Prof. Joulg and his group at the University of Marseille in the early sixties.
51. H. A. Bethe and J. Goldstone, Proc. Roy. Soc. A **238**, 551, (1957); R. K. Nesbet, Advances in Chem. Phys., Vol. **9**, 321, (1965).
52. E. Clementi, J. Chem. Phys. **38**, 2248, (1963); *ibid.* **39**, 175, (1963); E. Clementi and A. Veillard, *ibid.* **44**, 3050, (1966).
53. A. Rizzo, E. Clementi, and M. Sekiya, Chem. Phys. Lett. **177**, 477, (1991).
54. E. R. Davidson, S. A. Hagstrom, S. J. Chakravorty, V. M. Umar, C. F. Fisher, Phys. Rev. A **44**, 7071, (1991); S. J. Chakravorty, S. R. Gwaltney, E. R. Davidson, F. A. Parpia, C. F. Fischer, Phys. Rev. A **47**, 3649, (1993).
55. R. McWeeny, *Methods of Molecular Quantum Mechanics*, Academic Press, London, 1989.
56. P. Pyykko, Chem. Rev. **88**, 563 (1988).

57. *The effect of relativity in atoms, molecules and the solid state*. I.P. Grant, B. Gyorffy, S. Wilson, Eds., Plenum Press, New York, 1991.
58. The volume in honour of Prof. W. C. Nieuwport, New Challenges in Computational Quantum Chemistry, R. Broer, P. J. C. Aerts and P. S. Bagus Eds., Drukkerij Van Denderen B. V., Groningen, 1994, is presently an excellent source on modem research in relativistic quantum mechanics for atoms and molecules.
59. K. Balasubramanian, K.S. Pitzer Adv. Chem. Phys. **67**, 287 (1987).
60. Y.S. Lee and A.D. McLean, J. Chem. Phys. **76**, 735 (1982).
61. K.G. Dyall, P.R. Taylor, K. Faegri, Jr., and H. Partridge, J. Chem. Phys. **95**, 2583 (1991).
62. O. Visser, L. Visscher, P.J.C. Aerts and W.C. Nieuwpoort, J. Chem. Phys. **96**, 2910 (1992).
63. O. Matsuoka, J. Chem. Phys. **97**, 2272 (1992).
64. P. Grant and H. M. Quiney, Adv. in Atom. and Mol. Phys. **23**, 37 (1988).
65. Y. K. Kim, Phys. Rev. **154**, 17 (1967); *ibid.* **159**, 190 (1967).
66. T. Kagawa, Phys. Rev. A. **12**, 2245 (1975); *ibid.* **22**, 2340 (1980).
67. G. Malli and J. Oreg, J. Chem. Phys. **68**, 529 (1979).
68. O. Matsuoka, N. Suzuki, T. Ayoma, G. Malli, J. Chem. Phys. **73**, 1320 (1980)
69. T. Ayoma, H. Yamakawa, and O. Matsuoka, J. Chem. Phys. **73**, 1328 (1980).
70. R.E. Stanton and S. Havriliak, J. Chem. Phys. **8**, 1910 (1984).
71. Y. Ishikawa, R. C. Binning and K. M. Sando, Chem. Phys. Lett. **101**, 111 (1983).
72. Y. Ishikawa, R. C. Binning and K. M. Sando, Chem. Phys. Lett. **105**, 189 (1984); *ibid.* **117**, 444 (1985).
73. K. G. Dyall, I. P. Grant and S. Wilson, J. Phys. B**17**, 493 (1984); *ibid*. B**17**, L45 (1984); *ibid*. B**17**, 1201 (1984).
74. L. Pisani and E. Clementi, Chapter 8 in METECC-94, Vol A, E. Clementi, Ed., STEF Press, Cagliari, 1993; L. Pisani and E. Clementi, J. Comp. Chem. **15**, 466, (1994).
75. L. Pisani and E. Clementi, J. Chem. Phys. **101**, 3079, (1994).
76. L. Pisani and E. Clementi, J. Chem. Phys., in press.
77. P. Goldfinger and G. Verhaegen, J. Chem. Phys. **50**, 1467 (1969).
78. E. Clementi, J. Chem. Phys. **46**, 3851 (1967)
79. G. Corongiu, D. A. Estrin, G. Murgia, L. Paglieri, L. Pisani, J. Watts, and E. Clementi, Int. J. Quantum Chem. (in press).

HYPERSPHERICAL COORDINATES IN REACTIVE SCATTERING THEORY

ARON KUPPERMANN
Arthur Amos Noyes Laboratory of Chemical Physics
*Division of Chemistry and Chemical Engineering**
California Institute of Technology
Pasadena, CA 91125, USA

1. Introduction

In describing the asymptotic behavior of the solutions of the time-independent nuclear motion Schrödinger equation (SE) associated with reactive scattering processes occurring on a single electronically-adiabatic potential energy surface PES[1], different sets of Jacobi vector coordinates are used in different arrangement channel regions. Let us consider for example, a triatomic system whose nuclei are P_1, P_2 and P_3. In fig. 1 we display three such sets $\mathbf{r}'^{(1)}_\lambda$, $\mathbf{r}'^{(2)}_\lambda$ $(\lambda = 1, 2, 3)$. In each

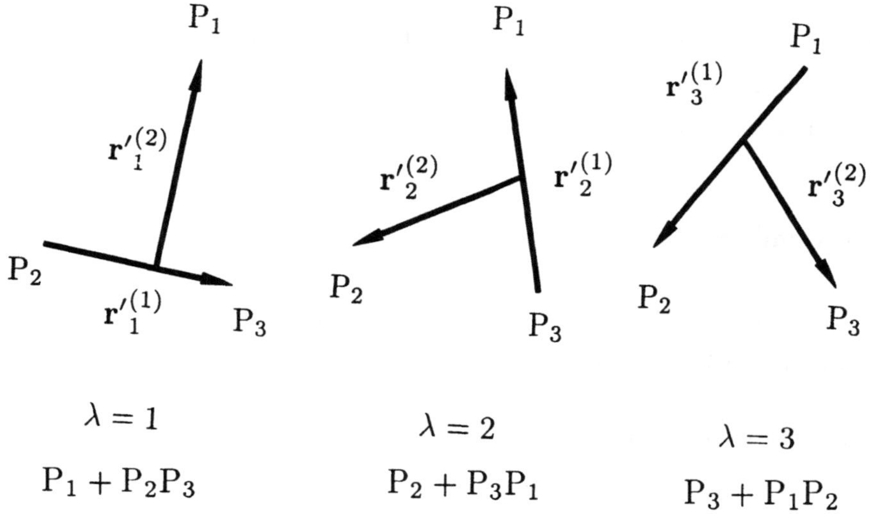

Figure 1. Jacobi coordinates for a three-particle system.

* Contribution number 9157

C. A. Tsipis et al. (eds.), New Methods in Quantum Theory, 501–532.

of these sets the vector $\mathbf{r}'^{(1)}_\lambda$ connects two of the nuclei and $\mathbf{r}'^{(2)}_\lambda$ connects their center-of-mass (CM) to the third nucleus. The $\lambda = 1$ vectors are convenient for decribing configurations of the system, designated $P_1 + P_2P_3$, in which P_2 and P_3 are the nuclei of a diatomic molecule; the third atom, with nucleus P_1, is distant from that molecule. Similarly, the $\lambda = 2$ vectors are convenient for describing $P_2 + P_3P_1$ configurations and the $\lambda = 3$ ones for $P_3 + P_1P_2$. One may partition the configuration space of the system into different arrangement channel regions λ and use in each such region basis sets in the corresponding coordinates $\mathbf{r}'^{(1)}_\lambda$, $\mathbf{r}'^{(2)}_\lambda$ to solve that SE. In this approach, the switch-over between the three pairs of such coordinates, the so-called matching problem[2], frequently leads to convergence problems. Alternatively, if basis sets in the union of these 3 sets of coordinates are used to solve the SE, problems of supercompleteness arise. For these reasons it is highly desirable to adopt a single set of coordinates in terms of which basis functions can be defined which are not supercomplete and which are symmetrized, *i.e.*, describe all regions of configuration space equally well, with equivalent regions described equivalently. For triatomic systems, a class of symmetrized hyperspherical coordinates satisfying these criteria has been described some time ago[3-8] and applied extensively to the calculation of state-to-state integral and differential cross sections in such systems. These coordinates will be reviewed in this paper and the results of their application to the $H + H_2$ reaction and its isotopomers will be summarized. Very recently, two sets of useful hyperspherical coordinates for tetraatomic systems were also developed[9]. One of them is symmetrized whereas in the other the system's nuclear motion hamiltonian takes on a particularly useful form. These coordinates will be described and their properties discussed.

2. Triatomic Systems

2.1 JACOBI MATRICES AND THEIR KINEMATIC ROTATION TRANSFORMATION PROPERTIES

Let us define the set of mass-scaled Jacobi vectors $\mathbf{r}'^{(1)}_\lambda$, $\mathbf{r}'^{(2)}_\lambda$ by the relations[10,11]

$$\mathbf{r}_\lambda{}^{(1)} = \left(\frac{\mu_\lambda^{(1)}}{\mu}\right)^{1/2}\mathbf{r}'_\lambda{}^{(1)}$$
$$\mathbf{r}_\lambda{}^{(2)} = \left(\frac{\mu_\lambda^{(2)}}{\mu}\right)^{1/2}\mathbf{r}'_\lambda{}^{(2)} \tag{1}$$

where

$$\mu_\lambda^{(1)} = \frac{m_{\lambda'}m_{\lambda''}}{m_{\lambda'}+m_{\lambda''}} \qquad \mu_\lambda^{(2)} = \frac{m_\lambda(m_{\lambda'}+m_{\lambda''})}{M}$$
$$\mu = \left(\frac{m_1m_2m_3}{M}\right)^{1/2} \qquad M = m_1+m_2+m_3 \tag{2}$$

In these expressions, $\lambda, \lambda', \lambda''$ is a permutation of 1, 2, 3 and m_i is the mass of nucleus $P_i (i = 1, 2, 3)$.

We now define the 3×2 Jacobi matrix ρ_λ^{sf} by [9]

$$\rho_\lambda^{sf} = (\mathbf{r}_\lambda{}^{(2)} \ \ \mathbf{r}_\lambda{}^{(1)}) = \begin{pmatrix} x_\lambda^{(2)} & x_\lambda^{(1)} \\ y_\lambda^{(2)} & y_\lambda^{(1)} \\ z_\lambda^{(2)} & z_\lambda^{(1)} \end{pmatrix} \tag{3}$$

where $x_\lambda^{(j)}, y_\lambda^{(j)}, z_\lambda^{(j)}$ $(j = 1, 2)$ are the components of $\mathbf{r}_\lambda{}^{(j)}$ in the cartesian frame Gxyz whose origin G is the CM of the nuclei and whose axes have space-fixed directions. A very useful property of these Jacobi matrices is that, under a $\lambda \to \nu$ change of arrangement channel coordinates they, transform as

$$\rho_\lambda^{sf} = \rho_\lambda^{sf}\mathbf{N}_{\lambda\nu} \tag{4}$$

where $\mathbf{N}_{\lambda\nu}$ is a 2×2 orthogonal matrix whose elements depend only on the masses of the nuclei. It is given by

$$\mathbf{N}_{\lambda\nu} = \begin{pmatrix} \cos\alpha_{\lambda\nu} & \sin\alpha_{\lambda\nu} \\ -\sin\alpha_{\lambda\nu} & \cos\alpha_{\lambda\nu} \end{pmatrix} \tag{5}$$

The angle $\alpha_{\lambda\nu}$ is in the $\pi/2$ to π range for $\nu = \lambda'$ and in the π to $3\pi/2$ range for $\nu = \lambda''$. The explicit expressions for this angle are [9,12]:

$$\sin\alpha_{\lambda\nu} = \left[\frac{m_{\lambda''}M}{(m_\lambda + m_{\lambda''})(m_{\lambda'}+m_{\lambda''})}\right]^{1/2}$$
$$\text{for} \quad \nu = \lambda' \tag{6}$$
$$\cos\alpha_{\lambda\nu} = -\left[\frac{m_\lambda m_{\lambda'}}{(m_\lambda + m_{\lambda''})(m_{\lambda'}+m_{\lambda''})}\right]^{1/2}$$

$$\sin\alpha_{\lambda\nu} = -[\frac{m_{\lambda'}M}{(m_\lambda + m_{\lambda'})(m_{\lambda'} + m_{\lambda''})}]^{1/2}$$

$$\text{for} \quad \nu = \lambda'' \qquad (7)$$

$$\cos\alpha_{\lambda\nu} = -[\frac{m_\lambda m_{\lambda''}}{(m_\lambda + m_{\lambda'})(m_{\lambda'} + m_{\lambda''})}]^{1/2}$$

Eqs. (1) through (4) can be easily generalized to N-atom systems, with the generalized $\mathrm{N}_{\lambda\nu}$ being an orthogonal square matrix of dimensions $N - 1$ whose elements still depend only on the masses of the nuclei[9]. As a result of this property, we call the $\lambda \rightarrow \nu$ mass-scaled Jacobi coordinate transformation a kinematic rotation[9,13,14]. The orthogonal nature of $\mathrm{N}_{\lambda\nu}$ is what makes the mass scaling of (1) such useful procedure. Two of its important consequences are[9]:

a. The hyperradius ρ, defined by

$$\rho = [Tr(\rho_\lambda^{sf}\ \tilde{\rho}_\lambda^{sf})]^{1/2} = [r_\lambda^{(2)^2} + r_\lambda^{(1)^2}]^{1/2} \tag{8}$$

is invariant under kinematic rotations, *i.e.*, is λ-independent.

b. The three-dimensional vector X_λ, defined by

$$\mathrm{X}_\lambda = \frac{1}{\rho}\begin{pmatrix} 2\ \mathbf{r}_\lambda{}^{(2)} \cdot \mathbf{r}_\lambda{}^{(1)} \\ 2 \mid \mathbf{r}_\lambda{}^{(2)} \times \mathbf{r}_\lambda{}^{(1)} \mid \\ r_\lambda^{(2)^2} - r_\lambda^{(1)^2} \end{pmatrix}, \tag{9}$$

where $r_\lambda^{(j)}$ is the magnitude of $\mathbf{r}_\lambda{}^{(j)}$ $(j = 1, 2)$, transforms under $\lambda \rightarrow \nu$ kinematic rotations according to

$$\mathrm{X}_\nu = \mathrm{M}_{\lambda\nu}\mathrm{X}_\lambda \tag{10}$$

with $\mathrm{M}_{\lambda\nu}$ defined by

$$\mathrm{M}_{\lambda\nu} = \begin{pmatrix} \cos2\alpha_{\lambda\nu} & 0 & \sin2\alpha_{\lambda\nu} \\ 0 & 1 & 0 \\ -\sin2\alpha_{\lambda\nu} & 0 & \cos2\alpha_{\lambda\nu} \end{pmatrix} \tag{11}$$

Let us now defined a set of internal configuration space mathematical axes $OX_\lambda YZ_\lambda$ in which the position vector of a general point P is $\mathbf{X}_\lambda$:

$$\begin{aligned} X_\lambda^{(P)} &= \frac{1}{\rho}\ 2\,\mathbf{r}_\lambda{}^{(2)} \cdot \mathbf{r}_\lambda{}^{(1)} \\ Y^{(P)} &= \frac{1}{\rho}\ 2 \mid \mathbf{r}_\lambda{}^{(2)} \times \mathbf{r}_\lambda{}^{(1)} \mid \geq 0 \\ Z_\lambda^{(P)} &= \frac{1}{\rho}\ (r_\lambda^{(2)^2} - r_\lambda^{(1)^2}) \end{aligned} \tag{12}$$

As a consequence of (10) and (11), OY and $Y^{(P)}$ are independent of λ. An important property of this space is that there exists a one-to-one correspondence between points P and nuclear configurations of the system[3,15]. In addition, a $\lambda \rightarrow \nu$ arrangement channel coordinate transformation is equivalent to a clockwise rotation of the $OX_\lambda YZ_\lambda$ frame around OY by the angle $2\alpha_{\lambda\nu}$.

2.2 JACOBI BODY-FIXED SYMMETRIZED HYPERSPHERICAL COORDINATES

Let, by definition, ρ, ω_λ and γ_λ be the polar coordinates of P in the internal configuration space $OX_\lambda YZ_\lambda$, where γ_λ is the angle in the 0 to π range between $\mathbf{r}_\lambda{}^{(2)}$ and $\mathbf{r}_\lambda{}^{(1)}$ and ω_λ is in that same range:

$$\begin{aligned} X_\lambda^{(P)} &= \rho \ \sin\omega_\lambda \ \cos\gamma_\lambda \\ Y^{(P)} &= \rho \ \sin\omega_\lambda \ \sin\gamma_\lambda \geq 0 \\ Z_\lambda^{(P)} &= \rho \ \cos\omega_\lambda \end{aligned} \tag{13}$$

The relations between $r_\lambda^{(1)}$ and $r_\lambda^{(2)}$ and the hyperspherical angle ω_λ are:

$$\begin{aligned} r_\lambda^{(1)} &= \rho \ \sin(\omega_\lambda/2) \\ r_\lambda^{(2)} &= \rho \ \cos(\omega_\lambda/2) \end{aligned} \tag{14}$$

It should be noted that the division of ω_λ by 2 in these expressions is essential for the one-to-one correspondence between $\rho, \omega_\lambda, \gamma_\lambda$ and nuclear configuration of the system to exist[3,15].

We now consider equipotentials of the electronically-adiabatic PES $V(\rho, \omega_\lambda, \gamma_\lambda)$ in the $OX_\lambda YZ_\lambda$ space. Under a $\lambda \rightarrow \nu$ coordinate transformation, these equipotentials are rotated clockwise by $2\alpha_{\lambda\nu}$ around OY without distortion and, as a result, display the permutation symmetry properties of the system of nuclei. By this we mean that if two of the 3 nuclei (which we will call B) are identical and the third one (A) is distinct, those equipotential maps have the symmetry of the S_2 permutation group of two objects, and for λ corresponding to $A + B_2$, OYZ_λ is a plane of symmetry. Simmilarly, if the 3 nuclei are identical, these maps have the symmetry of the permutation group of three objects, which is isomorphic with the C_{3v} point group. These properties of $V(\rho, \omega_\lambda, \gamma_\lambda)$ in the $OX_\lambda YZ_\lambda$ space justify calling $\rho, \omega_\lambda, \gamma_\lambda$ the internal λ-arrangement-channel symmetrized hyperspherical coordinates and OX_λ, OY, OZ_λ the corresponding (non-physical) axes. It should be kept in mind that this space is only the $Y \geq 0$ half-space of the full $OX_\lambda YZ_\lambda$ cartesian space. In fig. 2 we depict schematically an equipotential of $V(\rho, \omega_\lambda, \gamma_\lambda)$

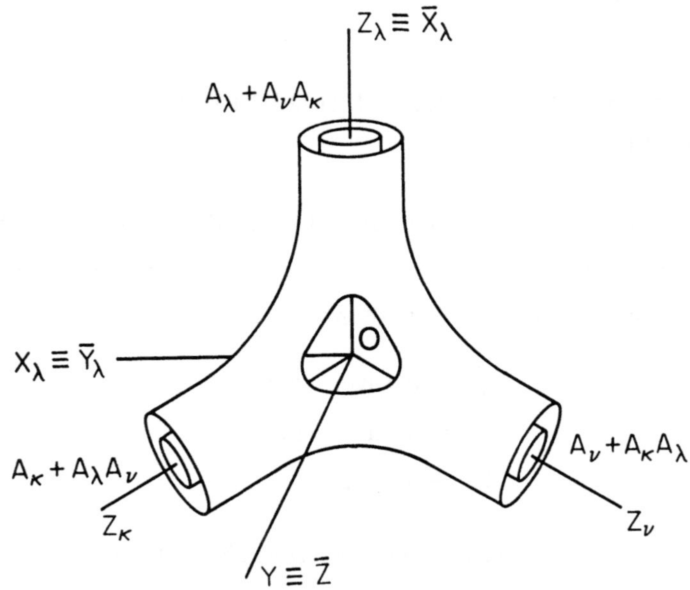

Figure 2. Equipotential energy surface for a system of three nuclei A_λ, A_ν and A_κ in symmetrized internal coordinates. The $O\bar{Y}_\lambda \equiv OX_\lambda$, $O\bar{Z} \equiv OY$ and $O\bar{X}_\lambda \equiv OZ_\lambda$ define two related non-physical internal nuclear coordinate frames. A point P has spherical polar coordinates $\rho, \omega_\lambda, \gamma_\lambda$ in the $OX_\lambda YZ_\lambda$ frame and $\rho, \vartheta, \varphi_\lambda$ in the $O\bar{X}_\lambda\bar{Y}_\lambda\bar{Z}$ frame.

including its $Y < 0$ mirror image for simplicity of display. The three nuclei are labelled $A_\lambda, A_\nu, A_\kappa$ rather than $P_\lambda, P_\nu, P_\kappa$, with λ, ν, κ being a permutation of 1, 2, 3. The set of λ-arrangement symmetrized hyperspherical coordinates is most useful in regions of the internal configuration space for which A_λ is sufficiently far from $A_\nu A_\kappa$ for this arrangement not to interact with either the $A_\nu + A_\lambda A_\kappa$ or the $A_\kappa + A_\lambda A_\nu$ arrangement. For those regions it is convenient to define a Jacobi body-fixed λ frame $(bf\lambda)$ $Gx^{bf\lambda}y^{bf\lambda}z^{bf\lambda}$ obtained from the physical space-fixed frame $Gxyz$ by a rotation through Euler angles $\phi_\lambda, \theta_\lambda, \psi_\lambda$ where $\theta_\lambda, \phi_\lambda$ are the polar angles of $\mathbf{r}_\lambda{}^{(2)}$ in that frame. ψ_λ is the angle in the 0 to 2π range between the $\mathbf{r}_\lambda{}^{(2)}, Gz$ and $\mathbf{r}_\lambda{}^{(2)}, \mathbf{r}_\lambda{}^{(1)}$ half-planes, whose common edge is $\mathbf{r}_\lambda{}^{(2)}$, and measured counterclockwise from the first to the second of these half-planes as viewed from the tip of the $\mathbf{r}_\lambda{}^{(2)}$ vector. The $Gz^{bf\lambda}$ axis lies along $\mathbf{r}_\lambda{}^{(2)}$ and the $Gx^{bf\lambda}z^{bf\lambda}$ half-plane with positive $x^{bf\lambda}$ contains $\mathbf{r}_\lambda{}^{(1)}$. As ψ_λ varies, the triatomic plane, which contains $\mathbf{r}_\lambda{}^{(1)}$ and $\mathbf{r}_\lambda{}^{(2)}$, rotates (*i.e.*, tumbles) around $\mathbf{r}_\lambda{}^{(2)}$, and is called the tumbling angle[2].

The set of 6 coordiantes $\rho, \Xi_\lambda \equiv (\phi_\lambda, \theta_\lambda, \psi_\lambda, \omega_\lambda, \gamma_\lambda)$[14] are called the Jacobi body-fixed λ arrangement channel symmetrized hyperspherical coordinates. The system's nuclear kinetic energy operator in those coordinates(after removal of the motion of the center-of-mass) is given by

$$\widehat{T} = \widehat{T}_\rho(\rho) + \frac{\widehat{\Lambda}^2}{2\mu\rho^2} \tag{15}$$

where $\widehat{T}_\rho(\rho)$ is the hyper-radial kinetic energy operator

$$\widehat{T}_\rho(\rho) = -\frac{\hbar^2}{2\mu}\frac{1}{\rho^5}\frac{\partial}{\partial\rho}\rho^5\frac{\partial}{\partial\rho} \tag{16}$$

and $\widehat{\Lambda}^2$ is the grand-canonical angular momentum operator

$$\widehat{\Lambda}^2(\Xi) = \widehat{L}^2(\omega_\lambda) + \frac{\widehat{j}_\lambda^2}{\sin^2\omega_\lambda} + \frac{\widehat{l}_\lambda^2}{\cos^2\omega_\lambda} \tag{17}$$

In this expression, $\widehat{L}^2(\omega_\lambda)$ is the angular momentum operator associated with the hyperangle ω_λ

$$\widehat{L}^2(\omega_\lambda) = -4\hbar^2\frac{1}{\sin^2\omega_\lambda}\frac{\partial}{\partial\omega_\lambda}\sin^2\omega_\lambda\frac{\partial}{\partial\omega_\lambda} \tag{18}$$

and $\widehat{j}_\lambda^2$ and $\widehat{l}_\lambda^2$ are the usual space-fixed angular momentum operators associated with $\mathbf{r}_\lambda{}^{(1)}$ and $\mathbf{r}_\lambda{}^{(2)}$ respectively, expressed in terms of the angles $\phi_\lambda, \theta_\lambda, \gamma_\lambda, \psi_\lambda$. In order to perform a partial wave expansion, it is convenient to express $\widehat{\mathbf{l}}_\lambda$ as

$$\widehat{\mathbf{l}}_\lambda = \widehat{\mathbf{J}} - \widehat{\mathbf{j}}_\lambda \tag{19}$$

where $\widehat{J}$ is the total nuclear orbital angular momentum vector of the system.

The ρ, Ξ_λ Jacobi symmetrized hyperspherical coordinates have been used in reactive scattering calculations for H_3 in its ground adiabatic electronic state PES, ignoring the presence of the conical intersection with its first electronically excited state PES[16].

2.3 PRINCIPAL AXES OF INERTIA SYMMETRIZED HYPERSPHERICAL COORDINATES

We now define another pair of angles, $\vartheta, \varphi_\lambda$ in the internal nuclear configuration space $OX_\lambda YZ_\lambda$ as the spherical polar angles of P with respect to axes $O\bar{X}_\lambda\bar{Y}_\lambda\bar{Z}$ where $O\bar{X}_\lambda \equiv OZ_\lambda$, $O\bar{Y}_\lambda \equiv OX_\lambda$ and $O\bar{Z} \equiv O\bar{Y}$[8], as indicated in fig. 2. The relation between ϑ, φ_λ and $\omega_\lambda, \gamma_\lambda$ is

$$\begin{aligned} \sin\vartheta\cos\varphi_\lambda &= \cos\omega_\lambda \\ \sin\vartheta\sin\varphi_\lambda &= \sin\omega_\lambda\cos\gamma_\lambda \\ \cos\vartheta &= \sin\omega_\lambda\sin\gamma_\lambda \\ 0 \le \vartheta \le \frac{\pi}{2} &\quad 0 \le \varphi_\lambda < 2\pi \end{aligned} \tag{20}$$

In addition, we define another body-fixed frame $I\lambda$ whose axes $Gx^{I\lambda}y^Iz^{I\lambda}$ are the principal axes of inertia of the three nuclei, and whose Euler angles with respect to $Gxyz$ are $a_\lambda, b_\lambda, c_\lambda$. The senses of those axes are chosen so that the $I\lambda$ system has the same handedness as the space-fixed one and in a manner which results in a one-to-one correspondence between $\rho, \Phi_\lambda \equiv (a_\lambda, b_\lambda, c_\lambda, \vartheta, \varphi_\lambda)$ and the space-fixed cartesian coordinates or $\mathbf{r}_\lambda{}^{(1)}$ and $\mathbf{r}_\lambda{}^{(2)}$. Furthermore, the $I\lambda$ axes are labelled so that the principal moments of inertia lie in the order

$$I_z^\lambda \le I_x^\lambda \le I_y^\lambda \tag{21}$$

The $\mathbf{r}_\lambda^{(1)}$ and $\mathbf{r}_\lambda{}^{(2)}$ vectors lie in the $Gx^{I\lambda}z^{I\lambda}$ plane. Under a $\lambda \to \nu$ arrangement channel transformation φ_λ changes according to

$$\varphi_\nu = \varphi_\lambda + 2\alpha_{\lambda\nu} \qquad \text{mod} \quad 2\pi, \tag{22}$$

ϑ and Gy^I are unchanged and the $Gx^{I\lambda}$ and $Gz^{I\lambda}$ axes are either unchanged or both change senses. This means that the Euler angles may at most change by $\pm\pi$ modulo 2π.

The coordinates ρ, Φ_λ are called the principal axes of inertia symmetrized hyperspherical coordinates. The nuclear motion kinetic energy operator in these coordinates is still given by (15) but $\widehat{\Lambda}^2$ is now expressed as

$$\begin{aligned}\widehat{\Lambda}^2 = & \frac{2}{1+\sin\vartheta}\widehat{J}_x^{I\lambda^2} + \frac{1}{\sin^2\vartheta}\widehat{J}_y^{I\lambda^2} + \frac{2}{1-\sin\vartheta}\widehat{J}_z^{I\lambda^2} \\ & + \frac{4}{\sin^2\vartheta}\widehat{L}^2 - 4\frac{\cos\vartheta}{\sin^2\vartheta}\widehat{L}\widehat{J}_y^{I\lambda} + \widehat{K}^2 - 8i\hbar\cot\vartheta\widehat{K}\end{aligned} \tag{23}$$

where

$$\widehat{L} = \frac{\hbar}{i}\frac{\partial}{\partial\varphi_\lambda} \qquad \widehat{K} = \frac{\hbar}{i}\frac{\partial}{\partial\vartheta} \tag{24}$$

are hyper-angular momenta associated with φ_λ and ϑ respectively and $\widehat{J}^{I\lambda}$, $\widehat{J}_y^{I\lambda}$ and $\widehat{J}_z^{I\lambda}$ are the components of the system's total nuclear orbital angular momentum operator $\widehat{\mathbf{J}}$ along the *bf* $I\lambda$ axes. In view of (22), $\widehat{L}$ is λ-independent. Given the transformation properties of the $I\lambda$ axes, every term in (23) is λ-independent, as is $\widehat{T}_\rho(\rho)$. As a result, not only is the nuclear motion hamiltonian $\widehat{H}$ independent of λ, but the same is true for every one of its terms. This is a particularly useful property, specially in the strong interaction region of nulcear configuration space, in which all three nuclei are close together, and none of the three λ-arrangement-channel sets of $I\lambda$ hyperspherical coordinates is particularly better suited than any of the other two.

2.4 METHOD OF SOLUTION OF THE NUCLEAR MOTION SCHRÖDINGER EQUATION

We consider a triatomic system under conditions such that it is confined to move on a single electronically adiabatic potential energy surface V and for which the Born-Oppenheimer approximation, in which the first and second deirvative coupling elements are neglected, is valid. Let us in addition assume that in the corresponding nuclear motion hamiltonian $\widehat{H}$ the spin-containing terms can be neglected, *i.e.*, that

$$\widehat{H} = \widehat{T} + V \tag{25}$$

We wish to solve the nuclear motion SE

$$\widehat{H}\Psi = E\Psi \tag{26}$$

for the orbital nuclear wave function Ψ, subject to the appropriate scattering asymptotic conditions and satisfying the Pauli principle for the exchange of identical nuclei. To that effect we define an auxiliary nuclear wave function $\Psi^{JM\Pi\Gamma}$ which is required to be a simultaneous eigenfunction of $\widehat{H}$, of $\widehat{J}^2$, of its space fixed z component $\widehat{J}_z$, and of the inversion operator $\widehat{\mathcal{I}}$ of the nuclei through their center-of-mass G according to the equations

$$\begin{aligned} \widehat{H}\Psi^{JM\Pi\Gamma} &= E\Psi^{JM\Pi\Gamma} \\ \widehat{J}^2\Psi^{JM\Pi\Gamma} &= J(J+1)\hbar^2\Psi^{JM\Pi\Gamma} \\ \widehat{J}_z\Psi^{JM\Pi\Gamma} &= M\hbar\Psi^{JM\Pi\Gamma} \\ \widehat{\mathcal{I}} &= (-1)^{\Pi}\Psi^{JM\Pi\Gamma} \end{aligned} \tag{27}$$

In these expressions, J and M are respectively the quantum numbers associated with $\widehat{J}^2$ and $\hat{J}_z$ and $\Pi = 0,1$ is a parity quantum number which specifies whether $\Psi^{JM\Pi\Gamma}$ is symmetric or antisymmetric with respect to inversion through G. In addition, $\Psi^{JM\Pi\Gamma}$ is required to transform according to the irreducible representation Γ of the permutation group of identical nuclei of the system. Once the $\Psi^{JM\Pi\Gamma}$ are obtained, Ψ can be expressed as a linear combination of these functions, with coefficients chosen so as to enforce the desired asymptotic conditions. This linear combination is called a partial wave expansion. To obtain the cross sections of interest, we never have to obtain Ψ explicitly. Instead, associated to $\Psi^{JM\Pi\Gamma}$ we calculate a partial wave scattering matrix $\mathbf{S}^{J\Pi\Gamma}$ in terms of which the differential and integral cross sections for the system can be obtained directly by standard methods[16], with the effect of the Pauli principle regarding the exchange of identical nuclei included.

In order to solve (27) we use the $I\lambda$ hyperspherical coordinates ρ, Φ_λ with a fixed λ in the strong interaction region of configuration space and the three sets of $bf\lambda$ hyperspherical coordinates $\rho, \Xi_\lambda(\lambda = 1,2,3)$ in each of the 3 corresponding weak interaction regions. Let $\boldsymbol{\alpha}_\lambda$ stand for either $\boldsymbol{\Phi}_\lambda$ or $\boldsymbol{\Xi}_\lambda$. We now expand $\Psi^{JM\Pi\Gamma}(\rho, \boldsymbol{\alpha}_\lambda)$ in a complete orthonormal basis set $\Psi^{JM\Pi\Gamma}_{\mathbf{n}_\lambda\Omega_\lambda}(\boldsymbol{\alpha}_\lambda; \bar{\rho})$ of the five hyperangules $\boldsymbol{\alpha}_\lambda$ which depends parametrically on $\bar{\rho}$, and are called local hyperspherical surface functions (LHSF). These are required to be eigenfunctions of $\widehat{J}^2$, $\widehat{J}_z$ and $\widehat{\mathcal{I}}$ and are defined by

$$\Phi^{JM\Pi\Gamma}_{\mathbf{n}_\lambda\Omega_\lambda}(\boldsymbol{\alpha}_\lambda; \bar{\rho}) = D^{J\Pi}_{M\Omega_\lambda}(\boldsymbol{\alpha}^{(1)}_\lambda)\phi^{\Pi\Gamma}_{\mathbf{n}_\lambda\Omega_\lambda}(\boldsymbol{\alpha}^{(2)}_\lambda; \bar{\rho}) \tag{28}$$

where $\alpha_\lambda{}^{(1)}$ stands for the set of three Euler angles contained in α_λ and $\alpha_\lambda{}^{(2)}$ stands for the remaining two hyperangles, *i.e.*, $\vartheta, \varphi_\lambda$ for the $I\lambda$ coordinate and $\omega_\lambda, \gamma_\lambda$ for the $bf\lambda$ ones. The $D^{J\Pi}_{M\Omega_\lambda}(\alpha^{(1)}_\lambda)$ are orthonormalized Wigner rotation function of definite parity[16] and the $\phi^{\Pi\Gamma}_{\mathbf{n}_\lambda\Omega_\lambda}(\alpha^{(2)}_\lambda)$ are eigenfunctions of an appropriately defined two dimensional reference hamiltonian in $\alpha_\lambda{}^{(2)}$[6,8] containing $V(\bar{\rho}, \alpha_\lambda{}^{(2)})$ and depending parameterically on Ω_λ. The quantum number $\mathbf{n}_\lambda$ is also two-dimensional.

In the strong interaction region of configuration space, where we use the internal coordinates $\vartheta, \varphi_\lambda$, $\phi^{\Pi\Gamma}_{\mathbf{n}_\lambda\Omega_\lambda}(\vartheta, \varphi_\lambda; \bar{\rho})$ is determined variationally by expanding it in a direct product of two orthonormal basis sets, $\{f_n(\vartheta; \bar{\rho})\}$ and $\{g^{\Pi\Gamma}_{p_\lambda\Omega_\lambda}(\varphi_\lambda; \bar{\rho})\}$, where n and p_λ are two quantum numbers which span those sets and $\mathbf{n}_\lambda \equiv n, p_\lambda$. The f and g functions may be chosen to be $\bar{\rho}$-independent linear combination of simple trigonometric functions in the angles ϑ and φ_λ respectively[6,8] or may be chosen instead to be functions reflecting the properties of $V(\bar{\rho}, \vartheta, \varphi_\lambda)$ so as to minimize the number of terms needed for this expansion to converge. The continuity conditions which the $g^{\Pi\Gamma}_{p_\lambda\Omega_\lambda}(\varphi_\lambda; \bar{\rho})$ must satisfy, are discussed in section 2.5. In the weak interaction region of configuration space, where we use the $bf\lambda$ coordinates ρ, Ξ_λ, we expand the $\phi^{\Pi\Gamma}_{\mathbf{n}_\lambda\Omega_\lambda}(\omega_\lambda, \gamma_\lambda; \bar{\rho})$ in the direct product of the basis sets $P^{j_\lambda}_{\Omega_\lambda}(\cos\gamma_\lambda)$ of the associated Legendre functions of $\cos\gamma_\lambda$, and the functions $t^{\Pi\Gamma}_{v_\lambda j_\lambda\Omega_\lambda}(\omega_\lambda; \bar{\rho})$ determined by the numerical solution of a one-dimensional eigenfunction in ω_λ which depends on V[8,16]. In this case, $\mathbf{n}_\lambda \equiv (j_\lambda, v_\lambda)$ and the quantum numbers $v_\lambda, j_\lambda, \Omega_\lambda$ correlate asymptotically with the vibration, rotation and helicity magnetic quantum numbers of the isolated $A_\nu A_\kappa$ diatom along the final velocity vector of A_λ with respect to $A_\nu A_\kappa$[2].

The expansion of the wave function $\Psi^{JM\Pi\Gamma}(\rho, \alpha_\lambda)$ in the LHSFs is

$$\Psi^{JM\Pi\Gamma\mathbf{n}_\lambda\Omega_\lambda}(\rho, \alpha_\lambda) = \frac{1}{\rho^{5/2}}\Sigma_{\mathbf{n}_\lambda,\Omega_\lambda} b^{J\Pi\Gamma\mathbf{n}'_\lambda\Omega'_\lambda}_{\mathbf{n}_\lambda\Omega_\lambda}(\rho; \bar{\rho})\Phi^{JM\Pi\Gamma}_{\mathbf{n}_\lambda\Omega_\lambda}(\alpha_\lambda; \bar{\rho}) \tag{29}$$

The coefficients $b^{J\Pi\Gamma\mathbf{n}'_\lambda\Omega'_\lambda}_{\mathbf{n}_\lambda\Omega_\lambda}$, once calculated, determine the $\Psi^{JM\Pi\Gamma}$ and therefore the partial wave scattering matrix $\mathbf{S}^{J\Pi\Gamma}$. The supercripts $\mathbf{n}'_\lambda$ and Ω'_λ in (29) were added to indicate that there as many linearly-independent solutions of (27) as there are LHSFs. Replacement of (29) into the first of (27) results in a set of coupled second order ordinary differential equations (called the coupled-channel hyperradial equations) in the **b** coefficients, which is cast as a differential equation in a matrix $\mathbf{b}^{J\Pi\Gamma}$ whose rows and columns are labelled by $\mathbf{n}_\lambda\Omega_\lambda$ and $\mathbf{n}'_\lambda\Omega'_\lambda$ respectively:

$$-\frac{\hbar^2}{2\mu}\frac{d^2\mathbf{b}^{J\Pi\Gamma}(\rho;\bar{\rho})}{d\rho^2}+\mathbf{V}^{J\Pi\Gamma}(\rho;\bar{\rho})\mathbf{b}^{J\Pi\Gamma}(\rho;\bar{\rho})=E\mathbf{b}^{J\Pi\Gamma}(\rho;\bar{\rho}) \qquad (30)$$

In this equation, $\mathbf{V}^{J\Pi\Gamma}(\rho;\bar{\rho})$ is an interaction potential matrix that produces coupling in both $\mathbf{n}_\lambda$ and Ω_λ. The Ω_λ part of this coupling is different for the $\rho, \mathbf{\Phi}_\lambda$ and ρ, Ξ_λ variables. For the former it is penta-diagonal, with Ω_λ coupled to $\Omega_\lambda \pm 1$ and $\Omega_\lambda \pm 2$, while for the latter it is trigonal, with Ω_λ coupled to $\Omega_\lambda \pm 1$ only. These different degrees of Ω_λ-coupling do not affect the numerical effort needed to solve the corresponding coupled hyperradial equations.

The numerical procedure used in association with the $\rho, \boldsymbol{\alpha}_\lambda$ coordinates is as follows[8]. The internal configuration space is divided into a number of spherical hyperradial sectors. The $\phi^{\Pi\Gamma}_{\mathbf{n}_\lambda\Omega_\lambda}(\boldsymbol{\alpha}^{(2)}_\lambda;\bar{\rho})$ are then determined at the center $\bar{\rho}$ of each sector and used to obtain the coupling matrix $\mathbf{V}^{J\Pi\Gamma}(\rho;\bar{\rho})$ over the entire sector. Eq.(30) is transformed into the first order non-linear Bessel-Ricatti matrix differential equation

$$\frac{d\mathbf{f}^{J\Pi\Gamma}(\rho;\bar{\rho})}{d\rho}+[\mathbf{f}^{J\Pi\Gamma}(\rho;\bar{\rho})]^2+\frac{2\mu}{\hbar^2}[E\mathbf{I}-\mathbf{V}^{J\Pi\Gamma}(\rho;\bar{\rho})]=\mathbf{0} \qquad (31)$$

where $\mathbf{f}^{J\Pi\Gamma}(\rho;\bar{\rho})$ is the logarithmic derivative matrix $[(d/d\rho)\mathbf{b}^{J\Pi\Gamma}(\rho;\bar{\rho})][\mathbf{b}^{J\Pi\Gamma}(\rho;\bar{\rho})]^{-1}$ and $\mathbf{I}$ is the identity matrix. This equation is integrated from the beginning to the end of each sector using the efficient fourth-order logarithmic-derivative method[17-19], and for which the initial $\mathbf{f}^{J\Pi\Gamma}(\rho;\bar{\rho})$ is obtained from the last such matrix at the end of the previous sector. The first sector starts at a hyperradius ρ_0 sufficiently small for the WKB solution to be applicable, and the ρ, $\mathbf{\Phi}_\lambda$ coordinates are used initially. In this manner, the $\mathbf{f}^{J\Pi\Gamma}$ equation is integrated from ρ_0 to a value ρ_s for which the interaction between the three different arrangement channels λ has become negligible. At this point one switches to the variables $\rho, \omega_\lambda, \gamma_\lambda$ and integrates that equation for each λ separately to a value ρ_a at which the interactions within the arrangement channels (associated with inelastic and elastic transitions) have also become negligible. It is also necessary to switch the $I\lambda$ principal axis of inertia frame to the Jacobi $bf\lambda$ frame, *i.e.*, from Euler angles $a_\lambda, b_\lambda, c_\lambda$ to Euler angles $\phi_l, \theta_\lambda, \psi_\lambda$. This can be done either at $\rho = \rho_s$ or $\rho = \rho_a$. At $\rho = \rho_a$ the $\Psi^{JM\Pi\Gamma}$ are matched to asymptotic atom-diatom wave functions expressed in the usual mass-scaled Jacobi coordinates $r_\lambda^{(1)}, \gamma_\lambda, \psi_\lambda, r_\lambda^{(2)}, \theta_\lambda, \phi_\lambda$. From this asymptotic analysis one obtaines the reactance matrix $\mathbf{R}^{J\Pi\Gamma}$ and from it the scattering matrix $\mathbf{S}^{J\Pi\Gamma}$. This is done for all Γ and both parities $\Pi = 0$ and 1, and a sufficient number of partial waves (*i.e.* values of J) for the resulting cross sections of interest to have converged to the desired degree of accuracy.

2.5 THE GEOMETRIC PHASE EFFECT

For systems displaying conical intersections between a pair of electronically adiabatic PESs, special consideration must be given to continuity conditions. When such systems are transported adiabatically around a closed loop in internal nuclear configuration space surrounding the conical intersection, the corresponding pair of adiabatic electronic wave functions, if required to be real and as continuous as possible, undergo a discontinuous change in sign[20,21]. In order that the system's electronuclear wave function be single-valued, as required by the postulates of quantum mechanics, the corresponding Born-Oppenheimer nuclear wave-functions must undergo a compensating discontinuous change in sign[7,8,22], called the geometric phase (GP). This change of sign must be included for such conically-intersecting systems in the calculation of the $\Psi^{JM\Pi\Gamma}$ and of the corresponding $\mathbf{S}^{J\Pi\Gamma}$. It can be accomplished as follows.

Let us consider a $\bar{Z} \geq 0$ hemisphere of radius $\bar{\rho}$ in the $\rho, \vartheta, \varphi_\lambda$ space of fig. 2 associated with a triatomic system in an electronically adiabatic state whose PES displays a conical intersection with a neighboring PES. Let the locus L of the conical intersection geometries intersect that hemisphere at a point $\bar{P}$. In order for the change in sign in $\Psi^{JM\Pi\Gamma}$ associated with the conical intersection to take place it suffices that for every $\bar{\rho}$ the products of basis functions $f_n^{\Omega_\lambda}(\vartheta; \bar{\rho}) g_{p_\lambda \Omega_\lambda}^{\Pi\Gamma}(\varphi_\lambda; \bar{\rho})$ should display such a change in sign. This means that these products should changes sign when a general point $P(\bar{\rho}, \vartheta, \varphi_\lambda)$ on this hemisphere executes a closed loop around $\bar{P}$. To indicate how this can be accomplished, let us consider first the H_3 system, which displays a conical intersection between its ground and first excited electronically adiabatic PESs for equilateral triangular configuration[23]. The locus L is for this case the $GY \equiv G\bar{Z}$ axis of that space, for which $\vartheta = 0$. For $\rho = \bar{\rho}$, and arbitary ϑ we can define a loop around $\bar{P}$ to be the circle of radius $\bar{\rho}\cos\vartheta$ on that hemisphere. As a point P traverses this loop, φ_λ varies from 0 to 2π. In order that the conical intersection boundary condition be satisfied, it suffices to impose on the basis functions $g^{\Pi\Gamma}_{p_\lambda\Omega_\lambda}(\varphi_\lambda)$ the geometric phase (GP) condition

$$g^{\Pi\Gamma}_{p_\lambda\Omega_\lambda}(\varphi_\lambda + 2\pi; \bar{\rho}) = -g^{\Pi\Gamma}_{p_\lambda\Omega_\lambda}(\varphi_\lambda; \bar{\rho}) \tag{32}$$

In the absence of such an intersection, the no-geometric-phase (NGP) continuity condition associated with the single-valuedness of $\Psi^{JM\Pi\Gamma}$ is instead

$$g^{\Pi\Gamma}_{p_\lambda\Omega_\lambda}(\varphi_\lambda + 2\pi; \bar{\rho}) = g^{\Pi\Gamma}_{p_\lambda\Omega_\lambda}(\varphi_\lambda; \bar{\rho}) \tag{33}$$

The formalism and methodologies for these two cases are otherwise exactly the same. As a result, using the symmetrized hyperspherical coordinates $\rho, \vartheta, \varphi_\lambda$, inclusion of the GP effect in scattering calculations for H_3 is extremely simple, and is accomplished by imposing a simple change of sign condition on the basis set in only one of the 6 coordinates which describe the nuclear motion of the system. This entails no increase in computational effort.

Let us now consider the isotopomers of H_3 of the AB_2 type, such as DH_2 and HD_2. The conical intersection line L is for these cases still a straight line, but no longer coincides with OY. It lies instead in the symmetry plane OYZ_λ of fig. 2 for $A_\nu \equiv A_\kappa$ and makes an angle ϑ_L with OY in the 0 to $\pi/2$ range, given by

$$\cos\vartheta_L = \frac{[3(1 + 2m_B/m_A)]^{1/2}}{2 + m_B/m_A} \tag{34}$$

For $m_A > m_B$, L tilts towards the positive OZ axis, whereas for $m_A < m_B$ it tilts away from it. For DH_2, ϑ_L is 11.5° and the tilt is towards OZ, whereas for HD_2 it is 14.5° and the tilt is away from OZ. A loop around OY with constant $\bar{\rho}$ and ϑ encloses the conical intersection point $\bar{P}$ if $\vartheta > \vartheta_L$ but not otherwise. At low energies, $\Psi^{JM\Pi\Gamma}$ does not sample regions for which ϑ is low, because $V(\bar{\rho}, \vartheta, \varphi_\lambda)$ is high for such regions. For example, for the DH_2 system and total energies $E \leq 1.8$ eV $\Psi^{JM\Pi\Gamma}$ only samples regions for which $\rho \geq 2$ bohr[24]. For such values of $\rho, \vartheta_L(\rho, \vartheta = 11.5°, \varphi_\lambda) > 3$ eV. As a results, for those E, regions of internal configuration space for which $\vartheta < \vartheta_L$ are highly classically forbidden, and $\Psi^{JM\Pi\Gamma}$ is essentially zero inside them. Under these conditions, loops around OY through regions for which $\Psi^{JM\Pi\Gamma}$ is not negligible always enclose conical intersection configurations and must change sign, just as for H_3. Similar considerations are valid for HD_2 for $E \leq 1.6$ eV[25]. We conclude that the use of basis functions

$g^{\Pi\Gamma}_{p_\lambda \Omega_\lambda}(\varphi_\lambda;\bar{\rho})$ which satisfy (32) consititutes an accurate way of including the effect of the geometric phase for DH_2 and HD_2 for $E \leq 1.8$ eV and $E \leq 1.6$ eV respectively.

For substantially higher energies, or for less symmetric systems, this methodology must be modified appropriately. One approach is to replace the hyperspherical coordinates $\vartheta, \varphi_\lambda$ by new variables $\bar{\vartheta}, \bar{\varphi}_\lambda$ orthogonal to and having the same ranges of definition as the old ones. We impose the condtion that loops generated by varying $\bar{\varphi}_\lambda$ over a zero to 2π range while keeping $\bar{\rho}$ and $\bar{\vartheta}$ constant always enclose the $\bar{P}$ conical intersection point, independently of the values of $\bar{\rho}$ and $\bar{\vartheta}$. It is always possible to obtain such $\bar{\vartheta}, \bar{\varphi}_\lambda$. To enforce the conical intersection boundary conditions we use a basis set $\bar{g}^{\Pi\Gamma}_{p_\lambda \Omega_\lambda}(\bar{\varphi}_\lambda;\bar{\rho})$ which satisfied (32) with φ_λ replaced by $\bar{\varphi}_\lambda$. Similarly, we replace the basis set $f^{\Omega_\lambda}_n(\vartheta;\bar{\rho})$ by another one $\bar{f}^{\Omega_\lambda}_n(\bar{\vartheta};\bar{\rho})$. The old variables $\vartheta, \varphi_\lambda$ are given functions of the new ones $\bar{\vartheta}, \bar{\varphi}_\lambda$. We need to evaluate matrix elements of differential operators in $\vartheta, \varphi_\lambda$ in the direct product of the new basis sets representation. The corresponding numerical quadratures can be performed using either $\vartheta, \varphi_\lambda$ or $\bar{\vartheta}, \bar{\varphi}_\lambda$ as integration variables. The hamiltonian is still expressed in the old variables. This approach to treating the geometric phase effect is conceptually sound not only for the H_3 isotopomers, but for other triatomic systems as well, but has not been tried out so far.

As seen in section 2.6, the GP can have a major effect on the cross sections of chemical reactions of systems displaying conical intersections.

2.6 RESULTS FOR H_3 AND ITS ISOTOPOMERS

Accurate quantum mechanical reactive scattering calculations have been performed over the last few years for the H_3, DH_2 and HD_2 systems, using the LSTH PES [26,27], including the GP effect, and compared with equivalent calculations in which this effect was omitted (NGP). Some of the more important results are summarized below.

For H_3 relatively low total energies E (in the 0.7 eV to 1.5 eV range with respect to the bottom of the isolated H_2 diatom well), para → ortho and ortho → para integral and differential cross sections are essentially not affected by the presence of the conical intersection [8]. However, para → para and ortho → ortho differential cross sections are strongly affected, as exemplfied in fig. 3. Inclusion of the GP effect results in a change of the phase of the oscillations of those cross sections with angle, even for E as low as 0.7 eV. This phenomenon results from

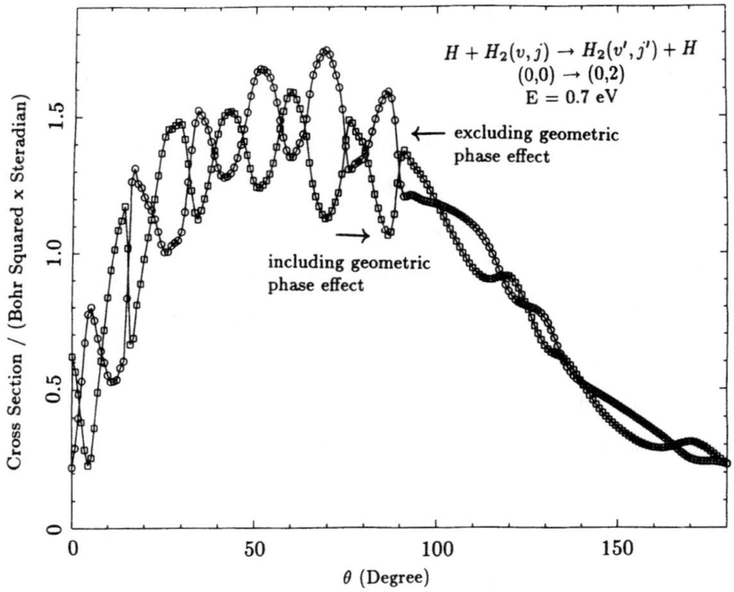

Figure 3. Degeneracy-summed differential cross sections for the $H + H_2(v = 0, j = 0) \rightarrow H_2(v' = 0, j' = 2) + H$ reaction, at a total energy of 0.7 eV, as a function of scattering angle. The squares (circles) correspond to the GP (NGP) calculations. (From ref. 8).

interferences between the direct and exchange scattering amplitudes. The calculations show that at this energy, the GP does not change the direct scattering amplitude, but changes the sign of the exchange one, in agreement with a simple semi-classical model [28]. These results are predictive of what will be found experimentally once measurements of such cross sections are made. These predictions differ from those of all other calculations for this system, since the latter were made without inclusion of the GP effect. Such measurements are highly desirable. The corresponding effect on the degeneracy summed-and-averaged integral cross sections is also significant, enhancing it by about 30 % for the $H + H_2(v = 0, j = 0) \rightarrow H_2(v' = 1, j' = 2) + H$ transition at $E = 1.2$ eV when compared with the corresponding NGP cross sections. The existence of such interference effects for triatomic systems between direct and exchange scattering amplitudes is restricted to the case in which the three atoms are identical, such as H_3, D_3 and Li_3.

It should be kept in mind that the lowest conical intersection energy for this system is 2.75 eV. As a result, the effect of the non-adiabatic coupling between the adiabatic ground and first excited electronic states is negligible for the geometries sampled by $\Psi^{JM\Pi\Gamma}$ at these relatively low energies because they are far from the conical intersection geometries. The effect of this conical intersection at such energies is entirely due to the geometric phase and the ensuant boundary conditions; it is present for real adiabatic electronic wave functions even when these couplings vanish identically, as has been assumed in the present calculations. As such, it is an effect that resembles the Pauli principle for electron exchange, which is also independent of the strength of the interaction between the electrons comprising a molecular system.

For E in the range of 1.8 eV to 2.3 eV [29], still significantly below the lowest conical intersection energy, large GP effects are also found in para $\rightarrow$ ortho and ortho $\rightarrow$ para integral cross sections. The origin of this effect is similar to the one which occurs in the DH_2 system and is discussed below.

The same equilateral triangle conical intersection which exists for the H_3 system exists also for its isotopomer DH_2. However, in the mass-scaled internal configuration space of fig. 2, the PES displays S_2 permutation group symmetry as discussed in section 2.2, with OYZ_λ being a plane of symmetry if $A_\nu \equiv A_\kappa$, rather than the S_3 symmetry (isomorphic with C_{3v}) exhibited by H_3. The conical intersection line is, as metioned in section 2.5, a straight line in that symmetry plane, making an angle of 11.5° with the OY axis and tilted towards the negative part of the OZ_λ axis. In addition, the angle between the positive OZ_λ axis, which is associated with the asymptotic $D + H_2$ $(A_\lambda + A_\nu A_\kappa)$ configuration and its $DH + H(A_\kappa + A_\lambda A_\nu)$ counterpart is no longer 120° as it was for the H_3 system, but 109° instead, because of the mass-scaling of the Jacobi coordinates given by (1). The Γ irreducible representations of S_2 are A for wave functions symmetric with respect to reflection through OYZ_λ and B for wave functions which are antisymmetric under that reflection.

Accurate GP and NGP calculations for the DH_2 system were performed in the total energy range 0.78 eV $\leq E \leq 1.8$ eV [24], using the LSTH PES [26,27]. Rotational state distribution measurements for the $D + H_2$ reaction were made by

Zare and coworkers [30-33]. For the $D + H_2(v = 1, j = 1) \rightarrow DH(v' = 1, j') + H$ process at $E = 1.8$ eV, there is a significant difference between the experimental results and the NGP calculations [31]. Experiment [32] and both GP and NGP theory [24] are compared in fig. 4. The NGP results agree with previous ones [31], but show a significant disagreement with the experimental ones, well outside experimental error. The GP results, on the other hand, agree quite well with experiment, displaying a maximum at $j' = 7$. The small remaining differences between the two may be due to residual systematic experimental errors, not reflected by the error bars. These results leave little doubt that the experiments are essentially correct and that the GP effect plays an important role for this process. The physical mechanism responsible for the GP effect in the DH_2 system, as well as in the para → ortho and ortho → para cross sections for the H_3 system can be explained with the help of fig. 5. This figure displays two kinds of exchange paths

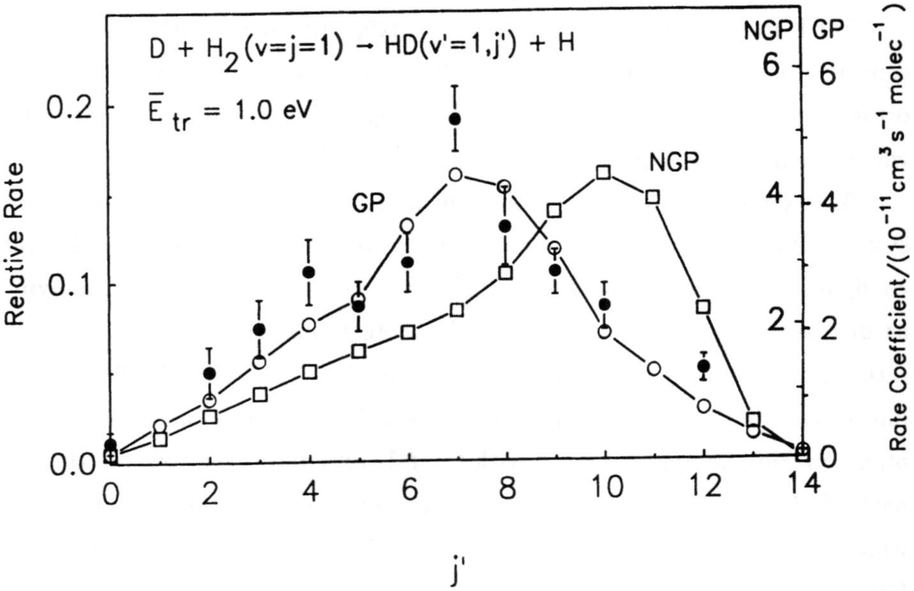

Figure 4. Comparison of theoretical and experimental rate (products of initial relative velocities by integral cross sections) as a function of rotational state quantum number j' for $D+H_2(v = 1, j = 1) \rightarrow DH(v' = 1, j')+H$ for an average initial relative translational $\bar{E}_{tr} = 1.0$ eV (average total energy $\bar{E} = 1.8$ eV). The solid circles display the experimental results of ref. 32. The open squares (circles) represent the NGP (GP) calculations. (From ref. 24).

going from the A + BC region of internal nuclear configuration space to the AB + C one. The first kind (full line) corresponds to a direct approach of A to B, formation

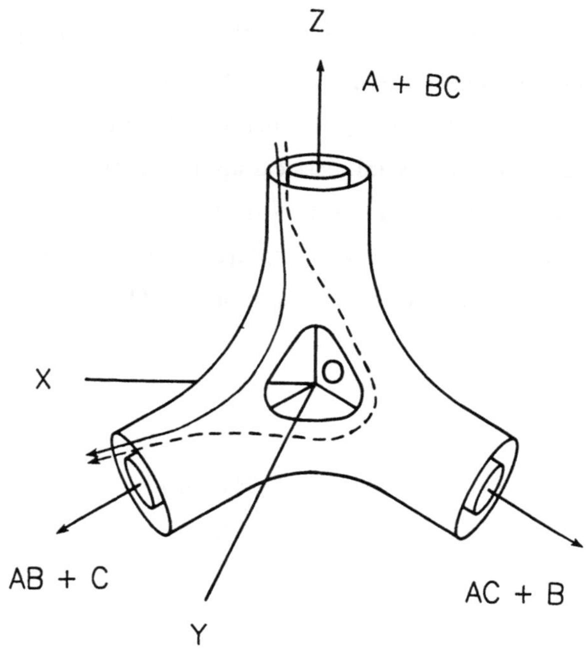

Figure 5. This figure depicts two kinds of semi-classical paths which contribute to the exchange scattering amplitude of a triatomic system ABC. An equipotential energy surface is represented. The $OXYZ$ frame is the same as the $OX_\lambda YZ_\lambda$ frame of fig. 2 with the λ subscript supressed. The dashed path partially encircles the conical intersection line, whereas the solid one does not. (From ref. 8).

of the A-B bond and concurrent rupture of the B-C bond, followed by the motion of C away from the newly formed AB molecule. The second kind (dashed line), which becomes more prevalent as the energy increases, moves around the locus of conical intersection configurations described in section 2.5. This corresponds to A first approaching the C end of the BC molecule, forming an incipient bond with that atom while relaxing the B-C bond, then strengthening the A-B bond while concomitantly breaking the B-C one, atom C finally receding from AB. The effect of the conical intersection on the semiclassical phases of each of these two kinds of paths is different, and in the case of the dashed path, depends on the detailed regions of internal nuclear configuration space traversed. As a results, there are interferences between these two kinds of exchange paths. The details of these interferences depend on whether the GP effect is included or not. It should be noted that this explanation is qualitative but that the calculations were done accurately (benchmark quality) using the methods described in sections 2.4 and 2.5, without any use of paths or of semi-classical approximations.

Differential cross sections for the $D + H_2(v = 1, j = 1) \rightarrow DH(v' = 1, j') + H$ reaction for j' equal 5 through 10 are displayed in fig. 6. As can be seen, large differences between the GP and NGP results occur. This is not surprising, given

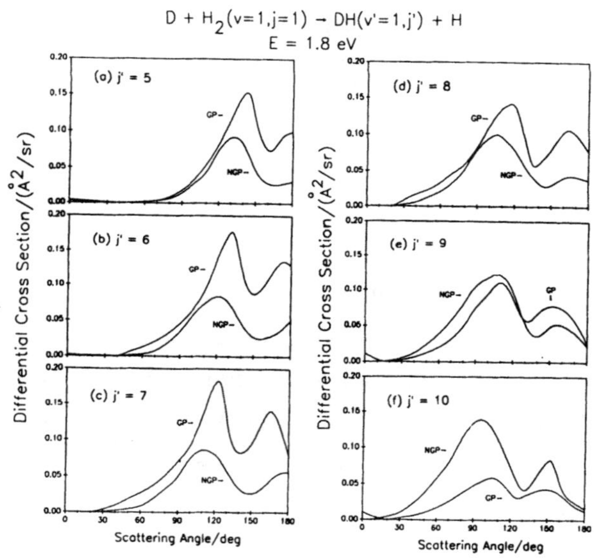

Figure 6. Theoretical GP and NGP differential cross sections at $E = 1.8$ eV for $D + H_2(v = 1, j = 1) \rightarrow DH(v' = 1, j') + H$. Panels (a) through (f) display the results (multiplied by 5 before plotting) for $j' = 5$ through 10. (From ref. 24).

the differences in the corresponding integral cross sections shown in fig. 4 and the fact that differential cross sections are usually more sensitive to the details of the dynamics than are integral ones. The GP curves in fig. 6 are predictive of what the results of measurements, not yet performed, should be. The only source of inaccuracy of the calculations, other than the small controlable computational errors associated with the numerical methods employed, is the inaccuracy of the LSTH PES used [26,27]. This PES is estimated to have an absolute accuracy of about 0.5 kcal/mole (0.022 eV) and an accuracy for differences of about 0.2 kcal/mole

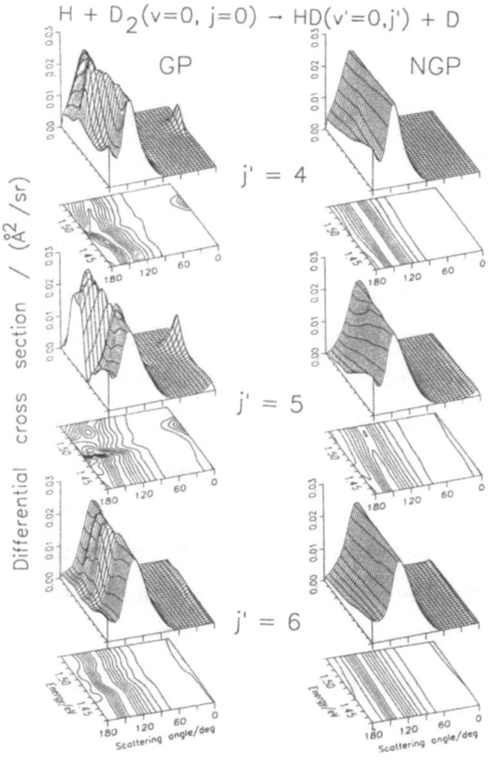

Figure 7. Degeneracy-summed GP and NGP differential cross sections for the $H + D_2(v = 0, j = 0) \rightarrow HD(v' = 0, j') + D$ reaction as a function of total energy and of the center-of-mass scattering angle of the HD product. Results are shown for $j' = 4$, 5 and 6. The contour lines displayed at the bottom of the plots range from 0.004 to 0.028 $Å^2/sr$ in step of 0.004 $Å^2/sr$. (From ref. 25).

(0.009 eV) over the entire range of nuclear configurations of interest. Barring the existence of a resonance in the neighborhood of $E = 1.8$ eV (see HD_2 results below), the GP curves of fig. 6 are expected be quite close to those which should result from good state-to-state differential cross section measurements. Such measurements would be very important for comparing experiment and theory.

Recent differential cross section measurements for the $H + D_2$ reaction [34,35] have stimulated accurate GP (and NGP) calculations on this system to be performed over the total energy range 1.42 eV $\leq E \leq$ 1.53 eV [25,36] which encompasses the energies of 1.47 eV [35] and 1.48 eV [34] at which the experiments were performed. The GP and NGP degeneracy-summed differential cross sections for the $H + D_2(v = 0, j = 0) \rightarrow HD(v' = 0, j') + D$ reaction is given as a function of E and of scattering angle [25] in fig. 7, for $j' = 4, 5$ and 6. The GP plots show significantly more structure as a function of E than do the NGP ones. In particular, for $j' = 5$ the GP DCS has a backward peak at $E = 1.484$ eV and scattering angle 163° with a fwhm of 41 meV and a forward one at $E = 1.502$ eV and scattering angle 0° with fwhm of 21 meV. By contrast, the NGP DCS shows near constancy or only a slow monotonic decrease with energy at all angles over the energy range considered. The energy dependence of the $j' = 5$ GP DCS is strongly suggestive of a resonance, whereas that of the NGP indicates none.

The degeneracy-summed resonance timedelays for the same processes of fig. 7 are given in fig. 8, also as a function of E and scattering angle. The $j' = 5$ GP plot displays a backward peak of 156 fs and a forward one of 98 fs, whereas the corresponding NGP resonance timedelay is everywhere zero within the accuracy of the plots. This confirms the resonance nature of the structure observed in the $j' = 5$ GP DCS. The absence of a resonance in the NGP calculations does not mean that it does not exist. It means, instead, that if it does exist, the corresponding resonance energy is shifted to an energy outside the range considered in these calculations and/or to a different transtion.

Some recent calculations have been done after the workshop ended[37] using an improved version of the BKMP PES for the ground adiabatic electronic state of H_3 [38]. We refer here to this as yet unpublished PES as the BKMP2 PES [39]. Its barrier height is 9.61 kcal/mole, which is 0.2 kcal/mole (0.009 meV) lower than that of the LSTH PES. This value is in excellent agreement with the barrier height of 9.61 kcal/mole, obtained from an exact quantum Monte-Carlo calculation having an absolute error of ± 0.01 kcal/mole. Although the BKMP2 barrier is only 9 meV lower than that of the LSTH PES, the corresponding resonance is shifted

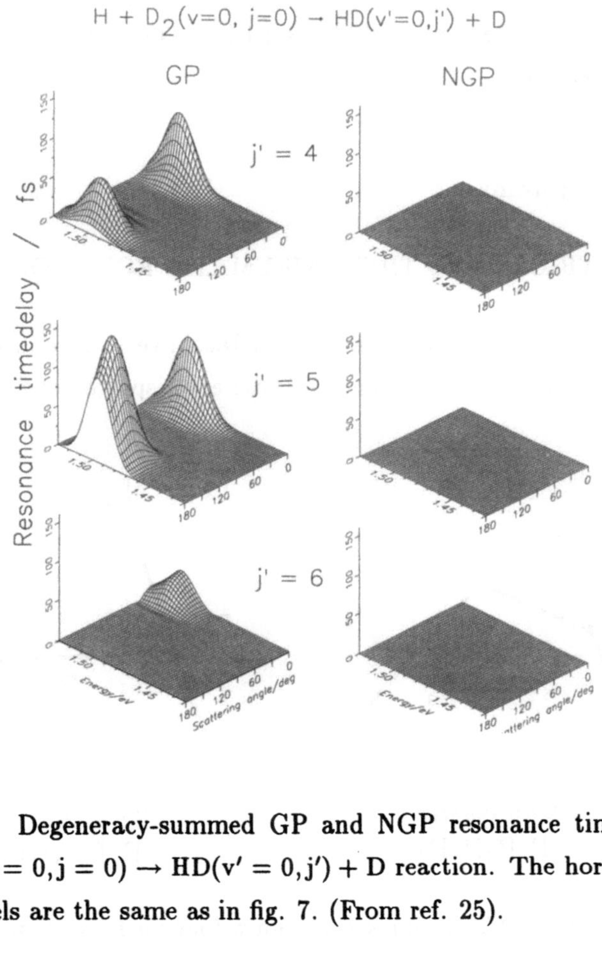

Figure 8. Degeneracy-summed GP and NGP resonance timedelays for the $H + D_2(v = 0, j = 0) \rightarrow HD(v' = 0, j') + D$ reaction. The horizontal axes and other labels are the same as in fig. 7. (From ref. 25).

downwards in energy by 55 meV, a decrease 6 times greater than the saddle point energy decrease. The reason is that other features of the PESs that are important for determining the resonance energy, such as curvatures in directions traverse to the energy path, are also different. The corresponding BKMP2 timedelay is a factor of 3 smaller than the LSTH one. This very great sensitivity of the resonance energy and timedelay to small changes in the PESs involved is indicative that these resonances can be used as spectroscopic-like information which may permit a determination of the topography of those surfaces. This is an exciting prospect. It also suggests that if, by saying that an *ab initio* PES is chemically accurate we mean that it is capable of predicting dynamic properties of the system within experimental accuracy, then

0.2 kcal/mole (9 meV) can no longer be considered a standard of chemical accuracy, at least for H_3 and its isotopic counterparts. A more appropriate standard for this system may be about 0.01 kcal/mole (0.4 meV), which is just being achieved by the exact quantum Monte-Carlo method [40] but not by presently existing variational quantum chemistry codes.

3. Tetraatomic systems

3.1 JACOBI MATRICES AND THEIR KINEMATIC ROTATIONS

For tetraatomic systems, there are two kinds of Jacobi vectors or clustering schemes used to locate the positions of the nuclei. These are displayed in fig. 9. As

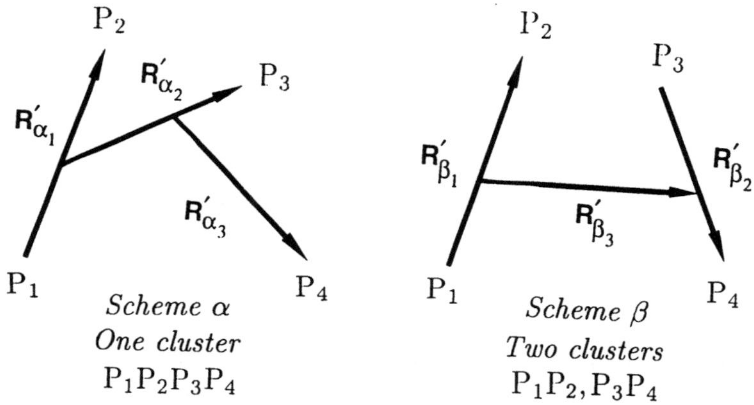

Figure 9. Clustering scheme for a four-nuclei system.

remarked after (7), eqs. (1) through (4) are easily generalized to N-atom systems. For a tetraatomic system, $\mathrm{N}_{\lambda\nu}$ is an orthogonal 3×3 matrix, and the corresponding versions of (8) and (9) are [9]

$$\rho = [Tr(\rho_\lambda^{sf} \; \tilde{\rho}_\lambda^{sf})]^{1/2} = [r_\lambda^{(1)^2} + r_\lambda^{(2)^2} + r_\lambda^{(3)^2}]^{1/2} \qquad (35)$$

and

$$\mathbf{X}_\lambda = \frac{1}{\rho}\begin{pmatrix} (1/2)(2r_\lambda^{(3)^2} - r_\lambda^{(2)^2} - r_\lambda^{(1)^2}) \\ (\sqrt{3}/2)(r_\lambda^{(2)^2} - r_\lambda^{(1)^2}) \\ \sqrt{3}\ \mathbf{r}_\lambda{}^{(3)} \cdot \mathbf{r}_\lambda{}^{(1)} \\ \sqrt{3}\ \mathbf{r}_\lambda{}^{(2)} \cdot \mathbf{r}_\lambda{}^{(1)} \\ \sqrt{3}\ \mathbf{r}_\lambda{}^{(3)} \cdot \mathbf{r}_\lambda{}^{(2)} \\ \rho X_6 \end{pmatrix} \tag{36}$$

where

$$\begin{gathered} X_6 = (-1)^{\chi_\lambda}\sqrt{3}(|\mathbf{r}_\lambda{}^{(3)} \times \mathbf{r}_\lambda{}^{(1)}|^2 + |\mathbf{r}_\lambda{}^{(2)} \times \mathbf{r}_\lambda{}^{(1)}|^2 + |\mathbf{r}_\lambda{}^{(3)} \times \mathbf{r}_\lambda{}^{(2)}|^2)^{1/2} \\ (-1)^{\chi_\lambda} = \operatorname{sgn}(\mathbf{r}_\lambda{}^{(3)} \cdot \mathbf{r}_\lambda{}^{(2)} \times \mathbf{r}_\lambda{}^{(1)}) \end{gathered} \tag{37}$$

For subsequent convenience, the order of the elements of $\mathbf{X}_\lambda$ in (36) is different than that in (9). χ_λ is a chirality quantum number whose value is 0 if the handedness of the set of vectors $\mathbf{r}_\lambda{}^{(1)}, \mathbf{r}_\lambda{}^{(2)}, \mathbf{r}_\lambda{}^{(3)}$ is the same as that of the space-fixed axes $Gxyz$ and 1 otherwise. Under kinematic rotations $\mathbf{X}_\lambda$ transforms as

$$\mathbf{X}_\nu = \mathbf{M}_{\lambda\nu}\mathbf{X}_\lambda \tag{38}$$

where

$$\mathbf{M}_{\lambda\nu} = \begin{pmatrix} \mathbf{V}_{\lambda\nu} & 0 \\ 0 & 1 \end{pmatrix} \tag{39}$$

The matrices $\mathbf{M}_{\lambda\nu}$ and $\mathbf{V}_{\lambda\nu}$ are related to $\mathbf{N}_{\lambda\nu}$ and have dimensions 6×6 and 5×5 respectively, their elements depending only on the masses of the four nuclei.

In analogy to the three nuclei case, we define a set of internal configuration space mathematical axes $OX_{\lambda_1}X_{\lambda_2}X_{\lambda_3}X_{\lambda_4}X_{\lambda_5}X_6$, in which the position of a general point P is $\mathbf{X}_\lambda$. These axes are the extension to the tetraatomic case of the $O\bar{X}_\lambda\bar{Y}_\lambda\bar{Z}$ axes introduced in section 2.3. As previously, there is a one-to-one correspondence between points P and nuclear configurations of the system, with some restriction on the ranges of the coordinates of P (similarly to the three nuclei case, where it was required that $Y \geq 0$) [9]. In addition, a $\lambda \to \nu$ kinematic rotation is equivalent to a 6-dimensional rotation of the $OX_{\lambda_1}X_{\lambda_2}X_{\lambda_3}X_{\lambda_4}X_{\lambda_5}X_6$ around the OX_6 axis, which is itself kinematic-rotation-invariant.

3.2 SYMMETRIZED HYPERSPHERICAL COORDINATES

As a result of these properties of $\mathbf{X}_\lambda$, it is useful to define the internal configuration space symmetrized hyperspherical coordinates $\rho, \mathbf{\Theta}_\lambda \equiv \vartheta, \varphi_\lambda^{(1)}, \varphi_\lambda^{(2)}, \varphi_\lambda^{(3)}, \varphi_\lambda^{(4)}$ by

$$\begin{aligned}
X_{\lambda_1} &= \rho\sin\vartheta\sin\varphi_\lambda^{(1)}\sin\varphi_\lambda^{(2)}\sin\varphi_\lambda^{(3)}\cos\varphi_\lambda^{(4)} \\
X_{\lambda_2} &= \rho\sin\vartheta\sin\varphi_\lambda^{(1)}\sin\varphi_\lambda^{(2)}\sin\varphi_\lambda^{(3)}\sin\varphi_\lambda^{(4)} \\
X_{\lambda_3} &= \rho\sin\vartheta\sin\varphi_\lambda^{(1)}\sin\varphi_\lambda^{(2)}\cos\varphi_\lambda^{(3)} \\
X_{\lambda_4} &= \rho\sin\vartheta\sin\varphi_\lambda^{(1)}\cos\varphi_\lambda^{(2)} \\
X_{\lambda_5} &= \rho\sin\vartheta\cos\varphi_\lambda^{(1)} \\
X_6 &= \rho\cos\vartheta
\end{aligned} \tag{40}$$

where ρ and ϑ are independent of λ. These coordinates are a generalization of the $\rho, \vartheta, \varphi_\lambda$ coordinates of section 2.3. Equipotentials of the electronically-adiabatic tetraatomic PES V in the internal configuration space spanned by $\mathbf{X}_\lambda$ are rotated around OX_6 without distorsion under a $\lambda \rightarrow \nu$ coordinate transformation, and as a result display the permutation symmetry properties of the system of nuclei, as in the triatomic case. The $X_{\lambda_i}(i = 1$ through 6 with $X_{\lambda_6} \equiv X_6)$ represent therefore all regions of configuration space equally well, and are very useful for visualizing the topological properties of V in a manner which is arrangement-channel independent. This is a very useful property. However, the nuclear motion hamiltonian in the 9 coordinates $a_\lambda, b_\lambda, c_\lambda, \rho, \mathbf{\Theta}_\lambda$ (where $a_\lambda, b_\lambda, c_\lambda$ are, as in section 2.3, the Euler angles which rotate the space-fixed frame $Gxyz$ into the principal-moment-of-inertia body-fixed frame $Gx^{I\lambda}y^{I\lambda}z^{I\lambda}$) is extremely complicated and very difficult to use in scattering calculations. This is quite different from the triatomic case, in which the hamiltonian in these coordinates was much simpler. As a result, it is convenient to introduce another system of hyperspherical coordinates, the row-orthonomal ones.

3.3 ROW-ORTHONOMAL HYPERSPHERICAL COORDINATES

It is possible to put the Jacobi matrix ρ_λ^{sf} in the form [9]

$$\rho_\lambda^{\text{sf}} = (-1)^\chi \widetilde{\mathbf{R}}(a_\lambda, b_\lambda, c_\lambda)\rho\mathbf{N}(\theta, \phi)\widetilde{\mathbf{R}}(\delta_\lambda^{(1)}, \delta_\lambda^{(2)}, \delta_\lambda^{(3)}) \tag{41}$$

In this expression, $\mathbf{R}(a_\lambda, b_\lambda, c_\lambda)$ is the 3×3 matrix which rotates $Gxyz$ into $Gx^{I\lambda}y^{I\lambda}z^{I\lambda}$ and $\mathbf{N}(\theta,\phi)$ is the diagonal matrix

$$\mathbf{N}(\theta,\phi) = \begin{pmatrix} \sin\theta\cos\phi & 0 & 0 \\ 0 & \sin\theta\sin\phi & 0 \\ 0 & 0 & \cos\theta \end{pmatrix} \tag{42}$$

where θ and ϕ are two angles in terms of which the system's nuclear principal moments of inertia are given by

$$\begin{aligned} I_1 &= \mu\rho^2\sin^2\theta \\ I_2 &= \mu\rho^2(1-\sin^2\theta\cos^2\phi) \\ I_3 &= \mu\rho^2(1-\sin^2\theta\sin^2\phi) \end{aligned} \tag{43}$$

with

$$0 \le I_1 \le I_2 \le I_3 \tag{45}$$

The quantities $\delta_\lambda^{(1)}, \delta_\lambda^{(2)}$, and $\delta_\lambda^{(3)}$ are three internal hyperangles in the range 0 to π. The quantities $\rho, \boldsymbol{\delta}_\lambda \equiv (a_\lambda, b_\lambda, c_\lambda, \theta, \phi, \delta_\lambda^{(1)}, \delta_\lambda^{(2)}, \delta_\lambda^{(3)})$ are called the row-orthonomal hyperspherical coordinates. In terms of them the system's nuclear motion hamiltonian is given by [41]

$$\begin{aligned} \widehat{H}(\rho,\boldsymbol{\delta}_\lambda) = {} & \widehat{T}_\rho(\rho) + \frac{\widehat{K}^2(\theta,\phi)}{2\mu\rho^2} - \frac{\hbar^2}{\mu\rho^2}[b(\theta,\phi)\frac{\partial}{\partial\theta} + \frac{c(\theta,\phi)}{\sin\theta}\frac{\partial}{\partial\phi}] \\ & + \frac{1}{2\mu\rho^2}\widehat{A}(a_\lambda, b_\lambda, c_\lambda, \delta_\lambda^{(1)}, \delta_\lambda^{(2)}, \delta_\lambda^{(3)}; \theta, \phi) \\ & + V(\rho, \theta, \phi, \delta_\lambda^{(1)}, \delta_\lambda^{(2)}, \delta_\lambda^{(3)}) \end{aligned} \tag{46}$$

The several quantities in this expression are defined below.

$$\widehat{T}_\rho(\rho) = -\frac{\hbar^2}{2\mu}\frac{1}{\rho^8}\frac{\partial}{\partial\rho}\rho^8\frac{\partial}{\partial\rho} \tag{47}$$

$$\widehat{K}^2(\theta,\phi) = -\hbar^2(\frac{1}{\sin\theta}\frac{\partial}{\partial\theta}\sin\theta\frac{\partial}{\partial\theta} + \frac{1}{\sin^2\theta}\frac{\partial^2}{\partial\phi^2}) \tag{48}$$

$$b(\theta,\phi) = \frac{N'_{\theta_1}N_1 - N'_{\theta_2}N_2}{N_1^2 - N_2^2} + \frac{N'_{\theta_1}N_1 - N'_{\theta_3}N_3}{N_1^2 - N_3^2} + \frac{N'_{\theta_2}N_2 - N'_{\theta_3}N_3}{N_2^2 - N_3^2} \tag{49}$$

$$c(\theta,\phi) = \frac{M_{\phi_1}N_1 - M_{\phi_2}N_2}{N_1^2 - N_2^2} + \frac{M_{\phi_1}N_1}{N_1^2 - N_3^2} + \frac{M_{\phi_2}N_2}{N_2^2 - N_3^2} \tag{50}$$

$$\widehat{A} = \frac{(N_2\widehat{J}_x^{I\lambda} - N_3\widehat{L}_{\lambda_1})^2 + (N_3\widehat{J}_x^{I\lambda} - N_2\widehat{L}_{\lambda_1})^2}{(N_2^2 - N_3^2)^2} + \frac{(N_3\widehat{J}_y^{I\lambda} - N_1\widehat{L}_{\lambda_2})^2 + (N_1\widehat{J}_y^{I\lambda} - N_3\widehat{L}_{\lambda_2})^2}{(N_3^2 - N_1^2)^2} + \frac{(N_1\widehat{J}_z^{I\lambda} - N_2\widehat{L}_{\lambda_3})^2 + (N_2\widehat{J}_z^{I\lambda} - N_1\widehat{L}_{\lambda_3})^2}{(N_1^2 - N_2^2)^2} \tag{51}$$

The $\widehat{J}_x^{I\lambda}, \widehat{J}_y^{I\lambda}, \widehat{J}_z^{I\lambda}$ are the components of the system's nuclear orbital angular momenta in the body-fixed $I\lambda$ frame whereas $\widehat{L}_{\lambda_1}, \widehat{L}_{\lambda_2}$ and $\widehat{L}_{\lambda_3}$ are space-fixed-type hyperangular momenta associated with $\delta_\lambda^{(1)}, \delta_\lambda^{(2)}$ and $\delta_\lambda^{(3)}$ according to

$$\begin{aligned}
\frac{\hbar}{i}\frac{\partial}{\partial\delta_\lambda^{(1)}} &= \widehat{L}_{\lambda_3} \\
\frac{\hbar}{i}\frac{\partial}{\partial\delta_\lambda^{(2)}} &= -\sin\delta_\lambda^{(1)}\widehat{L}_{\lambda_1} + \cos\delta_\lambda^{(1)}\widehat{L}_{\lambda_2} \\
\frac{\hbar}{i}\frac{\partial}{\partial\delta_\lambda^{(3)}} &= \sin\delta_\lambda^{(2)}\cos\delta_\lambda^{(1)}\widehat{L}_{\lambda_1} + \sin\delta_\lambda^{(2)}\sin\delta_\lambda^{(1)}\widehat{L}_{\lambda_2} + \cos\delta_\lambda^{(2)}\widehat{L}_{\lambda_3}
\end{aligned} \tag{52}$$

The remaining quantities in (49) through (51) are given by

$$\begin{array}{lll}
N_1(\theta,\phi) = \sin\theta\cos\phi & N_2(\theta,\phi) = \sin\theta\sin\phi & N_3(\theta) = \cos\theta \\
N'_{\theta_1}(\theta,\phi) = \cos\theta\cos\phi & N'_{\theta_2}(\theta,\phi) = \cos\theta\sin\phi & N'_{\theta_3}(\theta) = -\sin\theta \\
M_{\phi_1}(\phi) = -\sin\phi & M_{\phi_2}(\phi) = \cos\phi &
\end{array} \tag{53}$$

It can be proven that every term in (46) is invariant under kinematic rotations. This makes these coordinates very promising for accurate calculations. Recent results for tetraatomic systems have been published in which some of the degrees of freedom of the system were treated exactly and others approximately [42,43] or in which only the $J = 0$ partial wave was treated exactly and the others approximately [44]. In none of these treatments was the full permutation symmetry of identical nuclei considered. The use of the row-orthonomal hyperspherical coordinates together with massively parallel computers may permit an accurate treatment of all partial waves as well as the full inclusion of the effects of the Pauli principle.

4. Conclusions

The use of symmetrized hyperspherical coordinates for triatomic systems has permitted accurate bench-mark calculations of state-to-state differential and integral cross sections, including the effect of the geometric phase, up to relatively high energies. In particular, for HD_2 a pronounced reactive scattering resonance has been predicted, which is very sensitive to the details of the PES in the strong interaction region. A new system of row-orthonomal coordinates has been developed which, together with massively parallel computers, may permit the extension of such calculations to tetraatomic calculation.

5. Acknowledgements

This research was supported in part by NSF Grant CHE 9311401. The author wishes to thank Dr. Y.-S. M. Wu for help in the preparation of this manuscript.

6. References

1. Kuppermann, A. (1996) The geometric phase in reaction dynamics, in R.E. Wyatt and J.Z.H. Zhang (eds.) *Dynamics of Molecules and Chemical Reactions*, Marcel Dekker, Inc., New York, N.Y., in press.
2. Schatz, G.C. and Kuppermann, A. (1976) *Quantum Mechanical Reactive Scattering for Three- Dimensional Atom plus Diatom Systems. I. Theory*, J. Chem. Phys. **65**, 4596-4623.
3. a) Kuppermann, A. (1975) *A Useful Mapping of Triatomic Potential Energy Surfaces*, Chem. Phys. Lett. **32**, 374-375. b) Ling, R.T. and Kuppermann, A. (1975) Surface functions for three dimensional reactive scattering, in J.S. Risley and K. Geballe (eds.), *Electronic and Atomic Collisions, Abstracts of Papers of the IXth International Conference on the Physics of Electronic and Atomic Collisions*, University of Washington Press, Seattle, 353-354.
4. a) Johnson, B.R. (1980) *On hyperspherical coordinates and mapping the internal configurations of a three body system*, J. Chem. Phys. **73**, 5051-5058. b) Johnson, B.R. (1983) *The quantum dynamics of three particles in hyperspherical coordinates*, J. Chem. Phys. **79**, 1916-1925.
5. a) Pack, R.T. and Parker, G.A. (1987) *Quantum reactive scattering in three dimensions using hyperspherical (APH) coordinates. Theory*, J. Chem. Phys. **87**, 3888-3921. b) Pack, R.T. and Parker, G.A. (1989) *Quantum reactive scattering in three dimensions using hyperspherical (APH) coordinates. III Small θ behavior and corrigenda*, J. Chem. Phys. **90**, 3511-3519.

6. Launay, J.M. and Le Dourneuf, M. (1989) *Hyperspherical Close-Coupling Calculation of Integral Cross Sections for the Reaction $H+H_2 \rightarrow H_2+H$*, Chem. Phys. Lett., **163**, 178-188.
7. Lepetit, B., Peng, Z. and Kuppermann, A. (1990) *Calculation of bound rovibrational states on the first electronically excited state of the H_3 system*, Chem. Phys. Lett. **166**, 572-580.
8. Wu, Y.-S.M., Kuppermann, A. and Lepetit, B. (1991) *Theoretical Calculation of Experimentally Observable Consequences of the Geometric Phase on Chemical Reaction Cross Sections*, J. Chem. Phys. **186**, 319-328.
9. Kuppermann, A., (1994) A new look at symmetrized hyperspherical coordinates, *Advances in Molecular Vibrations and Collision Dynamics* (J. Bowman, ed.), JAI Press, Greenwich, CT, Vol. 2 B, 117-186.
10. Delves, L.M. (1959) *Tertiary and General-Order Collisions*, Nucl. Phys. **9**, 391-399.
11. Delves, L.M. (1960) *Tertiary and General-Order Collisions*, Nucl. Phys. **20**, 275-308.
12. Kuppermann, A., Schatz, G.C. and Baer, M. (1976) *Quantum mechanical reactive scattering for planar atom plus diatom system. I. Theory*, J. Chem. Phys. **65**, 4596-4623.
13. Smith, F.T. (1959) *Particpation of vibration in exchange reactions*, J. Chem. Phys. **31**, 1352-1359.
14. Smith, F.T. (1960) *Generalized angular momentum in many-body collisions* Phys. Rev. **120**, 1058-1069.
15. Kuppermann, A. (1981) Accurate Quantum Calculations of Reactive Systems, *Theoretical Chemistry-Theory of Scattering: Papers of Henry Eyring* (D. Henderseon, ed.) Academic Press, New York, NY, Vol. 6A, 79-164.
16. Cuccaro, S.A., Hipes, P.G. and Kuppermann, A. (1976) *Hyperspherical Coordinate Reactive Scattering Using Variational Surface Functions*, **154**, 4642-4667.
17. Johnson, B.R. (1973) *The Multichannel Log-Derivative Method for Scattering Calculations*, J. Comp. Phys. **13**, 445-449.
18. Johnson, B.R. (1979) The log derivative and renormalized numerov algorithms, *Alogrithms and Computer Codes For Atomic and Molecular Quantum Scattering Theory* (L. Thomas, ed.), NRCC Workshop, Lawrence Berkeley laboratory, Report No. LBL 9501, 86-104.

19. Manolopoulos, D.E. (1986) *An improved log derivative method for inelastic scattering*, J. Chem. Phys. **85**, 6425-6429.

20. Herzberg, G. and Longuet-Higgins, H.C. (1963) *Intersection of Potential Energy Surfaces in Polyatomic Molecules*, Discussions Farady Soc. **35**, 77-82.

21. Longuet-Higgins, H.C. (1975) *The Intersection of Potential Energy Surfaces in Polyatomic Molecules*, Proc. R. Soc. Lond. **344**, 147-156.

22. Mead, C.A. and Truhlar, D.G. (1979) *On the detremination of Born-Oppenheimer nuclear wave functions including complications due to conical intersections and identical nuclei*, J. Chem. Phys. **70**, 2284-2296.

23. Porter, R.N., Stevens, R.M. and Karplus, M. (1969) *Symmetric H_3: A Semiemperical and Ab Initio Study of a Simple Jahn-Teller System*, J. Chem. Phys. **49**, 5163-5178.

24. Kuppermann, A. and Wu, Y.-S.M. (1993) *The geometric phase shows up in chemical reactions*, J. Chem. Phys. **205**, 577-586.

25. Kuppermann, A. and Wu, Y.-S.M. (1995) *The quantitative prediction and lifetime of a pronounced reactive scattering resonance*, Chem. Phys. Lett. **241**, 229-240.

26. Siegbahn, P. and Liu, B. (1978) *An accurate three-dimensional potential energy surface for H_3*, J. Chem. Phys. **68**, 2457-2465.

27. Truhlar, D.G. and Horowitz, C.J. (1978) *Functional representation of Liu and Siegbahn's accurate ab initio potential energy calculations for $H+H_2$*, J. Chem. Phys. **68**, 2466; **71** (1979) 1514 (E).

28. Mead, C.A. (1980) *Superposition of reactive and nonreactive scattering in the presence of conical intersection*, J. Chem. Phys. **72**, 3839-3841.

29. Wu, Y.-S.M. and Kuppermann, A. (1993) *Prediction of the effect of the geometric phase on product rotational state distributions and integral cross sections*, Chem. Phys. Lett. **201**, 178-186.

30. Kliner, D.A.V., Rinen, K.D. and Zare, R.N. (1990) *The $D+H_2$ reaction. Comparison of experimental with quantum-mechanical and quasi-classical calculations*, Chem. Phys. Lett. **166**, 107-111.

31. Neuhauser, D., Judson, R.S., Kouri, D.J., Edelman, D.E., Shafer, N.E., Kliner, D.A.V. and Zare, R.N. (1992) *Sate-to-state rates for the $D+H_2$ (v=1, j=1)→HD(v', j')+H reaction. Predictions and measurements*, Science **257**, 519-522.

32. Kliner, D.A.V., Adelman, D.E. and Zare, R.N. (1991) *Comparison of experimental and theoretical integral cross sections for $D+H_2$ (v=1, j=1)→HD(v'=1, j')+H*, J. Chem. Phys. **95**, 1648-1662.
33. Adelman, D.E., Shafer, N.E., Kliner, D.A.V. and Zare, R.N. (1992) *Measurement of relative state-to-state rate constants for the reaction $D+H_2$ (v, j)→HD(v', j')+H*, J. Chem. Phys. **97**, 7323-7341.
34. Kitsopoulos, T.N., Buntine, M.A., Baldwin, D.P., Zare, R.N. and Chandler, D.W. (1993) *Reactive-product Imaging. The $H+D_2$ Reaction*, Science **260**, 1605-1610.
35. Schnieder, L., Seekamp-Rahn, K., Borkowski, J., Wrede, E., Welge, K.H. Aoiz, F.J., Bañares, L., D'Mello, M.J., Herrero, V.J., Sáez Rábanos, V. and Wyatt, R.E. (1995) *Experimental Studies and Theoretical Predictions for the $H+D_2$→HD+D Reaction*, Science **269**, 207-210.
36. Wu, Y.-S.M. and Kuppermann, A. (1995) *The importance of the geometric phase effect for the $H+D_2$→ HD+D reaction*, Chem. Phys. Lett. **235**, 105-110.
37. Wu, Y.-S.M. and Kuppermann, A. unpublished results.
38. Boothroyd, A.I., Keogh, W.J., Martin, P.G. and Peterson, M.R. (1991) *An improved H_3 potential energy surface*, J. Chem. Phys. **95**, 4343-4359.
39. The author thanks Boothroyd, A.I., Martin, P.G. and Keogh, W.J. for making their new BKMP2 H_3 PES available prior to publication.
40. Diedrich, D.L. and Anderson, J.B. (1994) *Exact quantum Monte Carlo calculations of the potential energy surface for the reaction $H+H_2 \rightarrow H_2+H$*, J. Chem. Phys. **100**, 8089-8095.
41. Kuppermann, A., unpublished results.
42. Wang, D.S. and Bowman, J.M. (1993) *Quantum calculations of mode specificity in reactions of H with HOD and H_2O*, J. Chem. Phys. **98**, 6235-6247.
43. Nyman, G. and Clary, D.C. (1993) *Quantum scattering calculations on $H_2O+H \rightarrow H_2+OH$ and isotopes. Rotational distributions and cross sections*, J. Chem. Phys. **99**, 7774-7786.
44. Zhang, D.H. and Zhang, J.Z.H.(1994) *Full-dimensional time-dependent treatment of diatom-diatom reactions. The H_2+OH reaction*, J. Chem. Phys. **101**, 1146-1156.

CONTRIBUTORS

Anderson J.B.

Department of Chemistry, The Pennsylvania State University, University Park, Pennsylvania 16802, USA

Aquilanti V.

Dipartimento di Chimica Universita di Perugia, I-06123 Perugia, Italy

Avery J.S.

Department of Chemistry, H. C. Oersted Institute, University of Copenhagen, DK-2100, Denmark

Baerends E.J.

Afdeling Theoretische Chemie, Vrije Universiteit, De Boelelaan 1083, 1081 HV, Amsterdam, The Netherlands

Barnovsky R.

Institute of Atomic Physics and Spectroscopy, Latvian University, Raina bulv., Riga LV-1586, Latvia

Billing G.D.

Department of Chemistry, H. C. Oersted Institute, University of Copenhagen, Denmark

Bleil R.

Department of Chemistry, Purdue University, West Lafayette, IN 47907, USA

Bohm J.H.

Department of Physics and James Franck Institute, University of Chicago, 5640 s. Ellis Avenue, Chicago, IL 60637 USA

Cavalli S.

Dipartimento di Chimica Universita di Perugia, I-06123 Perugia, Italy

Clementi E.

METECC Laboratoire, University L. Pasteur, 3, rue de l'Université, 67084 Strasbourg, France

Coletti C.

Dipartimento di Chimica Universita di Perugia, I-06123 Perugia, Italy

Corongiu G.

METECC Laboratoire, University L. Pasteur, 3, rue de l'Université, 67084 Strasbourg, France

Damburg R.

Institute of Atomic Physics and Spectroscopy, Latvian University, Raina bulv., Riga LV-1586, Latvia

De Fazio D.
Dipartimento di Chimica Universita di Perugia, I-06123 Perugia, Italy

Dunn M.
Department of Physics and Astronomy, University of Oklahoma, Norman, OK 73019, USA

Ermler W.C.
Department of Chemistry and Chemical Engineering, Stevens Institute of Technology, Hoboken, New Jersey 07030, USA

Fano U.
Department of Physics and James Franck Institute, University of Chicago,5640 s. Ellis Avenue, Chicago, IL 60637 USA

Germann T.C.
Department of Physics and Astronomy, University of Oklahoma, Norman, OK 73019, USA

Goodson D. Z.
Department of Chemistry, Southern Methodist University, Dallas, Texas, 752757, USA

Gritsenko O.V.
Afdeling Theoretische Chemie, Vrije Universiteit, De Boelelaan 1083, 1081 HV, Amsterdam, The Netherlands

Grossi G.
Dipartimento di Chimica Universita di Perugia, I-06123 Perugia, Italy

Herschbach D.R.
Department of Chemistry, Harvard University, Cambridge, MA 02138, USA

Hubac I.
Division of Chemical Physics, Faculty of Mathematics and Physics, Comenius University, 842 15 Bratislava, Slovakia

Kais S.
Department of Chemistry, Purdue University, West Lafayette, IN 47907, USA

Kryachko E.S.
Bogoliubov Institute for Theoretical Physics, Kiev, Ukraine 252143 and Instituto Estructura de la Materia, C. S. I. C., 28006, Madrid, Spain

Kuppermann A.
Arthur Amos Noyes Laboratoty of Chemical Physics, California Institute of Technology, Pasadena, 91 125 CA, USA

Linderberg J.
Department of Chemistry, Aarhus University, DK-8000 Aarhus C, Denmark

Loeser J.G.
Department of Chemistry, Oregon State University, Corvallis, OR 97331-4003, USA

Marino M.M.

Department of Chemistry and Chemical Engineering, Stevens Institute of Technology, Hoboken, New Jersey 07030, USA

Mercouris Th.

Theoretical and Physical Chemistry Institute, National Hellenic Research Foundation, 48 Vas. Constantinou Ave., 116 35 Athens, Greece

Morgan III J.D.

Department of Physics and Astronomy, University of Delaware, Newark, DE 19716, USA, and Cornell Theory Center, Cornell University, Ithaca, NY 14853, USA

Nicolaides C.A.

Theoretical and Physical Chemistry Institute, National Hellenic Research Foundation, 48 Vas. Constantinou Ave., 116 35 Athens, Greece

Popov V.S.

Institute of Theoretical and Experimental Physics, B. Cheremushkinskaya, 117259 Moscow, Russia

Proynov E.I.

Départment de Chimie, Université de Montréal, C.P. 6128, Succursale Centre-ville, Montréal, Québec H3C 3J7, Canada

Rakauskas R.J.

Department of Theoretical Physics, Vilnius University, Saulėtekio 9, 2054 Vilnius, Lithuania

Rost J.-M.

Facultät für Physik, Universität Freinburg, Hermann-Herder-Str. 3, D-79104 Freinburg, Germany

Ruiz E.

Department de Quimica Inorganica, Diagonal 647, Universitat de Barcelona, 08028 Spain

Sadovoy A.A.

Russian Federal Nuclear Center, All-Russian Research Institute of Experimental Physics, Arzamas-16, Russia

Salahub D.R.

Départment de Chimie, Université de Montréal, C.P. 6128, Succursale Centre-ville, Montréal, Québec H3C 3J7, Canada

Simonov Yu. A.

Institute of Experimental and Theoretical Physics, Moscow, Russia

Sulskus J.

Department of Theoretical Physics, Vilnius University, Saulėtekio 9, 2054 Vilnius, Lithuania

Themelis S.I.

Theoretical and Physical Chemistry Institute, National Hellenic Research Foundation, 48 Vas. Constantinou Ave., 116 35 Athens, Greece

Tsipis C.A.

Laboratory of Applied Quantum Chemistry, Faculty of Chemistry, Aristotle University of Thessaloniki, 540 06 Thessaloniki, Greece

van Leeuwen R.

Afdeling Theoretische Chemie, Vrije Universiteit, De Boelelaan 1083, 1081 HV, Amsterdam, The Netherlands

Vela A.

Departamento de Quimica, Universidad Autonoma Metropolitana-Iztapalapa, A.P. 55-534, México, D.F. 09340, México

Walcup J.R.

Department of Physics and Astronomy, University of Oklahoma, Norman, OK 73019, USA

Watson D.K.

Department of Physics and Astronomy, University of Oklahoma, Norman, OK 73019, USA

Wilson S.

Rutherford Appleton Laboratory, Chilton, Oxfordshire OX11 0QX, England

Index

MIX
Papier aus verantwortungsvollen Quellen
Paper from responsible sources
FSC® C105338

If you have any concerns about our products,
you can contact us on
ProductSafety@springernature.com

In case Publisher is established outside the EU,
the EU authorized representative is:
Springer Nature Customer Service Center GmbH
Europaplatz 3, 69115 Heidelberg, Germany

Printed by Libri Plureos GmbH
in Hamburg, Germany